CONTENTS

CHAPTER SIXTEEN

ATOMS AND NUCLEI....................................... 514

CHAPTER SEVENTEEN

SPECIAL RELATIVITY....................................... 550

APPENDICES

CHAPTER ONE

FORCES ON OBJECTS: STATICS

1-1 AN INTRODUCTION TO PHYSICS AS A SCIENCE

The subject of physics includes the study of a large number of diverse topics all related to things that go on in the world about us. Over the years many connecting links have been found among apparently different phenomena and these general ideas simplify the subject. For example, baseballs, planets and atoms all move according to the same laws describing force and motion. There is also much in common in the behavior of sound waves, light waves and seismic waves. The basic principles of physics are of philosophic value in the way they tell of the basic, simple plans of nature. But the basic ideas alone do not make the science of physics. The values of science are many. One of these is in how we become able to understand much of what goes on about us. This understanding not only enriches our life but teaches us what to be cautious of and what we need not fear. From this we learn to use our environment to our benefit and the benefit of others.

Science does not pretend that it teaches us everything. The working scientists probably know its limitations better than anyone else. We have to regard the small parts we know as a gain from no knowledge at all.

The plan of this book is to lay out the general principles, show to some extent how they were arrived at, and above all to show that they really do describe natural things as you too can see them. Science does not teach us that things are not as we see them, but it tells of how things are when we look very carefully. The general ideas arrived at by careful observation are then shown to explain things that happen about you. The ways in which the ideas of physics are used in a variety of studies are included. Examples are not only from aspects of physics but also from geology, architecture, biology, medicine and others. You need not dwell on all the examples but choose those that are of special interest.

You will find that you too can work problems in physics and in learning to do this the subject will become useful to you. Though you may not plan to become a working physical scientist, you will understand better how the scientist works and arrives at conclusions by participating in that process yourself.

1-1-1 THE PLAN OF PRESENTATION

The material begins with the study of force, for the concept of a force and of how it affects materials and motion is very basic in physics.

The units of force and other units needed in this study are then described. In any topic it is important to learn the units before going ahead. Specialized definitions of terms are also given early where necessary in a chapter because if you don't know what the words mean, you will not understand the later material.

The physical concepts are then presented, frequently with laws or other relations. Often these are in mathematical form. Generally, there are only a few formulas basic to a topic and then these are manipulated to describe various situations. It is largely in how you learn to do such manipulation that the science will be useful to you.

Applications to specific situations are then given, using numerical values. These are to show you how to approach the problems at the end of the chapters. These examples and problems teach you more about physics, as well as teach you how to work in it yourself. The answers to some of the problems are at the back of the book.

So, now we start with the role of force in the science of physics.

1-2 FORCE IN PHYSICS

When forces are applied to objects, several things may happen. First, they may be made to accelerate. It was Sir Isaac Newton who showed that an unbalanced force on an object would cause it to accelerate, or to change its velocity. The area of study dealing with force and acceleration is called **dynamics,** and is the subject of another chapter. More than one force may act on an object, and it may then not move at all, or it may not change its velocity. If the forces all "balance out," it is the same as if no force at all acts, and the object is in **equilibrium.** The area of study dealing with objects in equilibrium is called **statics,** and that is the subject of this chapter. Forces acting on an object of finite size (whether the object accelerates or is still) cause it to change its shape, to be distorted. The amount of deformation depends on the size of the forces and on the type of material of which the object is made. This area of study may be called the **strength of materials** or, since deformation is involved, it may be called **elasticity.** Too much deformation results in a permanent change in shape or in breakage.

1–3   UNITS FOR FORCES

The size of forces cannot be talked about unless the unit for the description of forces is known. Then the sizes involved cannot be deeply appreciated until those units become a part of your life experience. So the scientific units of force will be described and then, as introduction, those units will be compared to the everyday non-scientist description of forces. As your experience progresses, the scientific units will take on a more familiar meaning.

In physics the unit for description of a force is the **newton,** and its symbol is **N.** This unit has been adopted internationally and is the unit of force in the **Système International d'Unités** or SI units.

Large forces can be expressed in kilonewtons where 1 kilonewton (1 kN) is 1000 newtons. A million newtons is 1 meganewton or 1 MN.

Small forces may be expressed in millinewtons (1 mN), where 1 mN = 1/1000 N. For even smaller forces the micronewton (μN), which is a millionth of a newton, is used.

In the SI units a quantity is described by a **base unit,** in the case of forces it being the newton. Multiples and submultiples of a base unit are indicated by one of a set of standard prefixes, some of which were illustrated above for the newton. Below is a set of these prefixes that can be used with any base unit. In most cases the prefixes differ by a factor of 10^3 or 10^{-3}. Note the convenience of the use of powers of 10. The prefix centi is in the list because of wide usage, but the International Committee does not fully approve of its use.

Factor	Name	Symbol

$$1\ 000\ 000\ 000 = 10^9 = \text{giga} - \text{G}$$
$$1\ 000\ 000 = 10^6 = \text{mega} - \text{M}$$
$$1\ 000 = 10^3 = \text{kilo} - \text{k}$$
$$\text{base unit}$$
$$1/100 = 10^{-2} = \text{centi} - \text{c}$$
$$1/1000 = 10^{-3} = \text{milli} - \text{m}$$
$$1/1\ 000\ 000 = 10^{-6} = \text{micro} - \mu$$
$$1/1\ 000\ 000\ 000 = 10^{-9} = \text{nano} - \text{n}$$
$$1/1000\ 000\ 000\ 000 = 10^{-12} = \text{pico} - \text{p}$$

As examples of the size of a newton, the force required to pick up a 1 kilogram object on earth is about 9.8 N, or it may be expressed the other way, that the force of gravity of the earth on a kilogram mass is 9.8 N. The force of gravity on you is probably in the range 500 N to 1000 N. You could find your own mass in kilograms and multiply by 9.8 to get the actual force.

The force on one kilogram of mass varies a small amount with its position on the earth, and a few representative values are shown in Table 1–1. The moon is included for interest. The value shown, the number of newtons per kilogram, may be called the gravitational field strength at that place. The variation is due to altitude and to the non-sphericity of the earth (a latitude effect). There is also an effect resulting from the spin of the earth. Because of this latter effect, g as listed is not exactly the gravitational field strength but is more correctly referred to as

TABLE 1–1   Weight in newtons per kilogram at various locations

LOCATION	g (N/kg)
Sea level and 45° latitude	9.80665
Equator	9.78
N pole	9.83
San Francisco	9.80
New York	9.80
Washington, D.C.	9.80
Dublin	9.80
Glasgow	9.82
Johannesburg	9.78
Moscow	9.82
Melbourne	9.80
Rome	9.80
London	9.81
Tokyo	9.80
Bombay	9.79
Moon	1.62

the weight per kilogram. The difference is small and since the force is principally gravitation, g is often referred to as the gravitational field strength.

What we call the weight of an object depends on two things, the amount of mass m, and the gravitational field strength g. The weight is proportional to each of these independently. If you double the mass, the weight doubles; if you triple the mass, the weight triples, and so on. Also, if the gravitational field is increased, the weight will increase, double the gravity and you double the weight, and so forth. Considering both effects in turn:

$$W \alpha\, m\, (\alpha \text{ means "is proportional to")}$$

and $$W \alpha\, g$$

If both m and g are varied, the combined effect depends on the product (for example, if g is doubled and m is tripled, the total weight will be increased by six times), so

$$W \alpha\, mg$$

If g is in units of newtons per kilogram, m is in kilograms and W in newtons, the proportionality sign can be replaced by an equal sign to give

$$W = mg$$

Another unit of force, not an SI unit but one still widely used in science, is the **dyne.** The force of gravity on a 1 gram mass is 980 dynes, or g is 980 dynes per gram.

On 1000 grams or 1 kilogram the force of gravity is 980,000 dynes, and this is also 9.8 N. Consequently, 9.8 newtons are equal to 980,000 dynes or 1 newton is 100,000 or 10^5 dynes. A dyne could also be referred to as 10 micronewtons:

$$1 \text{ N} = 10^5 \text{ dynes}$$
$$1 \text{ dyne} = 10\,\mu\text{N}$$

It is common practice in scientific work to write numbers in powers of 10. This process is described in detail in Appendix 1.

Still another force unit sometimes encountered is the **kilogram of force.** The force of gravity on earth on 1 kilogram of mass is said to be 1 kilogram of force. On a 2 kilogram (kg) mass the gravitational pull is 2 kg of force, etc. The kilogram of force, then, is approximately 9.8 N. The force on 1 kg varies over the earth so the kilogram of force has no fixed value. You can see that it hardly qualifies as a scientific unit and should be avoided. The kilogram of force is sometimes referred to as a kilopond.

The word *kilogram* actually refers to mass. It is confusing to have the word also be a force unit. To avoid difficulty it is best to work in newtons in most of physics. The kilogram of force is not an SI unit and it can lead to confusion with mass. It has been introduced only because you will undoubtedly encounter it, in general use if not in scientific use, and you should at least know what it is.

EXAMPLE 1

A certain person has a mass of 50 kg. What is the force of gravity (or weight) of that person on earth? Express the force in N, in kg of force and in dynes.
The force on 1 kg is 9.8 N.
On 50 kg it is 50×9.8 N or 490 N.
The force is also 50 kg of force.
Since 1 N is 10^5 dynes, the 490 N are 490×10^5 dynes or 4.9×10^7 dynes.

In the English system of units the word *pound* is actually a mass unit, just as is the kilogram. The pound is often used as a force unit, one pound of force being the force of gravity on a one pound mass on earth. The pound of force should not be used as a scientific unit in physics and the conversion factor

$$1 \text{ pound of force} = 4.45 \text{ newtons}$$

can be used to convert to the basic unit of force, the newton.

1–4 EQUILIBRIUM

The subject of this chapter, statics, deals with objects that are under the influence of forces but do not accelerate. This includes objects at rest, and it also includes objects moving at a constant velocity. The two situations, from the standpoint of forces, are the same. You may be moving at 1000 km/h in an aircraft, but the muscle force required to lift a glass is the same as if you were stationary in your own living room. But wait! Is your velocity zero as you sit in your living room? Actually, you are rotating with the earth at a speed of over a thousand kilometers an hour and traveling around the sun at over 30 kilometers a second.

This subject of statics deals with objects either *at rest or moving with constant velocity in some reference frame.* This condition is called equilibrium with respect to linear motion, or more concisely, **translational equilibrium.** The motion of an object may involve translation or rotation. These two aspects will be considered in turn and then in combination. When objects are in translational equilibrium, the forces on them cancel and the result is the same as experiencing no force at all. *The state of translational equilibrium is a state of being either at rest or moving with constant velocity, and the cause of the state is that the forces, if any, cancel.* This book is probably in translational equilibrium. Gravity pulls it downward. Either your hands or the table exert an upward force on it to balance gravity. The net force is zero.

An automobile moving along a road at a constant speed is in translational equilibrium. Gravity pulls it down; the road holds it up. Air resistance and road friction hold it back; the motor turns the wheels and the result is a forward force to balance the frictional force. If the force from the motor exceeds the frictional force, the car accelerates forward or speeds up. If the motor force is less than the frictional force, the car slows down. Some other examples of objects in translational equilibrium are shown in Figure 1–1. These examples all involve forces in exactly opposite directions, so it is easy to cancel them out or sum them to zero. In these cases one might say that translational equilibrium results when "forces up" equal "forces down" and "forces to the right" equal "forces to the left."

Alternatively, the upward direction could be called positive so that downward is negative; then, considering signs, the sum

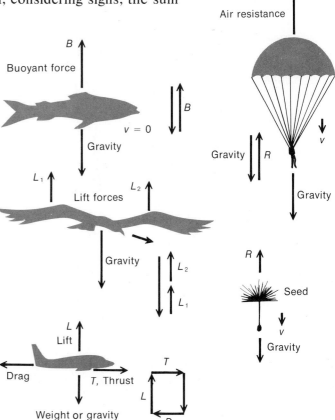

Figure 1–1 Some equilibrium situations in which the forces "balance" to zero. In each case the velocity is either constant or zero.

of the vertical forces is zero for equilibrium. Also, if right is chosen as positive so that left is negative, then the sum of the horizontal forces is zero when the signs are considered.

Forces belong to a class of quantities that are called **vectors.** This type of quantity needs not only a size, or magnitude, but also a direction for its complete description. Two forces of equal magnitude may act on an object in the same direction so they add to give an effect that is the sum of the two; or they may subtract to completely cancel if their directions oppose. If they are at some angle to each other (not $0°$ or $180°$), the effect of the two together will be somewhere between the sum and difference.

EXAMPLE 2

An elevator plus passengers with a total mass of 3000 kg is alternately at rest, then is moved upward at constant speed, stops, and then moves downward at constant speed. What is the force in the supporting cables in each of those states and how does the force differ from these values in the transition between the given states?

First, find the force of gravity on the elevator. It is 9.8×3000 N or 29,400 N (a downward force).

When it is at rest, the cables support it with a force of 29,400 N upward.

The force in the cables must exceed this value to start the upward motion, but when the upward speed is constant, the force is again 29,400 N, upward.

When the elevator slows to a stop, the force is reduced from 29,400 N, but when it is stationary, the force is again 29,400 N.

To begin the downward motion the force is reduced, but to hold it at a constant speed the force is again 29,400 N and still upward.

EXAMPLE 3

A boom as in Figure 1–2 is used in a vertical position to lift a mass. Find the force P at the base when $m = 1000$ kg and the weight W is 9800 N.

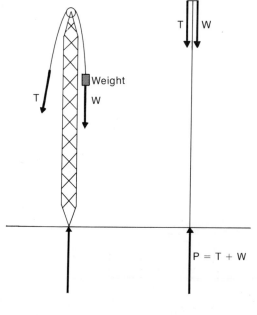

Figure 1–2 The forces applied when a boom is used in a vertical position to lift a weight. The total downward force is $T + W$, which is equal to $2W$, and the force at the ground is also $2W$.

$P = T + W$

To lift the mass the tension T must at least equal W. At the top of the boom there is a force W down on one side and T on the other, so the downward force is $2T$. The upward force P must be equal to $2T$ or 19,600 N.

The force P at the base is 19,600 N.

1–5   UNITS FOR LENGTH

The section to follow will require the use of length or distance as well as forces, so some words are in order on the scientific units for length.

The basic unit in the Système International is the **meter.** (For those of you schooled in English or American units, one meter is 39.37 inches). In the SI system there are prefixes for larger or smaller units. A thousand meters is a **kilo**meter, a thousandth of a meter is a **milli**meter. The **centi**meter is a hundredth of a meter (1 inch is 2.54 cm). The centimeter is not an official SI unit, but the committee responsible for the system has agreed that because of its wide use it will be retained. The meter and centimeter will be used in this work.

The full range of length units is given in Appendix 2.

The word meter is also commonly used to describe an instrument such as a speedometer, flowmeter, odometer, thermometer, etc. The International Committee has suggested that the unit of length be spelled *metre* to distinguish it from an instrument. To date this has been adopted by all countries except the United States.

A brief outline of the development of the meter (metre) may be of interest.

The meter was originally intended to be the length of a simple pendulum that swung with a half-period of 1 second. Such a pendulum, which takes 2 seconds to complete a swing, is called a **seconds pendulum.** After the French revolution in 1789 a commission of the National Assembly of France devised a new system of units, which later became the internationally adopted MKS units. In this system the meter was redefined as 1/10,000,000th the distance between the North Pole and the equator along the meridian through Paris. This was chosen because it would be an indestructible standard, but it was an inconvenient one. A standard meter bar was then constructed to be as close to this as possible, but as time passed it too was found to have some disadvantages. It lacked sufficient precision; its availability throughout the various laboratories of the world was not possible, and there was always the chance of its being damaged or destroyed. With developing technology, it became possible to measure the meter bar in terms of wavelengths of light. In 1961, the meter was redefined as 1,670,763.73 wavelengths of a certain spectral line of light from the element krypton. This is as it now stands. To go back to the original concept, in terms of the new meter a pendulum that swings with a half-period of 1 second is, at Paris, just 0.99390 meters.

The yard is defined as 36/39.37 meters.

EXAMPLE 4

The height of a certain person is 1.70 meters. Express this height in centimeters, in mm, and in the English system of feet and inches.
There are 100 cm in 1 m so 1.70 m = 170 cm.
There are 10 mm in 1 cm so 170 cm = 1700 mm.
1 m is 39.37 inches so 1.70 m = 1.70 × 39.37 inches, which is 66.9 inches or 5 feet, 6.9 inches.

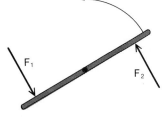

Figure 1–3 A situation in which two forces on an object are of equal size and act in opposite directions. Although in this case there will be an equilibrium for linear or translational motion, there will not be equilibrium for rotation.

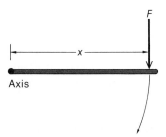

Figure 1–4 The force F produces a torque about the axis. The size of the torque is Fx, and it is clockwise.

1–6   ROTATIONAL EQUILIBRIUM

An object may not be in complete equilibrium under the action of two equal but oppositely directed forces if they do not act toward the same point. The two forces shown in Figure 1–3 would lead to equilibrium as far as linear or translational motion is concerned, but not for rotational motion.

To deal with rotational equilibrium, the concept of a "turning force," or "moment of force" must be introduced. This "turning force" depends not only on the force applied but also on the "lever arm." The **torque** is the product of the force and the perpendicular distance from the point of rotation to the line of action of the force as shown in Figure 1–4. The unit for torque is any unit of force times any unit of length. In SI units torque is expressed in newton meters or Nm.

Two forces of equal magnitude, in opposite directions but not acting along the same line, as in Figure 1–3, form what is called a couple. The size of the couple is the force F times the separation of the two forces. If that separation is x and $F_1 = F_2$, call it F in magnitude, the size of the couple is just Fx. If the rotation is considered to be about a point on the line between them, a distance a from F_1 and b from F_2 so that $a + b = x$, the torque due to F_1 is Fa and that due to F_2 is Fb. The total torque is $F(a + b) = Fx$. So the torque is independent of the position of assumed rotation. It is of magnitude Fx and this is what was given as the size of the couple.

EXAMPLE 5

If in a certain situation in using a wrench, a force of 200 N is exerted at 10 cm from the point of rotation, find the torque being produced. Also find the force that would be required if a wrench with a longer handle were used so that the force arm could be 40 cm rather than 10 cm.

The torque is 200 N × 10 cm = 2000 N cm. In SI units this would be 200 N × 0.1 m = 20 N m. If the new force arm is 0.4 m (40 cm) and the force is F, then for the same torque,

$$F \times 0.4 \text{ m} = 20 \text{ N m}$$
$$F = 20 \text{ N m}/0.4 \text{ m}$$
$$= 50 \text{ N}$$

The torque is 20 N m.
The force on a wrench with a 40 cm handle would be 50 N.

If more than one force operates, the resulting torques may cancel to produce no resulting torque at all, and rotational equilibrium will result. In both cases shown in Figure 1–5, the force F_1 alone would produce clockwise rotation, and F_2 alone would produce counterclockwise rotation. If the torques F_1x_1 and F_2x_2 were of the same size, then there would be no resulting torque. When there is no resulting torque, then there will be rotational equilibrium. This may mean no rotation at all or rotation at a constant speed.

EXAMPLE 6

To push a certain revolving door at a constant speed requires a force of 50 N perpendicular to the door and 0.60 m from the axis. This is balanced by friction, the frictional force acting at 1.5 cm from the axis, on the outside of the axle around which it rotates. Find the frictional force.

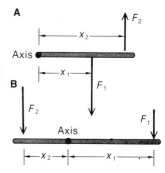

Figure 1–5 Two forces acting on an object to produce zero torque about the axis. The clockwise torque in each case is F_1x_1, and that counterclockwise is F_2x_2. Rotational equilibrium results if $F_1x_1 = F_2x_2$.

As in Figure 1–5, F_2 is 50 N and x_2 is 0.60 m. F_1 is the unknown and it operates at 1.5 cm. In any problem, the units must be consistent so x_1 is 0.015 m, not 1.5 cm. Then

$$50 \text{ N} \times 0.60 \text{ m} = F_1 \times 0.015 \text{ m}$$

or
$$F_1 = \frac{50 \text{ N} \times 0.60 \text{ m}}{0.015 \text{ m}} = 2000 \text{ N}$$

The force of friction is 2000 N.

1–7 SITUATIONS WITH TRANSLATIONAL AND ROTATIONAL EQUILIBRIUM

Equilibrium in a linear sense, which occurs when there is no net force, is called **translational equilibrium.** Translational equilibrium results from the net force being zero.

Rotational equilibrium results from the net torque or moments of the forces being zero. An object may be in both translational and rotational equilibrium, in which case both conditions must hold simultaneously.

The objects shown in Figure 1–5 may be in rotational equilibrium with the forces shown, but they are not in translational equilibrium. In A of the diagram the downward force exceeds the upward force and in B both forces are downward. If there is to be translational equilibrium for those objects, there must be an appropriate force at the axis. That force at the axis would contribute no torque because the force arm is zero, so the rotational equilibrium is not affected. But the force would have to be of such a size that the three forces would sum to zero.

An example of this situation would be the bar shown in Figure 1–6(A). If the upward force F is equal to the sum of the forces shown as W and P, then translational equilibrium will exist. If the point of rotation is at the same point at which P acts, the torques about that point could be zero. Then the clockwise torque produced by W, which is Wx_2, would be equal to the counterclockwise torque produced by F, which is Fx_1. The two simultaneous conditions for translational and rotational equilibrium in that case are

$$F = P + W$$

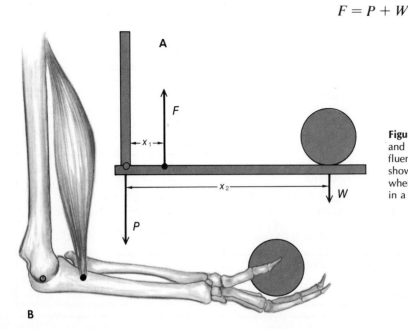

Figure 1–6   (a) A bar in both translational and rotational equilibrium under the influence of three forces. This situation is shown in (b) to represent the forearm when a weight is being held with the arm in a horizontal position.

and $$Fx_1 = Wx_2$$

This arrangement of forces is shown in Figure 1–6(B) to closely represent the lifting of a weight in the hand. The weight is the force shown as W in (A). The biceps pull upward with a force F. At the joint, the bone of the upper arm (the humerus) pushes down on the forearm (the ulna) with the force P.

You should hold a weight in your hand or push upward on the edge of a table or desk and feel for the tense muscle (biceps) and also the tendon and the position at which it is connected to the ulna.

EXAMPLE 7

Find the force in the biceps and between the bones at the elbow when a mass of 15 kg is being held in the hand with the forearm in a horizontal position as in Figure 1–6(B). Use $x_1 = 5$ cm and $x_2 = 35$ cm. Solve the problem using forces in newtons and again with kilograms of force.

The force of gravity on 15 kg is 147 N, and this is W of Figure 1–6. Use the conditions that

$$F = P + W$$

and $$Fx_1 = Wx_2$$

The second of these equations can be solved for the force F in the muscle, all other quantities being known. The known values are these:

$$W = 147 \text{ N}$$
$$x_1 = 5 \text{ cm}$$
$$x_2 = 35 \text{ cm}$$

Dividing both sides by x_1 and putting in the numbers gives

$$F = 147 \text{ N} \times \frac{35 \text{ cm}}{5 \text{ cm}}$$

Note that the unit cm cancels to give

$$F = 1029 \text{ N}$$

Then from the first equation, $P = F - W$:

$$P = 1029 \text{ N} - 147 \text{ N} = 882 \text{ N}$$

The force in the muscle is 1029 N and the force between the bones is 882 N. The muscular forces and the forces between the joints can be surprisingly large.

To solve the situation using forces in kilograms of force, use $W = 15$ kg of force and perform the same type of analysis. This will result in

$$F = 15 \text{ kg of force} \times \frac{35 \text{ cm}}{5 \text{ cm}}$$
$$= 105 \text{ kg of force}$$

and $$W = 105 \text{ kg of force} - 15 \text{ kg of force}$$
$$= 90 \text{ kg of force}$$

These answers are equivalent to the others.

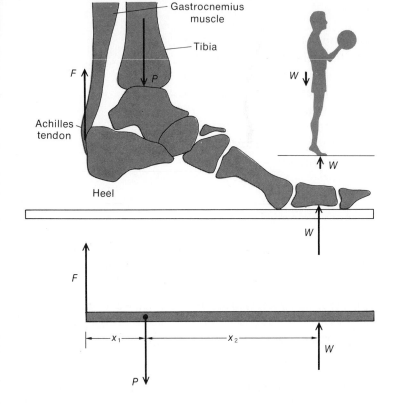

Figure 1–7 The forces involved in raising the body to stand on the ball of the foot.

A further example of such an application of statics as it applies to the body, an area of study often referred to as **biomechanics,** is the finding of the forces required for a person to stand on the ball of one foot. This is illustrated in Figure 1–7. The force between the ball of the foot and the floor is just the weight W of the person. The floor pushes up on the person at this point with a force W. At the joint in the ankle there is a downward force of P. Through the Achilles tendon in the heel, the gastrocnemius muscle pulls upward with a force F.

To have translational equilibrium, the upward forces $F + W$ must equal the downward force P. If W acts at a distance x_2 from the joint and F acts at a distance x_1, then (to have equilibrium against rotation) Wx_2 must equal Fx_1. The development of the heel in mammals has led to an increased lever arm for the muscle force F, allowing a smaller muscular force to lift the body onto the toes, or to jump, than would be necessary with a shorter distance x_1. The equations are

$$F + W = P$$

and
$$Fx_1 = Wx_2$$

Some typical values are $x_1 = 4.5$ cm and $x_2 = 13$ cm. If the mass of the person is 70 kg so the weight W is 685 N, then the second equation can be solved for F and this can be used in the first equation to find P, the force between the bones in the ankle. The results are that $F = 1980$ N (202 kg of force) and that $P = 2665$ N (272 kg of force). Note how the forces acting in the muscles and at the joints can greatly exceed the weight of the person, which in this case was only 685 N (70 kg of force). The muscle force was almost three times the weight and the force across the ankle was almost four times.

1–8 THE MOMENT OF A FORCE

In Figure 1–6(A), a bar was shown in equilibrium under the action of three forces. The force P was considered to be applied at a joint, which was a natural position about which to calculate torques. If there were no joint at that place, but the force P was acting at that position, there would still be equilibrium. The question arises: about what point should the torques be considered when there is no apparent axis of rotation? The answer is that if the object is in rotational equilibrium (consider zero rotation, for example), it is in rotational equilibrium about any point; the point that is considered as an axis is arbitrary. This statement will be justified in the following:

Figure 1–8 is similar to Figure 1–6(A), but the bar has been extended an arbitrary distance X to the point A. The forces P and W would each produce clockwise rotation about A, and F would produce counterclockwise rotation. If the point A was an actual axis of rotation, the **torque** about that axis could be calculated. If the point is not an actual axis, the product of force times force arm is usually referred to as the **moment** of the force about that point (or, frequently, the **first moment**). The word "torque" is usually reserved for an actual axis of rotation, but the term "moment of a force" can be used about any arbitrary point.

Equating clockwise and counterclockwise moments about A,

$$PX + W(x_2 + X) = F(x_1 + X)$$

Expanding this:

$$PX + Wx_2 + WX = Fx_1 + FX$$

Collecting the terms in X,

$$(P + W - F)X + Wx_2 = Fx_1$$

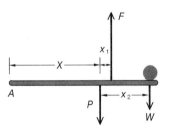

Figure 1–8 The moment of the forces about an arbitrary point A, for an object in equilibrium.

For translational equilibrium, $P + W = F$ or $P + W - F = 0$. The first term then drops out to leave

$$Wx_2 = Fx_1$$

Since the term in X dropped out, the choice of a point about which to take the moments of force is seen to be completely arbitrary in an equilibrium situation. If the object does not rotate about any one point, it is not rotating about any other point either.

EXAMPLE 8

Consider the forces acting on a bar as in Figure 1–9. The values shown in (A) are the same as were used in Example 7, but the point of rotation is to be considered at 10 cm to the right of the force of 147 N. Example 7 applied to the lifting of a weight, and the point of rotation was considered to be the elbow. In this example, the imagined point of rotation is somewhere around the finger tips, not a real axis so we should refer to moments not torques.

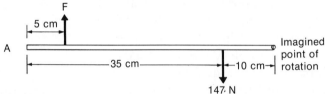

Figure 1–9 Calculation of moments about an arbitrary point of rotation.

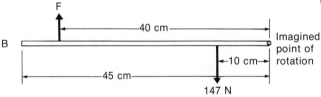

The problem is to find both P and F.

If the force F acted alone, it would cause clockwise rotation about R so it would have a clockwise moment. Both the force P and the force of 147 N would produce counterclockwise rotation about R. For rotational equilibrium the clockwise moments are equal to the counterclockwise moments. Using the numbers shown in Figure 1–9,

$$F \times 40 \text{ cm} = (P \times 45 \text{ cm}) + (147 \text{ N} \times 10 \text{ cm})$$

There are two unknowns so a second equation is required, and that can be based on the condition for translational equilibrium:

$$F = P + 147 \text{ N} \qquad \text{or} \qquad P = F - 147 \text{ N}$$

Substitute for P in the first equation,

$$F \times 40 \text{ cm} = (F - 147 \text{ N}) \times 45 \text{ cm} + 1470 \text{ N cm}$$
$$F \times 40 \text{ cm} = F \times 45 \text{ cm} - 147 \text{ N} \times 45 \text{ cm} + 1470 \text{ N cm}$$
$$-5 \text{ cm } F = -6615 \text{ N cm} + 1470 \text{ N cm}$$
$$F = 1029 \text{ N}$$

and it follows that $P = 882$ N.

These are exactly the same answers as were obtained in Example 7, so it is apparent that the choice of the point of rotation did not influence the answer. You cannot take the moments about the wrong point for any will do. However, in a practical problem a useful trick is to take the moments about the point of action of one of the unknown forces. That unknown does not appear in the resulting equation, and the algebra to obtain the solution is reduced.

1-9 CENTER OF GRAVITY

The previous examples and discussions have considered forces acting only at particular points. The problem now is how to deal with a force such as that of the force of gravity on an extended body. Gravity is pulling on all parts of the object and just how is such an extended force handled in equilibrium problems? The solution is in the idea that for equilibrium problems any extended body is equivalent to such an object with the total force of gravity acting at one point, that point being called the center of gravity, as in Figure 1-10. For a uniform solid such as a rod, sphere, etc., the center of gravity coincides with the geometrical center. If the object is not uniform, various techniques can be used to find the center of gravity. One of these is to suspend the object with one upward force. To get into equilibrium it must then move so the center of gravity is in the vertical line of the supporting force, as in Figure 1-11. Alternately, the supporting force must be made to act in the vertical line through the center of gravity.

Another technique is to apply two supporting forces, measure them and then calculate the position at which the total downward force would act to result in the measured upward forces.

As in Figure 1-12(A) or (B), the mass m of weight W is supported by forces F_1 and F_2 at distances x_1 and x_2 from one

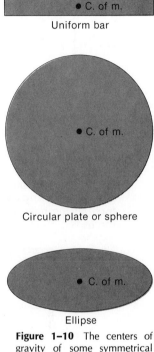

Figure 1-10 The centers of gravity of some symmetrical objects.

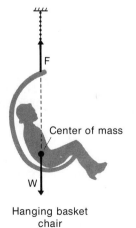

Hanging basket chair

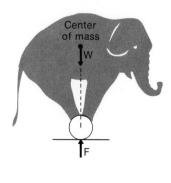

Figure 1-11 An object can be supported by one upward force if the center of mass is in the vertical line of the supporting force.

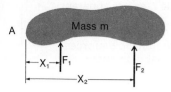

A

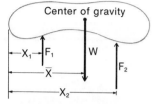

B

Figure 1-12 Using two up-ward forces to find the center of gravity of an object.

end. The total weight is considered to act at the unknown distance $\bar{x}$. Applying the equilibrium condition,

$$F_1 x_1 + F_2 x_2 = W\bar{x}$$

and

$$F_1 + F_2 = W$$

which combine to give

$$\bar{x} = \frac{F_1 x_1 + F_2 x_2}{W}$$

This method is easily adaptable to find the center of gravity of a person. Those of my friends who are swimmers tell me that this is a useful thing to know.

EXAMPLE 9

Find the position of the center of gravity of a person using two supporting forces and the following data. As in Figure 1–13, the person lies horizontally and the supporting force at the head is 366 N, while that at the heels is 300 N. The distance between the forces is 1.55 m.

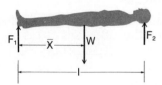

Figure 1–13 Finding the center of gravity of a person. A thin board may be used as a support, but the supporting forces for the board alone must be subtracted. This can be done with chairs and a "bathroom" scale.

The force W is the sum of the two upward forces or 666 N. (The mass of the person must be 68 kg.)

Let W act at a distance $\bar{x}$ from the heels, considering rotation about the heels:

$$666 \text{ N } \bar{x} = 360 \text{ N} \times 1.55 \text{ m}$$

$$\bar{x} = 0.84$$

The center of gravity is 0.84 m from the heels.

1–9–1   THE CENTER OF GRAVITY OF AN ARRAY OF OBJECTS

Any body of finite size can be considered to be made up of a very large number of smaller objects, and these ideas about centers of gravity will be illustrated using a number of separate objects. You will see that the ideas could be extended to continuous objects.

Consider four masses in a line on a weightless bar as in

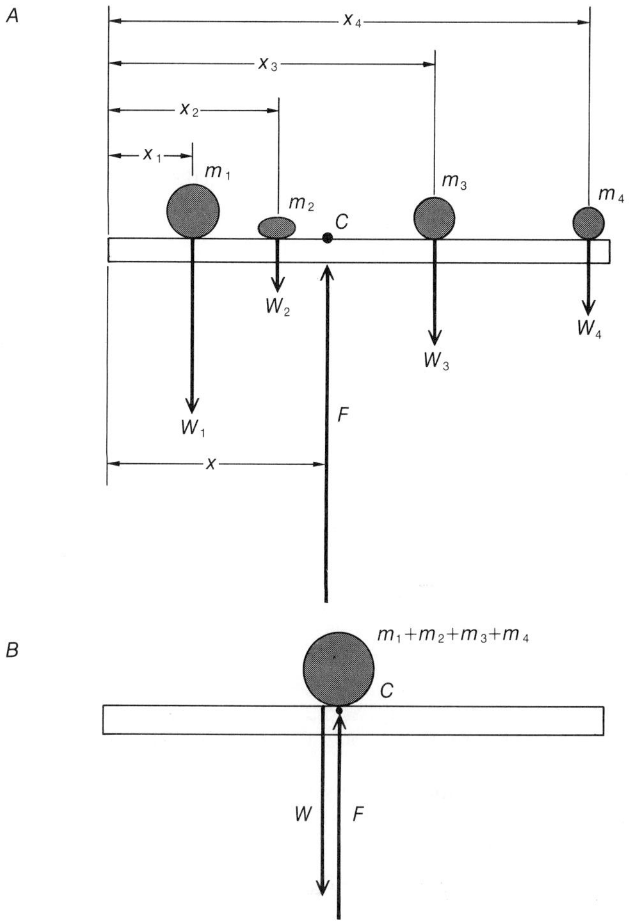

Figure 1–14 A large number of masses on a bar being balanced by one upward force at C. Equilibrium would also be achieved if all of the masses were concentrated at C, as in part (b).

Figure 1–14. The weights are W_1, W_2, W_3 and W_4. This situation is the same as if all the masses were concentrated at the position where F acts as in Figure 1–14(B). Considering the moments about the left hand end of the bar, F would produce a counterclockwise moment and the weights would each produce clockwise moments. For rotational equilibrium the net moment of all the forces must be zero.

So
$$F\bar{x} = W_1 x_1 + W_2 x_2 + W_3 x_3 + W_4 x_4$$

and for translational equilibrium

$$F = W_1 + W_2 + W_3 + W_4$$

Using this and solving for the position at which F would be applied, or the position of the center of gravity,

$$\bar{x} = \frac{W_1 x_1 + W_2 x_2 + W_3 x_3 + W_4 x_4}{W_1 + W_2 + W_3 + W_4}$$

In words,

$$\bar{x} = \frac{\text{sum of the moments of the weights}}{\text{sum of the weights}}$$

A notation often used is that of the upper case greek sigma (Σ). This is used to mean "the sum of," and a small roman letter, often i, is used as a subscript, which can take various values indicated.

Using this,

$$\bar{x} = \frac{\Sigma W_i x_i}{\Sigma W_i}$$

where i has the values 1 to 4 in our example, though in any given situation the range of i is whatever applies.

EXAMPLE 10

Consider a long bar of negligible weight with a 1 kg mass at 1 meter from the end, 2 kg at 2 meters and 3 kg at 3 meters. Find the position and the size of one upward force that would balance the bar horizontally, which will be the position of the center of gravity.

The weights, using g = 9.8 N/kg, are 918 N, 19.6 N and 29.4 N. The total weight is 58.8 N.

Using the relation $\bar{x} = \dfrac{\text{sum of the moments}}{\text{sum of the weights}}$

we see that $\quad \bar{x} = \dfrac{9.8 \text{ N} \times 1 \text{ m} + 19.6 \text{ N} \times 2 \text{ m} + 29.4 \text{ N} \times 3 \text{ m}}{58.8 \text{ N}}$

$$= \frac{137.2}{58.8} \text{ m} = 2.33 \text{ m}$$

The center of gravity is 2.33 m from the end indicated.

1–9–2 CENTER OF MASS AND MOMENT OF MASS

The term **center of mass** is often used interchangeably with the term **center of gravity,** and though they often coincide there is a basic difference between them. Because of a new concept to be introduced the distinction will be described. The center of gravity was given by the following relation, where any number of weights can be included as indicated by the lines of dots:

$$\bar{x} = \frac{W_1 x_1 + W_2 x_2 + W_3 x_3 + \ldots\ldots\ldots\ldots}{W_1 + W_1 + W_2 + W_3 + \ldots\ldots\ldots}$$

The forces called weights occur because of gravity acting on the masses, and if the gravitational field is g, then $W_1 = m_1 g$, $W_2 = m_2 g$, etc.
Using this,

$$\bar{x} = \frac{m_1 g x_1 + m_2 g x_2 + m_3 g x_3 + \ldots\ldots\ldots}{m_1 g + m_2 g + m_3 g + \ldots\ldots\ldots\ldots}$$

g occurs in all terms in numerator and denominator so it can be factored out of each as long as it is of the same strength at the position of each mass (herein is a fine distinction between centers of gravity and of mass). The g will then cancel to leave

$$\bar{x} = \frac{m_1 x_1 + m_2 x_2 + m_3 x_3 + \ldots\ldots\ldots}{m_1 + m_2 + m_3 + \ldots\ldots\ldots}$$

The quantity that is a mass times a distance is not like a torque. No force is implied or necessary. The quantity mx is referred to as the **moment of the mass** or more correctly, since the distance x is to the first power, the first moment of the mass. Then, in words, the center of a mass is at a distance $\bar{x}$ given by

$$\bar{x} = \frac{\text{sum of the moments of the masses}}{\text{sum of the masses}}$$

EXAMPLE 11

Repeat the previous example but find the center of mass.

$$m_1 = 1 \text{ kg}, \; x_1 = 1 \text{ m}$$
$$m_2 = 2 \text{ kg}, \; x_2 = 2 \text{ m}$$
$$m_3 = 3 \text{ kg}, \; x_3 = 3 \text{ m}$$

$$\bar{x} = \frac{1 \text{ kg} \times 1 \text{ m} + 2 \text{ kg} \times 2 \text{ m} + 3 \text{ kg} \times 3m}{1 \text{ kg} + 2 \text{ kg} + 3 \text{ kg}}$$

$$= \frac{14 \text{ kg m}}{6 \text{ kg}}$$

$$= 2.33 \text{ m}$$

The center of mass is at 2.33 m as was found for the center of gravity in the previous example.

1-10   FORCES AT ANGLES

In the preceding sections the situations chosen have been only those in which the forces are parallel to each other, and nicely perpendicular to the force arms. The basic ideas about

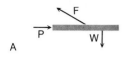

A

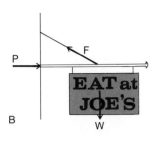

B

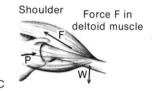

C

Figure 1–15 An object in equilibrium, with one of the forces at an angle. (a) The general diagram; (b) a sign on a building; (c) the forces at the shoulder necessary to hold the arm in a horizontal position.

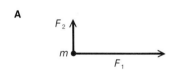

A

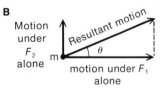

B

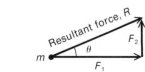

C

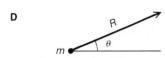

D

Figure 1–16 Two forces acting on a mass, and the one force R that is their equivalent.

equilibrium are all there, and it is only a small extension to consider forces not parallel to each other and not perpendicular to force arms.

Forces in that class are called **vectors,** and in adding them or otherwise considering their effects, the direction must be taken into account. Examples of forces acting at an angle are those exerted by the deltoid muscle (Fig. 1–15) in holding the arm out horizontally, and by the illustration of a sign being held outside a building. The same simplified diagram applies to both illustrations. It is not necessary that the force shown as P acts horizontally but it may.

Vectors are represented on a diagram by an arrow at the appropriate point. The length of the arrow is representative of the size (or magnitude) of the quantity, and the end is marked by an arrowhead. In print, vectors are usually indicated by bold face type or sometimes by an arrow over the letter, such as **F** or $\vec{F}$. The indication of a vector by the arrow over the letter is usually used in hand-written notes.

1–10–1   ADDING FORCES

No matter how many forces act on a body, the motion (if the body moves) could be duplicated if only one force was acting. An object cannot move in two or more directions at once! If the object does not move, it is as though no forces act on it. Consider the object shown in Figure 1–16(A); it is under the influence of two forces F_1 and F_2 which, in this case, are at right angles to each other. (Note that the ideas to be developed are not limited to perpendicular forces.) The motion under the influence of these forces individually is shown in Figure 1–16(B), along with the resultant motion. This resultant motion would be in the direction obtained by adding F_2 to F_1 as shown in Figure 1–16(C). The motion would be the same if the body was acted on only by one force R, shown in Figure 1–16(D). Two or more forces acting on an object can be added by this method of imagining or drawing a scale diagram and adding the forces by this "tail to point" method. It is not limited to forces that are perpendicular to each other. That configuration was chosen as an example because it is a simpler matter to solve right angle triangles than others. In the example of Figure 1–16(C), the value of R is found by the Pythagorean theorem:

$$R^2 = F_1^2 + F_2^2 \quad \text{or} \quad R = \sqrt{F_1^2 + F_2^2}$$

Also, the direction θ in the figure is given by

$$\tan \theta = F_2/F_1$$

The force R can be referred to as the **vector sum** of F_1 and F_2. The notation used, with the bold face type indicating that the quantities are vectors, is

$$\mathbf{R} = \mathbf{F}_1 + \mathbf{F}_2$$

In writing this longhand in your notes, arrows could be put over the quantities as a reminder that direction must be considered.

$$\overrightarrow{\mathbf{R}} = \overrightarrow{\mathbf{F}}_1 + \overrightarrow{\mathbf{F}}_2$$

EXAMPLE 12

A 10 kg mass is suspended by a long cord, and a force of 50 N is applied horizontally. Find the size and the direction of the resultant of the weight and the applied force, and hence the direction of the supporting cord and the tension in it.

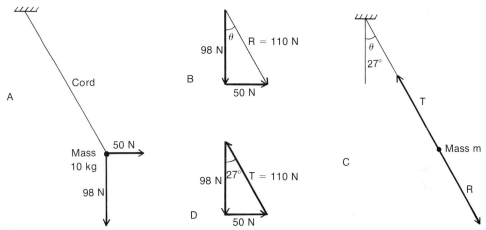

Figure 1–17 A mass hanging on a cord but held at an angle by a horizontal force.

First, the force of gravity on the 10 kg mass is 98 N downward, so the force diagram is as in Figure 1–17(A). The vector addition is done by the tail to point method as in (B) of the figure. The size of R is given from

$$R^2 = 50^2 + 98^2$$
$$R = 110$$

or

The resultant force is 110 N.
The direction of the force from the vertical, θ, of Figure 1–17(B) is found from

$$\tan \theta = 50/98 = 0.510$$
$$\theta = 27°$$

The cord must pull back along the same line as that resultant force and with a tension equal in size to R so equilibrium will result.
The tension in the cord is 110 N and the mass hangs at 27° from the vertical.
The three forces, the weight W (98 N), the sideways force F (50 N) and the tension T (110 N) could all be added by the tail to point method as in (D) of Figure 1–17. The three forces form a closed triangle so the resultant of all three is zero and equilibrium results.

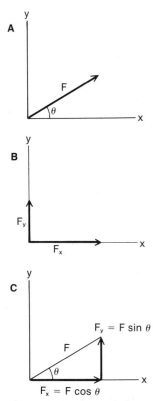

Figure 1–18 The single force in (a) is equivalent to the two forces in (b), which, when added as in (c), give the force in (a) as the resultant.

1–10–2 COMPONENTS OF FORCES

Let us consider now the reverse of adding forces. That is, the effect of any one force can be considered to be the same as if two other forces were acting, such that the two new forces sum to give the original one. In Figure 1–16(D), for example, if R is the force given in the situation, it could be replaced by the two forces F_1 and F_2. Those two forces, if they are perpendicular to each other, are called the **rectangular components** of R.

The procedure of breaking single forces into two would seem to complicate a situation, but this is not so; it can, in fact, simplify it. The equilibrium conditions dealt with in Figure 1–1 were limited to forces in two mutually perpendicular directions: forces up and down, or forces right and left. Any force acting at an angle can be broken into components in the two chosen directions. Then all of the previous ideas about balancing forces can be applied.

The components of a force F acting at an angle θ from some chosen direction, as in Figure 1–18, are obtained as follows. The directions in which the components are to be chosen are referred to as x and y, and the components are called F_x and F_y. They are given by:

$$F_x = F \cos \theta$$
$$F_y = F \sin \theta$$

The force of the shoulder muscle on the arm (the force F of Figure 1–15) can be broken into components, as shown in Figure 1–19. If F acts at an angle θ, it can be broken into $F \sin \theta$ upward and $F \cos \theta$ acting horizontally toward the shoulder. This example shows that by using components, a complex situation can be reduced to one in which the forces act only in two mutually perpendicular directions. If an object is in equilibrium, the forces in each of those two directions must cancel independently.

If the direction to the right is chosen as the positive x direction and that to the left is chosen as the negative x direction, then the sum of the forces (with signs) in the x direction must be zero at equilibrium. With similar attention to signs, the sum of the forces upward and downward, or in the positive and negative y directions, must also be zero at equilibrium. In symbols, with the upper case Greek (Σ) standing for "the sum of," this can be stated for translational equilibrium as:

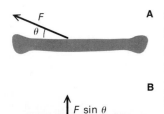

Figure 1–19 The force at an angle as in Figure 1–15, represented by two components: one along the arm and one perpendicular to it.

and

$$\Sigma F_x = 0$$

$$\Sigma F_y = 0$$

This is the standard way of expressing the condition for translational equilibrium.

EXAMPLE 13

Consider a horizontal object with three forces, all acting in line with the center of mass, as in Figure 1–20. This situation is similar to the shoulder muscle (the deltoideus) holding the arm horizontally, except that all the forces are considered to act at the same position so that there is no rotation. The force F shown acting at the angle θ above the horizontal can be broken into components, as in Figure 1–19. For equilibrium, $F \sin \theta$ must balance W, and P balances $F \cos \theta$. In the form of equations,

$$F \sin \theta = W \quad \text{or} \quad F \sin \theta - W = 0$$
$$F \cos \theta = P \quad \text{or} \quad F \cos \theta - P = 0$$

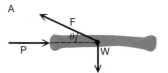

Figure 1–20   A set of forces almost the same as those in Figure 1–15; for simplification, all are assumed to act at a point. The three forces are shown in (a); in (b) they are shown with F broken into components. For a weight W of 3 lb and an angle θ of 15°, the values of F and P that lead to equilibrium are shown in (c).

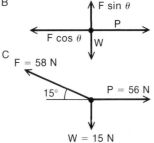

Using representative numbers, let the mass of the arm be about 1.5 kg, so W is 15 N and $\theta = 15°$. Then

$$F \sin 15° = 15 \text{ N}$$
$$F = 15 \text{ N}/\sin 15°$$
$$= 58 \text{ N}$$

Also

$$P = F \cos \theta$$
$$= 58 \text{ N} \cos 15°$$
$$= 56 \text{ N}$$

In Figure 1–19(C), the solution is illustrated diagrammatically. The forces F and P both greatly exceed the weight W.

1–10–3   MOMENTS OF FORCES NOT PERPENDICULAR TO THE FORCE ARM

Consider a force that acts on a lever arm to rotate it, but that is not perpendicular to the arm, as in Figure 1–21(A). The force is tipped to an angle θ from the perpendicular to the arm.

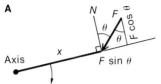

A

B

C

Figure 1–21 (a) The torque or moment of a force that is not perpendicular to the arm. (b) Only the component $F \cos \theta$ exerts a torque, which is then $(F \cos \theta)x$. If, as in (c), the perpendicular arm from the line of action of F is taken, the product $F(x \cos \theta)$ is again the torque.

That force can be broken into two rectangular components, as in Figure 1–21(B). The component $F \cos \theta$ tends to cause rotation; but the component $F \sin \theta$ pushes directly on the axis, and, having no lever arm, it cannot cause rotation. The torque or moment due to F is

$$torque = (F \cos \theta)x$$

Another way to handle this situation is shown in Figure 1–21(C). The line of action of the force is shown, and the perpendicular distance from the axis to that line is $x \cos \theta$. The force times this perpendicular distance is $F \cdot x \cos \theta$ or

$$torque = F(x \cos \theta)$$

This is exactly the same as the previous expression.

The conclusion is that if a force is acting at an angle θ from the perpendicular to the force arm, the torque (or moment) can be found either by multiplying the component of the force that is perpendicular to the arm ($F \cos \theta$) by the lever arm x, or by multiplying the force F by the perpendicular distance from the line of action of the force to the point of rotation. The results will be the same.

An example of this concept of the lever arm as the perpendicular distance from the axis to the line of action of the force is shown in Figure 1–22. The force W is shown acting at the center of mass of the leg; as the leg is lowered, the perpendicular force arm increases, and hence the torque about the hip joint also increases.

1–11   BIONICS

When you watch mechanical systems that lift, dig, scoop, etc, the similarity to biological systems such as arms, legs, hands, and so on is obvious. Mechanical systems can be devised with

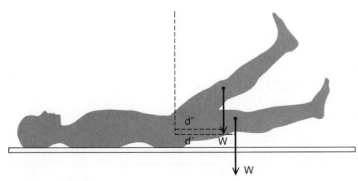

Figure 1–22 The use of the concept of moment or torque as the product of force times the perpendicular force arm. In doing the exercise shown, the perpendicular lever arm varies with the angle of the leg. (Adapted from M. Williams and H. Lissner, *Biomechanics of Human Motion*, W. B. Saunders Company, Philadelphia, 1962.)

Figure 1–23 Mechanical devices that are very similar to human limbs. (Part (b) courtesy of NASA.)

the aid of the study of biological systems. The technique of solving mechanical or other problems using the study of biological systems has earned the name **bionics,** this example of mechanical devices for lifting and digging being one example of it.

In this area there is one notable difference. Muscles can only pull, like cables. In making mechanical devices, hydraulic systems allow also the application of a push. In this way there is more scope in the design of the machines, but, nevertheless, the study of how the problems are solved in nature can be of great assistance in science and technology.

The devices referred to can be seen working about our cities, in construction and maintenance projects. More sophisticated examples are the instruments used for the remote control handling of radioactive materials and the machines that are sent to the moon or Mars to scoop soil samples, bring them back into the space vehicle, perform analyses and then radio the results back to earth. Some examples are shown in Figure 1–23.

1–12 FORCES IN PAIRS

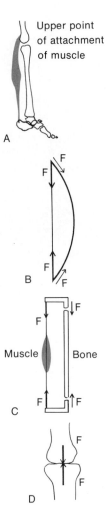

It is not possible to pull on the end of a rope that is in equilibrium without something exerting an equal force on the other end. Similarly, the muscle in the calf of the leg pulls up on the heel to raise it off the floor as in Figure 1–7, and to do this the upper end of the muscle must be fastened somewhere, at which point it also pulls down. In that example, the muscle is connected to the femur, the bone in the upper leg, just above the knee as in Figure 1–24(A). To be fully correct, the muscle divides and is fastened to each side of the femur. Not considering this detail, the situation is similar to a taut bow string as in Figure 1–24(B). The string pulls on each end of the bow and at those points the bow pulls back on the string. Muscles are like bow strings, but they also have the ability to shorten and exert a tension. This process at the cellular level is a further area of study.

The muscle tension force pulls along the bone, and the bone must be capable of withstanding that force as well as any weight already on it as in Figure 1–24(C). The joints are somewhat like slices through the bone as in the simplified diagram (C). The stress along the bone occurs across these joints, where it presses the two surfaces together. The bones push together as in (D) of Figure 1–24. The harder two surfaces push together, the more difficult it is to slide them. In order that joints can move under the large forces that occur, the friction must be extremely low. This is indeed the case in a normal joint, and it is one of the problems to be overcome in the design of artificial joints.

There is another important physical point here. *Whenever any force acts there is also acting another force of equal size but opposite in direction.* This is shown in the muscle or where the bow string meets the bow or across the joint as in Figure 1–24(D). This concept is one of the more important in physics and has earned the name of **Newton's third law.**

To illustrate Newton's third law further, hold your hands with two fingers touching as in Figure 1–25. Whether you push gently or hard the force of the left finger on the right is always exactly the same as that of the right finger on the left. Try it!

Figure 1–24 Forces always occur in pairs. The forces on the two ends of a muscle are equal in size. The force along a bone acts across a joint with equal forces as in (d). An analogy is made between the system of muscle and bone and the system of a bow with a taut string.

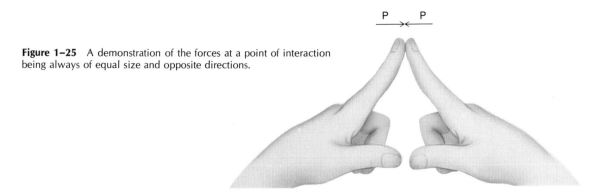

Figure 1–25 A demonstration of the forces at a point of interaction being always of equal size and opposite directions.

The two forces that occur in such situations act on different objects. The finger of your right hand exerts a force on the finger from the left hand and vice versa. In walking, your foot pushes on the earth: the earth pushes on your foot. You move forward and the earth moves backward (very, very little). At the knee joint the femur (the bone in the thigh) pushes on the ulna (the bone in the lower leg) and the ulna pushes back on the femur.

1–13 STABLE AND UNSTABLE EQUILIBRIUM

An object balancing on one or more points of support may be stable or unstable. If it is stable, small motion will cause the center of mass to move in such a manner that a restoring torque will be created to return the object to its original position. If it is unstable, a disturbance will result in an increase in the motion away from the equilibrium position. In the stable situation the disturbance raises the center of mass. In the unstable situation the disturbance lowers the center of mass. Between stable and unstable equilibrium is the situation called neutral equilibrium. A sphere on a level surface is in neutral equilibrium.

If there are two or more points of support for a linear object, the object will be stable as long as the center of mass is above the line between the points of support. For an extended object three or more points are necessary and the center of mass must be above the area enclosed by the supporting points. These cases, with more than one point of support, lead to stable equilibrium because the center of mass must rise if the object pivots about a support point.

Some particular stable situations with one point of support are of interest.

The common concept of a balance—two masses on a bar suspended at the center of mass as in Figure 1–26—is not a

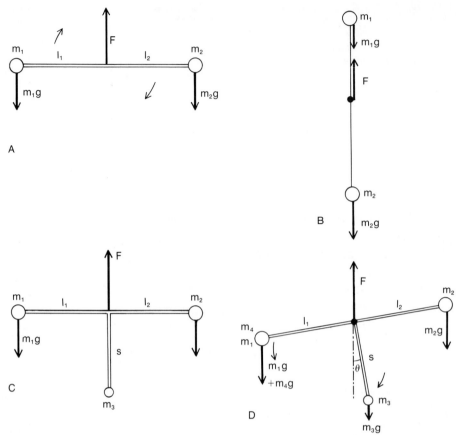

Figure 1–26 A simple equal-arm balance as in (a) is unstable, for any difference between m_1 and m_2 will result in movement to a position such as that in (b). If a bar s with a small mass m_3 is added as in (c), then a small imbalance as in (d) will lead to a slightly displaced but stable orientation.

stable situation. An initial rotation will continue because there is no restoring force created when a displacement is begun, or if a small unbalancing mass is added to m_1; the only stable position will be as in (B) of Figure 1–26. However, if a third mass (m_3) is added in (C) of Figure 1–26, a stable situation will occur. When a displacement is made, m_3 will move and create a restoring torque as in (D) of Figure 1–26. If there is imbalance caused by adding a small mass m_4 to m_1, the new equilibrium will occur when $m_4 l_1 \cos \theta$ is equal to $m_3 S \sin \theta$. The sensitivity of a balance is determined by the amount of restoring torque produced for a given displacement.

1–14 NUMERICAL ANSWERS

In science there is no such thing as an exact measurement. Every measurement is made to a certain degree of precision, which can be indicated approximately by the number of digits given. For example, if a common scale is used to measure the width of a piece of paper, the result could be given as

	0.213 meters
or	213 millimeters
or	213,000 micrometers
or	2.13×10^5 micrometers

These measurements are all to the same precision; there are three recorded digits. A term frequently used is three *significant figures*, which is corrupted to 3 sig figs. The sig figs are the key to this work. The "number of decimal places" has nothing to do with precision.

Had a more precise measuring instrument been used, the measurement could have added one or more significant figures. Perhaps it would have been

	0.2134 m
or perhaps	0.2127 m
or even	0.2130 m

The more precise measurement should be in the range from 0.2125 m to 0.2135 m if the rounded-off value is to be 0.213 m. A measured value is considered to have a possible variation of ± 5 in the first unrecorded digit. That is, $0.213 - 5$ is 0.2125 to give the lower end of the range, and $0.213 + 5$ is 0.2135 to give the upper end of the range.

If the last determined figure is 0 it must be recorded as such. 0.213 m is not the same as 0.2130 m.

The number written as 0.2130 m could also be written as 213,000 μm, which is the same as 0.213 m would be written. The style 2.130×10^5 μm would, however, indicate that four figure accuracy was obtained.

When measured numbers, which are always of limited accuracy, are used in calculations, the answer also has a limitation in accuracy. Perhaps the page measured above was 0.314 m long, so what was the area? Let A be the area, w the width and l, the length.

$$A = wl$$
$$= 0.213 \text{ m} \times 0.324 \text{ m}$$

This multiplies out to be 0.069012 m².

From this you would infer that the area is known correct to 5 sig figs. However, the width is only known to be in the range

0.2125 to 0.2135 m and the length is in the range 0.3235 to 0.3245 m. Multiplying these limiting values with

$$0.2125 \text{ m} < w < 0.2135 \text{ m}$$
$$0.3235 \text{ m} < l < 0.3245 \text{ m}$$

then $\quad 0.06874375 \text{ m}^2 < A < 0.06928075 \text{ m}^2$

Rounding the area to three sig figs,

$$0.0687 < A < 0.0693$$

With measured values to 3 figures the result is hardly good to three figures, but the answer does not warrant rounding off to two figures because the answer is closer than ± 5 in the third figure.

The general rule to follow, a rule which is only a rough guide to precision, is to give an answer with the same number of sig figs as were in the measured values used in the calculation. The answer can be no more precise than the values used in its calculation.

Numbers frequently occur in formulas, perhaps a 2 or 1/2. These numbers are considered infinitely precise at this stage.

The value π will frequently occur and it can be used to one more sig fig than the measurements, so it does not introduce error into the result. The approximation 22/7 as π is usually not acceptable. The ancient Chinese used 355/113 as an approximation to π, which is certainly better than 22/7 as shown below.

To ten sig figs: $\quad \pi = 3.141\ 592\ 654$
$$355/113 = 3.141\ 592\ 920$$
$$22/7 = 3.142\ 857\ 143$$

Draw your own conclusions!

In most of the problems in the book, three sig figs are called for, but sometimes two or four or more will be used. The answer should be given to about the same number of sig figs as were the data. You will usually not be faulted for putting in an extra one because they give only a guide to accuracy. However, if measured values are used and you write

$$\frac{4.56}{1.23} = 3.707317073$$

the answer should be marked wrong because it gives a completely incorrect idea of the precision.

$$\frac{4.56}{1.23} = 3.71 \text{ would be correct}$$

In the problems watch the significant figures!

DISCUSSION PROBLEMS

1-3   1. The word *mass* has not been defined in the text. Rather you have been expected to have some concept of what it is. Devise a definition of it, remembering that the weight of a given mass varies with its position and may even be zero, as in a spacecraft.

1-4   2. A person makes a trip in a car. Starting from rest the car accelerates to 20 km/h northward and travels 1 km. It then turns west and travels for 2 km still at 20 km/h. At that point the car stops. In what parts of the trip are the car and the person in it in translational equilibrium and in what parts are they not?

1-7   3. Push down on a table with your hand, keeping your forearm horizontal. Find the muscle that pulls on the main bone of the forearm (the tibia) and see where the tendon involved attaches to the ulna. Make a force diagram for this situation.

1-8   4. Make a diagram of the head supported on the spinal column, which is behind (posterior to) the center of mass of the head. Find and show the muscles that hold the head erect. These are also the muscles that pull the head back, and you can find them more easily by pushing back against something with your head while you feel for the tense muscles. Make a force diagram for this situation.

1-9   5. Ask a swimmer or a swimming instructor about the use of the knowledge of the center of gravity of the body.

1-9   6. Stand with your back against a wall, your heels touching the wall. Bend over to pick up an object on the floor in front of you. Usually men cannot do this without falling over, but women can. Explain the falling over in terms of the motion of the center of gravity of the body as the upper part moves outward from the wall. What difference between men and women does this show?

PROBLEMS

1-3   1. A certain book has a mass of 2.4 kg and contains 1240 numbered pages (620 sheets of paper).
(a) Find the force of gravity in newtons on the book.
(b) What is the mass of one page in kg and in g?
(c) What is the force of gravity in newtons on one page?

(d) What is the force of gravity in dynes on one page?

1-3   2. Consider a mass of 0.1000 kg of gold.
(a) What is the force of gravity on that mass of gold in Bombay, India?
(b) What is the force of gravity on that gold in Glasgow, Scotland?
(c) Gold is actually measured in Troy ounces, where 1 Troy ounce is 0.03110 kg. How many Troy ounces of gold are in that mass?
(d) Should the value of the gold be based on its weight or on its mass?

1-3   3. Find the force of gravity on a 165 pound mass. (There are 2.20 pounds in a kilogram.)

1-3   4. Find the mass that you can lift in one hand and calculate the upward force that you can exert. (If conventional weights are not available, use bags of sugar, flour, or cans of food—even a pail of water. Use the conversion factor that 2.2 lb = 1 kg if it is necessary.)

1-5   5. Convert the following lengths into the unit indicated:
(a) 395 miles = _____ km
(b) 6.00 feet = _____ m
(c) 36.0 inches = _____ cm
(d) 90 km = _____ m
(e) 1.53 m = _____ mm
(f) 620 sheets of paper are 5.3 cm thick
        1 sheet of paper is _____ mm
(g) 1 sheet of paper in (f) is _____ μm.

1-6   6. A cable is wound around a drum that has a radius of 0.15 m. The force in the cable is 4400 N. What torque is exerted on the drum?

1-6   7. Referring to Figure 1-5(A), if F_1 is 50 N and x_1 is 0.45 m, where must the force F_2 be applied to produce equilibrium if F_2 is to be of 15 N.

1-6   8. You catch a big fish, too big for the spring scale you have. To find its mass you get a pole 11 feet long, support it at one end, and lift upward with the scale at the other end. When the fish hangs 3 feet from the support, the scale reads 9 lb more than when the fish was not there. What is the mass of the fish? (Make a force diagram to assist in the solution.)
(Note: An 11 foot pole is also useful for touching things that you wouldn't touch with a ten foot pole. You might also want to convert the common expression referred to into its metric equivalent.)

1-6   9. If a force of 200 N pulls downward on your hand at 0.35 m from the elbow, what force must pull upward in the biceps (Fig. 1–6), which pulls at 4.0 cm from the elbow, to produce rotational equilibrium?

1–7 10. In problem 9, what must be the force at the elbow on the bone of the forearm, and in what direction does it act?

1–7 11. A bar is supported at one end, and at 0.5 m from the end there is a downward force of 3 N, at 1.0 m there is a downward force of 6 N and at 1.5 m, 9 N downward.
(a) Where would an upward force of 9 N be applied to put the system in rotational equilibrium?
(b) Where would an upward force of 18 N be applied to put the system in rotational equilibrium?

1–7 12. A pole 2 meters long is supported at both ends. A 20 kg mass is hung 0.5 m from one end, and at 1.30 meters is a 30 kg mass. Find the supporting force at each end. Neglect the weight of the pole. (Make a diagram to assist in the solution.)

1–7 13. A pole of negligible weight supports 20 kg at 0.5 m, and 30 kg at 1.3 meters, each distance measured from one end. Where would a single supporting force be applied to put the pole in equilibrium?

1–9 14. To the pole described in Problem 13 is added a third weight, 40 kg at 2 m from the end. Find where a single force would be applied to put this system in equilibrium. Where is the center of gravity?

1–9 15. A long board of mass 25 kg is supported at two points 3.0 m apart. A man walks along the board from one support to the other. Describe (using numerical values) the motion of the center of mass of the board plus man. The mass of the man is 75 kg.

1–10–1 16. A force of 25 N to the west and 35 N to the north act on an object. What is the resultant force and in what direction is it?

1–10–1 17. A child of mass 25 kg sits on a swing and a horizontal force P is applied so the ropes are at 20° from the vertical.
(a) Find the value of P.
(b) Find the tension in the ropes.

1–10–1 18. A boom as in Figure 1–27 is used to lift a mass of 500 kg. The force F must be such that the resultant of the weight and F is along the boom, which is 15° from the vertical. Find F.

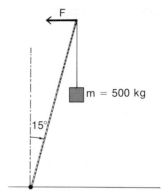

Figure 1–27 A boom used to lift a mass, as in Problem 18.

1–10–2 19. Find the components in the x and y directions of the following forces: The angles are all measured counterclockwise from the positive x axis. A table to show the force, the angle and the x and y components should be used to summarize the results.
(a) 55.0 N at 30°
(b) 22.2 N at 81°
(c) 66.0 N at 140°
(d) 92.0 N at 270°

1–10–2 20. Use the method of components to find the resultant of all the forces listed in Problem 19 acting together.

1–10–2 21. Two forces are to act on an object in such a way that the resultant is 45° from the x axis. One of the forces is of 15 N and along the x axis, while the other is 60° from the x axis.
(a) What is the magnitude of the second force?
(b) What is the magnitude of the resultant?

1–10–2 22. A mass of 50 kg is supported as in Figure 1–28 at the center of a cable be-

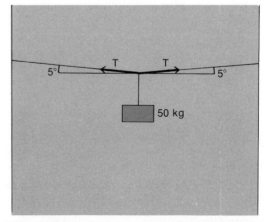

Figure 1–28 The forces in a cable supporting a mass at its center, as in Problem 22.

tween supports 3.0 m apart. The cable makes an angle of 5° from the horizontal. The upward components of the two tension forces together balances the weight.

(a) Find the tension in the cable.

(b) What is the ratio of the tension and the weight?

1-10-3   23.   Find the torque produced by a force F as in Figure 1–21 given the following values:

$$F = 200 \text{ N}, \quad x = 1.5 \text{ m}$$

and use values of θ equal to 30°, 45°, 60° and 80°.

1-10-3   24.   What torque is produced if you push vertically on the pedal of a bicycle when the pedal arm, which is 20 cm long, is at 30° from the vertical? You put half your weight on the pedal and your mass is 60 kg. A diagram to assist in the solution is recommended.

1-13   25.   Consider a balance as in Figure 1–26(D). l_1 and l_2 are equal at 0.15 m each. m_3 is of 2.9 g and $s = 0.30$ m. Find the value of

m_4 when $\theta = 1°$, 2° and 3°. Does the local value of g affect the result?

1-14   26.   Multiply the following numbers as given, keeping all digits in the product. Then, using the concept of there being an uncertainty of ±5 in the first unrecorded digit, find the range of the answer. Finally, give the answer with a reasonable number of significant figures.

(a) 2.31 × 4.14

(b) 3.1749 × 2.6

(c) π × 7.12

1-14   27.   Divide the following numbers as given, keeping at least six digits in the result. Then, using the concept of there being an uncertainty of ±5 in the first unrecorded digit, find the range of the answer. Finally, give the number with a reasonable number of significant figures. To obtain the range increase the numerator by 5 in that unrecorded place and decrease the denominator. Then reverse this for the other end of the range.

(a) 2.31/4.14

(b) 3.1749/3.6

(c) π/7.12

CHAPTER TWO

MOTION

The description of the motion of objects is a science that has challenges and rewards at every level. In this mobile society, situations that involve constant speed are commonplace and the solutions are not difficult. If you travel at a constant speed of 90 kilometers per hour, how long will it take to go 450 kilometers? The answer is 5 hours. That is not in the same category as the question "if you go from rest to 100 km/h in 5 seconds, what distance do you go?" To answer this question something must be known about the way in which the speed changes. If the increase is at a constant rate, the problem is quite simple and soon the way to handle it will be described. The situations that involve a constant rate of change of speed in a straight line, that is, constant linear acceleration, are quite common in physical situations. They result from a constant force in the same line as the motion.

If the acceleration is not in the same direction as the velocity, the direction of the motion will change. The resulting curved motion will also be described in this chapter.

A third distinctive type of motion is that which occurs in a gravitational field such as that around the sun or planets. The force of the sun on objects in the region of the planets is not constant but changes with distance from the sun though it is always directed toward the sun. This is the force that controls the path of the planets and space vehicles. The type of motion that results from such a force will be described in a later chapter.

2-1 THE DEVELOPMENT OF IDEAS IN KINEMATICS

Progress in physics had to await the precise description of motion. The ancient philosophers such as Aristotle did not use the concept of acceleration (rate of change of velocity), and it was, to a large extent, a precise description of this quantity that allowed Galileo Galilei to begin the science of motion. His ideas were set down in a book entitled *Two New Sciences* first published in about 1610. A translation into English is now available in paperback.* The two new sciences of which he writes are now referred to as "the strength of materials" and "the science of motion" or "kinematics." Strength of materials or "elasticity" will be dealt with in Chapter 7; in this chapter the study of motion is begun.

The style of Galileo's books is one no longer found in scientific writing. It is closer to that of a dramatic play than of a scientific text. The books are written as dialogues between several people who take different views. He, of course, presents his own ideas most forcefully and shows the ideas of many of his contemporaries to be rather ridiculous. It did not help his case when at least one of his characters, the one who was always wrong, could be identified as a local official. Though it was not the main reason, such tactics probably had more than a little to do with his being brought to trial and one of his books, *The Dialogues on the Great World Systems*, being ordered to be burned. Fortunately, some survived. Galileo, in his preface to that book, wrote:

> "Withal, I conceived it very proper to express these conceits by way of dialogue, which, as not being bound up to the rigid observance of mathematical laws, gives place, also, to digressions that are sometimes no less curious than the principal argument.
>
> I chanced, many years ago, as I lived in the stupendous city of Venice, to converse frequently with the Signor Giovan Francesco Sagredo, a man of noble extraction and most acute intellect. There came thither from Florence, at the same time, Signor Filippo Salviati, whose least glory was the eminence of his blood and magnificence of his estate, a sublime intellect that knew no more exquisite pleasure than elevated speculations. In the company of these two I often discoursed of these matters before a certain Peripatetic philosopher, who seemed to have no greater obstacle in understanding the truth than the fame he had acquired by Aristotelian interpretations."

In that book the characters are called Sagredo, Salviati, and Simplicio.

Galileo, in his *Two New Sciences,* introduced the methods and equations to work with accelerated motion. He showed that *freely* falling objects (objects falling without resistance) would *probably* have the same constant downward acceleration. Also, what is now called Newton's first law, that the natural, free

---

*Galilei, G. Dialogues Concerning Two New Sciences. New York, Dover Publications, 1914 (still available in reprint).

motion of objects would be in a straight line at constant speed, was put forward by Galileo. This was in contrast to the previously held ideas that natural motion would be circular. He almost, but not quite, related force to acceleration. The progress he had made was great; but it was left for Isaac Newton to add the concept of force and the concept of inertia to put the laws of motion in the forms that are still used. They are the laws used in most fields of physics even today. The only exceptions are in quantum mechanics, which deals with some aspects of atomic and nuclear physics, and with other cases in which the motion is so fast that it approaches the speed of light. In the latter instance, the relativistic mechanics of Einstein must be used.

Those who postulate reincarnation would be interested in the fact that Galileo died in Italy early in 1642, while Newton was born in England later the same year according to the calendar in use at the time.

While Galileo was in Italy formulating the laws of motion of objects on earth, Johan Kepler in Prague was studying the planets and describing the laws of planetary motion. These laws were the foundation for Newton's law of gravity. Though the motion of planets in the sky and of objects on the surface of the earth seem so different, they were shown by Newton to be all described by the same fundamental laws relating force and motion.

Galileo worked on earth where all moving bodies meet resistance, but from his observations he deduced the laws that would describe motion without resistance. These simple laws would then form the basis for the description of any movement, for other effects would only modify this basic motion. As a result of this extrapolation from observation, Galileo's work consists largely of theoretical derivations of what would occur in the ideal situation. Galileo understood that this was what he was doing for after a discussion of the effect of air resistance on objects of various sizes, shapes and densities he says

> ". . . in order to handle this matter in a scientific way, it is necessary to cut loose from these difficulties; and having discovered and demonstrated the theorems, in the case of no resistance, to use them and apply them with such limitations as experience will teach."

Kepler, on the other hand, worked with objects moving without resistance through space and sought to describe the observed motion with high precision. The basic laws that Kepler found for the motion of the planets do not have to be modified because of friction as must be done for laws of motion on earth.

The writings of Kepler are as unique in science as those of Galileo. Kepler did not describe only his final ideas and results, as is done in any modern scientific work, but told of all his wrong ideas and his mistakes. He talks of how he could not see the correct result even though he had it, but would throw it away and start again. His work contains not only mistakes in calculation but later mistakes that just happen to cancel the earlier ones. When students do this they say that they "did a Kepler." The reading of Kepler's works or well chosen excerpts, following his frustrations, seeing his blindnesses, is like watching a television show in which you want to scream at the performer about what he cannot see or is doing wrong. Finally when Kepler gets to the end, when he has found the laws describing planetary motion, with relief you share his joy of achievement.

2–2   THE QUANTITIES THAT DESCRIBE MOTION

The first step in this study is to define and describe the specialized words that will be used. Some of these, like speed or velocity, are in everyday use; but in this subject of mechanics they are given very precise meanings. The quantities to be used are as follows.

Path length, p, will refer to the actual distance traveled by an object as it moves between its initial and final positions.

Displacement, s, refers to the straight line distance between the two points in question. This will frequently differ from the path length, p.

Time, t, will usually refer to a time interval. In problems of motion, perhaps it will be the time required to make the displacement **s.** In terms of two clock times t_1 and t_2, the interval t is $t_2 - t_1$. If it is very small, it is often called Δt, where the symbol Δ is taken to mean "a small change in."

In this section there is the first encounter with time intervals, and an explanation of time standards is in order. The measurement of time intervals in seconds, minutes, hours, days or years is new to none of you, but precisely what is the basis for time? The day can be described as the time between successive passages of the sun across a line upward from the midpoint of the southern horizon (the line called the local meridian). Such a day is called a solar day but because of such things as the variable speed of the earth in its elliptic orbit about the sun the solar days are not the same throughout the year. Our ordinary time measurement is based on the average (or mean) length of a solar day and is called mean solar time. The time interval ordinarily called a second is 1/86,400 of a mean solar day.

The second defined according to the sun as above basically depends on the speed of rotation of the earth. The more sophisticated of modern timing techniques show that there are variations in the speed of rotation of the earth. The second has therefore been redefined in terms of an atomic standard. One second is the time required for 9,192,631,770 periods of a certain radiation from an atom of cesium 133. The type of equipment to measure a second in that way is found in few laboratories, but very precise time signals and frequencies governed by atomic clocks are available to anyone on the various scientific type of short wave broadcasts. These are on **WWV** in the United States, **CHU** in Canada and **GBR** or **MSF** in Europe, for example.

Speed, v, refers to what is ordinarily meant by how fast something is going. In scientific work, speed is the path traveled divided by the time taken. The average speed over a path p is given by $\bar{v} = p/t$. The bar over a quantity indicates that it is an average value. What is called instantaneous speed is measured over a very short path, Δp and is given by $v = \Delta p/\Delta t$.

Velocity, v, is not the same as speed; it includes the direction in its description. Average velocity $\bar{\mathbf{v}}$ over a time interval t is defined as displacement over time, $\mathbf{s}/t$. The direction of the average velocity is the same as the direction of the displacement. Velocity over a short displacement is $\mathbf{v} = \Delta\mathbf{s}/\Delta t$.

Acceleration, a, refers to rate of change of velocity, or alternatively to change of velocity per unit time. The average acceleration is found from

$$\bar{\mathbf{a}} = \frac{\text{change in velocity}}{\text{time interval}} = \frac{\mathbf{v}_2 - \mathbf{v}_1}{t}$$

Over a small time interval the acceleration is given by $\mathbf{a} = \Delta\mathbf{v}/\Delta t$. The direction of the acceleration is the same as the direction of $\Delta\mathbf{v}$ or $\mathbf{v}_2 - \mathbf{v}_1$.

In all cases, the bold face type is used to indicate directional quantities.

Displacement, velocity and acceleration are all **vector** quantities. Just as force is a vector and requires a direction for its complete description so do these. Path length and speed are not vectors and are called **scalars.** As was described for forces, these vectors are often designated by bold face type in books; in writing notes, they can be represented by an arrow over the quantity: $\vec{s}, \vec{v}$ or $\vec{a}$. These are reminders that they are directional. The notation will only be used when it is required to avoid ambiguity or to stress a point. In dealing with straight line motion, for instance, the special notation will sometimes be ignored.

An example of the difference between path and displacement and between speed and velocity is the case of a particle

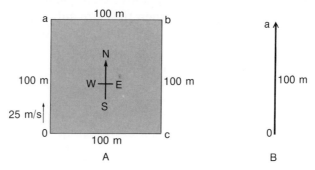

Figure 2-1 The path of a particle diffusing through matter. The path may be very long for a small displacement.

such as a molecule, an atom or a neutron diffusing through a material. The path of the particle may be like that shown in Figure 2-1. It suffers many collisons, and after a time t it will be displaced from its initial position by an amount **s**. The speed along the path may be high, even 500 meters per second for molecules in air at room temperature, yet the average rate of diffusion (s/t) may be only a few meters per second or less.

EXAMPLE 1

A vehicle moves around a square 100 m on a side as in Figure 2-2, starting at the corner marked 0 and proceeding past *A*, *B*, *C* and back to 0. The speed is 25 m/s.

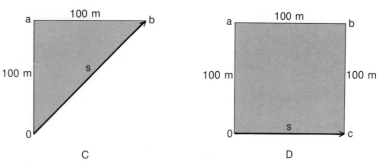

Figure 2-2 The path of the vehicle in Example 1.

Find the time that the object reaches points *A, B, C* and 0, as well as the path *p* traveled in that time and the displacement **s** from 0.

The vehicle reaches *A* in 4 seconds. The path to that position is 100 m and the displacement is 100 m north as in Figure 2–2(B).

B is reached in 8 seconds after following a path 200 m long. The displacement is then as shown in Figure 2–2(C). It is given by

$$\mathbf{s} = \sqrt{100^2 \text{ m}^2 + 100^2 \text{ m}^2}$$

$$= 141 \text{ m}$$

The direction is 45° east of north.

The displacement from 0 to *B* is 141 m, 45° east of north.

On reaching *C* the time is 12 seconds, the path is 300 m and the displacement from 0 is 100 m east. This is shown in Figure 2–2(D).

Finally, at 0 again, the path has been 400 m, the time was 16 seconds and the displacement is zero.

EXAMPLE 2

Situations in which the speed (or the velocity) is constant are not difficult to handle intuitively. If a car goes 270 kilometers in 3 hours, what is the average speed? Without resorting to equations, you can say 90 kilometers per hour. Using an equation, you would write

$$p = 270 \text{ kilometers}$$

$$t = 3 \text{ hours}$$

$$v = p/t = \frac{270 \text{ km}}{3 \text{ h}} = \frac{90 \text{ km}}{\text{h}}$$

Note in the above example how the units are included with the numbers when they are substituted into the equation.

To change the units write an equivalent in place of the word. For example km = 1000 m and h = 3600 s, so to convert 90 km/h to the same speed in m/s, write

$$v = 90 \frac{\text{km}}{\text{h}}$$

$$= \frac{90 \times 1000 \text{ m}}{3600 \text{ s}}$$

$$= 25 \frac{\text{m}}{\text{s}}$$

To express the above in English units use the conversion factor that 1 km = 0.6215 miles. Then

$$v = 90 \frac{\text{km}}{\text{h}}$$

$$= 90 \times \frac{0.6215 \text{ mi}}{\text{h}}$$

$$= 55.9 \frac{\text{mi}}{\text{h}}$$

EXAMPLE 3

Refer to Example 1 and Figure 2–2 and find the average velocity of the vehicle from the time it leaves the point 0 to the time that it is at the corners marked A, B, C and again at 0.

The vehicle reached A in 4 seconds and the displacement **s** was 100 m North. The velocity **v** is then

$$\bar{\mathbf{v}} = \frac{100 \text{ m North}}{4 \text{ s}} = 25 \text{ m/s North}$$

In 8 seconds the vehicle was at B where the displacement was 141 m at 45° East of North. The average velocity was

$$\bar{\mathbf{v}} = \frac{141 \text{ m at } 45° \text{ East of North}}{8 \text{ s}}$$

$$= 17.6 \text{ m/s at } 45° \text{ East of North}$$

The vehicle reached C in 12 seconds and the displacement was then 100 m East. The average velocity up to that point was

$$\bar{\mathbf{v}} = \frac{100 \text{ m East}}{12 \text{ s}}$$

$$= 8.3 \text{ m/s East}$$

The vehicle was back to 0 in 16 s but the displacement was then zero. The average velocity over those 16 seconds was then 0.

at A, $t = 4$ s	$\bar{\mathbf{v}} = 25$ m/s North	
at B, $t = 8$ s	$\bar{\mathbf{v}} = 17.6$ m/s at 45° East of North	
at C, $t = 12$ s	$\bar{\mathbf{v}} = 8.3$ m/s East	
at 0, $t = 16$ s	$\bar{\mathbf{v}} = 0$	

EXAMPLE 4

If a vehicle goes from rest to 90 km/h in 6.25 seconds, what was the acceleration?

In this problem, time occurs in two units, hours and seconds. The first step is to get them all the same. Also, in scientific work the unit of length is the meter. Fortunately for us, in Example 1 it was shown that 90 km/h = 25 m/s.

The data given are
the initial velocity v_1 is 0;
the final velocity v_2 is 25 m/s;
the time interval t is 6.25 s.

Using

$$a = \frac{v_2 - v_1}{t}$$

$$a = \frac{25 \text{ m/s}}{6.25 \text{ s}} = \frac{25 \text{ m}}{6.25 \text{ s} \cdot \text{s}}$$

$$= 4.0 \frac{\text{m}}{\text{s}^2}$$

Note how the unit second (s) occurs twice in the denominator and is written as s^2. The acceleration is 4.0 m/s². The units would be read as meters per second squared.

Another example of interest concerns the centrifuge. In a centrifuge, a sample is whirled around a circular path at a high speed. In a common bench type centrifuge, the speed of rotation may be about 6000 revolutions per minute. In an ultracentrifuge, a rotational speed of about 100,000 revolutions per minute may be used. The speed of the sample at a radius r is calculated from path length over time. In one revolution the path is $2\pi r$. If in one minute (60 seconds) there are n revolutions, the path is $2\pi rn$. For example, let r be 10 centimeters (0.1 m) and n be 6000:

$$v = \frac{n \times 2\pi r}{t}$$

$$v = \frac{6000 \times 2\pi \, 0.1 \text{ m}}{60 \text{ seconds}}$$

$$= 62.8 \text{ m/s} = 226 \text{ km/h}$$

In an ultracentrifuge rotating about 17 times as fast, the speed at a radius of 10 centimeters would be 17 times as great or about 1070 m/s. This is about 3840 kilometers per hour, much faster than an artillery shell. If the rotor of an ultracentrifuge should break while it is spinning, the broken part would no longer be held to the center but would fly off tangentially from its circular path; at this high speed it would behave in many ways like an artillery shell. The design of ultracentrifuges must be such that breakage is unlikely, and the rotors must be well balanced to keep stresses to a minimum. They must be operated only in their heavy protective casings.

2–2–1 ADDITION OF VELOCITIES

A velocity is a vector; that is, it has direction as well as size. When two velocities combine, the net motion is found by adding them, taking the direction into account. In Section 1–8 the methods used to add forces (which are also vectors) was described; velocities add or combine in a similar way. An example of combining or adding velocities is a soaring bird. A bird may glide without flapping its wings, but in still air it would slowly lose altitude, as shown in Figure 2–3(A). This must occur because there is always some air function to which energy is lost. This energy loss results in the loss of altitude. If there is also an upward movement of the air through which the bird glides, the actual path could be horizontal or even rising, as in Figure 2–3(B) and (C). Many birds glide for long distances in this way, even using air currents to carry them to high altitudes. Sailplane pilots make use of the same phenomenon, the trick being to find

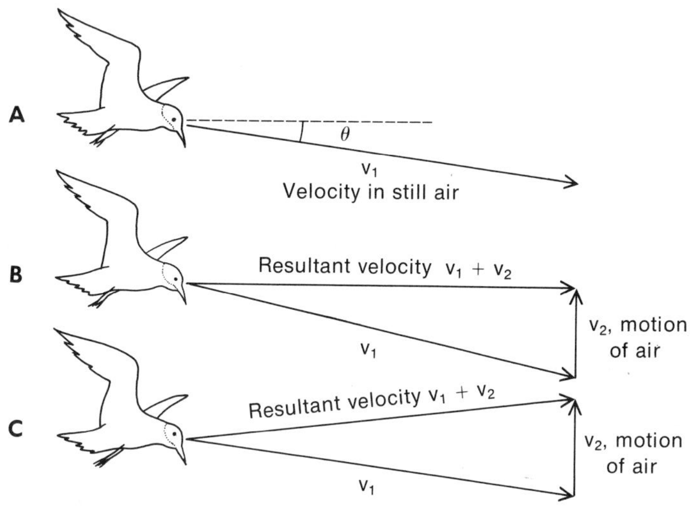

Figure 2–3 The glide path of a bird in still air (a), and in upward currents (b) and (c).

the upward air currents. The velocities add by the tail to point method, explained for forces in Chapter 1.

A further example is that of an aircraft flying in a wind. The speed of the plane over the ground is a combination of its speed in the direction in which it is pointed (and in which it would go if the air were still), and the speed of the moving air. This example is detailed below.

EXAMPLE 5

If a plane is going through the air at 200 km/h headed due north and there is a 40 km/h wind from the east, the combined velocity over the ground is not 240 km/h. In

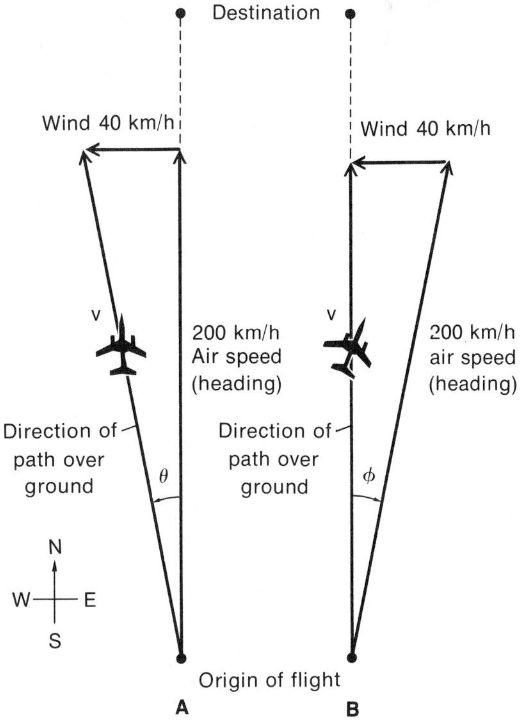

Figure 2–4 The path of an aircraft flying in a side wind. In (a) the craft heads due north, and the wind from the east carries it to the west. The resultant direction is found by the vector addition of the velocities. In (b), in order to fly due north, the aircraft heads slightly toward the east, so the vector addition of the velocity of the craft and the velocity of the wind gives a resultant velocity to the north.

any short interval of time as the plane goes northward, it is also carried toward the west at 40/200 or 1/5 of the northward speed. The resulting velocity is slightly toward the west, and is found as in Figure 2–4(A). The velocity vectors are shown as arrows with the lengths drawn according to some scale. The arrow head is put on the end of the vector. To add vectors in this case, the tail end of one is put at the point of the other. The *vector sum,* or *resultant,* is obtained by connecting the tail of the first vector to the head of the second; it is shown in the figure as *V.* The direction of the resultant velocity and its magnitude are found by solving for those quantities in the triangle formed by the three vectors. The angle θ is found in this right-angled triangle from $\tan \theta = 40/200$. Then $\theta = 11°19'$. *V* can be found by using the Pythagorean theorem or by using $\cos \theta = 200$ km/h/*V*. From this,

$$V = \frac{200 \text{ km/h}}{\cos \theta}$$

θ has been shown to be $11°19'$, so $\cos \theta$ is 0.981; then *V* is 204 km/h.

If a pilot wants to fly to a place that is due north from the starting point and if there is an east wind, he must head the aircraft in a direction such that the wind keeps pushing it back to a path heading due north, as in Figure 2–4(B). When migrating birds make long over-water flights, such as over the Gulf of Mexico, do they manage somehow to take the wind into account?

These examples were chosen to make use of right-angled triangles because they are relatively easy to solve. If vectors are not at right angles, they still add by this "tail to point" method. The solutions for the unknowns may be found mathematically if you can use trigonometry. Alternatively, the velocity vectors can be broken into perpendicular components as was explained for forces in Chapter 1. Finally, a scale diagram could be used, but this is the least exact method of solution.

EXAMPLE 6

A lighter-than-air craft (a blimp) is heading in a direction 30° to the east of north with a speed of 15 km/h through the air, while a wind of 20 km/h blows from 75° to the west of north. In what direction will the craft move over the ground?

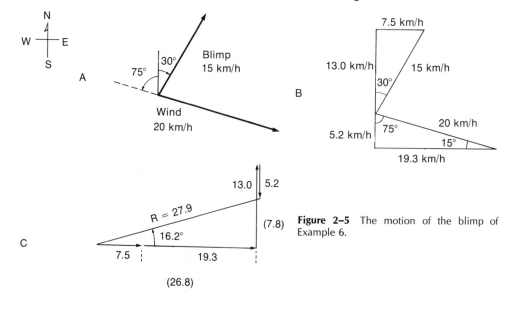

Figure 2–5 The motion of the blimp of Example 6.

This problem is most easily solved using components as shown in Figure 2–5.

The blimp moves northward at   15 cos 30° km/h = 13.0 km/h.
and eastward at   15 sin 30°   km/h =   7.5 km/h.
The wind moves it eastward at   20 sin 75°   km/h = 19.3 km/h.
and southward at   20 cos 75°   km/h =   5.2 km/h.
The net northward velocity is   (13.0 − 5.2) km/h =   7.8 km/h.
The net eastward velocity is   (7.5 + 19.3) km/h = 26.8 km/h.

The resultant velocity, v, as in (C) of Figure 2–5 is given by

$$v = \sqrt{7.8^2 + 26.8^2} \text{ km/h} = 27.9 \text{ km/h}$$

in a direction given by

$$\tan \theta = 7.8/26.8 \text{ or } \theta = 16.2°$$

The resultant velocity of the blimp is 27.9 km/h in a direction 16.2° N of E or 73.8° E of N.

2–3   ACCELERATED MOTION

If an object is not moving at a constant speed in a straight line (that is if the velocity is not constant), there is acceleration. Remember that the term *velocity* includes direction. There are two extreme categories of situations involving acceleration: motion along a straight line at a changing speed, and motion at constant speed but with changing direction. The two, of course, could be combined. In this section the movement of an object accelerating only in the same line as its motion will be considered, but not the acceleration that causes a change in direction.

In particular, the distances traveled by an accelerating object will be related to v_1, v_2, a and t. The relations that have been presented are

$$a = (v_2 - v_1)/t$$

and $s = \bar{v}t$ where $\bar{v}$ is the average velocity

The problem is that if the velocity changes from an initial value v_1 to a final value v_2, what is the average? If the average velocity is known, then the distance is just average velocity multiplied by time. It so happens that if the acceleration is constant over the interval, then the average is just the mean of the initial and final velocities:

$$\bar{v} = \frac{v_1 + v_2}{2}$$

This important relation, which will be justified in the work to follow, is one of the keys in dealing with accelerated motion.

EXAMPLE 7

A vehicle accelerates from 36 km/h (10 m/s) to 72 km/h (20 m/s) in a time of 10 seconds. What distance did it travel during that acceleration?
The average velocity was (10 m/s + 20 m/s)/2 which is 15 m/s.
It had this average velocity for 10 s, so the distance traveled was 150 m.
Using formulas

$$\bar{v} = \frac{v_1 + v_2}{2} = \frac{(10 + 20)\ m/s}{2}$$

$$= 15\ m/s$$

and $t = 10s$

using $s = \bar{v}t$

$$s = 15\ \frac{m}{s} \cdot 10\ s = 150\ m$$

The distance was 150 m.

2-3-1   LINEAR ACCELERATION

As an object moves along a line, the speed may be constantly changing; the graph of velocity versus time could appear as in Figure 2–6. Such a curve could result from putting a recorder on the speedometer of a car or from reading the speedometer frequently while traveling down the road. Alternatively, the position of the car or other object could be marked at regular intervals of time, the distance between marks measured, and the speed over each interval calculated. The time intervals would have to be short to give a realistic curve.

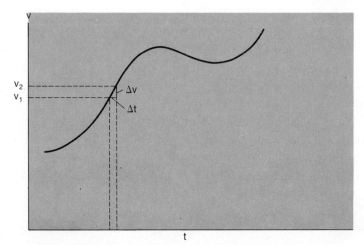

Figure 2–6 A plot of velocity against time in some particular situation. The acceleration at any given time is found by dividing the change in velocity, Δv, by the small time interval Δt in which it occurs. Then a, which is $\Delta v / \Delta t$, is also the slope of the curve.

The acceleration at any time is found by making note of the velocities at the beginning and end of a short interval Δt and then finding the change in velocity $\Delta \mathbf{v}$ that occurred in this time. The acceleration is then given by $\mathbf{a} = \Delta \mathbf{v}/\Delta t$. The time Δt should be short enough that the velocity curve is very close to a straight line across the interval. If the interval is very small (it is said then that Δt approaches zero), the calculated acceleration is called the *instantaneous acceleration*. The hypotenuse of the triangle defined by $\Delta \mathbf{v}$ and Δt is then tangent to the curve. Another way to express it is that the acceleration is the *slope* of the curve obtained when velocity is plotted as a function of time. The slope is defined in the usual way: the change in ordinate values on the graph divided by the corresponding change in the abscissa. When the curve is going down, $\Delta \mathbf{v}$ is negative and the slope and acceleration are negative quantities.

A special case is that in which acceleration is constant. This is a common situation and one for which further analysis will be worthwhile. In this instance $\Delta \mathbf{v}/\Delta t$, the slope of the line, is constant so the graph of velocity against time is a straight line as in Figure 2–7(A). In the even more restricted case shown in Figure 2–7(A), the initial velocity is zero. For the straight line graph, if Δt is expanded to one unit of time, the increase in velocity in this unit interval is equal to the acceleration a, as in Figure 2–7(B). After a time Δt, the total increase in speed is at and the resulting velocity is given by $v = at$. This says that for constant acceleration and an initial speed of zero, the speed is proportional to time.

Experiments have shown that for falling objects the acceleration is constant if the resistance is negligible. The acceleration of a freely falling object at the surface of the earth is measured to be about 9.8 meters/sec². This quantity is called g, the accel-

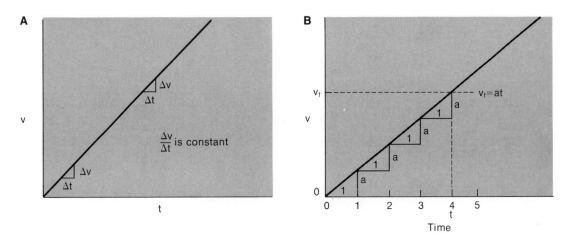

Figure 2–7 Graphs of velocity against time in a situation in which the initial velocity is zero and the acceleration is constant.

eration due to gravity, and it is numerically the same as what was referred to in Chapter 1 as gravitational field strength. The reason is left for the next chapter. The precise value depends on altitude and latitude. The units given are derived from the relation $\mathbf{a} = \Delta\mathbf{v}/\Delta t$. If $\Delta t = 1$ second, then $\Delta v = 9.8$ m/s, and

$$a = \frac{9.8 \text{ m/s}}{1 \text{ s}} = \frac{9.8 \text{ m}}{\text{s} \times \text{s}} = \frac{9.8 \text{ m}}{\text{s}^2}$$

In multiplying numerator and denominator by s, the quantity s × s appears on the bottom. Just as 3 × 3 would be written as 3^2, so s × s is written as s^2. Do not try to imagine a square second.

The variation of the downward velocity with time for a freely falling (i.e., resistanceless) object is shown in Figure 2–8(A). When an object falls through a resisting medium, the resistance force usually increases as the speed increases, so the acceleration decreases. The downward speed is then as illustrated in Figure 2–8(B). The velocity increases only until the resistance force balances the weight. This limiting velocity is called the **terminal velocity**. For a man jumping from a high aircraft, the terminal velocity before his parachute opens is about 50 m/s. After he opens his parachute, the terminal velocity drops to about 5 to 10 m/s. Some plant seeds, dandelions for instance, are like parachutes but have terminal velocities of only about 0.5 m/s. For a blood cell settling or "falling" through plasma under the force of gravity, the terminal velocity, referred to clinically as the sedimentation rate, is about 5 cm/h.

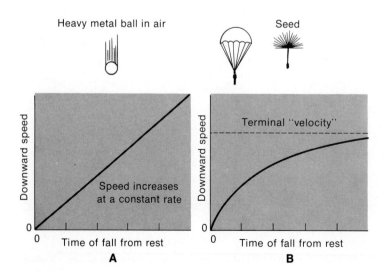

Figure 2–8 Acceleration is constant for falling objects in air if the air resistance is negligible, as for the heavy metal ball in (a). If the air resistance is large as in (b), the falling object approaches a terminal velocity.

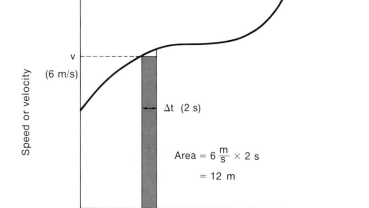

Figure 2–9 The distance moved by an object in a short time interval Δt is just the height times the width of the strip shown, which is its area.

The terminal velocity of particles depends on their weight*
in the medium in which they fall. After a nuclear explosion the
larger particles of radioactive "dust" settle more quickly from
the upper atmosphere where the explosion throws them than
do the very small ones. As a result of this the very small particles may be carried tremendous distances before falling to
earth.

Small particles of rock, clay or sand settle at different rates
in water, and separations of particles of various sizes can be
achieved by sedimentation.

How do you calculate the distance if the speed varies and
the acceleration is not constant? This is what will now be investigated.

Consider first a graph of velocity against time, as in Figure
2–9. Between the times t_1 and t_2, which define a short interval
Δt, the velocity changes very little and the distance moved in
that short interval could be calculated. Just take the average
velocity and multiply it by the time Δt. Let Δs be this short distance, and let v be the velocity for the interval Δt. Then in
symbols, $\Delta s = v\Delta t$. The quantity v is the height of the shaded
rectangle in Figure 2–9; the quantity Δt is the width. The area
of a rectangle is height times width, and the quantity $v\Delta t$ is, by
analogy, the area of that shaded strip. The units will not be the
usual units of an area but will be units of velocity multiplied by a
unit of time. On Figure 2–9 some representative numbers have
been placed. In this example, v is 6 m/s and Δt is 2 s.

---

*Weight here would be taken to mean the difference between the force such as due
to gravity and the buoyant force. Weight in this sense is also the negative of the force to
keep the object at rest.

Then $$s = \frac{6 \text{ m}}{\text{s}} \times 2\text{s} = 12 \text{ m}$$

The unit of seconds cancels just as a number would cancel. The resulting unit is just that of distance, which is the correct type of unit for Δs. We are happy when the units are correct as in this case. The area of the small shaded strip is the distance that the object went in the time interval Δt, or at least it is very close to it. Actually, the velocity did change by a small amount over that 2 second time interval. The area of the rectangular strip differs from the true distance by the area of the small triangle at the top of the strip. In the limit, as Δt gets very small, this small triangular area will be negligible.

To find the total distance moved in a long time interval, the graph of v against t is plotted and divided into a series of narrow strips, as in Figure 2–10(A). In each small time interval Δt, the distance moved is $\Delta s = v\Delta t$. The total displacement s is the sum of all the small displacements Δs, each of which is represented by the area of one strip. But adding the areas of all the strips gives the total area under the curve of velocity against time, as shown in Figure 2–10(B). Sometimes this area can be found very easily; at other times it is more difficult. If the relation between velocity and time can be expressed mathematically, the area may be found using calculus or by resorting to carefully drawing a graph and perhaps counting squares to get the area.

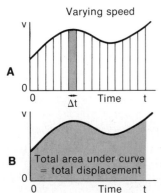

Figure 2–10 Over a long time interval, the total distance moved is the sum of all the short distances represented by the areas of the strips in (a). This amounts to the total area under the velocity-time curve as in (b).

If the velocity is constant, the graph of velocity against time is like that in Figure 2–11(A). The distance traveled in the time t is the area of the shaded rectangle shown. This is just vt, so the distance is given by

$$s = vt \qquad \text{if } v = \text{constant } (a = 0)$$

This is no surprise.

In graph (B) of Figure 2–11 is the case of constant acceleration, in which the velocity varies with time according to $v = at$. The distance covered up to time t is the area of the shaded triangle. The area of a triangle is half the height times the base, or

$$s = (at/2)\, t = (1/2)\, at^2$$

This is for the restricted case in which the acceleration is constant and the initial velocity is zero.

Sometimes the relation between final velocity, distance and acceleration will be needed. Frequently also, the initial velocity will be zero and the final velocity can be called v. Then use $v = at$, solve for t, and substitute for it in $s = at^2/2$. Carrying this out,

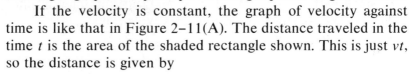

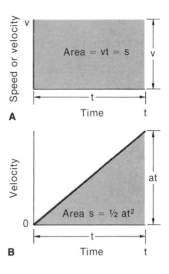

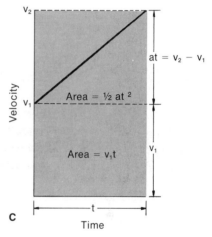

Figure 2–11 Finding the distance traveled over a time interval: (a) at constant velocity; (b) at constant acceleration; (c) with constant acceleration beginning with an initial velocity v_1.

$$t = v/a \quad or \quad t^2 = v^2/a^2$$

then $s = (1/2)\,at^2$ becomes $s = (1/2)\,a \cdot v^2/a^2$. Solving for v gives the result

$$v^2 = 2as \quad or \quad v = \sqrt{2as}$$

Another case of interest is that in which the object has an initial velocity v_1 and with constant acceleration achieves a velocity v_2 after a time t. The graph of velocity against time would be as in (C) of Figure 2–11. The distance traveled in the time t would be the shaded area. This can be found by the area of the rectangle shown plus the area of the triangle on top of it. This is

$$\text{Area} = s = v_1 t + \frac{(v_2 - v_1)}{2}\,t$$

which, with a bit of algebraic manipulation becomes

$$s = \frac{(v_1 + v_2)}{2} t$$

Using the definition of average velocity you can write

$$s = \bar{v}t$$

and comparing these two equations it is seen that

$$\bar{v} = \frac{v_1 + v_2}{2}$$

This was used earlier and the above shows that it applied if the acceleration is constant.

You can also show that the formula above for the area under the curve, which is the displacement, will take the form

$$s = v_1 t + (1/2) a t^2$$

EXAMPLE 8

A ball was thrown straight up and returned to the ground in 6 seconds. What was its initial speed? Consider that air resistance was negligible.

Letting the upward direction be positive, the acceleration, which is downward, will be negative, -9.8 m/s². When the ball gets back down, the displacement s is zero.

The knowns are
$$s = 0$$
$$t = 6s$$
$$a = -9.8 \text{ m/s}^2$$

The unknown is the initial velocity, v_1

Use
$$s = v_1 t + \tfrac{1}{2} a t^2$$

put $s = 0$ and solve for v_1

$$v_1 t + \tfrac{1}{2} a t^2 = 0$$

factor out a t: $t(v_1 + \tfrac{1}{2} at) = 0$
There are two solutions for the case of $s = 0$
either $t = 0$
or $v_1 + \tfrac{1}{2} at = 0$

putting in numbers $v_1 - \dfrac{9.8 \text{ m}}{2s^2} \cdot 6s = 0$

$$v_1 = \frac{9.8 \times 6 \text{ m}}{2 \text{ s}} = 29.4 \text{ m/s}$$

The initial upward speed was 29.4 m/s

EXAMPLE 9

An object falls freely (with negligible resistance) from a height h; what speed will it acquire? Replace s in the formula $v^2 = 2as$ by h and replace a by the acceleration due to gravity, g. Then $v^2 = 2gh$ or $v = \sqrt{2gh}$.

Consider, for instance, an osprey that is flying slowly 32 meters above a lake ($h = 32$ m) when it sees a fish. Then it folds its wings and may be considered to be a freely falling object ($g = 9.8$ m/s^2) as it hurtles toward the lake. With what speed will it strike the water? In $v = \sqrt{2gh}$, insert the appropriate values to get:

$$v = \sqrt{2 \times 9.8 \text{ m/s}^2 \times 32 \text{ m}}$$

$$= \sqrt{628 \text{ m}^2/\text{s}^2}$$

$$= 25 \text{ m/s}$$

$$= 90 \text{ km/h}$$

Note that the units are kept with the numbers and that the square root operation is performed on the units as well as on the numbers. Note also that if an automobile is driven off the roof of a building 32 meters high (about 10 stories) it will strike the ground at 90 km/h, a common highway speed. It corresponds to 55 mi/h.

EXAMPLE 10

An object was thrown upward with an initial velocity v_0 of 2 km/s or 2×10^3 m/s. How high will it rise?

The acceleration is due to the pull of gravity and is directed downward. Calling the upward direction positive, the acceleration a then takes the value $-g$, or -9.8 m/s^2. At the top of the trajectory, the speed is momentarily zero. The average speed on the way up is then $(v_0 + 0)/2$ or just $v_0/2$. Distance is average velocity multiplied by time; so the height is given by $v_0 t/2$. This expression contains t, which can be found from the knowledge that the initial speed is reduced by 9.8 m/s every second until it gets to zero. This time, t, is $v_0/(9.8$ m/s), or $t = v_0/g$. Substituting this for t gives the relation

$$h = v_0^2/2g$$

Putting in the numbers,

$$h = (2 \times 10^3 \text{ m/s})^2/2 \times (9.8 \text{ m/s}^2)$$
$$= 204{,}000 \text{ m}$$

The object will rise 204 km.

The initial speed was chosen to be approximately that of the V-2 rocket used by Germany in the last stages of World War II, and well into the 1950's by the U.S. research teams investigating the upper atmosphere. With much greater speeds, the calculations would have to take into account the decrease of gravity with height.

EXAMPLE 11

A passenger in an automobile traveling at 90 km/h is stopped in a distance of 1 meter in a collision. What is the acceleration?

In this case, the initial velocity is $v_1 = 90$ km/h, which is 25 m/s. The final velocity v_2 is zero, and the distance s is 1 meter.

The average velocity is $(v_1 + v_2)/2$, which is 12.5 m/s. To cover a distance of 1 meter while averaging 12.5 m/s would require 0.08 second. The change in velocity is from 25 m/s to 0, which amounts to −25 m/s. This occurs in 0.08 second, so the rate of change of velocity (the acceleration) is given by

$$a = \frac{-25 \text{ m/s}}{0.08 \text{ s}} = -313 \text{ m/s}^2$$

The minus sign indicates that the object is slowing down. The value −313 m/s² is 32 times as great as the acceleration of a falling object, and it could be said that the acceleration is −32 g.

2-3-2   THE TRAJECTORY

If an object is given an initial impulse and released, the path that it follows is called a trajectory. Baseballs, footballs or bullets follow trajectories. In a broader sense the paths of spacecraft are trajectories, though slightly more difficult to deal with than trajectories on earth. The trajectories on earth are special in that they occur in a gravitational field of constant strength. This condition applies with a reasonable degree of accuracy if we deal only with objects that do not go above about a hundred kilometers in altitude.

To analyze theoretically for the expected path, one more restriction must be put on; that is, air resistance must be negligible. Thus any results that are derived must be applied with caution to the real world. The most restricting condition is that of resistance, but there are still many situations that will be adequately described by the results. Resistance force increases with speed, and the paths of actual bullets will not be very accurately described. The results will tell only of what would happen if the air were not there. Baseballs, on which the air resistance is less than on a bullet, are more adequately described and so are rockets that spend most of their flight time above the main portion of the atmosphere.

The analysis of the trajectory will begin with a few simple ideas, set out in steps as follows:

1. Gravity pulls only down; it does not pull sideways.
2. The acceleration of an object is down because that is the direction of gravity.
3. There is no sideways acceleration due to gravity.

From these we would deduce that if an object is projected into the air, its horizontal velocity is unaffected by gravity; therefore, it is constant, while the vertical velocity is affected by gravity, the acceleration being 9.8 m/s² downward (or −9.8 m/s² upward). The object has a downward speed that is independent

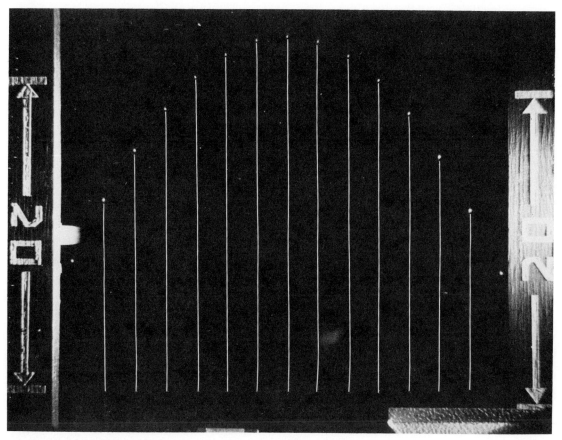

Figure 2–12 A stroboscopic photograph of a ball in a trajectory. The dots show the position of the ball every 1/60 of a second. The horizontal distance traveled in each 1/60 of a second is constant.

of its horizontal speed. It follows, for example, that if one object is dropped from a high building and at the same instant another is thrown horizontally from the same spot, the two will reach the ground at the same time.

Figure 2–12 is a stroboscopic photo of a ball thrown at an angle. The positions of the ball, the bright dots, were recorded by a strobe light flashing 60 times per second. Vertical lines have been drawn from each recorded position and the spacing is constant showing that the horizontal component of the velocity did not change. The vertical velocity, indicated by the spacing between the horizontal lines decreases steadily from its initial upward value till it reaches zero and then it increases downwards.

To analyze the trajectory, let the object be projected with a velocity v_0 at an angle θ above the horizontal as in Figure 2–13(A). This velocity is equivalent to the two components

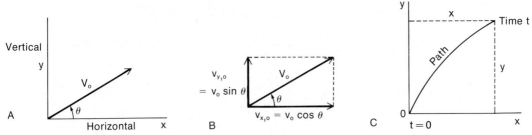

Figure 2–13 The analysis of a trajectory. (a) The initial velocity v_0 at an angle θ above the horizontal. (b) The vertical and horizontal components of the initial velocity. (c) The position of the object at time t.

shown in Figure 2–13(B). The horizontal component is v_{x_0} and is given by

$$v_{x_0} = v_0 \cos \theta$$

while the vertical component, v_{y_0}, is

$$v_{y_0} = v_0 \sin \theta$$

The horizontal component remains constant so after a time t the horizontal distance as in Figure 2–13(C) is given simply by $x = v_0 \cos \theta \cdot t$.

The vertical distance is found from the basic equation $s = v_1 t + (1/2) a t^2$, where s in this example is replaced by y, v is $v_0 \sin \theta$ and a is $-g$ so

$$y = (v_0 \sin \theta) t - (1/2) g t^2$$

EXAMPLE 12

An object is projected at 100 m/s at 60° above the horizontal. Find the position after 5 seconds and also the velocity.

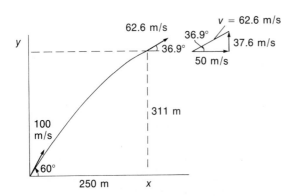

Figure 2–14 The velocity of a particle in a trajectory.

The initial horizontal velocity, $v_{x_0} = (100 \text{ m/s})\cos 60° = 50 \text{ m/s}$.
The initial vertical velocity, $v_{y_0} = (100 \text{ m/s})\sin 60° = 86.6 \text{ m/s}$.
The horizontal distance is $x = (50 \text{ m/s}) 5s = 250 \text{ m}$.
The vertical distance is $y = (86.6 \text{ m/s}) 5s - (9.8 \text{ m/s}^2) 25 \text{ s}^2/2$
$= 311 \text{ m}$.

The object will have gone 250 m horizontally and 311 m vertically as in Figure 2–14. The vertical velocity is reduced by 9.8 m/s every second for a total of 49.0 m/s in 5 seconds. The initial upward velocity was 86.6 m/s so it has been reduced to 37.6 m/s. The horizontal velocity is constant at 50 m/s. Using the vector diagram inset in Figure 2–14, the velocity is calculated to be 62.6 m/s at 36.9° above the horizontal.

The shape of the path of a projectile is found by solving for y in terms of x, that is by eliminating t.

Use
$$x = (v_0 \cos \theta)t$$

from which
$$t = \frac{x}{v_0 \cos \theta} \quad \text{and} \quad t^2 = \frac{x^2}{v_0^2 \cos^2 \theta}$$

substitute this in
$$y = (v_0 \sin \theta)t - (1/2)gt^2$$

to get
$$y = \frac{(v_0 \sin \theta)x}{v_0 \cos \theta} - \frac{g}{2} \frac{x^2}{v_0^2 \cos^2 \theta}$$

simplifying
$$y = (\tan \theta)x - \frac{g}{2 v_0^2 \cos^2 \theta} \cdot x^2$$

This equation is of the form
$$y = ax + bx^2$$

$$\left(\text{where } a = \tan \theta \text{ and } b = \frac{g}{2 v_0^2 \cos^2 \theta}\right)$$

which is the equation of a parabola. Therefore, the shape of the path of a projectile, on which the resistance is negligible and which is in a field of constant gravitation, is parabolic.

An interesting extension of this is to find the range of a projectile in terms of initial velocity and angle. The range is found by first finding the total time of flight. During this time the projectile moves horizontally at a constant speed and the distance it moves is the range.

The initial upward velocity is $v_0 \sin \theta$ if θ is measured from the horizontal. This velocity is reduced by an amount g each second. The time to reduce the upward speed to zero is then $v_0 \sin \theta/g$. The total time of flight, T, is double this or

$$T = 2 \, v_0 \sin \theta/g$$

The horizontal velocity is $v_0 \cos \theta$ and the horizontal distance traveled at this speed in the time T is the range R, which is given by

$$R = (v_0 \cos \theta)(2 \, v_0 \sin \theta/g)$$
$$= 2 \, v_0^2 \sin \theta \cos \theta/g$$

The expression can be simplified using the trigonometric relation

$$\sin 2\theta = 2 \sin \theta \cos \theta$$

Then $$R = v_0^2 \sin 2\theta/g$$

This gives the range on a horizontal surface for a projectile with an initial speed v_0 at an angle θ above the horizontal.

EXAMPLE 13

Find the angle θ necessary for a projectile having an initial speed of 2 km/s to have a range of 100 km. The angle is to be between 0° and 90°. This is a straight substitution type of problem. Use $R = v_0^2 \sin 2\theta/g$, and the problem is to find θ.

$$\sin 2\theta = Rg/v_0^2$$

$R = 200 \text{ km} = 2 \times 10^5 \text{ m}$
$v_0 = 2 \text{ km/s} = 2 \times 10^3 \text{ m/s}$
$v_0^2 = 4 \times 10^6 \text{ m}^2/\text{s}^2$
$g = 9.8 \text{ m/s}^2$

$$\sin 2\theta = \frac{2 \times 10^5 \text{m} \times 9.8 \text{ m}}{4 \times 10^6 \dfrac{\text{m}^2}{\text{s}^2} \text{ s}^2} \text{ (all units cancel)}$$

$$\sin 2\theta = 0.49$$

If θ is to be between 0° and 90°, 2θ may be between 0° and 180°. There are two angles between 0° and 180° for which the sine is 0.49. These are 29.3° and 150.7°.

Therefore
$$2\theta = 29.3° \text{ or } 150.7°$$
$$\theta = 14.7° \text{ or } 75.4°$$

The trajectories would be as in Figure 2–15, the possible angles of projection being 14.7° or 75.4°.

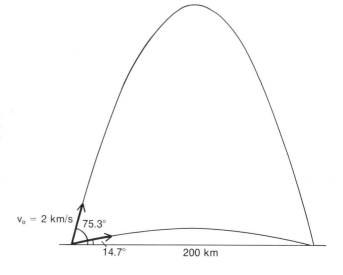

Figure 2–15 Two possible angles for the same initial speed, so that both trajectories have the same range. Note that the large angle is 90° minus the small angle.

The possible trajectories of an object thrown horizontally from a high point on a curved earth are interesting. In Figure 2–16(A) are some representative paths of projectiles with different initial velocities, v_0. The ranges are greater than for a flat earth. The scale is reduced in (B) of Figure 2–16 to illustrate the results of even higher initial velocities. The fate of the projectile on the path marked s is left for you to wonder about.

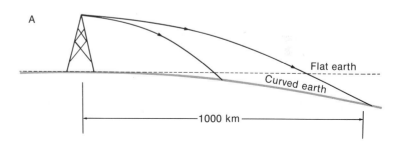

Figure 2–16 Trajectories on a curved earth. What would be the fate of the particle following the path marked s in part (b)?

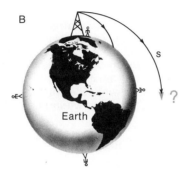

2–3–3 EFFECTS OF ACCELERATION ON THE HUMAN BODY

A constant speed, no matter how high it is, has no effect on the human body, but changes in velocity, or accelerations, do. The amount and the duration of the acceleration each contribute. Accelerations up to about 40 m/s² (about 4 g) for a long time have no detrimental effect, but as that value is exceeded the time duration for which it is possible to endure the acceleration is reduced. If the acceleration is along the axis of the body, "black-out" will occur if the acceleration is toward the head and a "red-out" (excessive blood in the head and the retina of the eye leading to a red visual sensation) if the acceleration is toward the feet. At 100 m/s² (about 10 g) the vertebrae will be crushed. In the anterior-posterior direction accelerations of 100 m/s² (about 10 g) are tolerable for short times and the limit of endurance is about 300 m/s² (about 30 g). Astronauts on lift-off and on re-entry may be subject to 100 m/s², and at those times they must be seated in form-fitting chairs oriented to direct the acceleration toward their back. Also, equipment designed for experiments in orbiting vehicles must be rugged enough to tolerate such accelerations.

The effects of various accelerations on the human body have been studied using high speed rocket sleds stopped quickly using a water braking system and also in large human centrifuges. The accelerations occurring in such circular motion will be considered in the next section.

EXAMPLE 14

A vehicle returning from the moon enters the atmosphere with a speed of 11 km/s and is brought to rest with an acceleration not to exceed 100 m/s² (about 10 g) in value. Allowing a constant −100 m/s², what length of path will be necessary?

The initial speed 11,000 m/s is to be reduced to zero at the rate of 100 meters per second every second, so 110 seconds would be required.

The average speed is (11,000 m/s + 0)/2 which is 5500 m/s or 5.5 km/s.

Traveling at an average of 5.5 km/s for 110s the distance traveled would be (5.5 × 110) km = 605 km.

The necessary distance is 605 km, which introduces the problem that the atmosphere is not that thick. Air friction is negligible above an altitude of 100 km. That is quite a problem! It means that the craft cannot enter vertically if disaster is to be avoided. The path must be gently sloping to allow sufficient time and distance to slow down.

The angle at which the spacecraft must re-enter can be estimated from these results. If the deceleration is to take place while the craft drops 60 km into the atmosphere, then the angle θ from the horizontal is as shown in Figure 2–17(A):

$$\sin \theta = \frac{60 \text{ km}}{605 \text{ km}} = 0.099$$

$$\theta = 5.7°$$

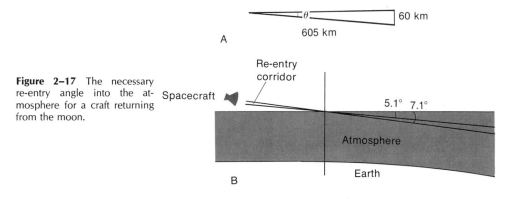

Figure 2–17 The necessary re-entry angle into the atmosphere for a craft returning from the moon.

The angle of entry into the atmosphere must be 5.7° off the horizontal. Some variation from this is possible but the allowable range of re-entry angle for the Apollo moon vehicles was only from 5.1° to 7.1° as in Figure 2–17(B). At an angle less than 5.1° the craft would "skip" back into space and at greater than 7.1° the acceleration forces would be too high. This is an extremely fine limit on the re-entry path and that to date all lunar vehicles have returned successfully is a tribute to those who designed the vehicles and controlled the flights.

2–4 RADIAL ACCELERATION WITH CONSTANT SPEED

If an object moves in a circular path, the velocity will always be changing even though the speed may be constant. This is because velocity was defined in a way that included the direction, and whenever the velocity changes there is acceleration. The problem is to relate the pertinent quantities: speed, radius of the path and acceleration.

In Figure 2–18, a rotating body is shown at two points on

Figure 2–18 A body moving on a curve at constant *speed*. The *velocity* changes between points 1 and 2 because the direction changes.

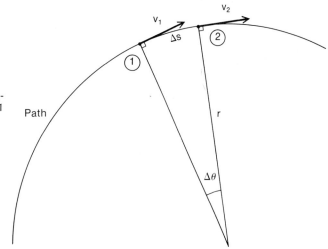

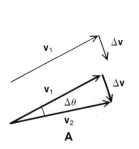

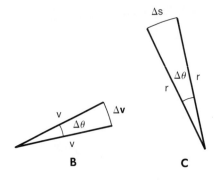

Figure 2-19 If v_1 is the velocity at point 1 in Figure 2-18, then the amount to be added to get v_2 is shown in (a) as Δv. The velocity diagram is shown in (b), and the displacement diagram is in (c). The triangles in (b) and (c) are similar.

its path. The velocity was initially v_1, and after traveling a short distance it was v_2. The speed was the same; only the direction of the velocity changed. To change the velocity vector from v_1 to v_2, something must have been added to v_1; call it Δv. The vector diagram is shown in Figure 2-19(A). This process of vector addition of velocities was explained in Section 2-2. The magnitudes of v_1 and v_2 are both the same, that is, the speed v around the circle. Showing only the magnitudes gives a diagram such as Figure 2-19(B). The angle between the velocity vectors is the same as in Figure 2-18. In Figure 2-19(C) is shown the distance moved between positions 1 and 2 of Figure 2-18, and the radii to those points. The triangles shown in Figure 2-19(B) and (C) are similar, so the ratios of corresponding sides are equal. That is, $\Delta v/v = \Delta s/r$. The change in velocity Δv is then given by

$$\Delta v = v\Delta s/r$$

Dividing the change in velocity by the time required for that change would give the acceleration. In this case of circular motion, we use a subscript c to distinguish circular from linear acceleration; then

$$a_c = \Delta v/\Delta t$$

where Δt is the time required to go from position 1 to position 2 in Figure 2-18. Now substitute the value already obtained for v to get

$$a_c = \Delta v/\Delta t = v\Delta s/r\Delta t$$

The part $\Delta s/\Delta t$ is just v, so the result is that

$$a_c = v^2/r$$

The direction of the acceleration is the same as the direction of the change in velocity, Δv. Examination of Figure 2–19(A) shows Δv to be toward the center of the circle. The acceleration toward the center of any circular path is called **centripetal acceleration,** and its magnitude is described by v^2/r.

EXAMPLE 15

Consider a centrifuge with a radius of rotation of 0.1 m and a rotational speed of 62.8 m/s, as was calculated in Section 2–2. The acceleration of a particle in the centrifuge is given by v^2/r. With $v = 62.8$ m/s and $r = 0.1$ m,

$$a_c = v^2/r = \frac{62.8^2 \, \text{m}^2/\text{s}^2}{0.1 \, \text{m}} = 39{,}400 \, \text{m/s}^2$$

This is a rather high acceleration. A falling body accelerates at only 9.8 m/s². The comparison is

$$a_c/g = \frac{39{,}400 \, \text{m/s}^2}{9.8 \, \text{m/s}^2} = 4020 \, \text{(the units cancel)}$$

and multiplying both sides by g yields

$$a_c = 4020 \, g$$

In this way centripetal accelerations are often given in terms of the acceleration of gravity, in g's. An ultracentrifuge may achieve over one million g.

EXAMPLE 16

What is the minimum radius for a turn in an aircraft moving at 900 km/h if the centripetal acceleration is not to exceed 4 g?

Use $\qquad\qquad\qquad\qquad\qquad\qquad a_c = v^2/r$

Solve for r, $\qquad\qquad\qquad\qquad\qquad\quad r = v^2/a_c$

The numerical values are $\qquad\quad v^2 = 900$ km/h
$\qquad\qquad\qquad\qquad\qquad\qquad\qquad = 250$ m/s

$\qquad\qquad\qquad\qquad\qquad\qquad a_c = 4 \, g$
$\qquad\qquad\qquad\qquad\qquad\qquad\quad = 39.2$ m/s².

Then $\qquad\qquad\qquad\qquad\qquad\qquad r = \dfrac{250^2 \, \text{m}^2/\text{s}^2}{39.2 \, \text{m/s}^2}$
$\qquad\qquad\qquad\qquad\qquad\qquad\quad = 1594$ m
$\qquad\qquad\qquad\qquad\qquad\qquad\quad = 1.59$ km

The minimum radius of the turn is 1.59 km. The maximum acceleration that a pilot in a sitting position can endure is 4 g. If a "g-suit" is worn to prevent too much blood leaving the brain, with a resulting blackout, up to 6 or even 8 g is possible.

2-4-1   THE LOW EARTH SATELLITE

Any object released near the surface of the earth has an acceleration directed toward the center of the earth, the acceleration g, which is due to gravity. A horizontal velocity is not affected by gravity. If the horizontal velocity is such that v^2/r is just equal to g, circular motion will result. This is the situation shown as S in Figure 2-16(B) and it is an earth satellite. To launch a satellite it is first raised by a rocket to an altitude where air friction is sufficiently low, at least 150 km. Another rocket is then fired in a horizontal direction to give the satellite sufficient speed to orbit the earth. The orbit will be circular if the speed is such that

$$v^2/r = g$$

where r is the radius of the orbit. At an altitude of 150 km the value of g is only 5 per cent less than it is at the surface of the earth, or 9.33 m/s². The necessary speed for a low earth satellite is calculated from the above relation.

EXAMPLE 17

Calculate the speed of a satellite orbiting 150 km above the surface of the earth where $g = 9.33$ m/s². The radius of the earth is 6380 km. Find the time for that satellite to orbit the earth.

The centripetal acceleration v^2/r is equal to g. The radius of the orbit, r, is 6380 km plus 150 km.

$$r = 6530 \text{ km} = 6.53 \times 10^6 \text{ m}$$
$$g = 9.33 \text{ m/s}^2$$
$$v^2/r = g \text{ from which } v = \sqrt{rg}$$

$$v = \sqrt{6.53 \times 10^6 \text{ m} \times 9.33 \text{ m/s}^2}$$
$$= 7.80 \times 10^3 \text{ m/s}$$

The speed of the satellite is 7.80×10^3 m/s or 7.8 km/s.

The circumference of a circle 6.53×10^6 m in radius is 4.10×10^7 m. The time required to go 4.10×10^7 m is given from $T =$ distance/speed.

$$T = 4.10 \times 10^7 \text{ m}/7.8 \times 10^3 \text{ m/s}$$
$$= 5260 \text{ s}$$
$$= 87 \text{ min } 40 \text{ s}$$

The satellite makes one orbit in 87 min 40 s.

DISCUSSION AND DERIVATION PROBLEMS

2-1 1. Do small bubbles in a liquid rise faster, slower or at the same rate as large bubbles? Justify your answer.

2-3-1 2. A ball thrown upward will take the same time to rise as to fall if there is no air resistance. In a situation in which there is air resistance and the greater the speed the greater the resistance, which will be the greater, the time to rise or the time to fall? Also, will the total time aloft be more or less than for the unresisted case?

2-3-2 3. A rifle is pointed directly at a target high above the ground. At the same moment the rifle is fired the target is released and falls freely. Will the bullet hit the target? Discuss the situation. Do not just give a yes or no answer.

2-3-2 4. Use the relation for the range of a projectile to find the angle that would give a maximum range for a given initial speed.

2-3-2 5. Derive a formula to give the horizontal range of a projectile that is given an initial horizontal velocity v at a height h above level ground.

2-4-1 6. Discuss the path of an intended low earth satellite that has an initial horizontal velocity just below the value given by $v^2/r = g$. Also discuss the path if the speed is slightly in excess of that needed for a circular orbit.

PROBLEMS

2-1 1. Find the average speed of a supersonic aircraft when it flies from London to New York, a distance of 5540 km, in 2 hours 52 minutes. Express the answer in km/h and in m/s.

2-2 2. A ball is thrown vertically, going to a height of 34.4 m. It returns to its original position after 5.3 s. Find for that 5.3 s:
(a) the total path,
(b) the displacement,
(c) the average speed,
(d) the average velocity.

2-2 3. A car goes around a circular track 0.40 km in radius at a constant speed of 25 m/s. Find the displacement and the average velocity in the first 25 seconds, the first 50 seconds, and in one revolution of the track.

2-2 4. Find the speed of a stylus along the groove of a record that turns at 33⅓ revolutions per minute when the stylus is in a groove 14.0 cm from the center, and also when the stylus is 7.0 cm from the center.

2-2 5. Find the average acceleration of a car in each of the intervals between the times shown in the following table. The speedometer reading at each time is shown in m/s and the direction is constant.

time (s)	velocity (m/s)
0	0
3	3.6
4	4.8
6	8.0
9	13.4
12	13.6
20	13.6
24	0

2-2 6. Find the speed of the car at each of the times listed in problem 5.

2-2 7. Find the final velocity of a mass moving constantly eastward if it accelerates at 8 m/s² for 2 s, then at 5 m/s² for 3 s and at 2 m/s² for 4 more seconds.

2-2 8. Consider that a certain playful parent runs at 10 m/s and is chasing a child who runs at only 5 m/s. The child is initially 100 meters ahead of the parent. How far will the parent run before catching up to the child? How long will it take?

2-2 9. Consider Problem 8 this way. How long will it take for the parent to run to the original position of the child? Where will the child be then? How long will it take the parent to run to that new position of the child, and then where will the child be? Continue the problem in this way. The parent is always running to the child's position but the child is still ahead. Add the times to get the total time for the parent to catch up to the child. Will the parent ever reach the point of running alongside the child, let alone passing? Reconcile this conclusion with the result of Problem 8 and with experience.

2-2-1 10. Calculate the downward speed of raindrops if they are seen from a vehicle moving at 15 m/s to apparently fall 30° from the vertical.

2-2-1 11. Find the heading (direction) an aircraft must take in order for it to fly due north when there is a wind of 60 km/h from the northeast. The airspeed of the craft is 190 km/h. Also, what is the resulting velocity over the ground?

2-2-1 12. A blimp that moves through

the air at 4.0 m/s is to travel to a point which is N.E. (45° east of north) of its starting point. A wind of 5.0 m/s is blowing from a direction 30° west of south.

(a) In what direction should the blimp head?

(b) What would be its speed over the ground?

2-3 13. An aircraft starting from rest moves 3300 m along a runway to achieve a take-off speed of 190 km/h. Assume constant acceleration.

(a) What is the average velocity?

(b) How long does the take-off run require?

(c) What is the acceleration?

2-3 14. A vehicle, starting from rest, goes on a straight track for a distance of 400 m in 5 s.

(a) What is the average speed?

(b) What assumption would be necessary in order that you could calculate a possible final speed?

(c) What would the final speed perhaps be?

2-3 15. Find the acceleration necessary for an elevator to move between two floors 3.0 m apart in 3.0 s. The elevator accelerates with an acceleration a for the first half of the journey and $-a$ for the second half. It starts and finishes at rest.

2-3 16. In what distance would a vehicle stop if the possible acceleration is -7.0 m/s^2 when the brakes are applied? Calculate this for speeds of 10, 20 and 30 m/s. These correspond to speeds of 22.4, 44.7 and 67.1 mi/h.

2-3 17. A ball is thrown upward at 49 m/s. Make a graph of the upward velocity against time for the first 10 seconds. In doing this consider air resistance to be negligible and that $g = -9.8$ m/s^2. Use the area under the graph to find the displacement in the first five seconds and then in the first ten seconds.

2-3-1 18. An object is projected at 76° above the horizontal with a speed of 30.3 m/s.

(a) Find the vertical and the horizontal components of the initial velocity.

(b) Calculate the horizontal and the vertical position of the object every second for six seconds.

(c) Make a graph of the trajectory using the data from part (b).

2-3-2 19. A ball in a trajectory was seen to have a range of 75 m and was in the air for 5.0 s. Find:

(a) the upward component of the initial velocity,

(b) the horizontal component of the initial velocity,

(c) the initial velocity (include the angle),

(d) the maximum height to which the ball rose.

2-3-2 20. Calculate theoretically the initial angle of projection of an object for which the maximum height is equal to the horizontal range.

2-3-2 21. Compare the speed on impact after falling 30 m on the earth to that after falling the same distance on the moon. On earth g is 9.8 m/s^2 and on the moon it is 1.57 m/s^2.

2-3-2 22. If you can throw a ball to a height of 15 m on earth, how high could you throw it on the moon where $g = 1.57$ m/s^2?

2-3-2 23. If you can throw a ball upward on earth so that it remains up for 4.0 seconds, how long would a ball thrown with the same speed stay aloft on the moon? ($g = 1.57$ m/s^2 on the moon).

2-3-2 24. An object is thrown horizontally at 15 m/s from a window which is 25 m above the ground.

(a) How long will it take to reach the ground?

(b) How far from the base of the building will the object land?

2-3-3 25. An orbiting spacecraft enters the atmosphere at 7.9 km/s and is to decrease its speed to effectively zero with a negative acceleration of 100 m/s^2 while it drops a vertical 70 km into the atmosphere. What will be the horizontal distance it travels between the time it enters the atmosphere and the time it has lost its speed?

2-4 26. What is the maximum speed at which a car can go round a curve of radius 400 m if the curve is level and frictional forces will allow a maximum centripetal acceleration of 0.4 g?

2-4 27. What centripetal acceleration occurs on a fly riding near the edge of a 33$\frac{1}{3}$ rpm record ($r = 14.0$ cm)?

2-4 28. What must be the radius of a turn to be made by an aircraft flying at 542 m/s if the centripetal acceleration is to be only 0.3 g? Express the answer in km?

2-4-1 29. Find the speed to orbit a spacecraft about the moon in a low orbit where $g = 1.47$ m/s^2 and the radius of the orbit is 1800 km. How long would it take the craft to orbit the moon?

CHAPTER THREE

FORCE AND ACCELERATION: DYNAMICS

3-1 INTRODUCING NEWTON'S SECOND LAW

On earth, everything that moves meets resistance. To maintain the motion, a force must be applied to balance the resistance force. It is no wonder that many, many years passed before it was realized that a uniform constant motion in a straight line requires no force at all, and that only a *force produces a change in motion or an acceleration.* This realization was one of the keys, perhaps the main one, that opened up to modern science the whole area dealing with force and motion. This includes the motion of planets and stars, the centrifuge, the car, the aircraft, the spacecraft, and major portions of the sciences of electricity and atomic physics. This seems rather formidable, but the basis of it all is summarized in a simple little equation:

$$\mathbf{F} = m\mathbf{a}$$

where $\mathbf{F}$ is the net force on the body; m is the mass, a scalar quantity; and $\mathbf{a}$ is the acceleration. It is important to note that $\mathbf{F}$ and $\mathbf{a}$ have direction, that is, they are vectors, and they are in the same direction.

This equation, simple as it seems, will require some explanation for its wide use. The implications that are buried in it will be revealed in discussion about it.

The equation itself is a form of what is called **Newton's second law.** It is not in its most basic form, nor is it just as Newton expresssed it. He wrote it out in Latin: "Lex II.

Mutationem motus proportionalem esse vi motrici impressae, et fieri secundum lineam rectam qua vis illa imprimitur." This translates as: "Law 2. The alteration of motion is ever proportional to the motive force impressed; and it is made in the direction of the right line in which that force is impressed."

The force that is used in Newton's second law is the *net force*. If two or more forces act on the body, the force to use is the *one* that would give the equivalent motion. This is just the vector sum of all those forces that act on the body. Frequently the forces will be acting along a single line and it will be an easy matter to find the net force.

If an object slides along a level surface, it slows down because of resistance or frictional forces, **R**, acting in the direction opposite to the motion. If a forward force, **P**, of the same size as the resistance force is applied to the object, it will move at a constant velocity because the net force, and thus the acceleration, is zero. This is illustrated diagramatically in Figure 3–1(A). If the applied force **P** exceeds **R** as in Figure 3–1(B), there will be a net forward force **F**, the size of which is the numerical difference between **P** and **R**. The object will accelerate in the direction of the net force.

The meaning of Newton's second law may be explained more by solving for the acceleration, **a**.

$$\mathbf{a} = \mathbf{F}/m$$

In this form the meaning is that the amount of acceleration is in direct proportion to the net force on the object: double F and a also doubles, etc. Also, the acceleration varies inversely as the mass m. If the same force is applied to two masses, one double the mass of the other, the larger mass will have just half the acceleration of the smaller. The object with the larger mass would, we say, have the greater **inertia**. It is by this property of inertia that mass can be detected and measured so it is sometimes called *inertial mass*. Mass can also be detected by the attraction to another mass. Two masses can be compared by

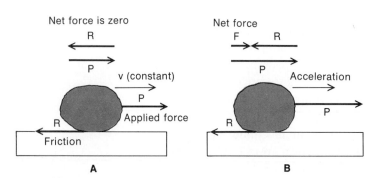

Figure 3–1 In (a), the forces on the object are balanced and there is no acceleration. In (b), the resistance R is less than P in magnitude, so there is a resultant force F to cause acceleration.

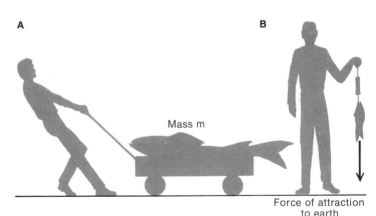

Figure 3-2 Mass is detected either by its inertia as in (a), or by its gravitational attraction to another mass, such as the earth, as in (b).

Mass m

Force of attraction to earth

the amount of attraction to another mass, to the earth for example. Mass detected in that way is called *gravitational* mass. The two methods of detecting mass are shown in Figure 3-2.

Oddly, gravitational mass and inertial mass are directly proportional, and masses compared using either property give the same comparative values. If one mass shows double the attraction to the earth compared to another, it will also exhibit double the inertia. This equivalence has been demonstated to the limit of our measuring ability. There is no philosophical or theoretical reason why these two properties of mass should be linked so closely, except that the concept is fundamental in the general theory of relativity.

3-2 UNITS IN DYNAMICS

The SI units have been devised as a consistent system for use in dynamics and that is their main advantage. Errors due to mixing of units from various systems will not occur if only SI units are used. For example, in the equation $\mathbf{F} = m\mathbf{a}$ it is not possible to use mass in kilograms and force in kilograms of force. The force must be in the SI unit, the newton. The newton, in fact, was chosen to fit Newton's second law expressed in that simple form.

The units for acceleration are that of length divided by the square of a time: meters per second squared (m/s^2) in SI units. The unit for mass has not yet been considered: it is the kilogram. There is a block of metal kept in the Bureau of Standards near Paris and that is defined as one kilogram. The original plan had been to use a cubic centimeter of water as one gram. That was found to be inconvenient as a precise standard. It was also inconveniently small, so a metal standard, a thousand grams or a kilogram, was constructed. There has been no practical way devised to define the kilogram in terms of atomic standards as has been done with the standards of length and of time.

A gram, g, is a thousandth of a kilogram: a thousandth of a gram or 10^{-3} gram is a *milligram,* (mg). A millionth or 10^{-6} of a gram is a *microgram,* (μg), and 10^{-9} gram is a nanogram. A thousand kilograms is a *tonne* or a *megagram,* Mg.

There is another standard of mass for use on the atomic scale. One atomic mass unit, symbol u, is one twelfth of the mass of an atom of carbon 12. The relation between atomic mass units and kilograms has been determined to be

$$1 \text{ u} = 1.660 \times 10^{-27} \text{ kg}$$

The construction of standard masses to which others are compared is not new. In Figure 3–3 is a photograph of some standards of mass, carefully made from stone and used in Ancient Egypt.

Throughout the years there have been many units devised for the measurement of mass. A list of many of them and how to convert them to SI units is included in Appendix 2. Many of these are still in use. The pound, or more correctly, the pound **avoirdupois** is the unit of mass still frequently used in some English speaking countries. One pound equals 0.454 kg. There are also the **troy** pounds and troy ounces used for measure of precious metals. In England the mass of a person is as often expressed in **stones** as in any other measure. One stone is 14 pounds or 6.36 kg.

In another system of units, referred to as the **British Engineering** (or **B.E.**) **system**, the word pound is used as a unit of force. It is defined as the force on a mass of one pound avoirdupois at sea level and 45° latitude.

The quantity called **density** refers to the mass per unit volume. Average density, especially of a non-uniform object, may also be used. It is the mass of an object divided by the volume. In symbols, if the mass M is in a volume V the density is given by

$$d = \frac{M}{V}$$

Figure 3–3 Mass standards used in ancient Egypt.

The total mass of an object is density times volume.

The SI units for density are kg/m³, though any unit of mass divided by any unit of volume will be a density.

The quantity called **relative density** is sometimes used. The relative density is the density of material or object relative to the density of a standard material. For solids and liquids the standard material is water; for gases, the standard for comparison is air at a standard temperature and pressure. Relative density has no units because it is a ratio.

$$\text{Relative density} = \frac{\text{density of material or object}}{\text{density of standard material}}$$

In Table 3–1 are shown a few densities and relative densities. A more extensive list is in Appendix 3.

The term relative density is similar to the term **specific gravity** used in the past. The term specific gravity is being discarded because the word specific had a variety of meanings. Now the term *specific* is defined to mean the quantity of anything per unit of mass (kilogram).

TABLE 3–1   Densities and relative densities*

MATERIAL	DENSITY kg/m³	RELATIVE DENSITY
Water	1000	1.000
Aluminum	2700	2.70
Copper	8890	8.89
Gold	19,300	19.3
Iron	7880	7.88
Lead	11,340	11.34
Magnesium	1740	1.74
Osmium	22,500	22.5
Silver	10,500	10.5
Uranium	18,700	18.7
Woods: Balsa	110– 140	0.11–0.14
Birch	510– 770	0.51–0.77
Oak	600– 900	0.6 –0.9
Ebony	1100–1330	1.1 –1.33
Mercury (liquid)	13,600	13.6
Bone	1700–2000	1.7 –2.0
Cement	2700–3000	2.7 –3.0
Feldspar	2550–2750	2.55–2.75
Hornblende	3000	3.0
Limestone	2680–2760	2.68–2.76
Quartz	2650	2.65
Earth (average)	5500	5.50
Sun (average)	1390	1.39
Moon (average)	3300	3.30

*A more extensive list is in Appendix 3.

For example, energy per kilogram would be called specific energy; volume per kilogram would be specific volume (the inverse of density).

EXAMPLE 1

Estimate the total mass of a slab of cement that is 8 m × 4 m and 0.20 m thick.
From Table 3–1 the density of cement may be in the range 2700 to 3000 kg/m³. The value used in a problem of this type is a personal choice. Let's use 2850 kg/m³.
The volume is

$$V = 8 \text{ m} \times 4 \text{ m} \times 0.2 \text{ m}$$
$$6.4 \text{ m}^3$$

The mass, density times volume, is

$$M = 2850 \, \frac{\text{kg}}{\text{m}^3} \times 6.4 \text{ m}^3$$

$$= 18{,}240 \text{ kg}$$

The mass is 18,240 kg or 18.2 tonnes.

The units for acceleration, which in SI units are m/s², are based on the definitions of the meter and the second.

The quantities on the right hand side of $\mathbf{F} \times m\mathbf{a}$ are all precisely defined, and there cannot be an independent force standard if the equation is to be correct in that form. The newton of force is in fact defined from Newton's second law. One newton is the force that would accelerate a mass of 1 kg with an acceleration of 1 m/s². In equation form,

$$1 \text{ N} = 1 \text{ kg} \cdot 1 \text{ m/s}^2$$
$$\text{or } 1 \text{ N} = 1 \text{ kg m/s}^2$$

The word newton is equivalent to and interchangeable with kg m/s².

The acceleration of an object at the surface of the earth is close to 9.8 m/s². If the mass is 1 kg, the force in newtons is found from $F = ma$ to be

$$F = 1 \text{ kg} \cdot 9.8 \text{ m/s}^2$$
$$= 9.8 \text{ N}$$

If the mass is m, the force of gravity in newtons is

$$F = m \cdot 9.8 \text{ m/s}^2$$

This is the origin of the conversion factor given in Chapter 1.

Over the surface of the earth the acceleration of a freely falling object varies slightly. If that acceleration is called g, then the force of gravity, or what we call weight, W, is given by

$$W = mg$$

The quantity g is open to two interpretations. It can be an acceleration (m/s^2) or the equation $W = mg$ can be solved for g to give $g = W/m$ where W is in newtons and m in kilograms. Then g has the units N/kg. This is interpreted as the gravitational field strength: the force in newtons on a unit mass, 1 kg, placed at that point in space. It must be pointed out that the acceleration in a system may be due to factors other than gravity so this interpretation must be used with caution.

Simply stated, the foregoing has shown that Newton's second law in the form $\mathbf{F} = m\mathbf{a}$ can be used with the SI units: m in kg, a in m/s^2 and F in N.

There is another system of units based on the unit of mass being the gram, g, of length being the centimeter, cm, and time, the second, s. The product ma then has units g cm/s^2. The unit for force in this system, the force that would accelerate 1 g at 1 cm/s^2 is called 1 **dyne**. The unit 1 dyne is 1 g cm/s^2.

Care must be taken not to confuse the symbol for the unit grams, g, with the gravitational acceleration, g.

The acceleration due to gravity in the centimeter–gram–second, or cgs system, is about 980 cm/s^2 so the force of gravity on 1 gram would, by $W = mg$, be 980 dynes. This factor was also used in Chapter 1.

3–3 USING F=MA

Newton's laws are very simple, but a few examples of how to use them will be in order.

EXAMPLE 2

Find the force of gravity in newtons and in dynes on a 150.2 g mass at Chicago where a freely falling object has an acceleration of 9.803 m/s^2.

The units are not consistent; change 150.2 g to 0.1502 kg.

$$\begin{aligned} W = F &= 0.1502 \text{ kg} \times 9.803 \text{ m/s}^2 \\ &= 1.472 \text{ kg m/s}^2 \\ &= 1.472 \text{ N} \end{aligned}$$

In the cgs units:

$$\begin{aligned} W = F &= 150.2 \text{ g} \times 980.3 \text{ cm/s}^2 \\ &= 147,200 \text{ g cm/s}^2 \\ &= 147,200 \text{ dynes} \end{aligned}$$

EXAMPLE 3

A rocket of mass 5000 kg accelerates upward from the earth at 25 m/s^2. What is the upward thrust?

The net force found from $F = ma$ is

$$F = 5000 \text{ kg} \times 25 \text{ m/s}^2$$
$$= 125,000 \text{ N}$$

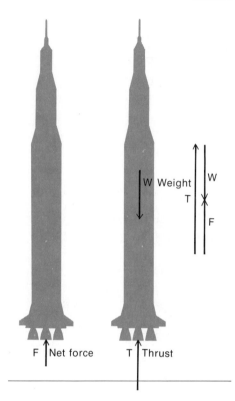

W Weight

T

F Net force

T Thrust

Figure 3–4 The value of the net force that accelerates a rocket upward is the difference between the thrust and the gravitational force.

This is the net force as in Figure 3–4, but the pull of the earth on 5000 kg is 5000 × 9.8 N or 49,000 N downward. The thrust must be enough to overcome this and still have 125,000 N left over. The upward thrust must therefore be 125,000 N + 49,000 N or 174,000 N.

EXAMPLE 4

A 70 kg kangaroo makes a leap for which the required initial speed is 7.0 m/s. This speed is acquired by pushing the hind feet on the ground while the body moves only 0.6 m along the path of the leap. Find the force on the ground (or, by Newton's third law, of the ground on the kangaroo) that produced the necessary acceleration. The situation is illustrated in Figure 3–5.

The equation **F** = m**a** cannot be used directly because **a** is not known. The acceleration can be found using the relations developed in kinematics. The distance (0.6 m), the initial speed (assumed to be 0) and the final speed (7.0 m/s) are known. Use $v^2 = 2as$ in which v and s are known.

Solving for a: $a = v^2/2s$

$$= \frac{49 \text{ m}^2}{2 \times 0.6 \text{ m s}^2} = 41 \frac{\text{m}}{\text{s}^2}$$

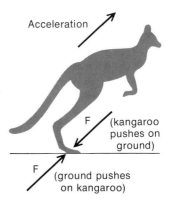

Acceleration

F (kangaroo pushes on ground)

F (ground pushes on kangaroo)

Figure 3–5 The force of the ground on a kangaroo (the reaction to the kangaroo pushing on the ground) gives it an acceleration for a leap.

Then use

$$F = ma$$
$$= 70 \text{ kg} \times 41 \text{ m/s}^2$$
$$= 2870 \text{ kg m/s}^2$$
$$= 2870 \text{ N}$$

The force would be 2870 N.

The force on the ground must be 2870 N in excess of that needed only to support the weight of the kangaroo. At the point of contact with the ground, the kangaroo pushes on the ground and the ground pushes back with an equal force. The force of the ground on the kangaroo accelerates the kangaroo. The force of the kangaroo on the earth pushes the earth backward, but not very much because of the large mass of the earth.

EXAMPLE 5

Two masses, one of which is of 5 kg and the other which is unknown, are connected together by a compressed spring as in Figure 3–6. The spring is released and the 5 kg mass accelerates at 3 m/s², while the unknown mass accelerates at only 0.2 m/s². What is the size of the unknown mass? There are two ways to do this. The first is to solve for the force and by Newton's first law that is also the force on the second mass. Then that mass can be found. The second method is to solve for the second mass directly.

$m_2 = ?$

$m_1 = 5$ kg

$a_2 = 0.2$ m/s² $a_1 = 3$ m/s²

Figure 3–6 Two masses being pushed apart by a compressed spring.

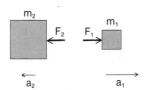

m_2

F_2 F_1

m_1

a_2

a_1

For the first mass: $F_1 = m_1 a_1$
For the second mass: $F_2 = m_2 a_2$
By Newton's second law $F_1 = F_2$ in magnitude so

$$m_1 a_1 = m_2 a_2$$

and solving for m_2

$$m_2 = m_1\, a_1/a_2$$

Substituting the values

$$m_2 = \frac{5\ \text{kg}}{0.2\,\dfrac{\text{m}}{\text{s}^2}}\ 3\,\frac{\text{m}}{\text{s}^2}$$

$$= 75\ \text{kg}$$

The second mass is 75 kg.

EXAMPLE 6

What would be the mass of a spacesuit and associated equipment that would result in a person of mass 65 kg having the same total weight on the moon as the person without the space suit on earth? On the moon g is 1.64 m/s² and on earth 9.8 m/s².

The weight of the person alone on earth is

$$W = 65\ \text{kg}\ 9.8\ \text{m/s}^2$$
$$= 637\ \text{N}$$

The total mass of the person wearing a suit of mass m on the moon is $m + 65$ kg. The total weight on the moon is

$$W = (m + 65\ \text{kg})\ 1.64\ \text{m/s}^2$$

and this must be 637 N or 637 kg m/s²

$$(m + 65\ \text{kg})\ 1.64\ \text{m/s}^2 = 637\ \text{kg m/s}^2$$
$$m + 65\ \text{kg} = 388\ \text{kg}$$
$$m = 323\ \text{kg}$$

The astronaut of mass 65 kg with a suit and equipment of 323 kg would be the same weight on the moon as without the suit and equipment on earth. This explains the ability of the astronauts to carry those large packs of equipment as in Figure 3–7, and still be able to walk on the moon.

Figure 3–7   An astronaut on the moon with his massive pack.

3–4 MOMENTUM

What we call momentum was once referred to as "quantity of motion." Momentum takes into account both the velocity of an object and its mass. A fast-moving object with a small mass could have the same momentum as a slower but more massive object. The momentum is defined as the product of mass and velocity; often the symbol $\mathbf{p}$ is used for it. That is, in symbols

$$\mathbf{p} = m\mathbf{v}$$

The direction of the momentum, which is a vector quantity, is the same as the direction of the velocity.

Newton's second law is intimately connected with momentum, as can be shown in the following way. Use the form

$$\mathbf{F} = m\mathbf{a}$$

But acceleration $\mathbf{a}$ is a rate of change of velocity. Let the initial velocity be $\mathbf{v}_1$ and the final velocity be $\mathbf{v}_2$, so the change is $\mathbf{v}_2 - \mathbf{v}_1$. If this change occurs in a time interval t, the rate of change of velocity, or acceleration, is

$$\mathbf{a} = \frac{\mathbf{v}_2 - \mathbf{v}_1}{t}$$

Putting this into $\mathbf{F} = m\mathbf{a}$ gives

$$\mathbf{F} = \frac{m(\mathbf{v}_2 - \mathbf{v}_1)}{t}$$

or

$$\mathbf{F} = \frac{m\mathbf{v}_2 - m\mathbf{v}_1}{t}$$

$m\mathbf{v}_1$ is the initial momentum of the object.

$m\mathbf{v}_2$ is the final momentum of the object.

$m\mathbf{v}_2 - m\mathbf{v}_1$ is the change in momentum that occurred in the time interval t. The change per unit time, or rate of change, of momentum is $(m\mathbf{v}_2 - m\mathbf{v}_1)/t$. That last equation expressed in words is

Force equals rate of change of momentum.

Another way of expressing this is that net forces cause changes in momentum. This *momentum form of Newton's second law* is actually more basic than the form $\mathbf{F} = m\mathbf{a}$ because, for example, mass changes can also be taken into account.

In the above derivations, the force is the average net force; to indicate this, a bar could be placed over the **F**, as in $\bar{\mathbf{F}}$.

EXAMPLE 7

How fast would a 1500 kg car have to travel to have the same momentum as a 5000 kg truck moving at 12 km/s in the same direction?
Let the masses be m_1 and m_2 and the speeds be v_1 and v_2.

$$m_1 = 1500 \text{ kg}$$
$$m_2 = 5000 \text{ kg}$$
$$v_2 = 12 \text{ km/s}$$
$$v_1 \text{ is unknown}$$

We want to find the value of v_1 which makes

$$m_1 v_1 = m_2 v_2$$

$$v_1 = \frac{m_2}{m_1} v_2$$

$$= \frac{5000 \text{ kg}}{1500 \text{ kg}} \times 12 \text{ km/s}$$

$$= 40 \text{ km/s}$$

The vehicle of mass 1500 kg would have to move at 40 km/s to have the same momentum as a 5000 kg vehicle moving at 12 km/s.

EXAMPLE 8

An object of mass 0.5 kg falls onto the floor with a speed of 3 m/s and rebounds with a speed of 2 m/s. What is the change in momentum?
The initial momentum was 0.5 kg × 3 m/s downward or 1.5 kg m/s downward. After the impact the momentum was 0.5 kg × 2 m/s upward or 1.0 kg m/s upward.
In changing from 1.5 kg m/s downward to 1.0 kg m/s upward, the total change is 2.5 kg m/s, for remember that momentum is a vector quantity.

EXAMPLE 9

If a 2 kg object is seen to change its speed from 5 m/s to 12 m/s in 3.5 seconds, what was the average force?
Initial momentum is 2 kg × 5 m/s = 10 kg m/s
Final momentum is 2 kg × 12 m/s = 24 kg m/s
The change in momentum is 14 kg m/s
This occurred in 3.5 s so the change per second is

$$\frac{14 \text{ kg m/s}}{3.5 \text{ s}} = 4 \text{ kg m/s}^2$$

Note that the units are the same as for newtons, so the average force is 4 N.

3-4-1   AN IMPULSE OR BLOW

There are some situations in which the initial and final momentum may be the only measurable quantities. For example, a golf ball is initially at rest; after being hit, it has a measurable speed (using proper techniques). The force is very large and acts for a very short interval. The mass of the golf ball is easily found, so the change in momentum can be found. Another situation is that of a ball that strikes a person's head and bounces off. The change in momentum of the ball can be found. In each of these situations we say that there has been a blow or an **impulse**.

The word impulse is defined in physics as the product of force and the time force acts. If the force varies in that time, the average force is used. The common word blow is a synonym.

An impulse causes the object on which it acts to change its momentum, and the amount of change in momentum is found from Newton's second law. Use it in the form

$$F = \frac{mv_2 - mv_1}{t}$$

Multiply through by t to get

$$Ft = \text{impulse} = mv_2 - mv_1$$

The left hand side of this equation is the impulse and the right hand side is the change in momentum of the object on which the impulse acted. In words, the equation says that *impulse equals change in momentum*. This is called the **momentum-impulse theorem.** An impulse may be measured by the change in momentum caused by it. This can be useful in situations in which it is not possible to measure the size of the force (and it may vary) or the time. One example is that of a ball being struck by a bat.

The momentum-impulse theorem shows that the units of momentum must be the same as the units of force multiplied by time. Momentum can therefore be expressed in newton seconds (N s) as well as in kg m/s.

EXAMPLE 10

In measuring the impact strength of a bone, a mass of 2.4 kg was made to strike the bone at 3.0 m/s. After the bone broke, the mass continued moving at 0.5 m/s. What impulse was delivered?

The initial momentum was 2.4 kg × 3.0 m/s or 7.2 kg m/s.
The final momentum was 2.4 kg × 0.5 m/s or 1.2 kg m/s.
The change in momentum was 5.0 kg m/s.

The impulse was therefore 5.0 kg m/s or in the alternate units (multiplying by s/s), the impulse was 5.0 N s.

3–4–2 CONSERVATION OF MOMENTUM

Science has at its beginning a collection of laws, but it is in the combination of these laws that it becomes an imposing structure, just as the proper collection of bricks makes an impressive building or a few transistors make a radio.

Consider a combination of some of the laws already worked with, laws that are useful on their own. Newton's second law expressed in the form that force equals rate of change of momentum, or its variation that impulse equals change in momentum, can be combined with the third law, that for every force occurring, there is an equal sized force acting in the opposite direction. If two objects interact, the force of the first on the second is the same size as the reaction force of the second on the first. These forces are at all times equal in size so the products of force and time, or the impulses on each other, are equal in magnitude though opposite in direction.

The impulse of object number 1 on object number 2 is equal to the negative of the impulse of object number 2 on object number 1. Then, since impulse equals change in momentum, the change in momentum of object number 2 (caused by the force from number 1) is equal to the negative of the change in momentum of object number 1 (caused by the force from number 2).

To simplify this let Δp_1 be the change in momentum of object number 1 and Δp_2 be the change in momentum of object number 2; then

$$\Delta p_2 = -\Delta p_1$$

or $$\Delta p_2 + \Delta p_1 = 0$$

The changes in momentum of the objects are in opposite directions and the total change is zero. If the change in the total of a quantity is zero, that means the amount remains the same or is conserved.

The above statement can be written as the *total momentum in any system of interacting bodies is constant* or that *momentum is conserved.*

Whenever objects interact, the total momentum remains the same. In the whole universe, since its beginning, there can have been no change in momentum. Objects interact, gaining and losing momentum but the total is always the same. On a smaller scale, in any isolated system the total momentum is constant.

The concept of **conservation of momentum** is one of the most basic in all of physics.

As an example of an isolated system and conservation of

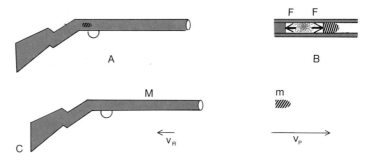

Figure 3–8 A system in which the initial momentum is zero (a). In (b) the high-pressure gases push the bullet forward and the rifle backward. In (c), the momentum of the rifle to the left is the same as the momentum of the bullet to the right. When direction is considered, the momentum of the whole system is still zero.

momentum consider a rifle with a bullet in it. The momentum of the system may be considered to be zero if the rifle is being held still. When it is fired the bullet is given momentum in one direction and the rifle acquires the same amount of momentum but in the opposite direction so the total is still zero. If the mass of the bullet is m, of the rifle is M, and if the bullet is propelled with a velocity v_P to the right while the rifle recoils to the left with a velocity v_R, as in Figure 3–8, then the momentum mv_P is equal in magnitude to the momentum Mv_R, or $mv_P = Mv_R$.

Solving for the recoil velocity:

$$v_R = (m/M)v_P$$

EXAMPLE 11

Calculate a recoil velocity in three situations. First, a 2 gram bullet is fired at 300 m/s from a rifle of mass 4 kg. The rifle recoils. Then a similar bullet is fired from the rifle while it is held tightly by a person sitting in a boat. The recoiling mass in the second case is the rifle plus person, plus boat—a total of 150 kg. In the third case, 250 bullets are fired from the boat.

1. For the rifle recoiling:

$$v_R = (m/M)v_P \text{ where } m = 2 \text{ g} = 0.002 \text{ kg}$$
$$M = 4 \text{ kg}$$
$$V = 300 \text{ m/s}$$
$$\text{Therefore } v_R = (0.002 \text{ kg}/4 \text{ kg}) \, 300 \text{ m/s}$$
$$= 0.15 \text{ m/s}$$

The rifle recoils at 0.15 m/s.

2. For the situation in the boat, $m = 0.002$ kg
$$M = 150 \text{ kg}$$
$$V = 300 \text{ m/s}$$

$$v_R = (0.002 \text{ kg}/150 \text{ kg}) \, 300 \text{ m/s}$$
$$= 0.004 \text{ m/s} = 0.4 \text{ cm/s}$$

The boat and person recoil at 0.004 m/s or 4 mm/s.

3. When one bullet is fired from the boat it acquires a speed of 0.004 m/s. A second bullet will increase the speed by a similar amount. It might be expected that after firing 250 bullets the speed will be 250 × 0.004 m/s or 1.0 m/s. This is not an insignificant speed.

Figure 3–9 A propulsion system based on the conservation of momentum.

In the latter example a boat was made to move in one direction, by firing bullets in the opposite direction as in Figure 3–9. The boat moved because of the reaction force when the bullets were propelled away. An engine built to move a vehicle by propelling mass in the opposite direction is called a **reaction engine.**

3–4–3 REACTION ENGINES

The most familiar form of reaction engine is the rocket. Inside the rocket is a mass of propellant, which is of a type that can be burned to produce a gas. The gas is ejected at a high velocity from a nozzle, that is, the gas is given momentum outward from the back of the rocket. The rocket acquires an equivalent amount of momentum forward. The momentum given to the propellant depends on both the propellant velocity and the mass of propellant ejected. The propellant velocity depends on the type of propellant, and with modern chemical rockets ejection velocities of the order of 2 km/s are obtained. The mass of propellant needed to give a satellite an orbiting speed of 8 km/s is many times the mass of the satellite. For each kilogram put into orbit about 1000 kg of propellant plus engine leave the ground. The exact figures vary from one rocket type to another.

The rocket engine is characterized by its carrying the total mass of propellant with it. This must include each part necessary for combustion. In a solid fuel rocket the solid must be such that the oxidant is incorporated into the solid. A liquid fuel rocket will have two tanks, one for oxidant and one for fuel, so when the propellants are brought together in the combustion chamber, they react to produce the high temperature gas that is ejected. A rocket will operate better outside the atmosphere than in it because there will be no air to slow the propellant gases.

A toy balloon that is blown up and then released without closing the neck is much like a rocket. The air comes out at a

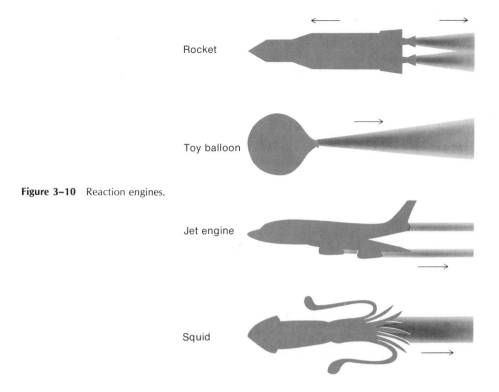

Rocket

Toy balloon

Figure 3-10   Reaction engines.

Jet engine

Squid

high speed and the balloon gains momentum in the opposite direction. The balloon and other reaction "engines" are shown in Figure 3–10.

The jet engine is a type of reaction engine but it does not carry the total mass of propellant in the vehicle. Air is taken in the front of the engine, compressed and burned with fuel in a combustion chamber. The hot gases produced are ejected backward. The additional momentum given to the air results in momentum being given to the craft.

Some animals move using a jet principle. The squid is an example. Water is drawn into a body cavity. Muscle tension forces eject the water at a high speed to propel the animal in the opposite direction.

A reaction engine gives an amount of momentum Δp to the propellant in a time Δt. The force given to the vehicle is rate of change of momentum or $\Delta p/\Delta t$. Reaction engines are rated by the force, or thrust, they produce rather than by the power, as is done with internal combustion engines such as are used in cars.

A rocket produces constant thrust if the propellant is ejected at a constant rate. If a mass Δm is ejected in a time Δt, the rate of ejection is $\Delta m/\Delta t$. The thrust is $\Delta p/\Delta t$, and since $\Delta p = \Delta m \cdot v_P$, then the thrust (of force) is

$$F = \Delta p/\Delta t = v_P \, \Delta m/\Delta t$$

EXAMPLE 12

At what rate (in kg/s) would propellant have to be ejected from a rocket of mass 10^5 kg to lift it off the ground? Consider a propellant speed of 2 kilometers per second (2×10^3 m/s).

The necessary force or thrust is mg or 9.8×10^5 N on the rocket which has a mass of 10^5 kg.

The thrust F is the momentum given to the propellant divided by the time, which gives the momentum per unit time:

$$F = \frac{\Delta m v_P}{\Delta t}$$

From which
$$\frac{\Delta m}{\Delta t} = \frac{F}{v_P}$$

$$= \frac{9.8 \times 10^5 \text{ N}}{2 \times 10^3 \text{ m/s}} \quad (\text{write N} = \text{kg m/s}^2)$$

$$= 4.9 \times 10^2 \text{ kg/s}$$

The propellant must be ejected at a rate of 490 kilograms per second to lift it off the ground. That is a lot of material to be ejected per second, and it illustrates one of the problems in the design of large rocket engines.

As propellant is ejected, the accelerating mass decreases and with a constant thrust the acceleration increases. The equations for motion under constant acceleration do not apply to rockets if the mass change is significant. If the mass of the propellant is a tenth of the total mass, an error of 5 per cent will occur if constant acceleration if assumed or if the propellant is considered to be ejected as a single mass m_P with a propellant velocity v_P.

3–5 FORCES DEDUCED FROM ACCELERATIONS

Newton's second law in the form

$$\text{net force} = \text{mass} \times \text{acceleration}$$

leads to some interesting deductions. When acceleration of a mass is detected, it is said that a force is acting. Force could even be defined as a cause of accleration. The significant thing here is that when an acceleration is seen, it is said that a force is acting to cause it: if there is no acceleration, it is said that there is no force, or no unbalanced force. The forces that cause accelerations can also cause distortions of objects — stretching of springs, for instance. Throughout this discussion, keep in mind that *net* forces or *resultant* forces are being referred to.

One origin of forces is called gravitational attraction. An object dropped near the surface of the earth falls with an acceleration that is called *g*; it is said then that a force, a gravitational force, is acting on the object to cause this acceleration. By Newton's second law, the size of this force is given by *mg*. To keep the object from falling, an upward force is required. If the object is held by a spring, the spring is stretched or distorted. The amount of stretch depends on the mass and also on the value of *g*. If the object is moving at constant speed upward or downward, or even sideways as long as there is no acceleration, the same analysis will apply: the forces would balance to zero, since the upward force would be of a size *mg*, and so would the downward force.

3–5–1 INERTIAL EFFECTS

To accelerate an object upward requires that the upward force be greater than the weight. To start an elevator in motion and accelerate it upward, the force in the cables has to exceed the weight. To start the elevator in motion downward, the tension in the cables must be less than the weight. When the elevator moves upward or downward at constant speed, the tension in the cables is just equal to the weight. Similarly, if a person in the elevator is holding a mass on a spring, as in Figure 3–11, and the elevator is moving at constant speed upward, downward or remaining still, the weight of the object according to the spring scale is the same. When the elevator *accelerates* upward, the object must accelerate also, so there is an added force on the spring. The object weighs more, be-

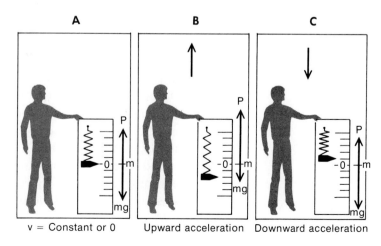

Figure 3–11 Forces on an object inside an elevator. In (a) there is no acceleration. In (b) there is an upward acceleration, and an apparent increase in weight. In (c) there is downward acceleration, and an apparent decrease in weight.

v = Constant or 0   Upward acceleration   Downward acceleration cause it requires a greater force to keep it stationary in the elevator. Similarly, during a downward acceleration the weight decreases. If the elevator was windowless, padded and completely isolated from the outside, the person inside would detect an acceleration from the change in weight; but the effect would be the same as would be seen if gravity were changing. This cause is ordinarily ruled out because experience has shown that gravity is constant, but the observer in the elevator would not be able to tell for sure that gravity had not changed. An observer outside the elevator would have no such decisions to make, because when the spring showed an increased upward force he would also see an upward acceleration of the object. The idea here is that *what is detected as weight depends on the system in which the observation is made.* In any place or system the weight is measured either by finding the supporting force necessary to keep a mass at rest or by removing the supporting force and measuring the acceleration.

Consider again the mass in the accelerating elevator. If the elevator accelerates upward with an acceleration a and the mass of the object is m, the force required is ma. The person in the elevator sees this as an extra downward pull on the spring required to hold the mass in equilibrium in the elevator. The upward acceleration produces, *in the system of observation connected to the elevator,* a downward force. This downward force appears on all masses in that accelerating system; that is, on all objects that have inertia. These forces are called **inertial forces**, or sometimes **fictional forces.** But to the observers in that system, the forces are real. There may be acceleration, or springs may be stretched because of the inertial forces. It is because of inertial forces that the human body can withstand only limited acceleration. **Inertial forces act opposite to the direction of the acceleration of the system.** The inertial force on a mass m is given by $-m\mathbf{a}$, where $\mathbf{a}$ is the acceleration of the reference system. These forces are illustrated in Figure 3–12.

Inertial forces occur with horizontal accelerations also. Imagine that a person is seated at a table in a closed railway car. On the table is a ball. As the train accelerates forward, the ball suddenly moves backward in this system. The person knows that a force is necessary to set that ball in motion, so he deduces that a force has acted on the ball. An observer outside the system sees the train and table accelerate forward, while the ball remains stationary. The outside observer does not detect the inertial force that was real to the person in the railway car. Weight and force are points of view.

Inertial forces are of more than academic interest, for they can assist in the understanding and analysis of some physical situations.

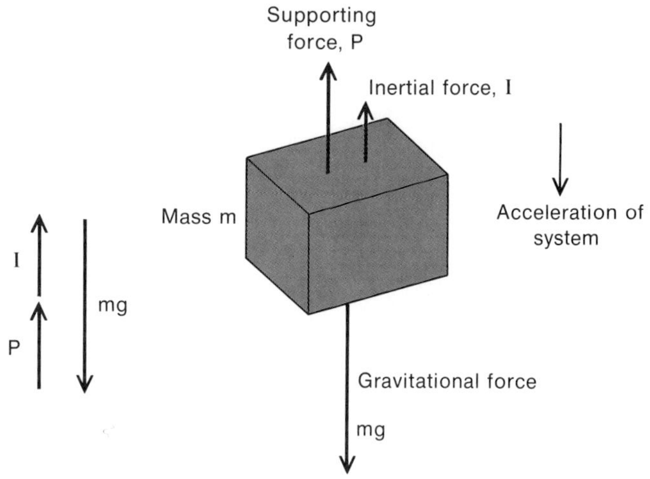

Figure 3–12 The forces on a mass in an accelerating system. The inertial force results from the acceleration of the system in which the measurements are made.

Supporting force, P

Inertial force, I

Mass m

Acceleration of system

Gravitational force

mg

EXAMPLE 13

When an automobile is in an accident, the accleration may be very large. To use numbers, assume that a car going 90 km/h (25 m/s) collides with a wall, and the people in it are stopped in a distance of 1 meter. The acceleration is backward, and the inertial force is forward. What would this force be on a 70 kg person in that car? The acceleration, found in Example 11 (page 55), is −313 m/s² (the negative sign indicates that the acceleration is backward). This is about 32 times the normal value of *g*. The weight is normally *mg*, but now the acceleration is 32 *g* so the forward inertial force is 32 times the normal weight.

Alternatively, the forward force given by *a* is 70 kg × 313 m/s or 22,000 N whereas the weight *mg* is just 686 N.

The force needed to stop a person in the car in the example above must be provided by something. It could be the windshield (but it would break), the steering column, the dashboard, or safety belts. If a seat belt is worn, the total backward force supplied by the seat belt must be 22,000 N or 11,000 N on each end, as in Figure 3–13. The belt must be sufficiently strong and anchored strongly enough to withstand forces of this magnitude. The belt must also be in such a position, across the pelvic bone and not the abdomen, so that internal damage will not be caused by the belt. Safety standards for automobiles stipulate that the seatbelts and anchors for them be able to withstand such large forces. The additional use of a shoulder belt distributes the force over a greater part of the body as well as preventing the upper part of the body from being thrown forward.

An astronaut experiences inertial forces whenever the rockets fire — whenever there is acceleration. On the launching pad he experiences only his normal weight; but as the rockets fire and he accelerates upward with it, the inertial forces press him back into his couch. This is of importance because the firing

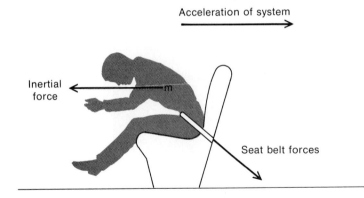

Figure 3–13 The forces on a seatbelt in a car stopping in 1 meter after going 90 km/h.

Acceleration of system

Inertial force

m

Seat belt forces

time for a rocket leaving earth is usually quite short; the shorter the firing time for a given amount of propellant, the higher the rocket will go. The acceleration must be kept within the limit that the astronaut can endure. With this limitation the rocket will eject its propellant as quickly as possible and then coast. In this coasting interval the rocket is like any object thrown upward. It has a downward acceleration of *g*, steadily reducing its speed. The rocket is a reference system with a downward acceleration, so on all objects in it there is an upward inertial force. But objects do not move upward in the cabin because the earth's gravity is still pulling them downward. The inertial force on a mass *m* is equal to *m* times the acceleration of the system, or *mg*, and is upward in this case. The downward force of gravity is also *mg*, which cancels the inertial force, leaving all objects in the vehicle in a state of apparent **weightlessness to an observer inside the vehicle.**

Whenever the rockets of a space vehicle are not firing, the objects in it are weightless. The firing of the rockets produces an acceleration of the system that introduces inertial forces, giving what appears as weight to everything in the vehicle. Use is made of this, for instance, to push fuel into the pipes before ignition of a large engine. Small thrusters are fired to momentarily give the liquid weight to settle it into the required part of the tank and pipes.

3–5–2 ROTATING SYSTEMS

Rotating or orbiting systems involve inertial forces if we consider what is going on within the rotating object (that is, if the reference frame is attached to the rotating object). Whether we deal with the forces acting in a centrifuge tube or those in an orbiting space vehicle, the analysis is similar, though in the first instance the weight is increased and in the second it is reduced to zero.

In Section 2–3–4 it was shown that a rotating object accelerates toward the center of rotation with a centripetal acceleration given by v^2/r. The speed is v, and r is the radius of rotation. The important thing is to transfer your thinking to the rotating system—to rotate with it. The rotating reference frame is accelerating toward the center, and objects in the system experience an outward force. An outside observer does not see this force at all; it is only to someone in the rotating system that the outward force appears. The size of the inertial force on a mass m is mv^2/r. This is called **centrifugal force** (Figure 3–14). When the motion of objects inside a rotating system is being described by an observer with his measuring frame attached to the rotating system, then the centrifugal force (the outward force) must be considered. An outside observer describing the motion in an external reference frame deals with a centripetal force, that is, a force toward the center. The use of a centrifugal force or of a centripetal force depends on the reference frame.

When an automobile goes around a curve, a passenger initially sitting at equilibrium in the car may be pushed against a door on the outside of the curve. The passenger inside the car experiences a force away from the center of the curve, a centrifugal force. An observer outside the car would see the door pushing inward on the passenger, a centripetal force, to move the passenger in the circular path that is observed in the external reference frame.

Inside a centrifuge tube there is a centrifugal force that moves particles to the bottom of the tube, toward the outside of the circle. In analyzing motion in the tube, with respect to the walls of the tube, a centrifugal force is used. An outside observer sees only circular motion that requires a force toward the center.

An astronaut in a satellite orbiting the earth detects no weight on himself or on objects in the craft (or outside it as in the

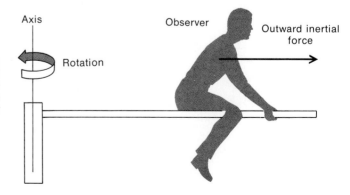

Figure 3–14 An observer in a rotating system experiences an outward force. This is an inertial force and is called the centrifugal force. The observer outside the system sees only the force toward the center that keeps the rotating observer in the circular path.

case of a space walk) when the measurements are with respect to the craft. The astronaut is weightless in the craft even though gravity is acting on him. At the orbit of a low satellite the gravitational field is only 5 per cent less than at the surface of the earth. The satellite is falling toward the earth but because of the sideways speed it never gets any closer. It "falls over the edge of the earth." The gravitational force mg pulls down, and inside the satellite there is an upward force, the inertial force, of mg. Therefore, to an observer inside the satellite objects are weightless.

3–5–3   INERTIAL FORCES AND PLANT GROWTH

The effect of inertial forces on plants is also of interest. Do plants "feel" inertial forces, or do they differentiate between inertial and gravitational forces? It is known that plants grow upward. If plants are grown on a slow centrifuge, in what direction will they grow? In Figure 3–15 are shown three possible situations. The correct answer to such a problem can be found only by performing the experiment. In Figure 3–16 is a photograph of wheat grown in the dark for 5 days on a centrifuge rotating at 1.3 revolutions per second. That is, $v = 0.026$ m/s at

Figure 3–15   Possible directions of growth of plants in a rotating system.

Figure 3–16 A photograph of plants growing in a rotating system. This answers the question posed by Figure 3–15.

a radius of 0.125 m. This picture shows which of *A*, *B* or *C* of Figure 3–15 describes the case.

The situation invites theoretical investigation. An observer riding with the plants would experience an outward force of mv^2/r and a downward gravitational force of mg. These forces are shown in Figure 3–17. The resultant describes the direction of the weight of an object in that rotating system. Small pendulums would hang in that direction. If the plants cannot distinguish between gravity and inertial forces they would grow opposite to the direction of this resultant force. This is illustrated by the single seed of Figure 3–18, in which the sloping line shows the calculated direction of the weight in the rotating system.

Such an experiment leads to many questions. Does the effectively increased gravity affect the rate of migration of hormones to the tips and hence the rate of growth? Does the increased *g* affect the size or strength of the plant? Would seeds sprout in weightlessness? Some of these questions have been investigated in space flight, but most are still unanswered!

3–5–4 ROTATION OF THE EARTH

The concept developed here is that in a rotating system the weight of objects is affected by the rotation. There is an outward

Figure 3–17 The forces experienced by an object such as a plant in a rotating system. The net force is downward and outward, and this is the direction of the apparent weight.

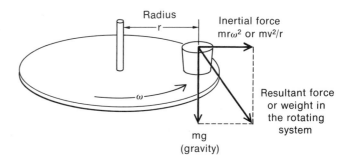

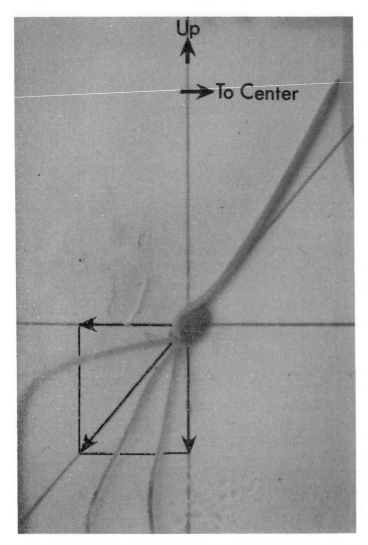

Figure 3–18 The vector diagram as in Figure 3–17 was drawn on a card, and it was put with the seed in the rotating system. The plant grew in the direction opposite to its weight in that system (at least, as close to it as do plants that ordinarily grow "straight up").

force (the inertial force) on objects because of that rotation, and if the objects are to be held still they must be supported by an inward force sufficient to balance the inertial force. If the objects are not supported, they will have an acceleration outward, or radially. Again it must be stressed that *these outward forces and accelerations are real only to an observer in the rotating system.* These inertial forces and subsequent accelerations are so similar to ordinary gravitational forces and accelerations that it can be said that, in working in rotating systems, ordinary dynamics can be used as they were for gravitational fields; but we must use a value of g appropriate to the rotating system. This will be outward (radially) and of a size v^2/r.

This analogy is useful, but there are a few areas in which it is not perfect. In rotating systems another apparent force arises, the **Coriolis force.** You may be familiar with this effect; it leads to counterclockwise circulation of air around low pressure areas in the northern hemisphere of the earth (and clockwise in the southern hemisphere).

The rotating earth is the kind of rotating system that is being considered. It turns around once in 24 hours, and has a radius of about 6380 km at the equator. There is an outward inertial force on all objects on the earth. This inertial force shows as an apparent small decrease in weight of an object, the effect being greatest at the equator and decreasing to zero at the poles. This decrease in gravity at the equator resulting from the earth's rotation was one of the early demonstrations of the spinning of the earth on its axis. Motion of the sun across the sky, and the motion of the stars or of the moon, can be explained as being due to either rotation of the earth or to the motion of the sun, moon and stars. Part of the decrease in g at the equator, however, cannot be explained in any way other than by the spinning of the earth. This spin, combined with the gravitational pull of the moon, has also caused the equator to "bulge out," so that the earth is not a perfect sphere. Thus, at the equator g is less than that at the poles for two reasons, the earth's spin and the greater distance from the center.

EXAMPLE 14

The measured value of g at sea level at the equator is 9.780 N/kg. The actual force due to gravity alone would be greater than this, the observed value being reduced because of the spin of the earth. The spin of the earth causes an observer at the equator to feel an outward force of mv^2/r on a mass or v^2/r per kilogram. The questions are: By how much does the spin of the earth reduce g at the equator, and what would the value of g be if it is considered to be only that due to gravity, not affected by spin?

It is required to calculate v^2/r. The speed v is the distance around the earth divided by the rotation time, 1 day or 86,400 seconds. The radius is 6.38×10^6 m. Then

$$v = 2\pi \times 6.38 \times 10^6 \text{ m}/86{,}400 \text{ s}$$
$$= 464 \text{ m/s}$$
$$v^2 r = (464^2 \text{ m}^2/\text{s}^2)/6.38 \times 10^6 \text{ m}$$
$$= 0.034 \text{ m/s}^2$$

Because of the spin of the earth the observed value of g at the equator is reduced by 0.034 m/s² or 0.034 N/kg. The value of g due to gravity alone is (9.780 + 0.034) N/kg or 9.814 N/kg.

3–6 FRICTION AND RESISTANCE FORCES

Any body on earth encounters resistance to motion. This is the case for solid objects sliding on solid surfaces, solid objects rolling, and objects moving through a gas or liquid. It is difficult to express any of these effects with high precision

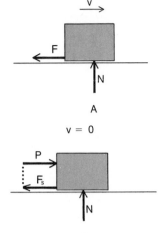

Figure 3–19 Sliding friction and static friction.

though there have been laws found that will give a close approximation to real situations. The in-depth study of various aspects of friction has expanded in recent years and that area of study was given the name **Tribology.**

3–6–1 SLIDING FRICTION

When one object slides over another as in Figure 3–19 there is a frictional force, **F**, opposite to the direction of the velocity. This frictional force is found to depend only on the force pressing the two surfaces together and on the type of material of the two surfaces. It does not depend on the area nor, in general, on the velocity. This discovery we owe to Leonardo da Vinci. The force pressing the surfaces together is shown as **N** in Figure 3–19. **N** will have to balance any other forces there may be: the weight of the object and any other forces in that direction. These others, because they may be of great variety are not shown.

The frictional force **F** is proportional only to the total force, **N**, pressing the bodies together or

$$\mathbf{F} \propto \mathbf{N}$$

A proportionality constant can be put in to make this an equation, and this constant, for which the symbol μ (the lower

TABLE 3–2 Coefficients of friction*

Materials in Contact	Coefficient of Sliding Friction, μ	Coefficient of Static Friction, μ_s
steel on steel	0.10	0.15
irridium-coated steel on steel	0.04	–
rubber on ashphalt, dry	0.7	1.0
rubber on ashphalt, wet	0.3	0.5
wood on ice, −5°C	0.04	–
wood on ice, −40°C	0.08	–
brass on ice, −5°C	0.06	–
brass on ice, −40°C	0.11	–
cartilage on bone**	0.015	–
bursa on bone**	0.012	–

*The values are representative. In practice there is considerable variation.
**Note how the coefficients of friction in biological situations are lower than in other situations. This is the source of one of the major problems in the design of artificial joints.

case Greek mu) is often used, is called the coefficient of sliding friction. Thus

$$\mathbf{F} = \mu \mathbf{N}$$

Some representative values of the coefficient of sliding friction are shown in Table 3–2.

EXAMPLE 15

A 200 kg container, perhaps of the massive lead type used for radioactive materials, is to be moved by sliding it away using several people and ropes. If each person could be expected to pull with a force of 250 N, and if the coefficient of friction is expected to be about 0.4, how many people would be needed?

The only downward force is the weight, *mg,* and this is balanced by the normal force *N* of the Figure 3–19.

$$N = mg$$

The frictional force is μN, so

$$F = \mu mg$$

The force, *P,* to pull the object must be at least equal to *F* so

$$\begin{aligned} P &= \mu mg \\ &= 0.4 \times 200 \text{ kg} \times 9.8 \text{ m/s}^2 \\ &= 784 \text{ N} \end{aligned}$$

Each person can be expected to pull with 250 N so three would hardly move it but four would move it with ease.

3–6–2 STATIC FRICTION

It may be in your experience that it takes a greater force to start an object sliding than to keep it moving. As in Figure 3–19(B) if a force **P** is applied to move an object, the static friction $\mathbf{F}_s$, will prevent that motion. $\mathbf{F}_s$ will exactly balance **P** until a limiting value of $\mathbf{F}_s$ is reached. Then any increase in **P** will start the object moving. This limiting value of $\mathbf{F}_s$ is related to the normal force **N** so

$$\mathbf{F}_s \leq \mu_s \mathbf{N}$$

where μ_s is the coefficient of static friction.

In general the coefficients of static friction are greater than those of sliding friction. (Looking back to the last example, a fourth person would be advisable to overcome static friction and start the object moving.)

It is of value to understand static friction in the acceleration or the stopping of a car. It is the force between the tires and the road that cause the acceleration or, if you put the brakes on, cause the car to stop. If the tires slide, the available force is the coefficient of sliding friction times the force between the tires of the driving (or braking) wheels that counts. If the tires do not slide, the force can go as high as the coefficient of static friction times that normal force, and this can exceed the force of sliding friction. So acceleration or braking is most effective if the wheels do not slide (or spin).

In drag racing the accelerating force is increased by increasing the weight on the driving wheels. To accelerate on a slippery road extra weight on the driving wheels is of great assistance but to steer on a slippery road, the extra weight should be on the wheels that turn.

3-6-3 RESISTED MOTION IN FLUIDS

When an object moves through a fluid, the resistance force does depend on the speed though the manner of that dependence varies from one situation to another. In general, the resistance force varies as a power of the speed ranging from $R = kv$ for low speed, streamline motion to $R = kv^2$ for high speed motion with turbulence produced. The constants k depend on the size and shape of the body. Also, let me hasten to say that this is an oversimplification though it does allow the derivation of some concepts that are close to reality.

Consider, for example, the case of an object falling through a gas or a liquid as in Figure 3-20. As the speed increases so does the resistance force and the net accelerating force decreases. The speed will approach a limit, called the **terminal speed.** At the terminal speed the resistance force is just equal to the force mg, the weight. If the situation is such that $R = kv$, then at the terminal speed v_t,

$$kv_t = mg$$

or

$$v_t = \frac{mg}{k}$$

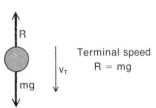

Figure 3-20 An object falling in a resisting medium. The speed approaches a limiting value.

From this, the terminal speed is seen to be proportional to the weight, at least for objects that have the same resistance force constant k. This is reminiscent of what Aristotle is said to have taught; that objects fall at speeds proportional to their weight. It is true for a limited situation!

DISCUSSION PROBLEMS

3–1   1. Devise a possible system to find the mass of a loaf of bread in a space vehicle where everything is weightless.

3–4   2. A "sand reckoner" is a timing device having an upper compartment from which sand slowly drains through a small hole to a lower compartment. They are often made to require 3 minutes to empty the upper chamber and are used to time boiling eggs or long distance phone calls.
Discuss the variation in weight registered by a scale or a balance when a "sand reckoner" is put on the pan. Will the weight be reduced when the sand is falling? What about the brief time when it starts to fall or when the last bit has just left the upper chamber and is still on the way down?

3–4   3. Discuss the problem of finding the mass of a fly in a closed bottle. Would the apparent weight of the system depend on whether or not the insect was flying?

3–4–2   4. Discuss how the conservation of momentum applies at each stage of the following: A car is initially at rest, it accelerates to 25 m/s and then hits a concrete wall.

3–4–2   5. Consider that you are in the center of a completely frictionless piece of ice as lunch time approaches, and the food is being served at the edge of the ice.
(a) How could you get to the edge of the ice?
(b) Devise a method by which you could get to be stationary in the center of a large piece of perfectly frictionless ice in the first place (this problem is related to moving about in space).

3–5   6. You are driving a car and want your passenger to move closer to you. How could you achieve this using an inertial effect?

3–5–3   7. Using your knowledge, look again at Problem 1.

3–9–3   8. At the North Pole $g = 9.932$ N/kg, while at the equator $g = 9.814$ N/kg (Example 15) after the effect of earth rotation is taken into account. Why is there a difference between the two values?

PROBLEMS

3–2   1. Find the density of a small bar of metal, supposedly gold, but of mass 0.666 kg

and dimensions 2.00 cm × 3.00 cm × 6.00 cm. Compare the calculated density to that in Table 3–1 to find whether or not the bar is of pure gold.

3–2   2. Lead bricks supplied for radiation shielding are often alloyed with antimony, which is of low density but hardens the lead so that it will be less susceptible to damage. Find the density of a lead brick which is 5.0 cm × 7.5 cm × 15.0 cm and has a mass of 5.85 kg. Use Table 3–1 to see if any antimony added has significantly reduced the density from that of pure lead. Express the decrease as a per cent.

3–2   3. Calculate the specific volume (volume per kilogram) of the following materials and then, for each, the length of the edge of a cube that would be a kilogram. Express the size of the cube in centimeters.
water, aluminum, lead, osmium, cement, quartz

3–2   4. Find the mass of a coin that is made of pure nickel, is 2.00 cm in diameter and has a thickness of 1.75 mm. The density of nickel is 8800 kg/m³.

3–2   5. Find the mass of the mercury in the bulb of a thermometer, which is a cylinder 3.0 mm in diameter and 8.0 mm long, with hemispherical ends added. The density of mercury is 13,600 kg/m³.

3–2   6. Find the average density of the material of a mass of small pieces of rock given the following data: A cylindrical container, mass 4.0 kg is 0.30 m in internal diameter and 0.45 m in height. It is filled with the pieces of rock and the mass rises to 74.0 kg. Water is poured in to fill the spaces between the rocks and the mass rises to 84.6 kg.

3–2   7. Find the average density of a proton that seems to be concentrated in a sphere 1.3×10^{-15} m in radius and has a mass of 1 amu.

3–3   8. Find the net force on an aircraft of mass 6.5 tonnes when it accelerates at 1.5 m/s².

3–3   9. Find the frictional force acting when an automobile of mass 1.5×10^3 kg is brought to rest from 25 m/s in a distance of 300 m.

3–3   10. Find the tension in the cables required to accelerate a 2.5 tonne elevator upward at 2.0 m/s².

3–3   11(a) A man of mass 85 kg is going to lower a box of bricks from the third floor of a building. The mass of the bricks is 90 kg. A rope is tied to the box, put over a pulley and down to the man on the ground. The box is pushed into space and it begins to fall. The man accelerates upward. Find the acceleration of each. Assume a frictionless pulley so the tension T in the rope is the same at each end.

Hint: Draw a diagram showing the forces on each object. Then write Newton's second law for each object separately. The accelerations of each are the same in magnitude. Eliminate T from the equations.

(b) When the box hits the ground, the bottom falls out so the mass remaining on the rope is only 20 kg. What is the acceleration of the man then?

3–3   12.   A toy car accelerates from rest down an incline in such a way that it moves 1.5 m in 4.5 s. If the mass of the car is 0.055 kg, what is the accelerating force?

3–3   13.   Find the speed of an electron (mass 9.1×10^{-31} kg) after it accelerates for a distance of 2.0 cm under a force of 8.0×10^{-15} N.

3–3   14.   A car constrained to move on a track is pulled with a rope that is at an angle of 30 degrees from the track. What must be the force in the rope to accelerate the car at 1.5 m/s^2 if its mass is 275 kg.

3–3   15.   A 75 kg box on ice (friction is negligible) is pulled using 2 ropes. One pulls directly east with a force of 300 N and the other pulls at 15 degrees east of north with a force of 600 N. What will be the acceleration of the box? Acceleration is a vector so include the directions.

3–4   16.   Find the momentum of a 35,000 tonne tanker moving at 5 m/s and compare it to the momentum of a 2 tonne earth satellite moving at 8 km/s.

3–4   17.   Find the momentum with which a 0.10 kg mass strikes the ground after falling 2.0 meters.

3–4   18.   Find the force due to the impact of a stream of sand falling from a height of 2.0 m onto a flat plate. The sand falls at a rate such that 6 kg falls in 1 minute. The sand comes to rest on impact.

3–4   19.   Use momentum ideas to find the force on a 25 kg object that changes velocity from 35 m/s to 20 m/s in the same direction after being acted on by the force for 15 seconds.

3–4   20.   If a ball initially moves at 17.1 m/s to the left and after being struck moves at 40 m/s to the right, what impulse was given to it? The mass of the ball is 0.35 kg.

3–4–2   21.   If a bullet of 5.0 grams is fired from the back of a small boat that, with the occupant, is 160 kg in mass, what velocity would be acquired by the boat? The muzzle velocity is 330 m/s backward. What speed would be reached after firing 100 bullets?

3–4–2   22.   Find the total height reached by the rocket described as follows. The rocket of 5.0 kg is loaded with a 0.50 kg of propellant that, on ignition, is ejected with the speed of 1000 m/s. The propellant is ejected in 5 seconds and during this time the thrust is constant. The accelerating mass may be approximated as just 5.0 kg. The rocket moves upward from the earth. Find:

(a) the thrust,
(b) the net upward force,
(c) the acceleration,
(d) the speed acquired during the acceleration,
(e) the height achieved while the acceleration is occurring (h_1),
(f) the additional height that the rocket will rise without power (h_2),
(g) the total height reached by the rocket ($h_1 + h_2$).

3–4   23.   Repeat Problem 22 but with the propellant being ejected in just one second. Comparing this result with that of Problem 22, what do you conclude about what should be done to get the rocket as high as possible with a given propellant mass and ejection velocity?

3–5–1   24.   Find your own mass in kg and then your weight in newtons. If you stand in an elevator that accelerates upward at 2.0 m/s^2, what is the total force between you and the floor of the elevator?

3–5–1   25.   Find the acceleration of a person of mass 70 kg who, while standing on a scale in an elevator, finds that it indicates a force of 637 N. (The scale may be graduated in kg but that would be on the basis of surface gravity which is 9.8 N/kg. The scale reading is basically a force.)

3–5–1   26.   While sitting in an aircraft on a runway you place on a table a mass of 0.45 kg connected to some springs. As the aircraft accelerates along the runway the 0.45 kg mass pulls backward on a spring showing a force of 0.30 N. This force persists for 100 s at which time the aircraft becomes airborne.

(a) What was the acceleration along the runway?
(b) What was the take-off speed?

3–5–2   27.   The basic idea of an inertial navigation system is that a mass is connected to some force measuring devices (springs perhaps). Forces on these springs indicate accelerations of the vehicle. Forces in line with motion are the result of changing speed. Forces perpendicular to the motion indicate motion along a curve. Find the final displacement of the vehicle that shows the following inertial forces on a mass of 1.00 kg. The initial state of the vehicle is stationary, facing west. A backward force of 1.2 N persists for 20 s. There are no more forces for an interval of five minutes at which time a force of 3.0 N to the right is seen and it persists for 50.3 s. There is no force for 30 s and then a forward force of 0.601 N for 40 s.

3–5–2   28.   A pilot dives his plane toward the ground at 45° and wants to pull out of the dive in such a way that the total forces experienced at the bottom (gravity plus centrifugal) is 4 times the normal gravitational force. If the speed is 250 m/s what must the radius at which the plane turns be? At what altitude must the vertical turn begin?

3–5–2   29.   Prolonged weightlessness is accompanied by certain physiological changes, so it may be necessary on long space flights to simulate gravity by inertial forces. For example, the crew could live on the inside of a rotating cylinder (Figure 3–21). If the spacecraft is a rotating cylinder 12 meters in radius, how fast will it have to rotate to give the sensation of one g at the outer edge?

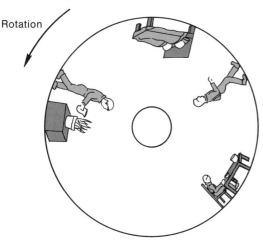

Rotation

Figure 3–21 In a rotating spacecraft, an outward inertial force could be a substitute for gravity on very long voyages.

3–5–3   30.   If you wish to raise animals or plants in a field of 2 g on earth, using a horizontal rotator of radius 2.5 meters, what rotation speed must be used? At what angle from the vertical will the forces in the system act?

3–5–4   31.   The mammoth planet Jupiter has the surface gravity usually given in tables of data about planets as 25.9 m/s² or N/kg. This value does not take into account the rapid spin of the planet which rotates once in 9.8 hours. The radius of the planet is 7.15×10^7 m. What would be the observed value of g on the surface of Jupiter at its equator?

3–6–1   32.   Find the stopping distance for a car going at 12 m/s (27 mi/h) on a slippery road if the coefficient of friction is 0.05.

3–6–1   33.   After an accident there was debate about whether or not one of the vehicles involved was traveling above the speed limit, which was 12 m/s. The road was dry so the coefficient of sliding friction between the tires and the road was at least 0.7. All four wheels skidded and the skid marks were 7.3 meters long to the position at which the car came to rest. Was the driver speeding?

3–6–2   34.   Find the maximum possible deceleration of a car, first when the wheels are rolling but are just on the verge of sliding, and then when they slide. The coefficient of static friction is 0.75 and of sliding friction is 0.65. How do the stopping distances compare in the two situations?

3–6–2   35.   You push a box, mass 25 kg, which is stationary on a floor. You increase the force on it until it moves, and then it accelerates as it slides. The coefficient of static friction is 0.75 and of sliding friction is 0.65.
(a) What force will start the box moving?
(b) With that force what will be the acceleration?

CHAPTER FOUR

ORBITS AND ANGULAR MOTION

The preceding work on force and motion has been concerned principally with situations involving constant force and constant acceleration. The analysis was in terms of linear quantities. The motion considered was not restricted to straight lines for it was shown that a trajectory in a constant force field is parabolic in shape and that circular motion at a constant speed results from a centrally directed, constant force. When the force field decreases with distance, as it does around the sun or planets, the basic motion of objects is in paths referred as conic sections. These may be ellipses, parabolas or hyperbolas. Objects that enter the solar system from deep space, pass by the sun and go out again to deep space, never to return, follow either parabolic or hyperbolic paths. An object that is in orbit round the sun or a planet moves in an elliptic path. This is the next situation to be discussed.

Motion around a central point can also be described in angular quantities rather than in linear quantities. In place of linear displacement, angular displacement is used. Similarly, angular velocity and acceleration replace linear velocity and acceleration. Kinematics can be developed in terms of angular quantities. Newton's laws have their counterpart in rotational motion and this leads to the dynamics of rotation. These topics will also be considered. They lead to the law of gravity, the force law that gives rise to the elliptic orbits of planets and satellites.

4–1   ORBITAL MOTION

The description of the motion of planets and satellites, and which now includes the motion of space vehicles, historically occurred at the same time as the description of accelerated motion in a straight line and the parabolic trajectory of an object on earth. The first two laws describing the motion of planets were published by Johann Kepler in 1609, and the third law was delayed until 1619. The laws describing accelerated motion on earth were published by Galileo Galilei in 1610. The physical laws to find the time for a space ship to go to Mars were known at about the same time as the equations to find the time for a stone to fall off a building. The laws that are basic for planetary travel, though known, were not applied until recently. Those laws describing the motion of the planets or other objects that orbit the sun are considered to be the first of all the physical laws, using the term "law" in its modern scientific sense.

A physical law is a description of an observed relationship between quantities. Laws may be derived directly from measurement and analysis or they may be suggested by theories or thought experiments, but even then the laws must be shown to correctly describe natural occurences. For example, one of the laws describing free fall is that the speed after falling from rest for a distance h is given from $v = \sqrt{2gh}$ (Section 2–3–1), where g, the acceleration due to gravity, is a constant. We derived this from a lot of other concepts but it could have been found from measurement of v and h. It has been shown to correctly describe the speed acquired by a falling body and can be called a law.

The laws describing planetary motion were found by J. Kepler directly from the analysis of observed data. One of the exciting sagas of science is of how Kepler worked for years and years, following ideas that proved false, making calculation errors, and more errors, and eventually found the descriptions that fit. The laws he found still apply, and not only to planets but to satellites and to space vehicles as well.

4–1–1 KEPLER'S LAWS

Kepler's laws are these:

I Each planet orbits the sun in an elliptical path with the sun in the plane of the orbit and at one of the foci of the ellipse.

II The speed of any planet varies in its orbit in such a way that the radius vector to it from the sun sweeps out equal areas in equal times.

III The ratio of the cube of the semi-major axis of the orbit divided by the square of the period for the planet to orbit the sun is the same for all planets.

These laws will be described in more detail.

The ellipse is only one of many types of closed curves, and Kepler tried a number of others before finding that it was an ellipse and no other that described the orbits. An ellipse is shown in Figure 4–1(A). It is described by a major axis and a minor axis. The comparative sizes of these axes is related to a quantity called the eccentricity, e. A circle can be called an ellipse with zero eccentricity. The foci of the ellipse are displaced from the center by an amount ea where a is half of the long axis of the figure or the "semi-major axis." An orbit is shown in Figure 4–1(B). The nearest approach of the planets to the sun, the perigee, is shown as r_1 and the farthest distance, at the apogee, is r_2. The semi-major axis is $(r_1 + r_2)/2$ and could be loosely referred to as a mean radius or effective radius.

Since the time of Kepler it has been found that any object, be it a planet or a comet orbiting the sun, a satellite (natural or artificial) or a spaceship orbiting a planet, if there is an orbit about some object, that orbit is an ellipse.

The orbits of the planets, though they are ellipses, are ac-

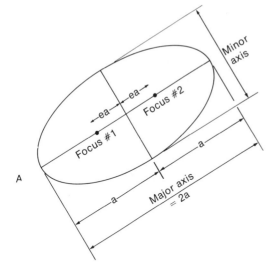

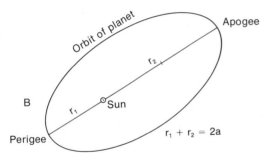

Figure 4–1 The ellipse and the orbit of a planet.

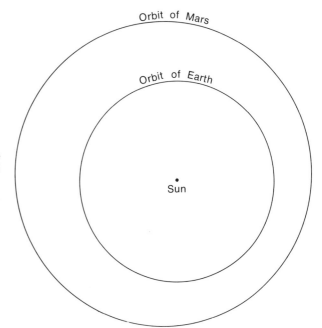

Figure 4–2 The orbits of the earth and Mars. They are both very close to circles, but with the centers displaced slightly from the sun.

tually very close to circles. In Figure 4–2 is a scale drawing of the orbits of earth and Mars; the orbits do appear very close to circles. In some of the work to follow, the planetary orbits will be treated as circles, and the results of the calculations based on that will be very close to what would be the result if ellipses were used.

The second law is explained by reference to Figure 4–3. If the same time is taken to go the distances AB, CD and EF, the shaded areas shown are equal. Consider the speeds to be v_1 at the perigee and v_2 at the apogee, and the time to go the distance shown to be t.

$$AB = v_1 t \quad \text{and} \quad CD = v_2 t$$

Figure 4–3 The planet moves the distances AB, CD, and EF in equal times, and the areas A_1, A_2, and A_3 are equal. This illustrates Kepler's second law.

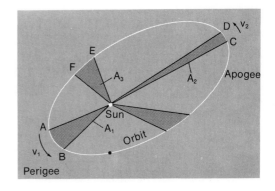

The shaded areas are very close to triangles with bases AB and CD and altitudes r_1 and r_2 so if the areas are A_1 and A_2

$$A_1 = (1/2) \, AB \, r_1 = (1/2) \, v_1 tr_1$$
$$A_2 = (1/2) \, CD \, r_2 = (1/2) \, v_2 tr_2$$

These areas are equal. Equating A_1 and A_2 the time t cancels and 1/2 cancels leaving

$$v_1 r_1 = v_2 r_2$$

The product of the speed times distance at the apogee is equal to the same product at the perigee. This law applies to the orbits of any objects, not only planets.

The third law, that the ratio of the cube of the semi-major axis, a, to the square of the period, T, is constant, applies also to more than planets. For all objects orbiting the sun, be they planets or space vehicles, the ratio a^3/T^2 is the same for all. Letting subscript 1 be for one orbiting object and 2 for another, then

$$\frac{a_1^3}{T_1^2} = \frac{a_2^3}{T_2^2}$$

The subsequent extension of this law is that for all objects orbiting the same central object the ratio a^3/T^2 is a constant, but that the constant is different for different central objects. The value of the constant depends, of course, on the units used for T and for a as well as on the central object.

To illustrate the law for the planets data are presented in Table 4–1. These data are made deceptively simple by using the time for each planet to orbit the sun expressed in years and the distance a expressed in a unit which is in terms of the orbit of the earth having a value of 1.00. The value of a for Mars, for example, is 1.52 times that for the earth and that is the value entered in the table. The value for a for earth is called 1 **astronomical unit**, the symbol for which is AU. 1 AU is equal to 1.49×10^{11} meters. The AU is commonly used in astronomy.

TABLE 4–1   Test of Kepler's Third Law

PLANET	a(AU)	T(YEARS)	a³	T²	a³/T²
Mercury	0.387	0.241	0.0580	0.0581	0.998
Venus	0.723	0.615	0.378	0.378	1.000
Earth	1.000	1.000	1.000	1.000	1.000
Mars	1.524	1.881	3.54	3.54	1.000
Jupiter	5.20	11.86	140.6	140.7	0.999
Saturn	9.54	29.46	868	868	1.000
Uranus	19.18	84.0	7056	7056	1.000
Neptune	30.06	164.8	27160	27160	1.000
Pluto	39.44	247.7	61350	61360	1.000

The mass of the planet did not seem to have any effect; the tiny Mars and enormous Jupiter fit equally well with the others.

EXAMPLE 1

Given data in SI units for two planets (Earth and Jupiter) calculate a^3/T^2 for each and compare them.

Given Data

	a(meters)	T(seconds)
Earth	1.495×10^{11}	3.156×10^{7}
Jupiter	7.78×10^{11}	3.743×10^{8}

Calculated Values

	$a^3(m^3)$	$T^2(s^2)$	$a^3/T^2(m^3/s^2)$
Earth	3.341×10^{33}	9.96×10^{14}	3.35×10^{18}
Jupiter	4.709×10^{35}	1.401×10^{17}	3.36×10^{18}

4-1-2 EARTH ORBITS

The period of a satellite orbiting the earth depends on the size of its orbit. Just as the value of a^3/T^2 is the same for all planets, the value of a^3/T^2 is the same for all earth satellites. However, the value of a^3/T^2 for earth satellites is not the same as for objects moving about the sun. In Example 2-17 it was shown that a low satellite in a circular orbit about the earth has a period of 5260 seconds. The semi-major axis (the radius in this case) is 6.53×10^6 m (the radius of the earth itself is 6.38×10^6 m).

Using these values, a^3/T^2 for this earth satellite, and hence for all others too, is

$$\frac{a^3}{T^2} = \frac{(6.53 \times 10^6 \text{m})^3}{(5.26 \times 10^3 \text{s})^2}$$

$$= 1.01 \times 10^{13} \text{ m}^3/\text{s}^2$$

EXAMPLE 2

Find the radius of a circular orbit for a satellite that will orbit the earth in 24 hours (86,400 s). If such a satellite is put over the equator, it will remain over the same point on the surface of the earth, because the point on the surface also goes around once in 24 hours. It is the "synchronous" satellite used for long distance communication on earth. Signals sent from point A on the earth as in Figure 4-4 can be re-transmitted from the satellite to point B.

For all earth satellites $a^3/T^2 = 1.01 \times 10^{13}$ m^3/s^2

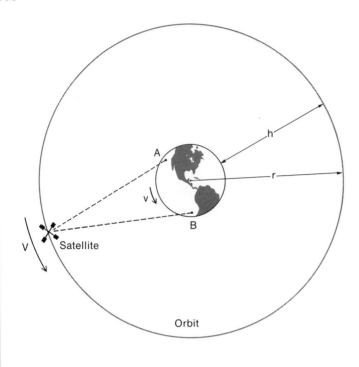

Figure 4–4 The synchronous satellite orbits in 24 hours, the same time as one rotation of the earth. It is thus always above the same point on the ground, and can be used for communication between points such as A and B.

Solve for a $a = \sqrt[3]{1.01 \times 10^{13}\ T^2\ \text{m}^3/\text{s}^2}$

Substitute $T = 8.64 \times 10^4\ \text{s}$, or $T^2 = 7.46 \times 10^9\ \text{s}^2$

Then $a = \sqrt[3]{1.01 \times 7.46 \times 10^{22}\ \text{m}^3}$

                   $= \sqrt[3]{75.3 \times 10^{21}\ \text{m}^3}$

                   $= 4.22 \times 10^7\ \text{m}$

The radius will be 4.22×10^7 m, or 42,200 km.
The altitude, h, above the surface of the earth is
         $h = (42{,}200 - 6380)$ km $= 35{,}800$ km

4–1–3 CIRCULAR AND ELLIPTIC ORBITS

A low circular orbit about the earth is achieved with a speed of 7.8 km/s. If the initial speed is less than this, the craft will fall to earth in what is actually an ellipse. If the initial speed is greater than that needed for a circular orbit, the object will move away from earth, slowing down as it does, and then it will fall back. The path will be an ellipse as in Figure 4–5.

The greater that the speed is in excess of the speed for the circular orbit the longer will be the ellipse. If the speed of a low earth orbiting craft is increased to 11.2 km/s, the ellipse will be so long that the craft will never return as shown in Figure

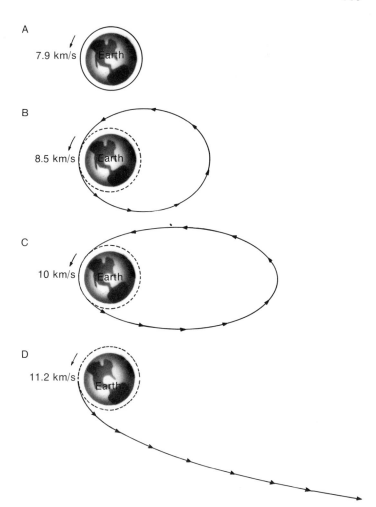

A

7.9 km/s Earth

B

8.5 km/s Earth

C

10 km/s Earth

D

11.2 km/s Earth

Figure 4–5 If a low satellite is given extra speed, it will go into an elliptic orbit such as (b) or (c). The higher the speed, the longer the ellipse. At the escape speed, as in (d), the ellipse becomes infinitely long.

4–5(D). This is called the escape velocity and the escape velocity from the earth is about 11 km/s.

4–1–4 A TRIP TO THE MOON

To go to the moon the craft is usually put into a low circular orbit about the earth. Then at a point on the opposite side from which it is planned to rendezvous with the moon, the rockets are fired to increase the speed and go into a long elliptical orbit as in Figure 4–6(B). This is referred to as "injection" into the orbit to the moon. The long elliptical orbit as in Figure 4–6(C) must have the high point at the distance of the moon. The semimajor axis of the lunar craft will be very close to half that for the moon and T^2/a^3 for the craft will be the same as for the moon.

Though the path is basically elliptical when the craft nears the moon, it will be attracted to the moon and deviate from that path.

A

Low circular orbit

B

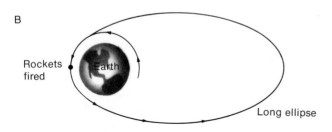

Rockets fired

Long ellipse

Figure 4–6   The concept of the moon trip.

C

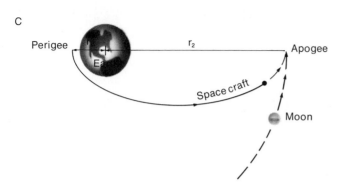

Perigee

r_2

Apogee

Space craft

Moon

EXAMPLE 3

Given that the moon orbits the earth in 27.4 days find the approximate time for a journey to the moon.

Let the orbiting time for the moon be T_m and its semi-major axis be a_m. Let the corresponding quantities for the craft be T_c and a_c, where T_c would be for a complete elliptical orbit of the craft.

By Kepler's third law:

$$\frac{T_c^2}{a_c^3} = \frac{T_m^2}{a_m^3}$$

from which

$$T_c = T_m \sqrt{\frac{a_c^3}{a_m^3}}$$

$$T_m = 27.4 \text{ d}$$

$$a_c = a_m/2 \text{ (very close)}$$

$$a_c^3/a_m^3 = 1/8$$

$$T_c = 27.4 \text{ d } \sqrt{1/8} = 27.4 \text{ d}/\sqrt{8}$$
$$= 9.7 \text{ d}$$

The complete elliptic orbit of the craft would require 9.7 days.

The craft requires only half this time to reach the far point in its orbit so the time to go to the moon is about 4.8 days. The same time would be required for the return trip.

Space travel is achieved through knowledge of the laws of nature and a recognition that they are not arbitrary rules to be fought against but that they describe how things happen. We do not "submit" to these laws; we "use" them to achieve our goals.

4–1–5 INTERPLANETARY SPACECRAFT

A space vehicle may be given enough speed as it leaves earth to escape from the gravitational pull of the earth. If the speed it acquires in moving out of earth orbit is 11.2 km/s (the escape speed from earth), the craft will not return, but as it gets farther and farther away its speed away from the earth will slowly reduce to zero. However, such a craft would still be held by the sun and would take up a position with an orbit similar to the earth. It would be an artificial planet.

If the spacecraft is given more than the escape speed, it will have some speed left over when it gets away from earth. If the launch was in the direction of earth motion as in Figure 4–7,

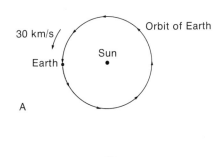

Figure 4–7 The launch of a spacecraft into orbit around the sun.

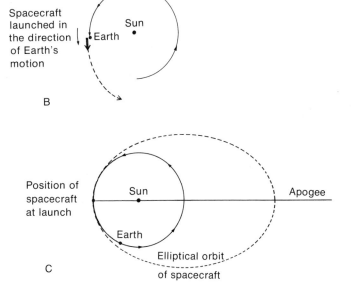

the space craft, which before launch had the same orbital speed as the earth (30 km/s), then will be moving at greater than 30 km/s, too fast to stay in earth orbit. The sun still pulls on the craft and it will take up an elliptic orbit as in Figure 4–7 (B) and (C). The craft then becomes another body orbiting the sun: it is like a planet, except for its size, and its motion is described by the same laws. For example, if the time for the craft to make a complete ellipse is T_c and the semi-major axis of its orbit is a_c then a_c^3/T_c^2 is the same as for any other planet. Using years and astronomical units that value was numerically just 1, so in those units $a_c^3/T_c^2 = 1$ or $T_c^2 = a_c^3$.

The speed of the craft when it is injected into the planetary orbit can be chosen such that at the apogee (the farthest point from the sun) it will be at the same distance from the sun as another planet. For example, to go to Mars, the apogee must be at 1.52 AU, to go to Jupiter, 5.2 AU.

EXAMPLE 4

Find the time for a craft to orbit the sun if the distance r_1 at the perigee is 1.00 AU (at earth) and at the apogee it is 1.52 AU (at the orbit of Mars).

For planet earth, a^3/T^2 is 1 AU³/1 y²

For the craft, a_c^3/T_c^2 is also 1 AU³/y²

and $a_c = (1.00 + 1.52)$ AU/2 = 1.26 AU.

If a^3/T^2 is numerically equal to 1, so is the inverse T^2/a^3, then

$$T_c^2/1.26^3 \text{AU}^3 = 1 \text{ y}^2/\text{AU}^3$$
$$T_c^2 = 1.26^3 \text{ y}^2$$
$$T_c^2 = 2.00 \text{ y}^2$$
$$T_c = 1.41 \text{ y}$$

The time for the spacecraft to make the complete orbit is 1.41 y. Note that the earth will have made 1.41 orbits about the sun, and though the craft will return to the same place in space at which it was launched, the earth will no longer be there.

The time for a craft to go to Mars can now be calculated, and some of the particulars about such a trip can be discussed. The craft is launched into the elliptic orbit that will take it as far out as the orbit of Mars. The timing of the launch must be chosen such that it will meet Mars at the apogee. Then it will reach Mars after traversing only half of its orbit. The time for a complete orbit of that type was shown in the last example to be 1.41 y. The time for half of the orbit is then 0.705 y or 8.5 months. This is the time for a spacecraft to go from earth to Mars.

Mars requires 1.88 y (22.6 months) to orbit the sun and in the 8.5 months that the craft spent in its journey, Mars moved through about 135° in its orbit. The trip is illustrated in Figure 4–8. The craft is launched from earth with the intent of meeting

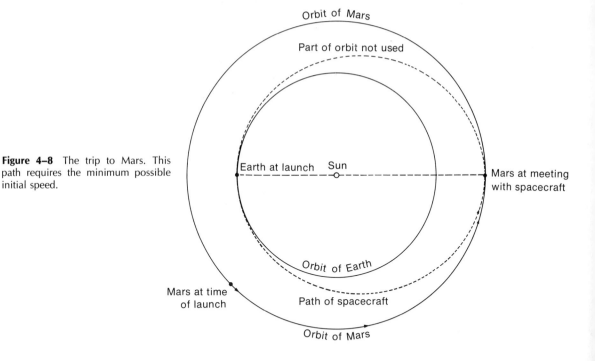

Figure 4–8 The trip to Mars. This path requires the minimum possible initial speed.

with Mars at a position on the opposite side of the sun from where it is launched. Mars at launch is 135° back in its orbit at that time. The craft in its ellipse then moves slowly outward from earth orbit and Mars moves along also. As the craft pulls away from the sun, it gradually slows and 8.5 months later it reaches the orbit of Mars, and Mars will then be in that position too. The craft may then use its rockets to go into orbit about the planet or even to land on it.

A Mars trip may be made only when the earth, sun and Mars are in the correct positions as in Figure 4–8. This occurs every 780 days (just over 2 years) and this is the interval between space shots to Mars. There is only a short span of time then, at which the craft can be sent. It is referred to as the launch "window."

To bring back a craft from Mars, the other half of the ellipse is the path. The craft must wait on Mars till earth is in the correct position to be met at the low point of the spacecraft orbit. The time for a complete trip (to Mars, waiting on Mars and return to earth) would require just over two years.

The craft in its journey to Mars merely coasts through space with no power. The rockets would be used only to make small course corrections. The velocity required is given to the craft as it leaves earth. Doing it in this way requires the least propellant.

It is possible to go to Mars in a slightly shorter time than described above by putting the craft into an orbit that would go

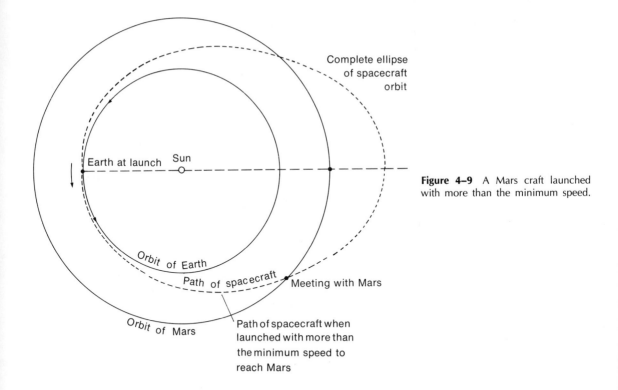

Complete ellipse
of spacecraft
orbit

Earth at launch Sun

Figure 4–9 A Mars craft launched
with more than the minimum speed.

Orbit of Earth

Path of spacecraft · Meeting with Mars

Orbit of Mars

Path of spacecraft when
launched with more than
the minimum speed to
reach Mars

beyond the orbit of Mars as in Figure 4–9. This would require
a higher launching speed.

To go to an inner planet such as Venus or Mercury, the
rocket is launched against the direction of the earth's motion in
its orbit, and with greater than escape velocity. When the craft
has escaped from earth, it is still following the earth in its orbit.
(The earth is moving at 30 km/s and if the craft had 2 km/s left
after escaping, it would still be moving in the same direction as
the earth but at 28 km/s.) The craft will then not have enough
speed to remain in earth orbit but will fall closer to the sun,
again in an elliptical path. The first craft to Venus, sent by the
U.S.S.R. in 1961, is shown on a stamp in Figure 4–10. The
craft shown was given a higher speed from earth than would
have been required to just meet Venus opposite the sun from
launch point on earth. The relative positions of Venus and
earth at launch time, closest approach and on arrival at Venus
are also shown.

Newton's first law of motion states that objects on which
there are no forces travel in a straight line at a constant speed.
This is the state of equilibrium. The interplanetary craft do not
operate their rocket engines during the flight but there is a force
on them anyway, the gravitational pull of the sun. Because
there is a force, straight line motion does not occur, rather, the
path is elliptical. Though the highways on earth may at times be
straight, the "highways" in interplanetary space are elliptical.

Figure 4–10 The path of a spacecraft to Venus, depicted on a Hungarian postage stamp.

4–2 ANGULAR MEASURE

The further description of rotation or of orbital motion requires the use of angular quantities. The systems used for the measurement of angles in scientific work will therefore be described.

Angles may be measured in terms of revolutions. This is nothing new; a wheel could turn half a revolution or a hundred revolutions—you would understand. In this system angular velocity would be in units such as revolutions per second, revolutions per minute, etc.

Degrees are also used in angular measure. One degree is defined as 1/360 of a complete circle. A minute of angle is 1/60 of a degree. Degrees are not new to you. The system of angular measure in degrees dates from at least 3000 years ago in Babylonia.

The angular measure most commonly used in scientific work, the unit adopted in the SI units, is the **radian**, the symbol for which is *rad*. One radian is an angle of a size such that when an arc is drawn as in Figure 4–11 the arc length is equal to the radius. If the angle were half a radian, the arc length would be just half the radius. The angle θ in radian measure is that given by the ratio of arc length, p, to the radius, r, or

$$\theta = p/r.$$

The arc length is referred to as p (defined as path) rather than s because s was defined as a straight line distance between two points.

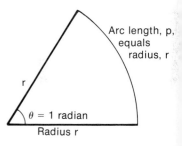

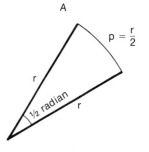

Figure 4–11 The radian. One radian is an angle for which the arc length is equal to the radius as in (a). In (b) is shown an angle of half a radian.

EXAMPLE 5

An angle is measured by drawing an arc across it at a radius of 15.0 cm, and with a curved ruler measuring the length of the arc. The arc length between the arms of the angle is found to be 6.0 cm. What is the size of the angle?

$$\theta = p/r \text{ where } p = 6.0 \text{ cm and } r = 15.0 \text{ cm}$$
$$= 6.0 \text{ cm}/15.0 \text{ cm}$$
$$= 0.40$$

The angle is 0.40 radians.

Note that the units of length cancel and the radian is a dimensionless quantity. It does not depend on the units used for the length measurement.

In one complete circle the angle in radians, arc length/radius, is $2\pi r/r$, which is 2π. That is, there are 2π radians in a circle. This gives the conversion factor between degrees and radians:

$$2\pi \text{ rad} = 360°$$
$$1 \text{ rad} = 360°/2\pi$$
$$= 57.3°$$

One radian is 57.3° and similarly, one degree is $2\pi/360$ radians or 0.0175 radians.

EXAMPLE 6

In the previous example, the angle was calculated to be 0.40 rad. To how many degrees does this correspond?

$$1 \text{ rad} = 57.3°$$
$$0.40 \text{ rad} = 0.40 \times 57.3° = 22.9°$$

4-2-1 ANGULAR VELOCITY

A rotating object turns through a certain angle in a given time, and one can talk about a *rotation speed* or, equivalently, **angular velocity**. Angular velocity is often represented by the lower case Greek omega (ω), the angle by the lower case Greek theta (θ). If the time to turn through an angle θ is t, the average angular velocity $\bar{\omega}$ is

$$\bar{\omega} = \theta/t$$

Similarly, at an angular velocity $\bar{\omega}$ the angle turned through in a time t is given by $\theta = \bar{\omega}t$. Note how these equations are similar to those for linear motion where **v** is replaced by $\bar{\omega}$ and **s** is replaced by θ.

Angular displacement or velocity are vector quantities, the direction being described by the direction of the axis in space.

Figure 4-12 The angular velocity vector. The direction is given by the right hand rule, as illustrated.

There are two ends to an axis, and by convention the positive direction for the vector is the direction given by the thumb when the right hand is curled around the axis with the fingers in the direction of the rotation as in Figure 4–12.

EXAMPLE 7

What is the angular velocity of the earth on its axis?
The earth turns through one turn or 2π radians in one day ($t = 24$ hours, or 1440 minutes or 86,400 seconds). (Actually, with respect to the stars it takes only 86,164 seconds for one rotation and you may have to think about that!)

Use $\qquad\qquad \omega = \dfrac{\theta}{t}$ where $\theta = 2\pi$ radians

$$\text{and } t = 86{,}164 \text{ seconds}$$

Then $\qquad\qquad = \dfrac{2\pi \text{ rad}}{86{,}164 \text{ s}} = 0.0000729 \text{ rad/s}$

$$= 7.29 \times 10^{-5} \text{ rad/s}$$

4–2–2 ANGULAR ACCELERATION

Angular rotation may not necessarily be at a constant speed. If there is acceleration, the angular velocity may change from ω_1 to ω_2 in a time t. The angular acceleration, often represented by the lower case Greek alpha (α) is given by

$$\alpha = \frac{\omega_2 - \omega_1}{t}$$

Note how this is similar to the equation for linear acceleration

$$\mathbf{a} = \frac{\mathbf{v}_2 - \mathbf{v}_1}{t}$$

Angular acceleration is a vector in the direction of the axis, with the positive direction given by the right hand rule as for angular velocity.

The equations describing angular motion are of the same form as those for linear motion and the same steps are followed to derive the specialized relations. Some of these follow:

Linear Motion	*Angular Motion*
$s = \bar{v}t$	$\theta = \bar{\omega}t$
$\bar{v} = \dfrac{v_1 + v_2}{2}$	$\bar{\omega} = \dfrac{\omega_1 + \omega_2}{2}$

$$a = \frac{v_2 - v_1}{t} \qquad\qquad \alpha = \frac{\omega_2 - \omega_1}{t}$$

$$s = v_1 t + at^2/2 \qquad\qquad \theta = \omega_1 t + \alpha t^2/2$$

$$v^2 = 2as \qquad\qquad \omega^2 = 2\alpha\theta$$

EXAMPLE 8

A wheel turns at 3600 rev/min. Express this in degree/s and in rad/s. Use the method of substitution of the equivalent thing for a unit.

$$1 \text{ rev} = 360°$$
$$1 \text{ min} = 60 \text{ s}$$

$$3600 \frac{\text{rev}}{\text{min}} = \frac{3600 \times 360°}{60 \text{ s}}$$

$$= 21,600°/\text{s}$$

That is, 3600 rev/min = 21,600°/s

or

$$1 \text{ rev} = 2\pi \text{ rad}$$

$$3600 \frac{\text{rev}}{\text{min}} = 3600 \times \frac{2\pi \text{ rad}}{60 \text{ s}}$$

$$= 377 \text{ rad/s}$$

That is, 3600 rev/min = 377 rad/s

EXAMPLE 9

A wheel acclerates from rest to 3600 rev/min in 30 seconds. What was the angular acceleration?

From the previous example 3600 rev/min = 377 rad/s. The initial angular speed was zero and the final was 377 rad/s. The time required was 30 s so the change per second, which is accleration, was

$$\alpha = \frac{377 \text{ rad}}{30 \text{ s} \cdot \text{s}} = 12.6 \text{ rad/s}^2$$

EXAMPLE 10

A limb moves from rest through 45° in 0.50 s. Find the acceleration in rad/s² assuming that it is constant.

$$\theta = 45° = (45/57.3) \text{ rad} = 0.785 \text{ rad}$$

Use $\qquad \theta = \alpha t^2/2$, solved for α

$$\alpha = 2\theta/t^2$$
$$= 2 \times 0.785 \text{ rad}/0.50^2 \text{s}^2$$
$$= 6.28 \text{ rad/s}^2$$

4–3 MOTION ON A CIRCUMFERENCE

The distance that an object moves on a circular path is readily expressed in terms of the angle at the center when radian measure is used. The quantities are illustrated in Figure 4–13. By definition of the angle, $\theta = p/r$, from which the path is given by $p = r\theta$.

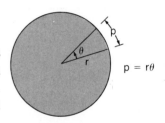

$p = r\theta$

If the distance p is moved through in a time t, the speed on the circular path can be expressed in terms of angular speed

$$v = \frac{p}{t} = r\frac{\theta}{t} = r\omega$$

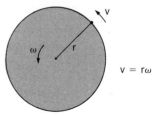

$v = r\omega$

where the substitution $\omega = \theta/t$ has been made. An initial speed may be v_1 and a final speed v_2. Then

$$v_1 = r\omega_1 \quad \text{and} \quad v_2 = r\omega_2$$

The acceleration along the path is

$$a = \frac{v_2 - v_1}{t} = \frac{r\omega_2 - r\omega_1}{t}$$

$$= r\frac{(\omega_2 - \omega_1)}{t}$$

$$= r\alpha$$

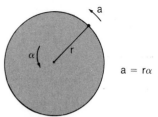

$a = r\alpha$

Figure 4–13 Displacement, speed, and acceleration on a circumference related to the corresponding angular quantities.

Summarizing, when radian measure of angles is used, the relations between the linear and angular quantities are given by

$$p = r\theta$$
$$v = r\omega$$
$$a = r\alpha$$

EXAMPLE 11

In an object moving in a path of radius 0.40 m accelerates at 6.28 rad/s², what is the linear acceleration along the path?

Given

$\alpha = 6.28$ rad/s²
$r = 0.40$ m
$a = r\alpha = 2.51$ m rad/s²

The units do not seem to be the units of acceleration but an angle in radians actually has a dimension of unity. This is because the units of length cancel when an angle is calculated in radians. The word radian is inserted to indicate that an angle is being referred to. The unit can be discarded when it is no longer needed.
The acceleration is therefore 2.51 m/s².

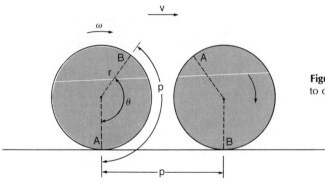

Figure 4–14 A rolling object. The diagram shows how to obtain the relations that $p = r\theta$ and $v = r\omega$.

4–3–1 ROLLING

The relation between the linear speed of the axis of a rolling object and the angular speed is often of interest. If there is no slipping between the flat surface and the curved, the path p (as in Figure 4–14) is the same as the distance on the circumference. The axis also moves through the distance p. If the time interval is t the linear speed v is given by $v = p/t$. The angular speed is $\omega = \theta/t$, if the angle turned through is θ. Writing $p = r\theta$, the linear speed is $v = r\theta/t$ but $\theta/t = \omega$ so $v = r\omega$ for rolling without slipping.

4–3–2 CENTRIPETAL ACCELERATION

Centripetal acceleration can also be expressed in terms of angular quantities through the relation that $v = r\omega$ or $v^2 = r^2\omega^2$ where ω is angular velocity. The units used for ω in this relation must be radians per second. Substituting for v in the expression for centripetal acceleration, $a_c = v^2/r$, the new form of the relation is

$$a_c = r\omega^2$$

EXAMPLE 12

What angular speed in radians per second and in revolutions per second would give a centripetal acceleration of 10 g if $r = 5$ meters?

Solve $a_c = r\omega^2$ for ω to get
          $\omega = \sqrt{a_c/r}$ where $a_c = 10 \times 9.8$ m/s²
                              $= 98$ m/s²
                              $r = 5$ m

$$\omega = \sqrt{\frac{98}{5}\frac{m}{m}\frac{m}{s^2}}$$

$$= 4.4/s$$

Note the units of just s^{-1}, but remember that the unit rad can be inserted where applicable, so the answer is that $\omega = 4.4$ rad/s.

To change this to revolutions per second substitute the equivalence 1 rad $= 1/2\pi$ revolutions.

Then
$$\omega = (4.4/6.28) \text{ rev/s}$$
$$= 0.71 \text{ rev/s}$$

The answers are that the angular speed is

4.4 rad/s   or   0.71 rev/s

4–4 TORQUE AND ANGULAR ACCELERATION

If the net torque on an object is zero, that object will not rotate or it will rotate at constant angular velocity. If the net torque is not zero, the angular velocity will not be constant, or there will be angular acceleration. Torque is related to angular acceleration in a manner similar to the relation between force and linear acceleration. Torque is a vector as is angular acceleration, and the angular acceleration is in the same direction as the torque.

To analyze rotational motion consider a mass m constrained to move in a circular path and with a force always along the path as in Figure 4–15.

The mass will have a linear acceleration along its path determined from $\mathbf{F} = m\mathbf{a}$. The symbol $\mathbf{L}$ is often used for torque, where $L = Fr$. The vector $\mathbf{L}$ lies along the axis of rotation, and is formally given by the "vector cross product" $\mathbf{r} \times \mathbf{F}$; here we will deal only with the magnitudes of the vectors. To obtain the product Fr, multiply both sides by r to give

$$Fr = mar$$

and substituting $a = r\alpha$ where α is angular acceleration

$$Fr = mr^2\alpha$$

Finally, since $L = Fr$,

$$L = mr^2\alpha$$

In linear motion the ratio force to acceleration is called

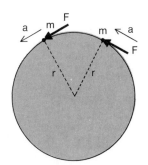

Figure 4–15 Acceleration along a circular path.

inertia (or mass). In rotational motion the ratio of torque to angular acceleration gives the corresponding quantity, called **moment of inertia**, often designated by the symbol I.

$$\text{In linear motion } \frac{F}{a} = m$$

$$\text{In circular motion } \frac{L}{\alpha} = mr^2 = I$$

For a small mass at a radius r the moment of inertia is given by mr^2. This quantity can also be called the second moment of the mass (the first moment, mr, was used in the calculation of centers of mass).

If there are several masses at different radii, the total moment of inertia is the sum of the moments of each of the masses.

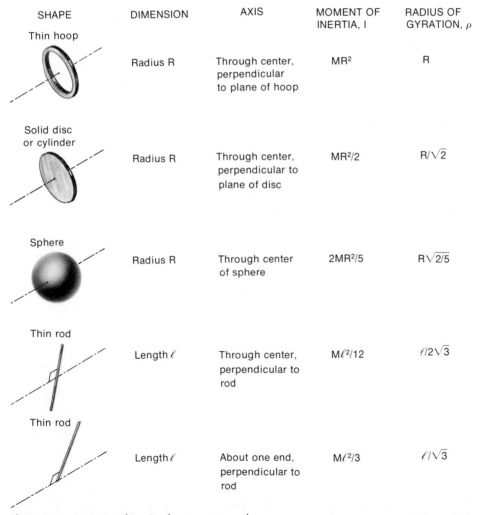

SHAPE	DIMENSION	AXIS	MOMENT OF INERTIA, I	RADIUS OF GYRATION, ρ
Thin hoop	Radius R	Through center, perpendicular to plane of hoop	MR^2	R
Solid disc or cylinder	Radius R	Through center, perpendicular to plane of disc	$MR^2/2$	$R/\sqrt{2}$
Sphere	Radius R	Through center of sphere	$2MR^2/5$	$R\sqrt{2/5}$
Thin rod	Length ℓ	Through center, perpendicular to rod	$M\ell^2/12$	$\ell/2\sqrt{3}$
Thin rod	Length ℓ	About one end, perpendicular to rod	$M\ell^2/3$	$\ell/\sqrt{3}$

Figure 4–16 Moments of inertia of some common shapes.

The moment of inertia of an extended body, like a rod or disc, cannot be calculated simply, because its parts rotate with different radii. The one exception is the thin hoop for which all the mass is at the same radius so $I = mr^2$. The moment of inertia of objects of various geometrical shapes and with different axes of rotation has been calculated, but for odd shapes it has often to be found experimentally using the ratio L/α. Moments of inertia can also be expressed in terms of what is called the **radius of gyration** often designated by the Greek lower case rho (ρ). For any object of mass m the radius of gyration is the quantity which is found from

$$m\rho^2 = I$$

The quantity, ρ, may be called an effective radius for the calculation of a moment of inertia. The expression for moments of inertia of some simple shapes and the radii of gyration are shown in Figure 4–16.

EXAMPLE 13

A solid disc with axis through the center is to be made to rotate by means of a force applied to a cable around it as in Figure 4–17. Find the angular acceleration if the force is 20 N, the radius is 0.30 m and the mass is 150 kg.

The torque, $\qquad L = 20 \text{ N} \times 0.30 \text{ m} = 6.0 \text{ Nm}$

The moment of inertia of a solid disc is given by

$$I = mr^2/2 = 150 \text{ kg} \times 0.30^2 \text{m}^2/2$$
$$= 6.75 \text{ kg m}^2$$

The angular acceleration is given by

$$\alpha = L/I$$
$$= \frac{6.0 \text{ N m}}{6.75 \text{ kg m}^2} \quad \text{(write, 1 N} = 1 \text{ kg m/s}^2)$$
$$= 0.89/\text{s}^2$$

The angular acceleration is 0.89 rad/s².

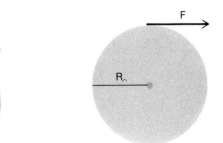

Figure 4–17 A disc with angular acceleration.

4-4-1 ANGULAR MOMENTUM

To arrive at the quantity called momentum in linear motion, Newton's second law was expressed in terms of initial and final velocity over a time interval t. Doing a similar analysis in rotational motion write

$$L = I\alpha \qquad \text{(compares to } F = ma\text{)}$$

and α, the angular acceleration is rate of change of angular velocity

$$\text{or} \quad \alpha = \frac{(\omega_2 - \omega_1)}{t} \qquad \text{compares to} \quad a = \frac{(v_2 - v_1)}{t}$$

$$\text{so} \quad L = \frac{I(\omega_2 - \omega_1)}{t} \qquad \text{compares to} \quad F = \frac{m(v_2 - v_1)}{t}$$

$$= \frac{I\omega_2 - I\omega_1}{t} \qquad \text{compares to} \quad = \frac{mv_2 - mv_1}{t}$$

The quantity $I\omega$ is called **angular momentum** by analogy to linear momentum, mv.

Linear motion	*Angular motion*
m = inertia	I = moment of inertia
v = linear velocity	ω = angular velocity
mv = linear momentum	$I\omega$ = angular momentum

Just as there is conservation of linear momentum in a closed system, even in the whole universe, there is also conservation of angular momentum.

The angular momentum of a mass m moving with a speed v in a circle of radius r as in Figure 4–18 is a simple case worth looking at in more detail. In this case, $I = mr^2$ and $\omega = v/r$.

$$\text{Then} \qquad I\omega = mr^2 \cdot v/r$$
$$= mvr$$

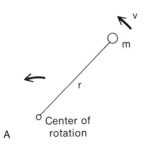

A

Center of rotation

That is, the angular momentum is linear momentum times radius.

As an example of conservation of angular momentum consider a small object on a string as in Figure 4–18(B). If the object is swung around in a circle and then the string is allowed to wind on your finger so r decreases, then in order that mvr be constant, the speed v must increase. This is indeed the case and this example is worth doing yourself.

B

Figure 4–18 Angular momentum.

A similar example is that of a diver. The diver leaves the board with enough angular velocity to turn through half a revolution before entering the water so the entry will be head first. However, if the diver curls up during the dive as in Figure 4–19 the radius of gyration, and hence the moment of inertia, I, is reduced. For $I\omega$ to be constant, ω, the angular velocity, must increase. The diver may then perform a complete revolution and straighten out again before hitting the water.

A further example concerns the tides and the earth-moon system. The tidal bulge as in Figure 4–20 moves around the earth, and because of frictional effects is gradually slowing the speed of rotation of the earth. That would reduce the angular momentum of the earth and to compensate, the moon must move further out from the earth. That is, the orbit of the moon is slowly increasing in size. The effect is very small, and the interest is in such thoughts that a few billion years ago the moon would have been much closer to the earth with the earth spinning much faster. In the distant future the earth will have slowed till it always has the same side to the moon and the moon will no longer recede. That this has not yet happened allows a limit to be set on the beginning of the earth-moon system. It cannot have been there forever.

The moon has already been slowed to the extent that it keeps the same side always toward the earth. The far side of the moon had never been seen until *Luna 1* orbited it in *January 1959* and sent back photographs of the far side. Since then the various lunar vehicles have mapped the far side of the moon in great detail.

An orbiting planet also has constant angular momentum. If the nearest distance to the sun (the perigee) as in Figure 4–21 is r_1 where the planet moves at a speed v_1 perpendicular to the radius line r_1, angular momentum is mv_1r_1. At the apogee, a distance r_2, the speed is reduced to v_2. Again it is perpendicular to the radius line and the angular momentum is mv_2r_2. Because angular momentum is constant

$$mv_1r_1 = mv_2r_2$$

or

$$v_1r_1 = v_2r_2$$

This is what was actually found by Kepler before the development of this type of theory—it is Kepler's second law. So Kepler's second law is really a statement of conservation of angular momentum.

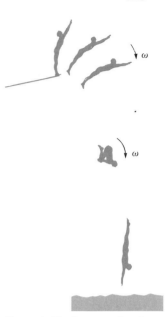

Figure **4–19** An example of constant angular momentum but changing radius of gyration and hence changing angular velocity.

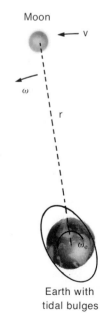

Moon

Earth with tidal bulges

Figure **4–20** The effect of tidal friction on the earth-moon system. Angular momentum in the system is conserved.

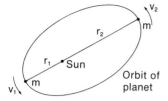

Figure 4–21 The varying speed of a planet in its orbit is such that angular momentum, *mvr*, is constant.

4–4–2   THE CORIOLIS EFFECT

If a steel ball is dropped from a high tower on the earth will it land directly below the point from which it is dropped? The rotation of the earth, which is toward the east, will have an effect so the answer is not obvious. The ball could land to the east of the point directly below its points of release, to the west, or it could fall straight down. The experiment was conducted by C. Flammarion in Paris in 1898. The result was that in a fall of 60 meters small steel balls were consistently deflected eastward by 8 mm from the point directly below their point of release. The 8 mm deflection in a fall of 60 m seems to be very small, but in large scale phenomona such as the movement of air masses or ocean currents this effect, called the **Coriolis effect**, is of prime importance.

The eastward deflection of the falling ball can be explained on the basis of conservation of angular momentum. The ball at the top of the tower, at a distance r from the axis of the rotation of the earth as in Figure 4–22, moves with the angular velocity of the earth, ω_E. Angular momentum is $I\omega$ and $I = mr^2$ so the angular momentum of the ball is $mr^2\omega_E$. As the ball falls toward the earth r decreases. The angular momentum $mr^2\omega$ must be constant so ω must increase above the value ω_E. The angular velocity of the ball will increase and an eastward deflection will result.

Motion upward from the earth, outward from the axis, would result in a westward deflection.

As a mass moves on the surface of the earth it also tends to approach the axis or to recede from it. In the northern hemisphere a mass moving northward approaches the axis and to conserve angular momentum its angular velocity must increase. An eastward deflection or a deflection to the right will occur as

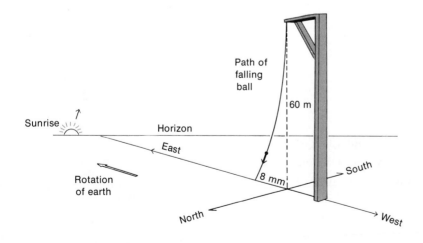

Figure 4–22 The eastward deflection of a ball falling toward the earth.

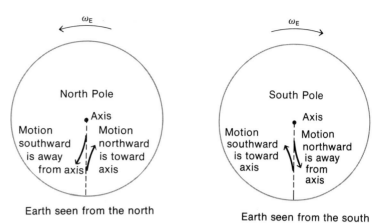

Figure 4–23 The deflection to the right of a moving mass in the northern hemisphere and to the left in the southern hemisphere. This is an example of the coriolis effect.

in Figure 4–23. Southward motion would be away from the axis so angular velocity must decrease to conserve angular momentum and westward motion would result. Again this is a deflection to the right. In the southern hemisphere the situation is reversed. Northward motion there is away from the axis so westward deflection, a deflection to the left, results. Southward motion is toward the axis so there is an eastward deflection, increasing the angular velocity. Again the deflection is to the left. Summarizing; in the northern hemisphere of the earth a mass moving on the surface is deflected to the right and in the southern hemisphere is deflected to the left. This is the Coriolis effect.

Figure 4–24 In the northern hemisphere, the coriolis effect causes a counterclockwise motion of air about a low pressure area and a clockwise motion about a high pressure area. In the southern hemisphere, the rotations of the air are in the opposite sense.

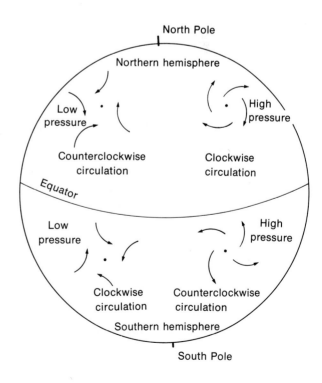

One of the consequences of the Coriolis effect is the counterclockwise motion of air as it moves toward a low pressure area in the northern hemisphere (cyclone) and the clockwise motion of the air (anticyclone) about a low in the southern hemisphere. These are illustrated in Figure 4–24 along with the anticyclonic motion about a high in the north and cyclonic motion in the south.

Another consequence of the Coriolis effect is the breaking up of weather patterns on the earth into bands or cells. If the earth did not rotate, the warm equatorial air would rise being replaced by a southward flow of polar air. In the upper atmosphere the air would flow northward, move down in the polar region and move southward. There would be one cell of circulating air in each hemisphere as in Figure 4–25(A). In reality, as the upper air moves away from the equator toward the north pole it is deflected eastward until at about 30° the motion is due east. The air moving from the north toward the equator is deflected westward. So there is a band of circulating air about 30° wide near the equator with vertical as well as horizontal circulation. The motion on each side of the equator is similar. This cell near the equator is called the Hadley cell. A similar effect produces another band of circulation or a vertical cell from about 30° to 60° called the Ferrel cell. There is a third region, the polar cell between about 60° and the poles. The three cells are illustrated in Figure 4–26.

The patterns described above are the basic structures on which changes due mainly to uneven heating of the atmosphere are superimposed to give our changing local weather.

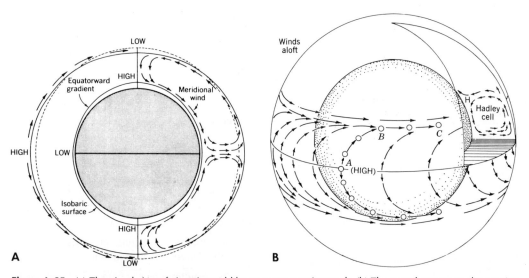

Figure 4–25 (a) The circulation of air as it would be on a non-rotating earth. (b) The actual pattern on the rotating earth. (From A. N. Strahler, *Planet Earth: Its Physical Systems Through Geologic Time,* Harper & Row, New York, 1972.)

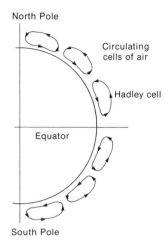

North Pole

Circulating
cells of air

Hadley cell

Equator

South Pole

Figure 4–26 The rotation of the earth causes the atmospheric circulation to occur in three distinct cells.

4–5 CIRCULAR ORBITS AND THE LAW OF GRAVITY

An object will move in a circular path only if there is a force toward the center of the circle as in Figure 4–27. In the case of orbits of planets or satellites the force is gravitational attraction to the central body. We will consider only cases in which the central body is far more massive than the revolving object so that the central body is effectively still.

For the general case let the central force be F_c, then for a circular orbit F_c must be such that it provides the centripetal force mv^2/r. So

$$\frac{mv^2}{r} = F_c$$

Unless this condition is met the orbit will not be circular.

Putting this in terms of the time to make a complete orbit, (the period T), use

$$v = 2\pi r/T$$

or
$$v^2 = 4\pi^2 r^2/T^2$$

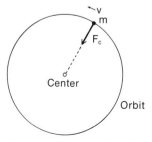

Center

Orbit

Figure 4–27 The circular orbit.

and $\qquad$ $\dfrac{mv^2}{r}$ becomes $\dfrac{m4\pi^2 r^2}{rT^2}$

or $\qquad$ $\dfrac{4\pi^2 mr}{T^2} = F_c$

or $\qquad$ $\dfrac{r}{T^2} = \dfrac{F_c}{4\pi^2 m}$

The relation between the orbit size r and the period T depends on the way that the force varies with factors such as distance r and the mass of the body. (The radius r of the circular orbit is equivalent to the semi-major axis, a, used for elliptic orbits). Various relations between r (or a) and T will be obtained with different force functions, F_c.

In the case of orbits under gravity, the form of the force function that describes the behavior of gravity must be such that when put into that last equation r^3/T^2 must be a constant (Kepler's third law) independent of the mass m of the orbiting body, though it will depend on the central body. The problem in finding the law of gravity was to find a form of F_c that would do just that. Without pursuing details of the search, the result is that if the law of gravity is of the form

$$F = \frac{GMm}{r^2}$$

where G is a constant, now called the **universal gravitation constant,** and M is the mass of the central body; then r^3/T^2 will be constant.

To show this use $\quad \dfrac{r}{T^2} = \dfrac{F_c}{4\pi^2 m} \quad$ and put $F_c = \dfrac{GMm}{r^2}$

$$\frac{r}{T^2} = \frac{GM\cancel{m}}{4\pi^2 \cancel{m} r^2}$$

multiply by r^2 $\qquad$ $\dfrac{r^3}{T^2} = \dfrac{GM}{4\pi^2}$

The right hand side will be the same (constant) for all objects orbiting the same mass M. The constant depends on the central mass M.

Thus, Kepler's third law showed Newton the correct form of the law of gravity and the formulation of this law $F = GMm/r^2$ is considered one of Newton's greatest achievements.

The mathematical form of the law of gravity has been given, but a complete statement of it would be that "Every body in the universe attracts every other body with a force that is proportional to the product of their masses and inversely proportional to the square of the distance between them." If the bodies have masses M and m then, as in Figure 4–28,

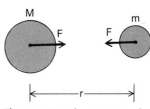

$$F = \frac{GMm}{r^2}$$

Figure 4–28 The gravitational attraction between two masses.

The force between masses in the laboratory has been measured (first published by Cavendish in 1798). In that case M, m, r and F would all be measured. The gravitation constant G would be the only unknown. The value of G has been found from the laboratory experiments to be

$$G = 6.67 \times 10^{-11} \frac{N \, m^2}{kg^2} \text{ (or } m^3/kg \, s^2)$$

G is numerically the force between two unit masses (a kilogram each) separated by a unit distance (1 meter).

EXAMPLE 14

Find the force of gravitational attraction between two lead spheres, one with a radius of 0.3 m and the other, 0.03 m. The relative density of lead is 11.3. The balls are so close that they touch. The separation of their centers is the sum of their radii, 0.33 m.

The mass of the large sphere, M, is the volume times density (and water is 1000 kg/m^3) so

$$M = \frac{4}{3} \pi \times 0.3^3 m^3 \times 11.3 \times 1000 \frac{kg}{m^3}$$
$$= 1278 \text{ kg}$$

The mass of the small ball, m, is found in a similar way to be

$$m = 1.278 \text{ kg}$$

also $r = 0.33$ m and $r^2 = 0.1089$ m²
$G = 6.67 \times 10^{-11}$ N m²/kg²

Use

$$F = \frac{GMm}{r^2}$$
$$= 6.67 \times 10^{-11} \frac{N \, m^2}{kg^2} \times \frac{1278 \text{ kg} \times 1.278 \text{ kg}}{0.1089 \text{ m}^2}$$
$$= 1.00 \times 10^{-6} \text{ N}$$

The force is 1.00×10^{-6} N, 1 μN or 0.1 dyne.

EXAMPLE 15

Find the mass and the average relative density of the earth.

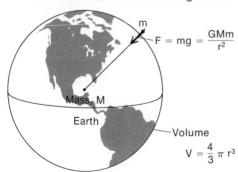

Figure 4-29 The law of gravity applied to the earth and an object on the surface.

A mass m on the surface of the earth is attracted to it with a force mg but that force is also described by the law of gravity $F = GMm/r^2$ where M is the mass of the earth and r is its radius as in Figure 4-29.

So

$$\frac{GM\not{m}}{r^2} = \not{m}g$$

This equation can be solved for the mass of the earth, M:

$$M = \frac{r^2 g}{G}$$

where r is 6.38×10^6 m
g is 9.8 N/kg
G is 6.67×10^{-11} N m^2/kg^2

So

$$M = \frac{6.38^2 \times 10^{12} \text{ m}^2 \times 9.81 \text{ N}}{6.67 \times 10^{-11} \text{ N} \dfrac{\text{m}^2}{\text{kg}^2} \text{ kg}}$$

$$= 5.99 \times 10^{24} \text{ kg}$$

The mass in the whole earth must be 5.99×10^{24} kg. To find the volume of the earth, consider it to be a sphere of radius 6.38×10^6 m. The volume of a sphere is given by $4\pi r^3/3$ which comes to 1.09×10^{21} m^3.

The average density of the earth is therefore

$$\frac{5.99 \times 10^{24} \text{ kg}}{1.09 \times 10^{21} \text{ m}^3} = 5.50 \times 10^3 \text{ kg/m}^3$$

The relative density of the earth (compared to water which has a density of 10^3 kg/m^3) is therefore 5.5.

Surface rocks on earth have a relative density in the range 2.6 to 4. It is apparent that the density deep in the earth must be greater in order that the average density be 5.5.

Another example will be given to show the power of the law of gravity.

Consider two sets of orbiting bodies, one going round a large mass M_1 and the other round another mass M_2.

For the first set $\qquad (a^3/T^2)_1 = GM_1/4\pi^2$
For the second set $\quad (a^3/T^2)_1 = GM_2/4\pi^2$

The ratio of the two values of a^3/T^2 is

$$\frac{(a^3/T^2)_1}{(a^3/T^2)_2} = \frac{GM_1/4\pi^2}{GM_2/4\pi^2} = \frac{M_1}{M_2}$$

The ratio of the values of a^3/T^2 gives the ratio of the masses. It has been shown (Section 4–1) that for the planets orbiting the sun a^3/T^2 is 3.35×10^{18} m³/s² and that for objects orbiting the earth a^3/T^2 is 1.01×10^{13} m³/s² (Section 4–1–2). The ratio of the masses of the sun and the earth is then

$$\frac{M_{\text{sun}}}{M_{\text{earth}}} = \frac{3.35 \times 10^{18} \text{ m}^3/\text{s}^2}{1.01 \times 10^{13} \text{ m}^3/\text{s}^2}$$

$$= 3.32 \times 10^5$$

The total mass of the sun is 332,000 times the mass of the earth.

This law of gravity can be extended to find the masses of planets, stars or galaxies, but it is limited to orbiting systems when used in this way, that is, to planets with satellites, stars that rotate about each other, rotating galaxies, etc.

DISCUSSION PROBLEMS

4–1–1 1. Kepler's laws arose from his finding and describing regularities or order in the orbits of the planets. How would you define a law in science?

4–1–2 2. If the laws governing the putting up of satellites were known a couple of hundred years ago, why was it not till 1957 that a satellite was put into orbit? Can you find out why the effort was put in to do it at that time?

4–1–3 3. Would it be possible to orbit the earth in one hour? If so, how?

4–1–3 4. Describe how to move a satellite from a low circular orbit to a high circular orbit with two rocket firings in the process. (The expression "a kick in the apogee" applies here.)

4–2–2 5. In the text it was stated that a final angular speed ω was related to the angular acceleration α, and an angle turned through θ by $\omega^2 = 2\alpha\theta$. Derive this relation. There is also the requirement that the system be initially at rest.

4–3 6. Show that centripetal acceleration, v^2/r, is described also by $r\omega^2$.

4–4–1 7. Use angular momentum ideas to explain the circular motion of the water as it goes down the drain of a sink or tub.

4–4–1 8. Use angular momentum ideas to explain the extremely high wind velocity in a tornado (twister).

4–4–1 9. Is it true that in the northern hemisphere the water going out of a bathtub usually swirls in a counterclockwise direction, while in the southern hemisphere it goes the opposite way? Why would such a difference occur?

4-5   10.   Kepler's laws were stated before the law of gravity was found. In fact, Newton's form of the law of gravity was found using Kepler's laws. Would you argue with the statement that the law of gravity can be used to prove Kepler's laws? In the text the relation between the law of gravity and the third law was discussed in detail.

PROBLEMS

4-1-1   1.   At the closest approach to the sun the earth is at 147 million km from it and at the farthest it is at 152 million km. What is the change in speed of the earth, between these two positions? Express it as a per cent of the lowest speed.

4-1-1   2.   The planet Mercury moves at 39 km/s when it is at the apogee of its orbit (at 70 million km from the sun). The perigee is at 46 million km from the sun. How fast does Mercury move at the perigee?

4-1-1   3.   If a new planet were discovered and measured to be at 60 A.U. from the sun, how long would it take that planet to orbit the sun?

4-1-1   4.   What must be the ratio of the distances of two planets from the sun if one is to make a complete orbit in twice the time of the other?

4-1-1   5.   The following data are for earth satellites. Show whether or not Kepler's third law applies to these satellites. Evaluate the constant in S.I. units.

Satellite	Mean Radius of Orbit in km	Period in Minutes
Sputnik I	6952	96.2
Explorer I	7582	109.6
Vanguard III	8504	130.0
Vanguard I	8671	133.9
Explorer VI	27,640	762
Moon	384,400	39,343

4-1-1   6.   Halley's comet is in a long elliptical orbit about the sun. After the first recorded sighting in 467 B.C. it has returned to the vicinity of the sun with an average period of 76.5 years. It was last seen in 1910 and is expected to be visible again in 1985. At the closest approach to the sun it is at 0.6 AU. How far is it at the farthest distance from the sun? Where is that in relation to the planets?

4-1-2   7.   Find the speed of a satellite in an orbit that has a radius of 41,200 km. The orbiting time is 24 hours. Express the answer in km/s.

4-1-3   8.   A satellite in an orbit 150 km above the surface of the earth (r_1 = 6530 km) fires its rockets to increase its speed to 10.2 km/s. It then goes into a long elliptical orbit, which at the high point, the apogee, is at 41,200 km. Find the speed at the apogee.

4-1-4   9.   (a) An earth satellite in a long elliptical orbit with perigee, r_1, at 6530 km and the apogee, r_2, at 385,000 km has a speed of 10.9 km/s at r_1. What is the speed at the apogee? (r_1 is for a low satellite and r_2 is at the distance of the moon).
(b) Calculate the speed of the moon in its orbit. The orbiting time is 27 days 7 hours. Compare the result to that obtained in (a).

4-1-5   10.   Find the time for a spacecraft to go to Jupiter in the path requiring the least speed from earth.

4-1-5   11.   Find the time for a spacecraft to go to Venus.

4-2   12.   Convert the following angles to radians, expressing the results both as a fraction or multiple of π and as a number in the decimal system. 180°, 90°, 45°, 60°.

4-2   13.   Make a table with the angles from 1° to 10° in 1° steps in the first column, the corresponding value in radians in the second column, the sine of the angle in the third and the tangent of the angle in the fourth column.

4-2   14.   When viewed from earth the angle between lines to opposite sides of the sun is very close to ½°. The sun is 1.50×10^8 km away. What is the diameter of the sun?

4-2   15.   Find the angular velocities of the second hand and the hour hand of a clock.

4-2-1   16.   What is the angular velocity of the earth in its daily rotation? Express it in rad/s.

4-2-1   17.   What is the angular velocity in rad/s of a centrifuge that is spinning at 6000 rev/min?

4-2-2   18.   Find the angular acceleration of a centrifuge when it requires 30 seconds to go from rest to 6000 rev/min.

4-2-2   19.   Through how many radians does a wheel turn when it goes from rest to 400 rev/min in 20 seconds?

4-2-2   20.   An elevator moves upward at 0.8 m/s pulled by a cable winding on a cylinder of radius 0.12 m. What is the angular velocity of the cylinder?

4-3   21. Find the speed of a point on the rim of a wheel rotating at 14 rad/s with a radius of 0.28 m.

4-3-1   22. Find the angular speed of the wheel of a car as it goes down the road at 90 km/h. The radius is 0.35 m.

4-3-1   23. Find the angular acceleration of a point on the outside of the tire of a car that is accelerating at 3.0 m/s² along a road. The radius of the wheel is 0.34 m.

4-4   24. Find the angular acceleration that occurs when a force is applied along the edge of a wheel. The wheel is a solid disc of mass 12 kg, radius 0.25 m, and the force is of 12 N.

4-4   25. Compare the final speeds of two wheels each of the same mass and radius, and to which the same torque has been applied for the same length of time. One wheel has the mass concentrated on the edge while the other is a solid disc.

4-4   26. Find the angular acceleration of a rod pivoted at one end and pulled by a force of 250 N at 5 cm perpendicularly from the pivot as in Figure 4–30A. The rod is of mass 6.5 kg, uniform, and of length 0.60 m. This is not unlike the acceleration of the lower leg as it makes a kicking motion. The force is applied by a muscle in the upper leg and a tendon over the knee to pull on the ulna. The function of the knee cap is to increase the torque by increasing the distance to the line of action of the force as in Figure 4–30B.

4-4-1   27. (a) Find the angular momentum of a wheel of mass 22 kg and of radius of gyration 0.33 m when it rotates at 4.4 rad/s.

(b) What torque would stop the wheel in 5.5 s?

4-4-2   28. (a) What is the perpendicular distance from the axis of the earth to objects at 30° latitude and at 40° latitude? The radius of the earth is 6380 km.

(b) Find the angular momenta of masses size m, one at 30° and one at 40°, both stationary on the earth. (Express it in terms of m).

(c) What eastward speed on the surface of the earth would a mass at 40° latitude have to have in order that its angular momentum would be the same as that of a similar mass stationary on the earth at 30° latitude?

4-5   29. Calculate the force of attraction due only to gravity between two people, one of mass 80 kg and the other of 50 kg with 0.60 m between their centers. Because of the complex shapes involved, the answer will be only an approximation when the law of gravity as given in the text is used.

4-5   30. Compare the mass of Mars to the mass of the earth given the following data. The earth has a satellite (called moon) that orbits with a radius of 3.84×10^8 m in a period of 27 days 7 hours. One of the satellites of Mars, Deimos by name, orbits at 2.35×10^7 m and with a period of 1 day 6.3 hours.

4-5   31. Calculate g on the surface of the moon given that the radius of the moon is 1740 km and its mass is 7.36×10^{22} kg.

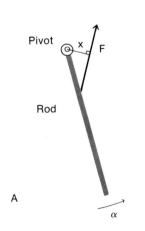

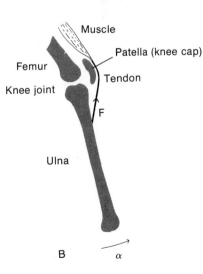

Figure 4–30 The angular acceleration of the lower leg in a kicking motion.

CHAPTER FIVE

ENERGY

5-1 DEFINITION AND UNITS

Energy is one of the more abstract of the concepts in physics, yet it is one of the most useful. To begin, consider energy as the ability to do work; by **work** is meant the movement of a force through space. For instance, an upward force on an object, which moves it up and away from the earth, does work on it. *The amount of work is given by the magnitude of the force multiplied by the distance moved in the direction of the force.* On the atomic scale, work is done in moving an atom out of a solid; work may be done in moving an atom out of a molecule; and work is done in moving an electron out from an atom. Work is done when the heart forces blood into the arteries, and when sodium ions are moved out of cells through the membranes.

What is now called energy was once referred to as **vis viva,** translated as **force of life.** Perhaps this older term is more expressive, more picturesque. In our age, life is associated with exuberance and, too often, work with drudgery.

Energy appears in many forms; when at first sight in some situations it is thought that some energy was lost, a closer look will show that it merely changed to another form. As you push a box along the floor you do work on it, and this work or energy is transformed into heat by friction; the heat is transferred partly to the box and partly to the floor. In the combination of you, the box, and the floor, the total energy is constant; this is called a *closed system.* In fact, by a closed system in this case is meant one from which no energy escapes and to which no energy comes from outside.

5-1-1 THE JOULE AND THE ERG

Energy in its many forms is measured in a variety of units, all of which are interchangeable by the use of the appropriate conversion factors. The most basic unit is derived from the idea that energy is related to work. Since energy is required to do work, the amount of energy used is measured by the amount of work done. Work is equal to force times distance, so the units of energy also are force times distance. The unit of force usually used in physics is called a *newton,* and the unit of distance is the *meter.* Energy is measured, then, as the product of these units, or *newton meters.* This unit of energy is called a *joule.* Now, these words are possibly strange to you. To give this concept more reality, let us note that *power* is defined as the rate of doing work, or equivalently the rate of dissipation of energy. If this rate is one *joule per second,* it is referred to as one *watt.* Watts are familiar for measurement of electric power, and lurking behind this term is energy in joules.

In the CGS system, the unit of force is the *dyne* and that of distance is the *centimeter.* Work, or energy, then has the unit *dyne centimeter,* and this quantity is called an *erg* of work or energy. The conversion between ergs and joules is found from substitution in the equations that define them as follows:

$$1 \text{ joule} = 1 \text{ N} \times 1 \text{ m}$$

but $\qquad 1 \text{ N} = 10^5 \text{ dynes}$

and $\qquad 1 \text{ m} = 10^2 \text{ cm}$

Substituting: $\qquad 1 \text{ joule} = 10^5 \text{ dynes} \times 10^2 \text{ cm}$

$$= 10^7 \text{ dynes cm}$$

$$= 10^7 \text{ ergs}$$

Therefore, $1 \text{ joule} = 10^7 \text{ ergs}.$

5-1-2 CALORIES, SMALL AND LARGE

The SI unit for energy is the joule, and more than one unit for the measurement of the same quantity can only make things more complicated than necessary. The calorie, a common unit for energy till now, is therefore going to be phased out. A further reason for this is that there were many kinds of calories, each with a slightly different definition.

The idea of the calorie (with a lower case c) was that it would be the energy needed to raise 1 g of water by 1° C. However, the energy needed to raise 1 g of water from 2° C to 3° C is not the same as from 15° C to 16° C or from 90° C to 91° C. The differences are small, but they can be of importance in precise work. A *mean calorie* was defined as a hundredth of the energy required to raise 1 g of water from 0° C to 100° C. A *15° calorie* was the energy needed to raise 1 g of water from 15° C to 16° C.

The conversion factor that is usually of sufficient accuracy is 4.18 J = 1 calorie, or 1 calorie = 0.239 J.

There is also the large Calorie (with a capital C), often called a kilogram calorie or kilocalorie. One Calorie is the energy needed to raise 1 kg of water by 1° C. Thus, 1 Calorie is 1000 calories at the appropriate temperature. There are 4180 J in 1 large Calorie. In rating foods by their caloric content, it is the large Calorie that is used.

5-1-3   THE ELECTRON VOLT

Another energy unit is frequently used when considering individual particles of atomic size. In these cases, even the erg is a large unit; the one that has been devised is the *electron volt* or eV. The derivation of this unit will be dealt with more fully in Chapter 14, but briefly it is that if a particle bearing a unit electronic charge passes across a voltage difference of one volt, the amount of energy it acquires is one electron volt. The electron volt is related to the erg and the joule very closely by

$$1 \text{ eV} = 1.60 \times 10^{-19} \text{ joule per particle}$$
$$= 1.60 \times 10^{-12} \text{ erg per particle}$$

The electron volt differs from the joule in that it usually refers to the energy of only one particle. Common multiples of the electron volt are the keV, MeV, and GeV, which are:

$$1 \text{ keV} = 1000 \text{ eV} = 10^3 \text{ eV}$$
$$1 \text{ MeV} = 1,000,000 \text{ eV} = 10^6 \text{ eV}$$
$$1 \text{ GeV} = 1 \text{ BeV} = 1,000,000,000 \text{ eV} = 10^9 \text{ eV}$$
$$1 \text{ GeV} \equiv 1 \text{ BeV}$$

The letter "B" stands for billion, and in North America billion means 10^9. In Europe, however, billion means 10^{12}. The letter "G" stands for the prefix giga, which has been assigned on all continents to have the meaning of 10^9.

5-1-4   POWER

In non-scientific conversation the words force, energy, and power are often used interchangeably, but in science the words have different and specific meanings. Force is that quantity which causes a mass to change its velocity. Energy is the product of force and distance. Power is the rate of doing work, of using energy, or of dissipating energy. In other words, power is work per unit time or energy dissipated per unit time. The energy is not used up or destroyed, but is converted from one form to another, perhaps in a useful way.

The units of power are, basically, any unit of work or energy divided by any unit of time. The SI unit is a *joule per second* and is called a *watt*.

A unit commonly used in biology and medicine is the kcal/day; 20.7 kcal/day is equal to one watt.

In the British engineering system, the unit of power is called the horsepower. One *horsepower* is equal to 746 watts, or very nearly three-quarters of a kilowatt.

Power is Energy/Time

Using P for power and E for energy,

$$P = E/t$$

With E in joules and t in seconds, P is in J/s, which is called watts. Energy is given by $E = Pt$. From this we see that energy used can also be measured in terms of power used multiplied by time. If P is in watts and t is in seconds, E is in joules. That is,

1 watt second equals 1 joule

1000 watts for one hour would be

$$E = 1000 \text{ watts} \times 3600 \text{ s} = 3.6 \times 10^6 \text{ J}$$

The quantity 1000 watts (1 kw) for 1 hour is given the name *kilowatt hour,* and one kwh is 3.6 million joules.

5–1–5 MASS AND ENERGY

Popularly, the name Einstein and the equation $E = mc^2$ are associated with the relation between mass and energy. E is the energy produced when the mass m is converted by some process to a more familiar form of energy; c is the speed of light (which is a constant), so c^2 is merely a proportionality constant. The equation actually states that mass and energy are two forms of the same thing and the conversion factor is c^2. The numerical value of the conversion factor depends on the units used. In English units, for example, c is 9.84×10^8 ft./s. If a stick 300,000 km long were used for a unit of length, then c would have the numerical value of one. The equation in such units would take the form $E = m$. This emphasizes that energy and mass are two forms of the same thing. Theoretical physicists frequently use units in which c has the numerical value of 1 to simplify the equations and to clarify the basic concepts.

The equation $E = mc^2$ tells how much energy in the familiar form is released when a mass m is converted.

A calculation of the energy (in joules) that is equivalent to 1 kilogram of mass gives a result that can be described only as fantastic. Using $E = mc^2$ with $m = 1$ kg and $c = 3 \times 10^8$ m/sec gives

$$E = 9 \times 10^{16} \text{ kg m}^2/\text{s}^2 = 9 \times 10^{16} \text{ joules}$$

Mass is another form of energy, and the conversion factor to joules is 1 kg $= 9 \times 10^{16}$ joules. The quantity 9×10^{16} joules is a large amount of energy. If this energy were released in 1 second. it would amount to 9×10^{16} or 90,000,000,000,000,000 watts (1 joule per second is 1 watt). Even if this energy were released in one day (86,400 seconds), it would amount to 9×10^{16} joules/8.64×10^4 seconds or about 1×10^{12} watts, a million million watts; a million megawatts for a whole day, all from the conversion of only 1 kg of matter to energy! Is it any surprise that after Einstein found the way to calculate this result, those who understood it became excited and many exerted terrific effort to achieve such a conversion?

5–2 MECHANICAL ENERGY

Energy appears in many forms, and in mechanics there are two important forms: the energy that an object has by virtue of its position in space and the energy that an object has because of its motion. These are called potential energy and kinetic energy, respectively.

5–2–1 POTENTIAL ENERGY

One form of energy of an object is due only to its position; this is a "mechanical" form of what is called *potential energy*. If there is a high platform with a massive object on it, that object can do work on something else merely because of its elevated position. This process is illustrated in the series of diagrams in Figure 5–1. In (a) the mass m is shown at the elevation h above the ground. In (b) an equal mass m is on the ground; a rope is tied to it, put up over a pulley, and tied to the elevated mass. In (c) the mass on the platform has been pushed over the edge and given a slow velocity downward. If the friction in the pulley is zero, the mass on the ground will be raised up to the level of the top of the platform as in (d). The force required to lift it at constant speed is numerically the same as its gravitational weight, mg. The work done to lift it the height h is the force times the distance moved, or mgh, and this is the same as the work that had to be done to raise the mass that was initially at the top of the tower to that position. The mass at the top of the tower was able to do work on the other mass because of its position. That is, because of its position in the gravitational field around the earth, it had energy; this form of energy is called *gravitational potential energy*. The amount of such potential energy is a function of height, which would imply that at zero height the potential energy is zero. The second mass could be raised even higher if the first were allowed to fall into a hole, as in part (e). The zero level for potential energy is arbitrary. In any specific problem, the zero level to be used for that problem must be specified. Sometimes the zero level, often ground level, will be implied. Then negative potential energy also has to be allowed. In any problem, what is usually important is the energy difference between two levels, and this does not depend on zero level.

In the previous example, if the zero level is at the level of the ground, the potential energy of the mass on top of the tower is mgh. At the bottom of the hole of depth d, as in Figure 5–1(e), the potential energy is $-mgd$. The difference in the potential energy of the mass between the bottom of the hole and the top of the tower is then $mg(h+d)$, and this is the work that the mass can do in moving between those two levels.

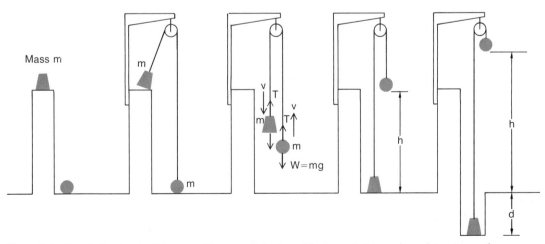

Figure 5–1 Gravitational potential energy. The mass at the top of the tower in (a) can do work to raise another mass, as in (b) and (c), to the top of the tower (d). By allowing the first mass to fall into a hole as in (e), the second mass may be raised even higher.

5-2-2 KINETIC ENERGY

Kinetic energy is energy due to motion. To set an object in motion, work must be done on it; for a moving object to stop, it must exert a force on something else (which also pushes back), and it does work on this other object. To find how the kinetic energy of an object depends on the various physical factors, the amount of work done on an object to give it a certain speed must be found. The energy it has due to its motion is the same as the work done on it to give it that motion.

Let a mass m be initially at rest, and let a force F be applied to it over a distance s, as illustrated in Figure 5-2. The object accelerates (according to $F = ma$), and at the end of the distance it has a speed v. The work done on it is Fs, which must be related to the resulting speed of the body and its mass. The acceleration will be constant since F is constant, and it has been shown that if an acceleration is constant over a distance s, the velocity acquired is given by $v^2 = 2as$. Then from Newton's second law ($F = ma$), a can be replaced by F/m to yield $v^2 = 2sF/m$. This has the quantity Fs in it, so we solve for that quantity and get $Fs = mv^2/2$. This is the work done on the body, which is also the kinetic energy it has, so the result is

$$KE = \frac{1}{2} mv^2$$

The expression $KE = \dfrac{1}{2} mv^2$ is applicable to many situations, and it will be of value later in dealing with the motion of electrons, of molecules in gases, and so forth; but how it is useful can be shown immediately in some more tangible situations.

Consider an object raised to height h, as in Figure 5-3. Its potential energy is mgh, and its kinetic energy is zero. If the object falls freely back to ground level ($PE = 0$), all the PE (mgh) is changed to KE, so $mv^2/2 = mgh$ or $v^2 = 2gh$, so that $v = \sqrt{2gh}$. This is the same expression that was derived in Section 2-3 for

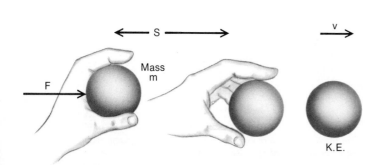

S

v

Mass
m

F

K.E.

Figure 5-2 Work being done on a mass to give it a velocity.

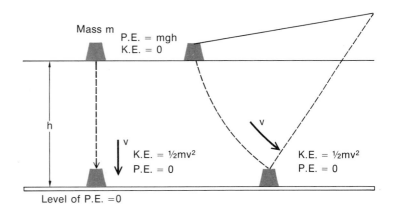

Figure 5–3 A mass at a height h has potential energy; when it falls to a lower level, that potential energy is transformed into kinetic energy. The path between the two levels does not matter, as long as there is no friction. In both cases shown, the P.E. at the higher level is transformed into K.E. at the lower level, so the speeds are the same.

the speed of an object after falling from a height h. This time the relation was derived entirely on the basis of energy considerations.

The potential energy change depends only on the difference in height between two ends of a path. If there is no friction on the object which would change some of the energy to heat, the potential energy will be all changed to kinetic energy, no matter what the shape of the path.

The final direction of motion does not matter either. When an object falls through a height h from rest it will have a speed given by $v = \sqrt{2gh}$. The path may be on a curve, as in Figure 5–3(b), or it may be along an inclined plane.

5–2–3 ROTATIONAL KINETIC ENERGY

Work has been defined as force times distance. In the case of rotation, as in Figure 5–4, a force F may move a mass along a circular path for a distance p. The angle turned through will be θ. The work is given by Fp, for everywhere F is along the direction of p.

$$\text{Work} = Fp$$

Multiply the right side by r/r to get

$$\text{Work} = (Fr)(p/r)$$
Fr is torque, L
p/r is the angle θ

The work in a rotation is then torque times angle, $L\theta$, or

$$\text{Work} = L\theta$$

If a torque L is applied to an object of moment of inertia I, which is initially at rest, the object will begin rotating; it will

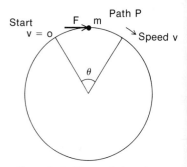

Figure 5–4 The work done in a rotation through an angle θ.

have an angular acceleration. If it turns through an angle θ while the torque is applied, the final angular velocity will be given by

$$\omega^2 = 2\alpha\theta$$

From Newton's second law for rotation, $\alpha = L/I$, so

$$\omega^2 = 2L\theta/I$$

Solving for $L\theta$, which is the work done on the object or the energy acquired by it, gives

$$\text{Work} = \text{Energy} = L\theta = I\omega^2/2$$

This is similar to the equation for linear kinetic energy, with I replacing m and ω replacing v.

> angular kinetic energy $= I\omega^2/2$
> linear kinetic energy $\;= mv^2/2$

EXAMPLE 1

Two objects move down a path having a vertical height h, from the top to bottom, one sliding and one rolling. How would the speed of the rolling object, if it is a thin hoop, compare to the speed of the sliding object? In both cases assume that none of the energy is turned to heat by friction. The situation is shown in Figure 5–5.

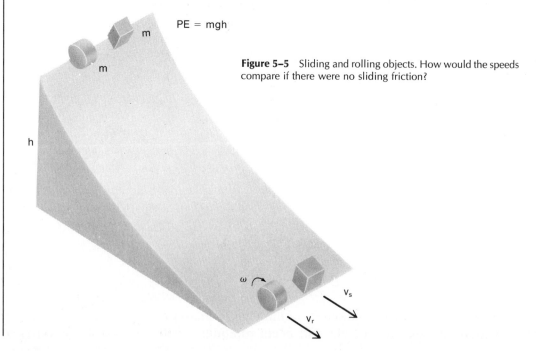

PE $=$ mgh

Figure 5–5 Sliding and rolling objects. How would the speeds compare if there were no sliding friction?

For the sliding object, the PE at the top is *mgh,* and this is changed to linear KE at the bottom. Letting v_s be the speed of the sliding object:

$$mv_s^2/2 = mgh$$

from which

$$v_s = \sqrt{2gh}$$

For the rolling object, the PE at the top is also *mgh*; but this energy goes partly into rotational kinetic energy ($I\omega^2/2$) and partly into linear kinetic energy ($mv_r^2/2$), where v_r is the linear speed of the rolling object. Then

$$mgh = mv_r^2/2 + I\omega^2/2$$

For rolling, $\omega = v_r/r$ or $\omega^2 = v_r^2/r^2$

For a hoop, $I = mr^2$

so $mgh = mv_r^2/2 + mr^2\, v_r^2/2r^2$
$$= mv_r^2/2 + mv_r^2/2 = mv_r^2$$

from which $v_r = \sqrt{gh}$

The problem was to compare the speed of the rolling object to that of the sliding object: that is,

$$\frac{v_r}{v_s} = \frac{\sqrt{gh}}{\sqrt{2gh}} = \frac{1}{\sqrt{2}} = 0.707$$

The rolling hoop will have only 0.707 of the speed of the sliding object. Surprised?

EXAMPLE 2

It is possible to store energy in a rotating wheel. The flywheel found in many machines stores energy for short intervals and is often necessary for the continued operation of the machine. Investigate the feasibility of storing enough energy in a flywheel to operate a 100 watt light bulb for one hour.

Consider that a speed of 2 revolutions per second can be used, and find the mass necessary if the wheel is in the form of a solid cylinder. The radius is to be 0.7 meter.

Solution: The energy is found by noting that 100 watts is 100 joules per second, and in 1 hour there are 3600 seconds. The energy is therefore 3.6×10^5 joules.

Rotational kinetic energy is $I\omega^2/2$, where $I = mr^2/2$ for a solid disc, so

$$\text{Energy} = E = mr^2\omega^2/4$$

Solve for *m*, $m = 4E/r^2\omega^2$

The numerical values are $r = 0.7$ m
$$\omega = 2 \text{ rev/s} = 12.6 \text{ rad/s}$$

$$E = 3.6 \times 10^5 \text{ J}$$
$$= 3.6 \times 10^5 \text{ kg m}^2/\text{s}^2$$

Then

$$m = \frac{4 \times 3.6 \times 10^5 \text{ kg m}^2}{0.7^2 m^2 \times 12.6^2 \frac{\text{rad}^2}{s^2} s^2} \quad \text{(note: rad}^2 = 1)$$

$$= 18{,}500 \text{ kg}$$

The mass would have to be 18,500 kg, 18.5 metric tonnes. If such a wheel were made of iron in the form of a solid cylinder of radius 0.7 meter, the length would have to be about 1.5 meters.

The frictional problems have been neglected.

5–3 HEAT, ENERGY, AND TEMPERATURE

When an object is heated, when energy is put into it, of what form is the energy in that material? Heat might behave like a fluid as it moves around through an object or from one object to another; but it is not a fluid at all. The energy put in is there in the form of either kinetic or potential energy. If a metal is heated, it expands. The molecules move further apart, requiring that work be done against the interatomic forces that hold the metal together. Some of the energy put into an object exists in a potential form because of this; the remainder of the heat energy put in increases atomic or molecular motion (that is, it is in the form of kinetic energy). In solids, in which the atoms have fixed positions, the energy is in the form of vibration.

In gases, the kinetic energy of the particles (atoms or molecules) is increased when heat is put in. For monatomic gases, all of the energy is in the form of kinetic energy of linear motion in the form discussed: each particle (atom or molecule) has a kinetic energy $\frac{1}{2}mv^2$. If the gas molecules consist of two atoms they may, as shown in Figure 5–6, spin or vibrate in and out. The energy put into the gas divides into three forms: kinetic energy due to translational motion in each of the three dimensions, kinetic energy due to rotation, and kinetic energy of vibration of the atoms in the molecule.

If two objects in which the atoms have different average kinetic energies are put in contact, the faster moving atoms of one bump against the atoms of the other and transfer energy to them. This process goes on until the average kinetic energy of the atoms is the same in both objects. It is known that when two objects at different temperatures are put together, heat (energy) moves from the one at a higher temperature to the one at a lower temperature until they finally reach the same temperature. This means that they reach the state at which the average kinetic

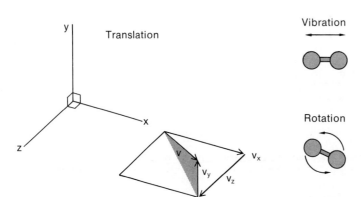

Figure 5–6 Monatomic molecules distribute their energy into motion in each of the three directions shown in (a). Diatomic molecules (b) may have energy of vibration and rotation, as well as the three directions or degrees of freedom for translational kinetic energy.

energy of the particles (atoms or molecules, as the case may be) is the same in both. So what is being detected when temperature is measured is actually the average kinetic energy of the particles. In the case of a gas it is translational KE; in a solid it is vibrational energy of the atoms. The situation is a little more complex in a liquid, but the basic idea that temperature is related to average kinetic energy of the particles still holds true. In a given sample of material, the individual particles will always occupy a wide range of energies, of course, ranging from zero to many times the average.

The word heat is usually (but not always) reserved for the energy associated with random motion of molecules. However, heat is just energy and there is actually little need for the word. In SI units, heat is measured in the same unit as any other form of energy, in joules.

5-3-1 TEMPERATURE SCALES

Temperature is one of the *base quantities* of the Système International. The unit for temperature is the kelvin, for which the symbol is K. Temperature is related to kinetic energy, and at the zero of the kelvin scale the kinetic energy of the particles of the medium (atoms or molecules) would be zero. That is called absolute zero, and at one time the kelvin scale was called the absolute temperature scale.

The kelvin scale had its origin in the temperature scale now called the Celsius scale (formerly Centigrade). On the Celsius scale the freezing point of water is zero degrees (0°) and the boiling point of water at normal atmospheric pressure is one hundred degrees (100°C). It has been established that absolute zero is at -273.15°C. The unit intervals on the kelvin scale are the same size as those on the Celsius scale. The equivalent points on the kelvin scale and the Celsius scale are:

absolute zero is 0 K $= -273.15$°C

water freezes at 273.15 K $= 0$°C

water boils at 373.15 K $= 100$°C

In terms of temperature intervals, 1 K $= 1$°C. The degree symbol or word is not used with the kelvin temperature, whereas it is used with the Celsius scale. For example, water freezes at 273.15 kelvins or 273.15 K. The freezing point is also zero degrees Celsius or 0°C.

The kelvin scale and the Celsius scale are both recognized as SI units. There have been other temperature scales, and one still encountered is the Fahrenheit scale. The fixed points on the Fahrenheit scale are the freezing point of water at 32°F and the boiling point, 212°F. The interval between freezing and boiling is 180 Fahrenheit degrees, which correspond to 100 Celsius degrees. The Fahrenheit degrees are therefore smaller than Celsius degrees, one Fahrenheit degree being 100/180 or 5/9 of a Celsius degree. To convert from Fahrenheit to Celsius, first convert to degrees above or below the freezing point (subtract 32°F from the reading), and then multiply by 5/9 to adjust for the size of the degree. For example, 68°F is 36°F above the freezing point of water; 36 Fahrenheit degrees correspond to 36 × 5/9 Celsius degrees or 20 Celsius degrees. Therefore, 68°F = 20°C.

The liquid thermometer was invented by Galileo, but it was put in the present mercury form by G. D. Fahrenheit, a Prussian scientist (1686–1736), and the scale he used was found to have the fixed points given above at 32° and 212°. The Centigrade scale with the fixed points at 0° and 100° was devised by a Swedish astronomer, Anders Celsius (1701–1744). The term "Centigrade scale" was used until 1948 when, by international agreement, it was changed to honor Celsius, its inventor. With the establishment of an absolute zero, an absolute temperature scale was devised, now called the kelvin scale in honor of the British physicist, Lord Kelvin.

5–4 SPECIFIC HEAT CAPACITY

When energy is absorbed by a material, part of the energy will contribute to further molecular motion, which is seen as an increase in temperature; and part of the energy may do work in expanding the material, which is a form of potential energy. In different materials, the fraction of the energy that contributes to a temperature change will be different. As a result, the same amount of energy put into the same mass of different materials will result in different temperature changes. Another way of expressing this is that different amounts of energy must be put into the same mass of different materials to cause the same temperature change.

For example, 4180 J put into 1 kg of water raises its temperature by 1°C, but only 880 J will raise the temperature of 1 kg of aluminum by 1°C. The quantity of energy required to raise the temperature of 1 kg of a substance by 1°C is called the *specific heat capacity* of that substance. Specific heat capacities for a few common substances are listed in Table 5–1. Reference is sometimes made to the *heat capacity of an object*. This refers to the amount of energy needed to raise the temperature of the

TABLE 5–1 Specific heat capacity and relative heat capacity for a few substances

SUBSTANCE	SPECIFIC HEAT CAPACITY J/kg °C	RELATIVE HEAT CAPACITY (with respect to water)
water	4180	1.000
ice, −10°C to 0°C	2100	0.50
mercury	138	0.033
aluminum	880	0.211
steel	447	0.107
glass	670	0.16
marble	880	0.21
wood	1600	0.4
iron	500	0.120

whole object by 1°C. If the object is made of only one substance, its heat capacity is merely the product of the object's mass in kg and the heat capacity per kg. If the object consists of several parts of different materials, the total heat capacity is the sum of the products of the mass of each material and its heat capacity.

Reference can also be made to the ratio of the heat capacity of a given substance to the heat capacity of the same mass of the usual reference material—water. That would be the *relative heat capacity*. The new term, relative heat capacity, is the same as the old term *specific heat,* which you may still sometimes encounter. Some relative heat capacities are also shown in Table 5–1.

EXAMPLE 3

If an object consists of 5.5 kg of iron and 3.3 kg of aluminum, and contains 2.0 kg of water, how much heat (energy) will raise the whole thing by 20°C? From Table 5–1 the specific heat capacities are:

for the iron   500 J/kg °C
for the aluminum 880 J/kg °C
for the water   4180 J/kg °C

For a 20°C temperature rise, the iron requires

$$500 \ \frac{J}{kg \ °C} \times 20°C \times 5.5 \ kg = 55{,}000 \ J$$

The aluminum requires

$$880 \ \frac{J}{kg \ °C} \times 20°C \times 3.3 \ kg = 58{,}000 \ J$$

The water requires

$$4180 \ \frac{J}{kg \ °C} \times 20°C \times 2.0 \ kg = 167{,}000 \ J$$

The total is 280,000 joules.

EXAMPLE 4

If a piece of lead is moving at 330 m/s and if, on being brought to rest, the kinetic energy is all changed to heat that is all absorbed by the lead, by how much would the temperature rise? The specific heat capacity of lead is 126 J/kg °C.

The mass is not given; let it be m. The kinetic energy is

$$mv^2/2 = \frac{m \times 330^2 \text{ m}^2}{2 \text{ s}^2}$$

$$= 54{,}450 \; m \text{ m}^2/\text{s}^2$$

The energy will have to be left in terms of m.

Let the temperature rise be $T_2 - T_1$, the energy E, and the mass m. For lead

$$E = \frac{126 \text{ J}}{\text{kg °C}} \, m \, (T_2 - T_1)$$

Solving for $(T_2 - T_1)$,

$$T_2 - T_1 = \frac{E \text{ kg °C}}{126 \, m \text{ J}}$$

Insert

$$E = 54{,}450 \; m \text{ m}^2/\text{s}^2$$

and write kg m²/s² for the unit J. Then

$$T_2 - T_1 = \frac{54{,}450}{126 \, m} \frac{m \text{ m}^2 \text{ kg °C}}{\text{kg} \dfrac{\text{m}^2}{\text{s}^2} \text{s}^2}$$

The mass m cancels, as do all units but °C. Then

$$T_2 - T_1 = 432 \text{°C}$$

Note that it was not necessary to know the mass m.

Also, lead melts at 327 °C. The speed chosen was typical for a bullet. You will find that a bullet fired into something like a block of wood will be at least partially melted.

5–5 MEASUREMENT OF ENERGY— CALORIMETRY

One of the fundamental ways of measuring energy is to allow that energy to be absorbed by an object for which the heat capacity is known, and then to find the temperature change. The object that absorbs the energy (called a calorimeter) must be constructed so that no heat comes in from, or is lost to, the surroundings during the measurement. One type of calorimeter is the Dewar flask, which is very similar to a Thermos bottle. It is used with a liquid, usually water, in it. The mass of the liquid and its heat capacity are measured precisely. Then a measurement of the temperature rise when energy is added will allow calculation of that energy. A small correction will allow for the energy absorbed by the inner wall of the container, and even the small amount of energy lost to the surrounding room can be corrected for. A simple calorimeter is illustrated in Figure 5–7.

Consider, for example, that you have an object and in some project that you are doing you need to know its heat capacity, or the energy that would change its temperature by 1°C. You may be able to measure the masses of the various parts; knowing the specific heat of each, the heat capacity can be calculated. But this is not always possible. Another way to find the heat capacity would be to heat the object, say in hot water, and then put it into water in a cold calorimeter. The heat energy will go from the hot object to the cold calorimeter until the whole thing reaches the same temperature. The final temperature is measured. The energy that went into the calorimeter to raise it to that temperature can be calculated. This is the same as the energy that left the object as its temperature fell from its initial to its final value. The energy involved in that temperature change is then known, and the energy per degree, or heat capacity, is calculated.

In some measurements, even in precise calorimetry, the thermal capacity of a thermometer must be known. For example, you may have a small amount of warm liquid and you want to measure its temperature. You put a cold thermometer into it and read the temperature. But the thermometer absorbed some heat and cooled the object. You read the final temperature, but not the temperature that you wanted. If you knew the heat absorbed per degree by the thermometer, perhaps you could calculate how much it cooled the object.

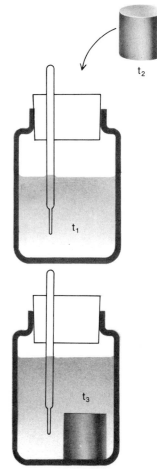

Figure 5–7 A simple calorimeter. A hot object put into the cool water in the calorimeter raises the water's temperature; from the temperature rise and the known mass of water, the energy that went into it can be found.

EXAMPLE 5

Find the heat capacity of a thermometer in the following way and with the following measurements.

A small calorimeter containing 12 g of water is in equilibrium at room temperature, 22.3°C. The thermometer is then put in boiling water, and reaches 98.9°C. (It is not at sea level, so the boiling point is not 100°C). The thermometer is then quickly transferred to the small calorimeter, and the final temperature reached is 22.9°C. The temperature rise was 0.6°C.

The heat energy gained by the 12 g of water in the calorimeter is $0.012 \text{ kg} \times 4180 \frac{J}{kg} \times 0.6°C \doteq 30$ joules.

The thermometer lost 30 J in falling from 98.9°C to 22.9°C, that is, by 76.0°C. The heat capacity is 30 J/76.0°C = 0.39 J/°C.

The temperature change was measured to only one figure accuracy, so the answer should be rounded off to one figure. It could be called 0.39 J/°C or 0.4 J/°C. Rather than argue about which it should be, you could accept either or choose between them with more precise measurement.

5–6   LATENT HEAT

When a liquid is vaporized, energy is required to move the molecules away from each other, to work against the inter-molecular forces. To vaporize a liquid requires heat without any change in temperature. This is called *latent heat*. Associ-ated with any change of state—from liquid to vapor, from solid to liquid, or from one crystalline form to another—there is an associated latent heat. The specific latent heat is the energy per kilogram required for a change of state.

When water in a container reaches the boiling point and more heat is put in, the temperature does not rise but the water is slowly evaporated. Each kilogram of water vaporized re-quires 2.26 million joules, and this quantity is referred to as the *specific latent heat of vaporization of water*. When water vapor condenses, a corresponding amount of energy, 2.26 million joules per kilogram, must be given to the surroundings. The specific latent heat of vaporization of water can also be referred to as 2,260 kJ/kg at the boiling point.

A similar process occurs at the freezing or melting point. To change 1 kg of ice at 0 °C to 1 kg of water, still at 0 °C requires 334,000 joules or 334 kJ. To change 1 kg of water at 0 °C to 1 kg of ice at 0 °C, 334 kJ must be removed. The *specific latent heat of liquefaction* (or of *fusion*) of water is 334 kJ/kg.

The latent heats play an important part in the weather on earth. Large bodies of water not only have a large capacity for heat storage; as they vaporize during hot weather, there is a cooling effect on the surroundings. In cool weather, freezing is slowed because of the necessity to remove the latent heat of fusion. Both processes have a stabilizing effect on the tempera-ture in the surrounding area. Also, when water vapor con-denses to form a mist, the latent heat goes to the surroundings and a slight warming effect can be noted. Similarly, the latent heat of fusion is released when snow flakes are formed, and a noticeable warming occurs just before snow falls on an other-wise cold winter day.

The large amount of energy released when steam con-denses also explains why burns caused by steam are often much more severe than burns caused by even boiling water.

The human body continually produces heat; even in the resting state the average person produces heat at a rate of about 90 J/s. When you are using energy to do work, even more heat is produced. This heat must somehow be carried to the environ-ment. If the surroundings are cool there is no problem in getting rid of the heat, but rather the contrary; it is difficult not to pass it to the surroundings too fast. If the surroundings are warm, perhaps there can be no heat transfer to them; yet the body heat produced must be dissipated. This can be done by vaporization of water. Just 2¼ grams of water turned to vapor per minute would be enough to take away the total body heat. Water can vaporize during the process of perspiration; also, with every breath vapor is transferred to the air in the lungs and exhaled. This exhaled vapor carries away some of the body heat. In dogs, which do not have glands for sweating and often have too thick a coat to allow sufficient heat transfer to the surroundings, the principal method of heat loss is through breathing. The dog pants in order to increase the amount of exhaled vapor and hence the rate of heat loss.

5-7 HEAT TRANSFER

Heat (energy) moves about, causing cooling of hot objects and consequently warming of cool ones. Energy does not spontaneously move from any object to another at higher temperature. This is what happens in a refrigerator, but work must be done to accomplish it. In this section we deal only with the natural movement of energy, from any object only to one at a lower temperature.

Heat moves in three ways, commonly referred to as *convection, conduction,* and *radiation.* A warm object in a room cools by all three processes. The general cooling effect can be handled in a manner called thought analysis—with the results compared to what really happens—or the three separate cooling processes can be analyzed separately. The results of the three together should correspond with the experimental result.

The analysis of the cooling of an object was carried out by Sir Isaac Newton, and the results, valid for only small temperature differences, are described by what is now known as *Newton's law of cooling.*

5-7-1 NEWTON'S LAW OF COOLING

It is common experience that the greater the temperature difference between an object and its surroundings, the more quickly will there be a transfer of heat. An object 20°C above room temperature will cool more quickly than one which is only 10°, 5°, or 1° above room temperature. Newton noted that the rate of cooling is directly proportional to the temperature difference. That is, if an object cooled at the rate of 2°C/min when it was 20°C above room temperature, at 10° above room temperature it would cool at only 1°C/min, and at 5° above room temperature the cooling rate would drop to 0.5°C/min.

The cooling per unit time would be a constant fraction of the excess temperature: in the example above, $0.1 \times$ temperature excess per minute.

If the temperature excess is x, then a change Δx occurs in a time interval Δt. The rate of cooling is described by

$$\frac{\Delta x}{\Delta t} = -kx$$

The negative sign is used because for cooling Δx is a negative temperature change. The quantity k is called a cooling constant, and k will be different for each different object.

If the temperature outdoors is $10\,°C$ below indoor temperature, your house will lose heat at a certain rate; in order to keep the indoor temperature constant, heat will have to be supplied at the same rate. When the outdoor temperature is $20\,°C$ below that indoors, the cooling rate would be doubled; to prevent a drop in indoor temperature, the rate of burning of fuel in the house would also have to be doubled.

These observations do not contradict experience in a general way, and yet Newton's law of cooling stands up to a numerical test only in very restrictive situations. Ordinarily it is good only for a rough guide. One of the restrictions is that temperature differences must be small; in fact, in practice the law is limited to only a $10\,°C$ or perhaps a $20\,°C$ temperature difference, and also it varies in degree of applicability from one situation to the next. Newton's law of cooling must be used with care!

EXAMPLE 6

Consider that Newton's law of cooling applies to your bowl of soup. It is $20\,°C$ above room temperature, and you note that in 1 minute it cools to $19\,°C$ above room temperature. How long will it take to reach $10\,°C$ above room temperature?

The data for this problem can be worked out in steps, and it will be convenient to arrange them in tabular form. In each one minute interval the temperature drops by 1/20 of the excess. The cooling constant k of Newton's law is 1/20 or 0.05 per minute. A large number of significant figures are used so the error does not accumulate excessively.

The time required for the soup to cool to $10\,°C$ above room temperature will be just over 12 minutes.

This type of problem raises several questions.

How long would it take to drop to room temperature? Would it ever get to room temperature?

Is there not a quicker way to do such a problem? A computer could be used, but also there is a theoretical way to handle problems in which the rate of change of a quantity is proportional to the size of the quantity. This type of situation is actually quite common; it will be met again and it will soon be necessary to learn how to analyze it theoretically.

Time, minutes	Temperature excess, °C	Loss of temperature in 1 min. = 0.05 × excess	Temperature after 1 min.
0	20	1	19
1	19	0.95	18.05
2	18.05	0.9025	17.1475
3	17.1475	0.8574	16.2901
4	16.2901	0.8145	15.4756
5	15.4756	0.7738	14.7018
6	14.7018	0.7351	13.9667
7	13.9667	0.6983	13.2684
8	13.2684	0.6634	12.6050
9	12.6050	0.6303	11.9748
10	11.9748	0.5987	11.3761
11	11.3761	0.5688	10.8073
12	10.8073	0.5404	10.2669
13	10.2669	0.5133	9.7536

A more precise description of cooling will result from a study of the three main processes involved, convection, conduction, and radiation.

5–7–2 HEAT TRANSFER – CONVECTION

Convection currents are actual movements of fluid material, thereby transferring heat from one place to another. The currents are due to uneven heating that causes local changes in density. There is the old saying that "hot air rises," but in general the warmer, less dense material, be it air, water, or whatever, will rise.

A warm object in air passes heat to the air close to it. This air expands, and its density decreases. It now is lighter than the surrounding air, so it rises and is replaced by cooler air, which in turn absorbs heat and then carries it away (Figure 5–8). The rate of heat transfer from an object resulting from its setting up convection currents in the air or liquid around it is a difficult

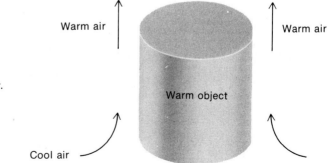

Figure 5–8 Heat transfer by convection in air.

Warm air

Warm air

Warm object

Cool air

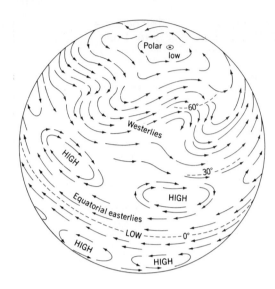

Figure 5–9 Global convection currents in the atmosphere. (From A. N. Strahler, *Planet Earth: Its Physical Systems Through Geologic Time,* Harper & Row, New York, 1972.)

problem to deal with in a quantitative way. It can, however, be accepted that the greater the temperature difference, the greater the heat transfer rate by convection. With natural convection the cooling rate is found to be close to the 1.25 power of the temperature difference; with forced convection it is closer to the first power of the temperature difference. For small temperature differences the heat transfer rate by convection is roughly proportional to the temperature difference.

Heat transfer by convection may be difficult to deal with theoretically, but that does not diminish its importance in natural phenomena. Global weather patterns are in part due to convection currents in the atmosphere (Figure 5–9). On a smaller scale, local upward convection currents are sought by sailplane pilots.

The drifting of the continents has been established, but the driving mechanism is still being investigated. Convection currents in the mantle have been suggested. The mantle is not liquid, but is material beyond its elastic limit; the rate of flow is measured in centimeters per year. Figure 5–10 illustrates two

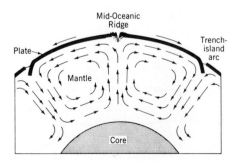

Figure 5–10 Possible patterns of convection currents in the mantle of the earth, which could be the cause of continental drift. (From A. N. Strahler, *Planet Earth: Its Physical Systems Through Geologic Time,* Harper & Row, New York, 1972.)

types of convection currents that could be responsible. These currents could also be responsible for heat transfer to the crust of the earth from the deep interior.

5–7–3 HEAT TRANSFER – CONDUCTION

The second method of heat transfer is the passage of energy from one molecule to another by direct contact. This is conduction. It is the method by which heat moves through a metal rod, through a windowpane or a wall, through a coat or a layer of fur, and through the crust of the earth from the mantle below.

Many situations involving conduction can be analyzed quite simply in a quantitative way. Doing this will result in a better understanding of some natural phenomena.

Consider a slab of material of thickness x and cross sectional area A, as in Figure 5–11. One surface is maintained at a temperature T_1 and the other surface at a higher temperature T_2. Energy moves by conduction from the hot side to the cool side. The rate of energy transfer is found to vary directly as the temperature difference $T_2 - T_1$ (that is, it is directly proportional to $T_2 - T_1$). Also, the greater the area, the greater the amount of energy carried through; again, it is a direct proportion. The thicker the object, the slower the heat transfer; the rate at which energy flows through is inversely proportional to x. In symbols, if E is energy in calories or in joules, and t is time so that E/t is a rate of energy transfer, then

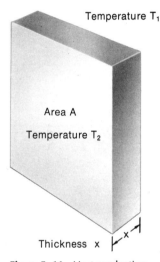

Figure 5–11 Heat conduction through a slab of material.

$$E/T \propto T_2 - T_1$$

$$\propto A$$

$$\propto 1/x$$

Combining these:

$$E/t \propto \frac{A(T_2 - T_1)}{x}$$

Making one measurement of all the quantities in a given situation would give a value on each side, and they could be made equal by multiplying one side by a constant, called a proportionality constant. Writing the equation with the constant k,

$$E/t = kA\frac{(T_2 - T_1)}{x}$$

In this case the constant k is different for different materials; it is called the *thermal conductivity* of the material. E is the amount of energy passing through the material in a time t, the cross sectional area is A, and thickness is x. The temperature on one side is T_1 and that on the other side is T_2; k is called the thermal conductivity of the material. If k is high for a given material, it conducts energy at a higher rate than another material having a lower value of k. Some values of the thermal conductivity k for a few materials are shown in Table 5–2. In general, the metals are good conductors. Those materials with low values of k are good insulators against heat transfer.

The units for k in the SI system are found by solving the conductivity equation for k and substituting the units for each quantity. In words, k will be in joules per second per square meter of surface per degree Celsius temperature drop per meter of thickness. Joules per second are watts, and one meter unit cancels to leave the result that k is in watts/m °C.

Insulation material used in the construction of homes and other buildings is rated according to its resistance to heat flow. The resistance value (R value) is related inversely to the thermal conductivity.

The R values quoted in the construction industry are based on the thermal conductivity in units of British thermal units per hour per square foot per degree Fahrenheit per inch of thickness. That is not an SI unit. The value of k in SI units is related to the R value used in the industry by $k = 0.144/R$.

TABLE 5–2   Thermal conductivities of a few materials

MATERIAL	THERMAL CONDUCTIVITY, k W/m °C
silver	421
copper	384
gold	293
aluminum	201
platinum	69
wrought iron	60
lead	35
rocks	2 to 4
ice	2
glass	1.0
water	0.63
woods	0.2 to 0.05
cork	0.050
glass wool	0.040

EXAMPLE 7

Find the rate of heat conduction through a pane of glass 3 mm thick and 1.2 m by 0.6 m in area. The inside temperature is 20°C and the outside temperature is −18°C. From Table 5–2, the thermal conductivity of glass is 1.0 W/m °C.

Use
$$\frac{E}{t} = kA \frac{(T_2 - T_1)}{x}$$

where
$$(T_2 - T_1) = 38°C$$
$$x = 0.003 \text{ m}$$
$$A = 1.2 \text{ m} \times 0.6 \text{ m} = 0.72 \text{ m}^2$$
$$k = 1.0 \text{ W/m °C}$$

Then
$$\frac{E}{t} = \frac{1.0 \text{ W}}{\text{m °C}} \frac{(0.72 \text{ m}^2)(38°C)}{0.003 \text{ m}}$$
$$= 9100 \text{ W}$$

The heat will be going out the window at 9100 watts. In these days of energy shortage a double or even triple window with an insulating air space should be used.

Whenever there is a temperature difference between two points, there must be a flow of heat from the high temperature to the low temperature. As one goes down into the crust of the earth, as in a mine for example, the temperature rises. This indicates that heat is flowing outward from the interior of the earth. The earth seems to be cooling.

EXAMPLE 8

As one goes into a deep mine, the temperature rises with depth. There is a temperature gradient of about 3°C per 100 m of depth. This indicates an outward heat flow, so the earth is cooling. From this measured temperature gradient and the measured thermal conductivity of rock, the rate of heat loss from the earth can be calculated.

The total mass of the earth is known. Estimating a specific thermal capacity and an initial temperature for the molten material of the earth in the past, the total energy that has been lost can be calculated.

For example, melted rocks and metals would perhaps have been 2000°C hotter at some time in the past. On the basis of the present measured heat loss, the time that the earth has been cooling can be calculated. The time will be a rough estimate, because when it was hotter it would have cooled faster. The cooling time calculated will therefore be longer than the answer that a more complicated calculation would give, but it is instructive in telling about the history of the earth.

Measured values are that temperature increases at 3°C per 100 m and the thermal conductivity of rock is about 2 W/m °C. The heat loss is then given by

$$E/t = kA \frac{(T_2 - T_1)}{x}$$

where
$$k = 2 \text{ W/m °C}$$
$$T_2 - T_1 = 3°C$$
$$x = 100 \text{ m}$$

$$E/t = 2 \frac{W \times 3°C}{(m\ °C)(100\ m)}\ A$$

$$= 0.06 \frac{W}{m^2} \times A$$

The rate of heat loss from the earth is therefore 0.06 watt per square meter.

The total area of the earth, a sphere 6.38×10^7 m in radius, is given by $A = 4\pi r^2$. From this, the area is 5×10^{16} m². Using this value, the rate of energy loss from the whole earth is

$$E/t = 0.06 \frac{W}{m^2} \times 5 \times 10^{16}\ m^2$$

$$= 3 \times 10^{15}\ W \quad or \quad 3 \times 10^{15}\ J/s$$

That is, from the measured temperature gradient and thermal conductivity, the earth must be losing energy to space at 3×10^{15} J/s. In one year this is

$$3 \times 10^{15} \frac{J}{s} \times 86,400 \frac{s}{day} \times 365 \frac{days}{year} = 9.5 \times 10^{22} \frac{J}{y}$$

The mass of the earth is 6.0×10^{24} kg. The specific heat capacity of the material may be estimated to be about 1000 J/kg °C. To cool such a mass by 2000°C, the heat loss would have to be:

$$E = 1000 \frac{J}{kg\ °C} \times 6.0 \times 10^{24}\ kg \times 2000°C$$

$$= 1.2 \times 10^{31}\ J$$

The earth is losing heat at the rate of 9.5×10^{22} J/y, and to lose 1.2×10^{31} J would require a time given by

$$\frac{1.2 \times 10^{31}\ J}{9.5 \times 10^{22}\ J/y} = 1.3 \times 10^8\ y$$

Our estimate is that the earth would have cooled for about 100 million years. Such an estimate is much too long, for when it was hotter it would have cooled much faster. The total heat loss from a state of hot molten rock to the present state would more likely have occurred in no more than 15 or 20 million years.

A calculation such as this, but in more detail, was carried out by Lord Kelvin toward the end of the nineteenth century and was thought to give an upper limit to the possible age of the earth. Geological evidence such as the deposition of ocean sediments showed a much greater age, as much as 4000 million years. These data, the age calculated by physicists from cooling rate and that estimated by geologists from other evidence, differed by so much that the scientists in the two areas argued and began to mistrust each other. Only recently, after three quarters of a century, is trust again being established between the two groups.

The short cooling time did not take into account the fact that there is a source of heat in the earth. This source is the naturally radioactive materials, the principal ones being potassium, thorium, and uranium (with their assorted products). At the time the calculation was originally carried out, radioactivity was unknown. There was no imagined energy source in the earth, so theoretical physicists could not accept the long life claimed by the geologists. The age shown by geologists has told of the amount of heat production and hence the amount of radioactive material in the earth.

There is even debate about whether the earth was formed in a molten state or formed as a cold body and then melted by the energy given off by radioactive materials.

5–7–4   HEAT TRANSFER: RADIATION

Hot objects radiate heat to their surroundings in a form that is of the same basic nature as light: it is an electromagnetic radiation. The wavelength of the radiation associated with a hot object depends on its temperature. As an object such as a piece of metal is slowly heated, it begins to re-emit heat; it begins to glow red, and then orange; and perhaps it finally becomes white hot, giving off all colors of the spectrum. The sun, with a surface temperature of about 6000 K, gives its peak radiation in the yellow-green at about 550 nm or 0.55 μm. The wavelengths emitted range on both sides of that value, not only in the visible (0.4 μm to 0.7 μm) but also into the infrared and ultraviolet. An object that is not hot enough to glow red will emit in the infrared. The human body emits principally in the wavelength range from 4 to 20 μm, which is well into the infrared region.

An object isolated in outer space will radiate energy to its surroundings as long as it is not at absolute zero. The rate at which energy is radiated away depends on the temperature, though it is not a direct relation. In fact, it is found that if there are two objects, one twice as hot as the other, the hotter one will radiate at 2^4 or 16 times the rate of the cooler one. The rate of energy loss by radiation is found to increase as the fourth power of the kelvin temperature: this is known as the *Stefan-Boltzmann law*.

If an object is not isolated, it can absorb radiation that is being radiated by other bodies. On earth, an object at a kelvin surface temperature T_2 radiates to its surroundings at a rate proportional to T_2^4. If the environment is at a uniform temperature T_1, that environment radiates to the object at a rate proportional to T_1^4. The net radiation loss is proportional to $T_2^4 - T_1^4$.

If an animal has a skin temperature T_2 in an environment at a temperature T_1, it loses heat to the surroundings at a rate that depends on the difference between the fourth powers of the kelvin temperatures.

The heat loss also depends on the radiating area and is directly proportional to it. The rate of heat loss from an area A is described by

$$E/t \propto A\,(T_2^4 - T_1^4)$$

A proportionality constant could be inserted to make the relation into an equation, but the constant would depend on the surface material. For what is called a perfect radiator (which would also be a perfect absorber), the constant is called σ, known as the Stefan-Boltzmann constant. Its value is 5.70×10^{-8} W/m^2 K^4. For a real surface the radiation efficiency is a fraction f of the ideal value, and the heat loss by radiation from that surface is described by

$$E/t = f\sigma A\,(T_2^4 - T_1^4)$$

where f is an efficiency factor

σ is the Stefan-Boltzmann constant

T_2 is the kelvin temperature of the surface of the body

T_1 is the kelvin temperature of the surroundings.

For a perfect radiator, $f = 1$; for any other surface, f is less than one. Because good radiators are good absorbers, and because the best absorber is one that reflects no radiation and in visible light would appear black, a perfect radiator is referred to as a *black body*. The Stefan-Boltzmann law for a black body is written with $f = 1$.

The Stefan-Boltzmann law in which the difference between the fourth powers of the temperatures occurs does not seem to be consistent with Newton's law of cooling. It can be shown, though, that for small temperature differences they are compatible. The analysis will also show why Newton's law of cooling is valid only for small temperature differences.

The Stefan-Boltzmann radiation law can be put in a different form if the temperature of the radiating body is only slightly above that of the surroundings. Let the temperature excess of the body be ΔT, so $T_2 = T_1 + \Delta T$, or $T_2 = T_1[1 + (\Delta T/T_1)]$. By a "small temperature excess" we meant that $\Delta T/T_1$ is much less than 1, preferably well below 0.1. Then the binomial expansion can be used, and

$$T_2^4 = T_1^4[1 + (\Delta T/T_1)]^4$$

$$= T_1^4 \left(1 + 4\frac{\Delta T}{T_1} + \text{terms that will be small} \right)$$

$$\approx T_1^4 + 4\, T_1^3 \Delta T$$

The Stefan-Boltzmann law then is

$$E/t = f\sigma A \, (T_2^4 - T_1^4)$$
$$\approx -f\sigma A \ 4T_1^3 \Delta T$$

The rate of heat transport from the surface is proportional to the temperature excess, ΔT, which is also $T_2 - T_1$.

If the temperature of the radiating body is just a small amount ΔT above the surroundings, which are at an absolute temperature T_1, the Stefan-Boltzmann radiation law takes the form

$$E/t = -f\sigma A \ 4T_1^3 (\Delta T)$$

The quantities f, σ, A, and T_1^3 are all constants, so the energy loss by radiation is

$$E/t = -k\Delta T$$

That is, the energy loss is proportional to the temperature difference between the object and its surroundings, as in Newton's law of cooling. Remember, though, that an approximation was used, so this formula is valid only for small values of ΔT.

EXAMPLE 9

At the position of the earth in space, the radiation from the sun amounts to about 1300 watts/m². The earth also radiates energy into space. The whole problem is very complex, but we will make a calculation to see how that figure compares with the heat radiated into space from each square meter of the earth at 20°C or 293 K.

Use
$$\frac{E}{t} = f\sigma A \, (T_2^4 - T_1^4)$$

where
$$f = 1$$
$$\sigma = 5.7 \times 10^{-8} \text{ W/m}^2 \text{ K}^4$$
$$A = 1 \text{ m}^2$$
$$T_2 = 293 \text{ K}$$
$$T_1 = 0 \text{ K}$$

Then
$$\frac{E}{t} = 5.7 \times 10^{-8} \, \frac{\text{W } 293^4 \text{ K}^4}{\text{m}^2 \text{ K}^4} \, 1 \text{ m}^2$$

$$= 420 \text{ W}$$

This is not what is received from the sun, but consider that the whole surface of the earth, both the night and day sides, is radiating to space. On that basis, with twice as much area radiating as receiving energy, the amount radiated per unit area would be expected to be only half as much as is received per unit area on the sunlit side. Also, much of the sunlight is merely reflected back into space without being absorbed. So we might say that there is approximate agreement.

5–8 ENERGY SOURCES

Our technological society has grown by using large amounts of energy per person. As technology spreads and population also increases, the problem of providing enough energy becomes more and more acute. The fossil fuels—oil, coal, and natural gas—have been the basis for development until now, but these are becoming scarce. Our children's children will not thank us for using them up. It is now that alternate energy sources must be developed.

Many new energy sources have been suggested. Some of these sources are renewable; that is, they are not "used up," leaving none for the future. These include wood, solar energy, wind, waves, tides, and geothermal sources. Almost in the same category are the nuclear sources, which are of two types: the fission reactors fueled with uranium, thorium, or plutonium, and the fusion reactors based on combining light elements, mostly hydrogen and perhaps lithium. We don't know yet what a fusion power plant will use, for no one has yet managed to make one.

There are many aspects to consider in planning future effort in development. These include availability, safety, pollution, and above all whether or not the source under consideration could provide enough energy for the needs. To use a ridiculous example, chicken droppings can be used to make methane gas and that can be used for cooking. One needn't even resort to numerical analysis to expect that this source would not provide enough energy for a city. But would it be worth looking at for use in a restricted situation?

The whole situation is complex enough to warrant a book itself, and the data of today must be extrapolated into the future, an unreliable but necessary process. In these situations arguments not based on numerical data are of little value, so let us look just a bit at some of the sources in a very definite way. This will be on the basis of large sources of energy for cities and for industries, and smaller sources for individual homes. For large plants, a typical output of 1 GW (gigawatt, a thousand megawatts), or 10^9 watts will be used for calculation.

5-8-1   SOLAR ENERGY

Solar energy is one of the sources commonly suggested. At the position of the earth in space, about 1300 watts per square meter come from the sun. Much of this is absorbed by the atmosphere; when the sun is directly overhead, about 700 watts per square meter reach the ground. This figure is reduced when the sun's rays strike at an angle. The sun does not always shine directly overhead; in fact, the average per day in the U.S.A. is 230 watts per square meter. Using this optimistic figure (which assumes that clouds do not block the radiation), a 1 gigawatt plant would require the collection of the solar power from almost 10^7 square meters. That would require solar panels covering an area 3 km on a side. Considering that the numbers used are extremely optimistic, the required area for a practical plant would probably be in the range of 5 to 10 km on a side. The capital cost of even one such plant would be prohibitive, even if it could be constructed.

The numerical analysis shows that for large power plants solar energy is probably not going to be the source. It is estimated that, even with a lot of effort in that direction until the year 2000, only about 1 to 5 per cent of our energy needs could be obtained from solar power.

The use of solar energy to provide part of the heating in individual homes is another matter, and that is where the effort in development of solar power should be directed.

5-8-2   WIND, WAVES, TIDES, and GEOTHERMAL ENERGY

Rather than dealing with these sources in detail, we will note that when numerical analysis is applied it seems that together they could provide only about 6 to 8 per cent of our needs in 25 years. These data are based on an analysis carried out by Dr. J. K. Dawson and R. A. Bird of Harwell and presented at the Combustion Engineering Association Conference on "Energy for Tomorrow's World," Eastbourne, November 26, 1975. It must be pointed out that such figures, being projections, are open to question, but that they cannot be questioned on their own. The questions must be directed at the sources of those results: the projected needs, the efficiency assumed for various systems, the data on wind, wave, and geothermal energy, and so forth.

The low percentage of the contribution suggested for these sources is not to be taken as insignificant. Also, they may turn out to be better (or worse) than presently expected.

5-8-3 NUCLEAR FUSION

The nuclear fusion power plant, producing energy by the conversion of hydrogen to helium as is done in the sun, is still in the future. The experimental installations at present are just that, experimental and none has yet produced significant power. One of the problems is that the material must be at a temperature of almost 100,000,000 K to sustain the reaction. This is not easy to obtain and maintain. The surface of the sun is at only 6000 K. The temperature in the interior of the sun is about 40,000,000 to 100,000,000 K, and it is this that must be duplicated. The required temperatures have been obtained for short intervals and the outlook is promising, although the date of achievement of a power source based on nuclear fusion is unpredictable.

The reason for the effort is that a small amount of hydrogen theoretically will give a lot of energy. One metric tonne of hydrogen converted to helium would release the same energy as 10,000,000,000 tonnes of oil. Even with an efficiency as low as 0.01 per cent, a tonne of hydrogen would be equivalent to a million tonnes of oil.

5-8-4 NUCLEAR FISSION

The present-day nuclear power plants are based on the fission of uranium. In a nuclear reactor, uranium atoms split when they are hit by neutrons. The result is two "fission fragments" and two to four more neutrons, which can split other uranium atoms. With each such event 200 million electron volts of energy are released. This is a lot, and it means that the energy from 1 tonne of uranium is about the same as that available from 10,000,000 tonnes of oil. Nuclear fission power plants have been used for decades, and many are producing power at a lower cost than fossil fuel plants. It is also estimated that for each year of operation of a 2 gigawatt reactor, three million tonnes of oil are saved.

One of the problems is that the fission products are dangerously radioactive and they must be stored for a long time, perhaps for more than a thousand years. Many of these products are also useful in medical diagnosis, research, and therapy. The radioactive material itself is dangerous to handle, and because of this the nuclear plants are built to high standards. The data regarding the causes of deaths of workers at nuclear installations are therefore interesting. The nuclear power program in Britain

TABLE 5–3 Causes of deaths of employees associated with the British nuclear power program*

CAUSE	NUMBER OF DEATHS COMPARED TO THOSE EXPECTED ON THE BASIS OF THE NATIONAL AVERAGE
All neoplasms	0.75
leukemia	0.48
lymphatic system other than leukemia	0.76
carcinoma of bone	0.95
carcinoma of lung	0.72
Circulatory system	0.91
ischemic heart disease	0.90
cerebrovascular disease	0.71
Respiratory system	0.40
Digestive system	0.72
Genitourinary system	0.55
Accidents, violence	0.78
Road traffic	1.04

*Based on data presented by Sir John Hill, chairman of the UKAEA, in a lecture to the British Nuclear Energy Society, Nov. 6, 1975.

has been operating since 1957, and a survey of the causes of deaths of radiation workers from 1962 to 1974 was made in 1975. A summary of the results given in terms of the deaths from similar causes for the whole country is given in Table 5–3.

It is interesting to note that even though leukemia is a disease often associated with radiation, the workers in the nuclear power program have less than half the national average incidence of this disease. Diseases of the respiratory system are only 40 per cent of the national average. In fact, deaths from all causes but road traffic were less than the national average. These data are probably surprising to everyone. The excellent health record is probably due to the high standards of cleanliness and safety associated with a nuclear reactor. Such standards would have to be continued, but they are maintained because they are necessary on technical grounds. It could even be remarked that such standards are not applied in general industry on health grounds.

These data and remarks are presented here to emphasize that discussions such as those about the safety of workers in the nuclear reactor power program must be based on numerical data—not on the phrase "I think that."

Not all aspects of a nuclear power program are discussed here, but enough to show on what data the arguments pro and con must be based.

DISCUSSION PROBLEMS

5-1-2 1. Make a list of the various units for energy. Include those in this text and those from other books if you are able. Comment on the value of the SI units.

5-1-4 2. (a) Show that the power P used to keep a vehicle in motion at a constant speed v against a resistance force F is given by

$$P = Fv$$

Road friction and air resistance are included in F.

(b) If a vehicle has an engine with a given maximum power, what determines the maximum speed? What would have to be changed to increase the maximum speed if the power is limited?

5-2-2 3. Does the kinetic energy of an object depend on the reference frame in which it is measured? Just as there is no absolute value for potential energy (a reference level or position must be given), is there also no absolute kinetic energy? Discuss this with respect to energy observed in moving vehicles, even on the moving earth.

5-2-3 4. If you were given two spheres of the same diameter and the same mass, how could you tell whether or not one was hollow? The spheres could be made of materials differing in density.

5-8-1 5. Heat can be stored in a large mass for use when needed. For example, solar heat can be stored in water during the day and used to heat a home at night. Discuss the advantages of a material that melts in the temperature range available.

PROBLEMS

5-1-1 1. How much work is done when a 60 kg mass is raised 3.0 m at constant speed by an upward force?

5-1-1 2. A cart is pushed along a track for a distance of 50 m by a force directed 30° away from the direction of the motion on the track. The force is 25 N.

(a) Find the component of the force along the direction of the track and the component perpendicular to it.

(b) How much work is done in moving the cart?

5-1-2 3. A human requires about 1850 Calories (kilocalories) per day to maintain the basic body processes. Convert this to joules per day and then to the number of joules used per second.

5-1-2 4. How much energy in joules, in calories, and in Calories will raise 25 kg of water from 20°C to 35°C?

5-1-2 5. (a) If 90 joules are put into a mass of 70 kg of water each second for a day, how many kilocalories are put into the 70 kg per day?

(b) How many kcal go into each kg?

(c) If none of the energy is lost from the water, what temperature rise would result?

5-1-3 6. Find the energy produced per second when 3×10^{16} electrons go across an x-ray tube per second, each acquiring an energy of 100,000 electron volts.

5-1-3 7. Find the energy in joules produced when 1.8×10^{19} atoms of uranium-235 disintegrate, each releasing 200 MeV of energy. One gram of natural uranium contains 1.8×10^{19} atoms of the type that will disintegrate in a nuclear power reactor, releasing 200 MeV each.

5-1-4 8. Convert the following to watts:

(a) 1850 kcal/day (a human)

(b) 400 horsepower (an automobile engine)

(c) 50,000 BTU/hour (one BTU raises 1 pound of water through 1°F and is equivalent to about 252 calories) (a furnace).

5-1-4 9. What will be the temperature rise when a 1500 watt electric heater is immersed in 3 kg of water for 4 minutes? Assume that all the energy remains in the water.

5-1-5 10. Find the mass equivalent to one gigawatt for one hour.

5-1-5 11. (a) Find the energy in joules equivalent to the mass of the head of a pin. The pinhead is a hemisphere of steel, 0.9 mm in radius, with a density of 7.86 g/cm^3.

(b) If this energy were to be released at a steady rate over 24 hours, what would be the power?

5-2-1 12. How much energy is required to lift each kilogram in a satellite from the surface of the earth to an altitude of 160 km? Neglect the small change in g in that distance.

5-2-1 13. If a 1 kg mass sits on a table in an upper floor room in a building on a hill, find the potential energy of that mass:

(a) With respect to the floor if the table is 1 meter high.

(b) With respect to street level if that floor is 12 meters above street level.

(c) With respect to the valley floor 200 meters below street level.

5-2-2   14. Find the kinetic energy in each kilogram of a satellite orbiting the earth at 7.96 km/s. Combine this result with the result of Problem 12, if you did it, to find the total energy with respect to the earth.

5-2-2   15. A 25 kg cart is pushed across the floor, starting from rest, and a horizontal force of 200 N is used. After a distance of 20 meters it is moving at 7.0 m/s.
(a) What is the work done?
(b) What is the kinetic energy acquired?
(c) What fraction of the total work was converted to heat because of friction?

5-2-2   16. If a projectile has a speed of 33 m/s when it is thrown from ground level, what is the speed when it is 20 m above the ground? Does the mass matter? Does the direction matter?

5-2-2   17. (a) A 500 kg elevator moving downward at 2.5 m/s is brought to a stop in a distance of 0.32 m. How much energy is converted into heat by the braking mechanism while the elevator is being stopped?
(b) Repeat part (a) but with the initial motion of the elevator being upward.

5-2-2   18. (a) Consider two masses, m_1 moving at a speed v_1, and m_2 moving at a speed v_2. If the masses have equal kinetic energy, solve for the ratio of v_2 to v_1 in terms of m_2 and m_1.
(b) In order for a hydrogen molecule (mass 2 units) to have the same kinetic energy as an oxygen molecule (mass 32 units), how many times faster than the oxygen would the hydrogen have to move?
(c) If m_1 is 32 units (oxygen) and m_2 is 100,000 units (a very large molecule), what would the speed of the large molecule be compared with the speed of oxygen if they have the same kinetic energy?

5-2-2   19. Find the speed of an electron (mass 9.1×10^{-31} kg) if it has an energy of 100 electron volts.

5-2-2   20. When a 3000 kg vehicle moving at 25 m/s is brought to rest in 18.8 seconds, what energy in watts must be dissipated by the brakes?

5-2-3   21. Compare the speed of a rolling hoop to that of a rolling sphere after they have rolled down an incline of the same height. assume that no mechanical energy is lost because of friction.

5-2-3   22. Consider a solid disc rolling along a horizontal surface. The motion of each part is a combination of the horizontal motion at the speed v of the disc and the rotational motion about the axis. How much of the total kinetic energy is rotational and how much is translational?

5-3   23. (a) What is absolute zero on the Fahrenheit scale?
(b) Convert 98.6°F to °C and to kelvins.
(c) Convert −40°C to °F and to kelvins.

5-4   24. If an object containing 0.50 kg of water and 0.042 kg of aluminum cools at 1°C per minute, at what rate is energy being carried from it?

5-4   25. (a) When 4000 joules are put into 0.125 kg of water, by how much would the temperature rise?
(b) If the same energy was put into the same mass of steel, what would the temperature rise be?

5-5   26. (a) Find the heat capacity of an object if, after it is heated to 100°C, it is put into 0.400 kg of water at 10°C and the water temperature goes to 25°C.
(b) If the object has a mass of 0.37 kg, what is the specific heat capacity?
(c) Find the relative specific heat capacity.

5-5   27. Find the temperature that results when 12 kg of water at 90°C are put into 30 kg of water at 25°C. Neglect the heat absorbed by the container.

5-6   28. (a) How many joules are required to change 0.50 kg of ice at 0°C to water at 0°C?
(b) When 0.50 kg of ice is put into 1.5 kg of water at 20°C, what will the result be?

5-6   29. If 3.0 grams of steam are added to 0.150 kg of water in a calorimeter, resulting in condensation, what temperature rise will result? The water is initially at 15°C.

5-6   30. Find the energy required to change 1 kilogram of water from the ice form at −10°C to steam at 100°C. The specific heat capacity of ice is 2090 J/kg °C. The other data are in the text.

5-7-1   31. If you note that a certain object takes 2 minutes to cool from 15°C above room temperature to 14° above room temperature, how long will it take to cool from 5° above room temperature to 4° above room temperature?

5-7-1   32. Use the data in the first two columns of the table in Example 6 to investigate the cooling law.
(a) Plot a graph of temperature excess against time.

(b) Tabulate the logarithm of the temperature excess at each time shown in the table.

(c) Plot a graph of the logarithm of the temperature excess against time. Comment on the graph and, if you can, write an equation for the line shown.

(d) Subtract adjacent values of the logarithm of the temperature excesses shown in the table.

(e) From the material above, find how long it would take for the soup to cool to 5° above room temperature.

(f) Find how much longer it would take to cool to 2.5° above room temperature.

5-7-3   33. Estimate the thermal conductivity of the material of a sleeping bag, given the data that the total surface area is 3.5 m², the thickness is 6 cm, and the outside temperature is −22°C. The temperature inside is maintained at 25°C by a person producing heat at the rate of 90 watts.

5-7-3   34. The walls of buildings are often made with vertical wooden studs with glass wool between them. Compare the heat flow through a stud with the heat flow through the glass wool. Use a length of wood 4 cm wide, 9 cm thick, and 1 m long. The glass wool is 9 cm thick, 1 m long and 36 cm wide. For the wood, $k = 0.15$ J/s m °C, and for the glass wool, $k = 0.040$ J/s m °C.

5-7-3   35. Compare the rate of heat loss from two buildings in a cold climate, one with 30 cm thick concrete walls ($k = 2$ J/s m °C) and the other with walls of wood 5 cm thick ($k = 0.1$ J/s m °C). The areas of the walls and the inside and outside temperatures are the same. Note that concrete and wood vary widely in conductivity according to their sources, so the results are only an approximation to any real situation.

5-7-4   36. Compare the energy radiated from an object at 1190 K with what it would radiate at 1000 K. Consider the surroundings to be effectively at 0 K.

5-7-4   37. What is the rate of radiation of energy from a sphere of radius 0.050 m, which is at 127°C in surroundings at 27°C? The surface radiates at only 0.5 of the rate of a perfect radiator.

5-8   38. Coal gives about 3×10^7 J/kg when it is burned. How much coal would have to be burned per year to give a continuous energy production of 29 megawatts?

5-8-1   39. Over what area would solar energy have to be collected to warm 50 kg of water from 15°C to 40°C in 1 hour? The sun at that time and location is giving 700 watts/m² and a fifth of it is being absorbed to heat the water.

5-8-1   40. To replace a continuous 5000 watt output heating system with a solar heat collector, what area of collectors would be needed? Consider an average of 200 watts per square meter for 6 hours a day. The energy from the solar collectors is stored in batteries for continuous use.

5-8-2   41. What would be the necessary diameter of a windmill if it is to produce 1000 watts of power? It removes 10 per cent of the kinetic energy in a cylinder of air of the same diameter as the windmill, and the wind speed is 11 m/s (about 25 mi/h). The density of air is 1.3 kg/m³.

CHAPTER SIX

GASES AND ENERGY

6-1 THE GAS LAWS

If a gas is confined in a space of variable volume so that none escapes (the mass is therefore constant), there are three physical quantities that describe the gas: the pressure, P, the volume, V, and the absolute or kelvin temperature, T. When any of these is varied the others are affected. The early work with gases was done in the accepted scientific way — keeping one of the three variables constant and finding how the other two are related. In any situation it is possible to study the relationship between only two things at a time. Three laws resulted from this work, each of which is for a constant mass of gas. They are:

Boyle's law: If T is constant, the product PV is constant.

Gay-Lussac's law: If V is constant, the ratio P/T is constant.

Charles's law: If P is constant, the ratio V/T is constant.

When these laws were found, in the 17th to the early 19th century, the concept of absolute temperature did not exist. The quantity $t + 273°$, with t in what we now call Celsius degrees, was used in place of T in those laws. This work led to the concept of $-273°C$ being an absolute zero, with $t + 273°C$ being the absolute temperature T.

The three laws are now often combined in one, the *general gas law*. This is that for a constant mass of gas,

$$\frac{PV}{T} \text{ is constant}$$

This can also be written $PV = \text{constant} \times T$, where the value of the constant will depend on the mass and the type of gas.

6-1-1 PRESSURE AND ITS UNITS

Pressure refers to the force exerted by a fluid, be it a gas or a liquid. The total force exerted by a fluid on a surface depends on the total area over which the pressure acts. The quantity called pressure is the force exerted per unit of area. The SI unit for description of force is the newton and the unit for area is the square meter. Thus, the units for pressure are N/m^2. The newton per square meter is also given the name *pascal,* abbreviated Pa, so pressures are expressed in pascals.

Normal atmospheric pressure is 101,300 pascals; for pressures of this size the kilopascal, kPa, is used, where the prefix kilo- means 1000. Still higher pressures could be expressed in megapascals, MPa. Pressure can also be expressed in terms of normal atmospheric pressure; 1 atmosphere of pressure is 101.3 kPa.

The units for pressure are these:

$$1 \text{ N/m}^2 = 1 \text{ Pa}$$
$$1000 \text{ Pa} = 1 \text{ kPa}$$
$$1{,}000{,}000 \text{ Pa} = 1 \text{ MPa}$$
$$1 \text{ atmosphere} = 101.3 \text{ kPa}$$

Since pressure is the force on a unit area, the force on an area A is given by

$$F = PA$$

EXAMPLE 1

Consider a mass of gas put into a balloon. On the ground the pressure is 100 kilopascals and the temperature is 27°C (300 K). The balloon is to rise to 25,000 meters, where the pressure is only 2.5 kPa and the temperature is −50°C or 223 K. The balloon can expand only to a radius of 15 meters, so the maximum volume is $4\pi r^3/3 = 14{,}000 \text{ m}^3$. Find the necessary volume and radius when the balloon is filled on the ground.

At 25,000 meters
$\qquad\qquad\qquad\qquad\qquad P_2 = 2.5 \text{ kPa}$
$\qquad\qquad\qquad\qquad\qquad V_2 = 14{,}000 \text{ m}^3$
$\qquad\qquad\qquad\qquad\qquad T_2 = 223 \text{ K}$

At the ground
$\qquad\qquad\qquad\qquad\qquad P_1 = 100 \text{ kPa}$
$\qquad\qquad\qquad\qquad\qquad V_1 \text{ is unknown}$
$\qquad\qquad\qquad\qquad\qquad T_1 = 300 \text{ K}$

If PV/T is constant, that means that PV/T at 25,000 meters is the same as PV/T on the ground.

Using the subscripts indicated above,

$$\frac{P_1 V_1}{T_1} = \frac{P_2 V_2}{T_2}$$

Solving for V_1 $$V_1 = \frac{P_2 T_1}{P_1 T_2} \times V_2$$

Substituting values, $$V_1 = \frac{2.5 \text{ kPa}}{100 \text{ kPa}} \frac{300 \text{ K}}{223 \text{ K}} \times 14{,}000 \text{ m}^3$$

$$= 470 \text{ m}^3$$

The volume on the ground will be 470 m³.
Using $V = 4\pi r^3/3$ and solving for r, the radius on the ground will be only 4.8 meters. The balloon will be far from filled on the ground. In Figure 6–1 a stratospheric balloon is shown at the time of launch. It is designed to carry scientific instruments to a very high altitude. Note the apparently small volume of gas at the head of the balloon.

There is one type of scientist who is happy to find laws from experimental data. Of this type were Boyle, Charles, and Gay-Lussac. There is another type who looks at those laws and asks, "What is the reason that they are of that form, and what more can be learned about nature by finding out?" These people set up a model of the real situation (or a theory, if you prefer) and analyze the model using known physical laws. Ludwig Boltzmann (1844–1906) was of this type, and it is largely to him that we owe the successful model to explain the general gas law. It is often referred to the kinetic theory of gases.

6–2 THE KINETIC THEORY OF GASES

In developing the kinetic theory of gases, we will use a model of an enclosed mass of gas containing a certain number of particles. These particles are effectively points, which means that all of their energy is translational kinetic energy; there is no spin or internal vibration. Also, there is effectively no force between the molecules. This means that there is no potential energy change when the gas expands or contracts. The force on the container wall will be assumed to be due to bombardment by the particles. The collisions will be assumed to be elastic. If the particles lost energy to the walls when they hit them, the walls of a container of gas would slowly heat up. This does not happen, so the collisions must be elastic.

The particles that constitute the gas will have a wide range of speeds but there will be an average speed. This average speed will be referred to merely as v. The container will be considered to be a rectangular box of sides a, b, and c as in Figure 6–2(a). The box is shown in a coordinate system with axes x, y, and z. The particles may move in any direction; none is preferred. If a particle has a speed v as in Figure 6–2(b), the components of v in the three dimensions are v_x, v_y, and v_z. There are n particles

Figure 6–1 A stratospheric balloon at launch. The small volume of gas at surface pressure will expand to fill the whole balloon at the low pressure of the stratosphere. (From J. M. Pasachoff: *Contemporary Astronomy*, W. B. Saunders Company, Philadelphia, 1977.)

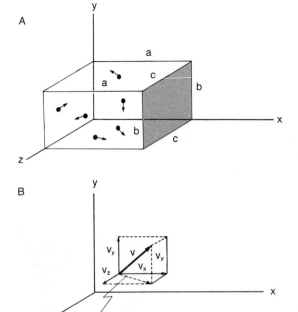

Figure 6–2 The model of a gas that is used in the kinetic theory of gases.

in the box, each of mass m. Each time a particle hits a wall of the box, it will be given a change in momentum so that an impulse will be given to the wall. The continual bombardment by the particles will result in a steady force, and this is the origin of the gas pressure. For example, a particle moves toward the wall of sides bc with a speed of v_x, as in Figure 6–3. In the collision the momentum in the x direction changes from mv_x toward the wall to mv_x away from it. The momentum parallel to the wall does not change. The total momentum change is $2mv_x$ in one collision, and this is the impulse given to the wall. The problem is to find the total number of collisions on one wall in each unit time (second); the momentum change per second is the force on the wall.

Each particle makes repeated collisions on the wall bc, the time between the collisions being the time required to travel across the box (a distance a) and back. The speed toward or away from the wall is v_x. The time between collisions with the wall bc is given by $t = 2a/v_x$. The number of collisions per second is $1/t$ or $v_x/2a$. (If the travel time across and back is $1/100$ s, then there are 100 collisions per second, for example.) The momentum change given to one particle per second because of its hitting the wall bc is the momentum change per collision ($2mv_x$) times the number of collisions per second ($v_x/2a$). This is the force exerted on that wall by only one particle.

From one particle, $F = (2mv_x)(v_x/2a) = 2mv_x^2/2a$.

From n particles, $F = 2nmv_x^2/2a$.

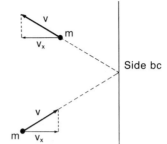

Figure 6–3 The bombardment of one wall of a container by one particle of gas.

The force is expressed in terms of one component of the speed rather than the speed. By applying the Pythogorean theorem to the velocity diagram in Figure 6–2(b),

$$v^2 = v_x^2 + v_y^2 + v_z^2$$

All directions of motion are equally probable, so on the average $v_x^2 = v_y^2 = v_z^2$. These three equal quantities must add to give v^2, so each must, on the average, be $v^2/3$. From this,

$$\overline{v_x^2} = \overline{v^2}/3$$

The bar is used to indicate that the average values are used; note carefully that by this we mean the average value of the squares of the velocities, *not* the square of the average values.

The total force exerted on the wall *bc* is then

$$F = \frac{2}{3} \frac{nm\overline{v^2}}{2a}$$

The pressure on the wall is force/area, the area A being *bc*. The pressure P is then given by

$$P = \frac{F}{A} = \frac{2}{3} \frac{nm\overline{v^2}}{2abc}$$

But *abc* is the volume *V*, so

$$P = \frac{2}{3}\frac{nm\overline{v^2}}{2V}$$

From this,

$$PV = \frac{2}{3}\frac{nm\overline{v^2}}{2}$$

The quantity $m\overline{v^2}/2$ in the above equation is the average translational kinetic energy, $\overline{E_t}$, of one particle. Again the bar indicates a mean value.

The equation derived from the particle model of the gas is therefore

$$PV = n \cdot \frac{2}{3}\frac{m\overline{v^2}}{2}$$

or

$$PV = n \cdot \frac{2}{3}\overline{E_t} \qquad (1)$$

This result can be compared to the general gas law as found from experiment, which is that the product *PV* is equal to a constant times the absolute temperature. The constant depends on the mass of gas, which in turn depends on the number of particles, *n*. The general gas law can then be written in the form

$$PV = nkT \qquad (2)$$

The constant *k* is known as Boltzmann's constant, and much effort has gone into its measurement. The result is that:

$$k = 1.38 \times 10^{-23} \text{ J/K particle}$$

Boltzmann's constant can also be expressed in terms of electron volts per kelvin degree, were 1 eV is 1.602×10^{-19} J/particle. In these units,

$$k = 8.61 \times 10^{-5} \text{ eV/K}$$

The inference is that absolute temperature is proportional to translational kinetic energy. Setting the right-hand sides of equations (1) and (2) equal to each other, the relation between *k* and the translational kinetic energy per particle per degree is

$$\frac{2}{3}\overline{E_t} = kT$$

or

$$\overline{E_t} = \frac{3}{2} kT$$

The translational kinetic energy per particle per kelvin degree is therefore $3k/2 = 2.07 \times 10^{-23}$ J/K particle, or 1.29×10^{-4} eV/K.

EXAMPLE 2

Find the average translational kinetic energy of air molecules at 20°C (293 K). Express the answer in joules and in electron volts.

Use

$$\overline{E_t} = \frac{3}{2} kT$$

where

$$k = 1.38 \times 10^{-23} \text{ J/K particle}$$

$$\overline{E_t} = \frac{3 \times 1.38 \times 10^{-23} \text{ J}}{2} \frac{}{\text{K particle}} \times 293 \text{ K}$$

$$= 6.07 \times 10^{-21} \text{ J/particle}$$

To express the answer in eV, use

$$k = 8.61 \times 10^{-5} \text{ eV/K}$$

Then

$$\overline{E_t} = \frac{3 \times 8.61 \times 10^{-5} \text{ eV}}{2} \frac{}{\text{K}} \times 293 \text{ K}$$

$$= 0.038 \text{ eV}$$

The average translational kinetic energy of each molecule of air at 20°C is therefore 6.07×10^{-21} J or 0.038 eV.

The theoretical analysis shows also what is measured by temperature — it is translational kinetic energy of particles. It shows what form the heat (energy) takes in the gas. The correlation between experimental and theoretical equations also shows that the particle model of a gas is valid and is worthy of further analysis.

6-2-1 KINETIC TEMPERATURE

The temperature of a gas depends on the energy per particle, and this depends on the speed of the particle. The relation between temperature and particle speed has some interesting applications.

The leading edges of high speed aircraft are subject to bombardment of air molecules at a speed equivalent to that of the aircraft. This can be the same as putting those leading edges in a furnace at a high temperature, where the air molecules would hit at an equivalent average speed. This is at the root of the heating problem in high speed craft. The energy must be taken away from those edges by refrigeration systems to prevent overheating. If the leading edges of supersonic aircraft were made of aluminum they would actually melt, which could be tragic.

In Figure 6–4 is a graph of the atmospheric temperature with altitude; beyond 25 km altitude the temperature rises. The region of higher temperature in the stratosphere results from absorption of the far ultraviolet light of the sun, principally by the ozone which is in the atmosphere at that altitude. This ozone is important in keeping the very high energy, short wavelength part of the ultraviolet solar spectrum from reaching the surface. There is considerable fear about anything (excessive use of spray cans with a fluorocarbon propellant, for example) destroying the protective ozone layer.

Beyond a hundred kilometers the temperature again rises, to about 600 °C at 300 km altitude. This does not mean that you could cook a chicken by holding it out the window of a satellite orbiting at 300 km. The chicken would lose heat by radiation faster than it would gain it by molecular bombardment. In such a sparse atmosphere, if it even warrants that name, the temperature of the gas can be expressed only as a description of the speed of the particles. It is referred to as kinetic temperature.

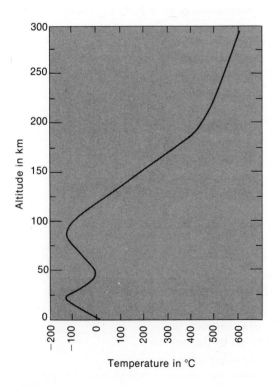

Figure 6–4 The temperature variation of the atmosphere of the earth with altitude.

The average speed of a particle of the gas is found by using the relations $\overline{E_t} = m\overline{v^2}/2$ and $\overline{E_t} = 3kT/2$:

$$\frac{m\overline{v^2}}{2} = \frac{3kT}{2}$$

$$\overline{v^2} = \frac{3kT}{m}$$

In this expression $\overline{v^2}$ is the mean (average) of the squares of the speeds of the individual particles. This is not the same as the square of the mean. The value that can be calculated for the speed from this expression is the square root of the mean of the squares of the speeds. It is often referred to as the *root mean square* speed or the r.m.s. speed. Let it be represented by just $\overline{v}$, with an understanding of the meaning.

$$\overline{v} = \sqrt{\overline{v^2}} = \sqrt{\frac{3kT}{m}}$$

EXAMPLE 3

Calculate the r.m.s. speed of oxygen molecules at 20°C. The mass of an oxygen molecule is 32 u, where 1 u = 1.66 × 10^{-27} kg.

Use

$$\overline{v} = \sqrt{\frac{3kT}{m}}$$

where

$k = 1.38 \times 10^{-23}$ J/K particle
$m = 32 \times 1.66 \times 10^{-27}$ kg/particle
$ = 5.3 \times 10^{-26}$ kg/particle
$T = 293$ k

$$\overline{v} = \sqrt{\frac{3 \times 1.38 \times 10^{-23} \text{ J} \times 293 \text{ K}}{5.3 \times 10^{-26} \dfrac{\text{kg}}{\text{particle}} \text{ K particle}}}$$

$\phantom{\overline{v}} = 478$ m/s

The r.m.s. speed of oxygen molecules at 20°C is 478 m/s. This is 1070 mi/hr.

6-2-2 BROWNIAN MOTION

The very small atoms of a gas cannot be seen, but it is possible to suspend tiny particles of smoke or pollen in a gas. These small particles will be bombarded by the atoms or molecules of the gas and will soon have the same average kinetic energy as the particles of the gas. The motion of these particles can be seen with a microscope.

The motions were first observed during the early nineteenth century, by Dr. Robert Brown, and are now given the name *Brownian motion*. The source of this motion was a puzzle at the time. It seemed to be so continuous, not decreasing because of frictional forces, that many considered it to be a basic "life" type of energy. This idea about a "life" energy was reinforced by the fact that the initial work was done with suspensions of pollen grains. It was Albert Einstein, in 1905, who showed that the average energy of these moving particles was the same as the energy of motion of the gas molecules. These visible particles moved because of molecular bombardment! Brownian motion is one of the most tangible demonstrations of the existence of molecules and atoms.

The atomic theories were first put forward by several ancient Greek philosophers (about 300 B.C.), and there is a common inclination to say that their ideas were entirely speculative. It is therefore somewhat surprising to read the words of Lucretius, the Roman poet who, in about 50 B.C., listed the evidence in favor of the atomic theory. Among the phenomena that he says gave rise to the concept of atoms is what can be interpreted as Brownian motion. He watched tiny particles of dust floating in a ray of sunshine in an otherwise darkened room. His descriptions so closely describe the phenomenon as it is seen in a microscope that I wonder if he really did see Brownian motion. In the translation by A. D. Winspear, his description is this:

> For think,
> When rays of sun pour through the darkened house
> Why then you'll see,
> A million tiny particles mingle in many ways,
> And dance in sunbeams through the empty space,
> As though in mimic war the particles wage everlasting
>   strife —
> Troop ranged 'gainst troop, nor ever call a halt;
> In constant harassment
> They're made to meet and part.
> So you can guess from this
> Just what it means
> That atoms should be always buffeted in mighty void.
> And so,
> A little thing can give a hint of big
> And offer traces of a thought
> And so it's very right
> That you should turn your mind to bodies dancing in the
>   rays of sun
> Movements like this will give sufficient hint
> That clandestine and hidden bodies also lurk

In atom stream
For if you watch the motes,
When dancing in a sunbeam, you will often see
The motes by unseen clashings dashed to change their
    course,
Sometimes turn back, when driven by external blows
And whirl, now this way, then now that
Dancing every way at once.
So you may know
They have this restlessness from atom stream.*

6-2-3 THE MOLE

It is usually not possible to work with individual gas particles; a sample of gas will have, to use an understatement, a
large number of particles. A number has been agreed on as a
standard; this is 6.02×10^{23} particles and is called a *mole*. Note
that 6.02×10^{23} particles of *anything* is a mole: the definition is
not restricted to atoms. This number is called *Avogadro's
number* and is actually defined as the number of atoms in 0.012
kg (12 g) of carbon-12.

EXAMPLE 4

Find the number of atoms in a diamond of mass 0.1 g. This is a size frequently found
in a ring. A diamond is pure carbon.
In 12 g of carbon there are 6.02×10^{23} atoms. In 0.1 g there are

$$\frac{0.1}{12} \times 6.02 \times 10^{23} = 5.0 \times 10^{21}$$

That diamond will have 5.0×10^{21} atoms

6-2-4 COMPARING THE MASSES OF THE
PARTICLES OF GASES

In terms of kinetic theory, using the form of the equation
$PV = nkT$, the form $PV/T = nk$ can be derived. Since k is a constant, the implication is that *samples of different gases for which
the product PV/T is the same have the same number of particles, n.* The quantities that have the same values of PV/T

---

*Alban D. Winspear: Lucretius and Scientific Thought, Harvest House, Montreal, .
1963, pages 84 and 85.

would have different masses because the individual particles would have different masses. Comparison of the masses that have the same value of PV/T would show how the masses of the individual particles compare.

EXAMPLE 5

Compare the masses of the particles that constitute argon gas to those that constitute hydrogen gas. Compare the result to the ratio of the accepted value for the atomic mass of argon (39.94 u) to that for a hydrogen molecule, H_2 (2×1.008 u).

Use the measured values for the density of each gas at the standard temperature and pressure: 1 atmosphere or 101.3 kPa and 273 K.

1 m^3 of argon gas at s.t.p. has a mass of 1.781 kg.

1 m^3 of hydrogen gas at s.t.p. has a mass of 0.0899 kg.

The total mass, M, is the number of particles times the mass of each, or nm. Using the subscripts A for argon and H for hydrogen,

$$M_A = nm_A$$
$$M_H = nm_H$$

n is the same for each because P, V, and T are the same for each. Then $\dfrac{M_A}{M_H} = \dfrac{m_A}{m_H}$.

The ratio of the masses of the particles of the gas is the same as the ratio of the mass per cubic meter at s.t.p.

Using numbers, $m_A = 1.781$ kg and $m_H = 0.0899$ *kg, so*

$$\frac{M_A}{M_H} = \frac{1.781}{0.0899} = 19.81$$

Compare this to the ratio of the accepted masses of A and of H_2. This is 39.94/(2×1.008) = 19.81. The two values are the same, so the theory seems to be valid for this case.

6–2–5 SPEED AND PARTICLE MASS

Another topic of interest is the comparison of speeds of particles of different masses at the same temperature. For example, in a gas that is a mixture of several species of molecules, the particles bump into each other and come to the same average kinetic energy. How do the mean speeds of the various types of particles compare?

Consider two sets of particles of masses m_1 and m_2 with r.m.s. velocities $\overline{v_1}$ and $\overline{v_2}$ ($\overline{v}$ is taken to mean $\sqrt{\overline{v^2}}$).

Then
$$\frac{1}{2} m_1 \overline{v_1^2} = \frac{3}{2} kT$$

and
$$\frac{1}{2}\, m_2 \overline{v_2^2} = \frac{3}{2}\, kT$$

The right sides are the same, so

$$m_1 \overline{v_1^2} = m_2 \overline{v_2^2}$$

so
$$\overline{v_1^2}/\overline{v_2^2} = m_2/m_1$$

or
$$\overline{v_1}/\overline{v_2} = \sqrt{m_2/m_1}$$

This shows that the mean velocities of molecules at the same temperature are inversely proportional to the square root of the masses of the molecules. The ratio of the masses of the particles is the same as the ratio of their molecular weights. The particles of the lowest mass will have the highest speeds in order that all have the same kinetic energy.

EXAMPLE 6

How do the speeds of any hydrogen molecules in air compare to the speeds of the oxygen molecules?

The mass of a hydrogen molecule is 2 × 1.008 u or 2.016 u. An oxygen molecule is 2 × 16.00 u or 32.00 u.

The r.m.s. molecular speeds compare inversely as the masses, so

$$\frac{\overline{v_H}}{\overline{v_O}} = \sqrt{\frac{m_O}{m_H}} = \sqrt{\frac{32.00}{2.016}} = 3.98$$

The r.m.s. speed of the hydrogen molecules will be 3.98 times that of the oxygen molecules.

The relative amount of hydrogen on earth is much less than it is on the massive planets like Jupiter and Saturn. This is because on earth the hydrogen atoms in the upper air will frequently achieve escape velocity, while the slower moving oxygen and nitrogen will not. Over the years the hydrogen has slowly leaked away into space from earth. The escape speed from the more massive planets is much higher than that from earth, so the hydrogen has been retained. The gravitational pull of our moon is much smaller than on earth, the escape speed is less, and the moon has lost all of its atmosphere. If we were to try to generate an atmosphere on the moon, it too would leak off.

Another example of use of the fact that particles with higher masses have lower speeds is in the separation of atoms of different mass. Natural uranium consists of atoms of two different masses. The majority of it (99.27 per cent) has a mass of 238 u, while 0.72 per cent has a mass of 235 u. It is the U-235 that produces the power in most nuclear power plants, and the uranium for many plants must be made to have a higher percentage of U-235 than occurs naturally. This can be achieved by putting the uranium in a compound that can be made into a gas. Then the gas is allowed to diffuse through a porous plug from one chamber to another. The lighter U-235 will diffuse faster than the U-238, so the uranium gas in the second chamber will be enriched in U-235. As this process is done over and over, the result can be an enrichment in U-235 to the required level for a power reactor. This is called the *thermal diffusion* method of enriching uranium.

6–3   HEAT CAPACITY OF GASES

The specific heat capacity of a gas, the energy required to raise 1 kg of the gas by 1°C, can be measured directly and expected specific heats can be derived from kinetic theory. To do this, it will be convenient to put the general gas law in terms of the number of moles rather than the number of particles. Use the form

$$PV = nkT$$

In a sample of gas of mass M, the number of moles is $M/\mathcal{M}$ if $\mathcal{M}$ is the molar mass, the number of kilograms per mole.* One mole consists of 6.02×10^{23} particles, the number called Avogadro's number, A. Then

$$n = A \times M/\mathcal{M}$$

The gas law then becomes:

$$PV = \frac{M}{\mathcal{M}} AkT$$

The quantity Ak, with units of energy per mole, is given the name of the *universal gas constant,* for which the symbol R is

---

*The molar mass in kilograms per mole is used in the SI units. The term formerly used was molecular weight, which is grams per mole. The kilogram per mole is just 1/1000 of the molecular weight in grams per mole. Another way to look at it is that grams per mole are the same as kilograms per kilomole or kg/kmole.

used. R is equal to 8.31 joules per mole k. Using this universal gas constant, the general gas law becomes:

$$PV = \frac{M}{\mathcal{M}} RT$$

From kinetic theory, the relation using the translational kinetic energy, E_t, of each particle was

$$PV = \frac{2}{3} nE_t$$

From these equations it is apparent that

$$\frac{2}{3} nE_t = \frac{M}{\mathcal{M}} RT$$

or
$$nE_t = \frac{3}{2} \frac{M}{\mathcal{M}} RT$$

The quantity nE_t is the total kinetic energy of all the particles in the mass of M of the gas.

At a temperature T_1, let the total translational kinetic energy be nE_{t_1}, and at a temperature T_2 let it be nE_{t_2}; then

$$nE_{t_1} = \frac{3}{2} \frac{M}{\mathcal{M}} RT_1 \qquad \text{and} \qquad nE_{t_2} = \frac{3}{2} \frac{M}{\mathcal{M}} RT_2$$

The difference between nE_{t_2} and nE_{t_1} is the difference in the total kinetic energy of the particles at temperatures T_2 and T_1. The difference in energy is

$$nE_{t_2} - nE_{t_1} = \frac{3}{2} \frac{M}{\mathcal{M}} R(T_2 - T_1)$$

Let all the energy put into the gas go to translational kinetic energy. This means that the particles are effectively points and cannot have energy of rotation or internal vibration; this situation applies to monatomic gases. Also, the gas cannot be allowed to expand, or else some energy would be used to do work on the surroundings. This latter idea means that a different amount of energy would be required to raise the temperature of a mass of gas if it is kept at a constant volume than if it is kept at a constant pressure and allowed to expand. There are two specific heats that are commonly used for a gas:

specific heat at constant volume, called C_v
specific heat at constant pressure, called C_p

We are calculating C_v. C_v is $(nE_{t_2} - nE_{t_1})/M(T_2 - T_1)$, which has units of energy per kg degree. Solving for this,

$$C_v = \frac{3}{2}\frac{R}{\mathcal{M}}$$

EXAMPLE 4

Calculate the specific heat at constant volume for argon. The measured value is 311 J/kg °C.

Use

$R = 8.31$ J/mole °C
$\mathcal{M} = 0.0399$ kg/mole

$$C_v = \frac{3}{2} \times \frac{8.31 \text{ J/(mole °C)}}{0.0399 \text{ kg/mole}}$$

$$= 312 \text{ J/kg °C}$$

The calculated value of C_v for argon is 312 J/kg °C, while the experimentally measured value is 311 J/kg °C. The agreement is satisfactory.

6–3–1 MOLAR HEAT CAPACITY

Still working with monatomic gases, it is interesting to express the specific heat in terms of energy per mole per degree. Using constant volume and putting $M = \mathcal{M}$ as well as $T_2 - T_1 = 1°C$, then $nE_{t_2} - nE_{t_1}$ will be the energy needed to raise 1 mole of the gas by 1°C.

Use $$nE_{t_2} - nE_{t_1} = \frac{3}{2}\frac{M}{\mathcal{M}} R (T_2 - T_1)$$

$M/\mathcal{M}$ will be 1 and $T_2 - T_1$ will be 1°C. Then the molar heat capacity is $\frac{3}{2}R$. This should be the same for all monatomic gases at constant volume. That value, $3R/2$, is 12.5 J/mole °C. The measured values for two gases of very different molecular weight, helium with $\mathcal{M} = 0.004$ kg/mole and argon with $\mathcal{M} = 0.040$ kg/mole, are respectively 12.4 J/mole °C and 12.7 J/mole °C. Again the calculations and concepts of kinetic theory are validated.

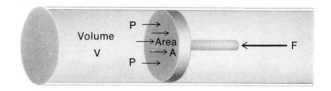

Figure 6–5 Finding the work done on the surroundings (on the piston) by a gas when it expands.

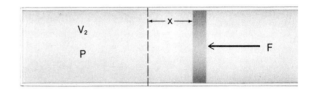

6–3–2 ENERGY IN AN EXPANSION

If a gas is heated and kept at constant pressure so that it expands, the pressure forces pushing out on the sides do work. Some of the energy put into the gas goes to do this work while the remainder increases the internal energy and, with monatomic gases, shows as a temperature rise. The problem is to find the amount of energy that is used in expanding the gas.

To do this, consider one mole of the gas so that $M = \mathscr{M}$ and $PV = RT$. Then consider an initial volume V_1 at a temperature T_1; the heated gas has a volume V_2 and temperature T_2. In calculating the molar heat, $T_2 - T_1$ will be 1°C. To allow the calculation, consider the gas to be in a cylinder as in Figure 6–5; as it expands it does work in pushing back the piston. The force on the piston is pressure times area or

$$F = PA$$

As in Figure 6–5(b) and (c), in the expansion the piston moves a distance x in going from a volume V_1 to a volume V_2 at the constant pressure P.

Using the general gas law,

$$PV_1 = RT_1 \ (M = \mathscr{M}, \text{ i.e. one mole of gas is used})$$
$$PV_2 = RT_2$$

Subtract:
$$P(V_2 - V_1) = R(T_2 - T_1)$$

The change in volume $V_2 - V_1$ is also given by Ax (see Figure 6-5). Also, $T_2 - T_1$ is 1°C, and $P(V_2 - V_1)$ becomes

$$PAx = R \qquad \text{(with } T_2 - T_1 = 1°C\text{)}$$

Using $PA = F$, then $PAx = Fx =$ work done in the expansion. The work done is then just the gas constant R.

The conclusion of all this is that when one mole of an ideal gas is heated by one degree, the energy used in the expansion is just R, or 8.31 joules per mole degree. In the development of this there were no restrictions such as the gas being monatomic, diatomic, and so forth.

6-3-3   SPECIFIC HEAT AT CONSTANT PRESSURE

When energy is put into a gas that is kept at constant pressure while the temperature rises, the gas must expand. Part of the energy put into it increases the translational kinetic energy and part does work on the surroundings as the container boundary moves outward in the expansion. If one mole of a monatomic gas is used and the temperature rise is one degree, then the increased kinetic energy is $3R/2$ and the work against the expansion is R. Only the amount $3R/2$ increases the internal energy of the gas; the amount R is work done on the surroundings. As in Figure 6-6, $5R/2$ is put into the gas, $3R/2$ going to kinetic energy and R going to expansion.

This analysis suggests that the heat required to raise one mole of any monatomic gas by one degree should be $5R/2$ if the pressure is kept constant. Using $R = 8.31$ J/mole °C. the specific heat, C_p, per mole can be calculated:

$$C_p \text{ per mole} = 5R/2 = 20.78 \text{ J/mole °C}$$
or $\qquad C_p \text{ per mole} = C_v + R$

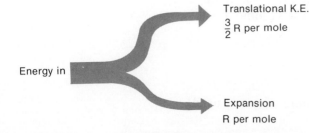

Translational K.E.
$\frac{3}{2}$ R per mole

Expansion
R per mole

Figure 6-6 The division of the energy put into a gas when it is allowed to expand.

TABLE 6-1   Measured and calculated specific heats
at constant pressure for some monatomic gases

Gas	C_p, per mole J/mole °C Measured	$5R/2$	$\mathcal{M}$ kg/mole	C_p, per kg J/kg °C Measured	$5R/2\mathcal{M}$
helium	20.74	20.78	0.00400	5180	5195
neon	20.75	20.78	0.02018	1028	1030
argon	20.69	20.78	0.03994	518	520
krypton	20.67	20.78	0.0837	247	248

(C_p and C_v are sometimes used to indicate heat capacity per mole rather than per kilogram. When they are used to indicate molar heat capacity, that use should be clearly indicated.)

If the molar mass is $\mathcal{M}$ (in kg/mole), the molar heat capacity at constant pressure is

$$C_p = \frac{5R}{2\mathcal{M}} \text{ (per mole)}$$

The correlation between the theory and measured values is shown for monatomic gases in Table 6-1.

6-3-4   RATIO OF HEAT CAPACITIES

The ratio of the specific heats at constant pressure and at constant volume, C_p/C_v, enters into some physical situations and it has been assigned the symbol γ (Greek lower case gamma):

$$\gamma = \frac{C_p}{C_v}$$

For monatomic gases, C_p per mole is $5R/2$ and C_v per mole is $3R/2$. Then γ, the ratio of these, is expected to be just $5/3$ or 1.667. The measured value of γ for helium is 1.63, and that for argon is 1.667. For gases that are not monatomic, γ is less than this

6-4   ISOTHERMAL AND ADIABATIC PROCESSES

When an enclosed mass of gas expands or contracts, energy may or may not be put in, and this affects the temperature during the process. The restriction is that PV/T remains constant.

There are two basic types of process.

In one type of process, as the gas changes pressure and volume it is kept at the same temperature. This is called an *isothermal* process. If a gas expands isothermally, energy must be put into it as the process occurs; otherwise the gas will cool. This is because energy is used to push out the walls during the expansion. If the gas contracts isothermally, energy must be taken out or the temperature will rise. This is because the surroundings do work on the gas as the container boundaries move in. Temperature and mass are constant in an isothermal process, so it is described by Boyle's law:

$$PV = \text{constant (isothermal)}$$

The second process is that in which, as the gas expands or contracts, no energy is allowed in or out. Such a process is called *adiabatic*. Adiabatic expansion results in cooling. Adiabatic compression results in heating. The relation between P and V for an adiabatic process is $PV^{\gamma} = \text{constant}$, where $\gamma = C_p/C_v$.

The relations between pressure and volume for isothermal and adiabatic processes are illustrated in Figure 6–7. The adiabatic curve is steeper than the isothermal because as the volume goes up, the pressure is reduced by the amount for the isothermal process as well as by an increased amount caused by the temperature drop.

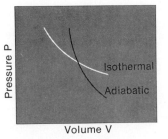

Figure 6–7 Isothermal and adiabatic processes for a gas.

6–4–1 WORK IN AN EXPANSION

In Section 6–3–3, it was shown that the work done in an expansion is pressure times change in volume. If the pressure changes, the work is the area under the pressure-volume curve as in Figure 6–8. The expansion may take place in a cylinder as in Figure 6–8, and the piston does useful work for us on the outside. It is not possible to get all the energy out of the gas unless the pressure were to drop to zero. This would occur only if the absolute temperature dropped to zero.

6–5 ORDER AND DISORDER

The temperature of a sample of gas depends on the average kinetic energy of the particles and hence the particles' speeds. The individual particles have speeds ranging from well above to well below the average. The speed distributions in samples of gas at two different temperatures are shown in Figure 6–9.

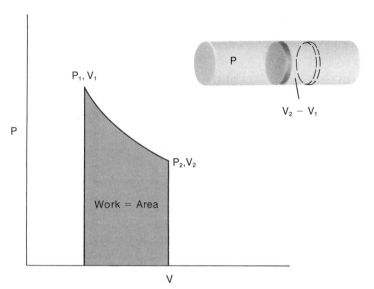

Figure 6–8 The work done in the expansion of a gas.

There are no numbers attached; it is only to illustrate that the particle speeds are distributed above and below a mean in each case. The distributions are quite random but are clustered around the mean in a manner described by what is called a Maxwell-Boltzmann distribution.

If two compartments of gas are arranged as in Figure 6–10 with a heat conducting barrier between them, the heat energy will be transferred from the hot gas to the cold until they both reach a temperature somewhere between the two initial temperatures. The initial and final speed distributions will be like those in Figure 6–11. The final state has less order, more disorder, than the initial state. At least initially there were two separate distributions, and in the final state there is only one distribution curve.

Figure 6–9 The distribution of the speeds of the particles in a low-temperature gas and in a high-temperature gas.

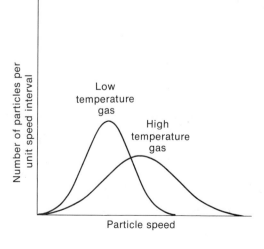

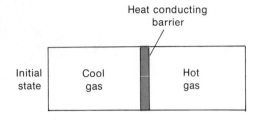

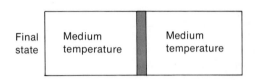

Figure 6–10   A hot gas and a cool gas reach an equilibrium temperature by the natural process of energy flow from the hot to the cool object.

In nature, heat energy flows from a hot to a cool object and a system goes naturally from a state of order to a state of disorder. Disorder is measured by a quantity called *entropy*. In natural processes the entropy increases, and order decreases.

There are examples of order increasing in a small portion of a system. For example, atoms may form an orderly crystal. In doing so, heat is given to the surroundings, increasing disorder there.

In an engine, order is given to a piston in which all the atoms move in the same direction, but this is done only by the expenditure of energy, perhaps by burning an orderly, complex hydrocarbon. Again the total order of the system decreases. An engine designed to convert heat energy to useful work will usually do so by the expansion of a gas to move a piston or a turbine blade. In this process there will be cooling, and the cooling cannot be to a temperature below that of the surround-

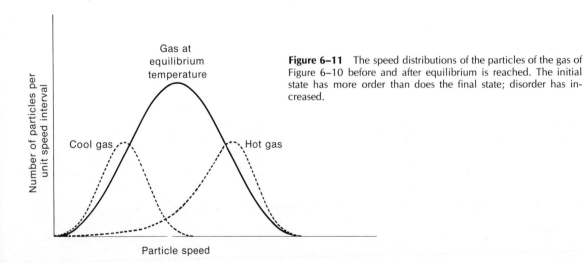

Figure 6–11   The speed distributions of the particles of the gas of Figure 6–10 before and after equilibrium is reached. The initial state has more order than does the final state; disorder has increased.

ings. Energy must be put into the gas to heat it in the first place, and also some of the expansion can be at a constant temperature as long as heat is being put into the gas. In fact, the energy put into any engine is always greater than the useful work gotten out; but the greater the temperature change in the process, the more efficient it is. Because the lower temperature is limited by the surroundings, it follows that the higher the maximum temperature, the greater the efficiency. Also, an engine cannot work unless the temperature can be made to exceed that of the surroundings.

All the energy put in cannot be extracted (the efficiency is not 100 per cent), so there will be some energy lost to the surroundings. This can lead to a problem of *thermal pollution* at the site of large power plants, where the energy lost may be warming a body of water. This may result in an ecological change. There are some cold areas of the world where the thought of thermal pollution when the outdoor temperature may be at −40°C is amusing. One problem is to find how to make use of the waste heat from a heat engine or power plant.

It is one of the basic laws of nature that entropy or disorder is increasing. The hot objects cool, the cool ones warm up. One even hears talk of the heat death of the universe when everything has reached one final temperature. This will, of course, never completely occur, for it will be approached asymptotically. Also, it will be well beyond our lifetimes. One of the values of the concept is that it causes us to think about the total plan of the universe.

6–6   THE LAWS OF THERMODYNAMICS

The subject of energy and the movement of energy (thermodynamics) has been developed without stress on the laws. Physics develops by working with a subject until the laws that govern it finally become obvious and then are stated. The laws have been used in the preceding part of the chapter, and their statement here is a summary. In thermodynamics there are four basic laws:

The zeroth law of thermodynamics was added after the others had been widely accepted and had been numbered. It is the important statement that heat flows naturally from a hot to a cold object.

The first law is a statement of the conservation of energy. In a closed system the total energy, in all its forms, is constant. This law includes the concept of the conversion between heat and mechanical energy.

The second law states that in any process involving energy conversion or natural transfer (heat to mechanical, for example, or flow of heat from a hot to a cold object) there is an increase in the entropy of the total system involved.

The third law is that absolute zero cannot be realized in a real system.

DISCUSSION PROBLEMS

6-2-1   1.   Discuss the idea of devising a thermometer scale graduated to m/s rather than in °C. Would it be limited to one gas, air for example?

6-2-1   2.   If a planet similar in size and mass to the earth was ten times as far from the sun as is the earth, what difference might be expected in the atmosphere?

6-4   3.   The gas laws and kinetic theory give insight into the nature of *laws* and *theories* in physics. Discuss the following statements, *using examples* from the material of this chapter to support your statements.
(a) The gas laws are proved by kinetic theory.
(b) The gas laws lend validity to the model of a gas consisting of particles (molecules) in which gas pressure is due to bombardment of the container by the particles.
(c) Laws in physics describe regularities observed in nature, and theories are devised to describe a part of nature that we cannot see.
(d) Theories stand as long as they are not in conflict with the laws, but they must be discarded or modified if they disagree with the laws.

6-4   4.   How can the measurement of the ratio of the specific heats of a gas tell of the structure of the molecules of the gas?

6-5   5.   A block slides across a floor and because of friction is brought to rest. Initially the atoms of the box all have motion in the same direction. Discuss the change in entropy in this case.

PROBLEMS

6-1   1.   What pressure would give a force of 1500 N on a surface 1.5 cm in radius?

6-1   2.   Find the net force on the side of a container outside of which the pressure is 101 kPa and inside of which it is 301 kPa. The side is rectangular, 0.15 m × 0.30 m.

6-1   3.   Find the pressure when air in the cylinder of a pump is compressed from a length of 35 cm to a length of 5.0 cm. The temperature remains constant during the compression. The initial air pressure is 101.3 kPa.

6-1   4.   Find the radius of a bubble of air at the surface if it is formed at a depth in water where the absolute pressure is 810 kPa. The pressure at the surface is 1 atmosphere. At formation the bubble has a radius of 0.10 mm. There is no temperature change.

6-1   5.   If a confined mass of gas is held at a constant volume and heated, the pressure will change. If at 273 K the pressure is 101 kPa, what would the pressure be if the gas is heated to 373 K?

6-1   6.   A sealed can containing air at a pressure of 1 atmosphere and room temperature (20°C) is put into an oven, where it is heated to 200°C. Find the amount by which the pressure in the can exceeds 1 atmosphere. Express the answer in kPa.

6-1   7.   A rectangular tin can is heated to the temperature of boiling water while it is open to the air. The lid is put on tightly and the can is allowed to cool to room temperature. In a situation like this the air pressure, which is then greater outside the can than in it, will sometimes crush the can. The problem is to find the total inward force on one of the large sides if the can did not crush. Use the following data. The room temperature is 20°C and the boiling point of water is 100°C. Atmospheric pressure is 101.3 kPa. The hot air in the can is also at a pressure of 101.3 kPa before it is sealed. The dimensions of the can are 20 cm by 15 cm by 9 cm.

6-1   8.   A mass of air at 20°C is confined to a length of 4.0 cm in a cylinder. The air is heated to 200°C and one end of the cylinder (a piston) moves outward. Find the final length of the cylinder of air if the pressure is constant.

6-1   9.   A balloon of radius 15 cm at room temperature (20°C) is cooled to −40°C and the pressure remains constant. Find the radius at the cooler temperature.

6-1   10.   If 35 cm^3 of gas is collected at a temperature of 35°C and a pressure of 90.3 kPa, what would the volume be at 0°C and 101.3 kPa?

6-1   11.   Find the temperature resulting when compressed hydrogen gas is suddenly expanded to four times its original volume and the pressure is reduced to a tenth of its initial value. The initial temperature is 20°C.

6-1   12.   Air in a pump is compressed quickly so that no energy escapes. Find the final temperature (in °C) if the air in a cylinder 35 cm long and 2.5 cm in diameter is compressed by a piston into a column only 5.0 cm long. The initial temperature is 22°C and the initial pressure is 1 atmosphere. The pressure when it is compressed is 1540 kPa.

6-1   13.   Find the average force exerted on a flat shield when it is hit by 200 tennis balls per minute, each of mass 0.10 kg. The balls approach at 20 m/s and rebound with the same speed.

6-2   14.   In order for a thermonuclear reaction to occur, a gas temperature of about 40 million K is required. At that temperature, what is the average kinetic energy of the particles of the gas expressed in electron volts?

6-2-1   15.   (a) What is the average translational kinetic energy of an oxygen molecule in air at 293 K?
(b) Oxygen molecules have a mass of 32 u. What is their "average" speed (the square root of the mean of the square of the speed) at 293 K?

6-2-1   16.   (a) If a particle 0.1 μm in diameter having relative density of 1.0 is suspended in a gas at room temperature, 18°C or 291 K, what would be the average speed as obtained from the average kinetic energy?

6-2-3   17.   Find the number of atoms per cubic meter of air in a chamber at what is considered to be a very good vacuum, 10^{-8} Pa. At 1 atmosphere and 273 K the density of air is 1.3 kg/m³. The average molar mass of the air molecules is 0.0288 kg/mole. The temperature of the evacuated chamber is also 273 K.

6-2-4   18.   When 1 microgram of water evaporates from a container, how many molecules have gone? The molar mass of water is 0.018 kg/mole.

6-2-3   19.   Using techniques involving radioactivity, it is possible to measure samples with as few as a million atoms of iodine per cubic centimeter of solution. The molar mass of the iodine in question is 0.131 kg/mole. What is the amount of iodine measured, expressed in g/cm³?

6-2-4   20.   Compare the expected densities of carbon monoxide (CO, 0.028 kg/mole) and carbon dioxide (CO_2, 0.044 kg/mole) to that of air (average 0.0288 kg/mole) when all are at 20°C and 100 kPa.

6-2-4   21.   Radon gas has a molar mass of 0.222 kg/mole. How would the density of radon gas compare to that of air (average molar mass, 0.0288 kg/mole) at the same temperature and pressure?

6-2-4   22.   Compare the speeds of molecules of heavy water (D_2O, molar mass 0.020 kg/mole) and of ordinary water (H_2O, molar mass 0.018 kg/mole) when they are mixed in the same vapor.

6-2-5   23.   How would the masses of the particles of two gases have to compare in order that the r.m.s. speed of one of the gases was just double that of the other?

6-3   24.   Find the molar mass, in kg/mole and also in g/mole, of a gas for which measured values are: $P = 101$ kPa, $V = 1 \ell = 10^{-3}$ m³, $M = 0.045$ kg, $T = 273$ K.

6-3   25.   (a) What would be the temperature rise when 50 J are put into 0.011 kg of argon gas that is kept at a constant volume?
(b) What would be the temperature rise when 50 J are put into 0.011 kg of argon gas that is kept at a constant pressure?

6-3   26.   Calculate the expected value of the quantity PV/T when 0.050 kg of helium (molar mass 0.004 kg/mole) is confined to 0.33 m³ at 293 K.

6-3-4   27.   (a) Calculate the molar heat capacity at constant volume for a diatomic gas. Of the energy put in, three parts go to translational motion in the three dimensions, and this energy shows as a temperature rise; one part goes to rotation of the system of the two atoms; and one part goes to energy of vibration of the pair of atoms.
(b) Calculate the molar specific heat at constant pressure for a diatomic gas. The energy put in is divided into three parts for translational motion, one for rotation, and one for vibration. Further, the work done in expansion is R per mole per degree.
(c) What is the theoretical value of γ for a diatomic gas?

6-4   28.   (a) Find the final pressure when argon in an insulated cylinder with a movable piston is compressed adiabatically to half its original volume. The original pressure is 101.3 kPa. Argon gas is monatomic and γ is 1.667, or 5/3.
(b) What would the final pressure be if the expansion was isothermal?
(c) Explain the reason, in terms of temperature, for the difference between the answers in (a) and (b).

CHAPTER SEVEN

FORCE AND DISTORTION

Forces acting on a solid object may stretch it, compress it, bend it, twist it, or break it. In short, they distort it. If all the forces on an object balance to zero, the object may be in equilibrium; but still, under the influence of those forces, the object will be distorted. It may be ever so small a distortion, but it will be there. When the forces are released the object may resume the shape and size that it had before the forces were applied. If it does, the material is said to be *elastic*. If the distortion is permanent, the object is *inelastic;* in fact, if there is no tendency at all to return to the original form, the object is referred to as being *perfectly inelastic*. If it goes back exactly to its original form it is *perfectly elastic*. **Elasticity** is not a measure of the ability to stretch or distort because of forces, but rather it is the property of returning to the original form (size and shape) when a distorting force is removed. Spring steel is highly elastic, while chewing gum is inelastic. It is obvious that there is a limit to the elasticity of any material. For example, a piece of steel can be bent so much that it will remain bent. However, the analysis of the behavior of materials inside the elastic limit is a very profitable area of study.

7–1 STRESS, STRAIN, AND YOUNG'S MODULUS

A typical example of elasticity is the stretching of a long narrow object by forces pulling on the ends, as in Figure 7–1(a). The forces F applied to a rod will stretch it by an amount shown as e. The force must be applied to both ends; if it is on only one end the result will be acceleration, not stretching. One of the important experimental observations on elastic materials is that

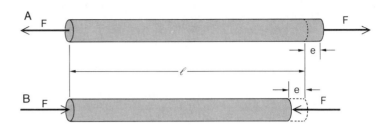

Figure 7–1   (a) A force *F* applied to a rod stretches it by the amount shown as e. (b) A compressional force causes a similar change in length.

the distortion, *e* in this case, is proportional to the distorting force. Double *F* and *e* also doubles — as long as what is called the elastic limit is not exceeded. If the forces produce compression, as in Figure 7–1(b), the amount of compression is usually the same as the elongation that would be produced by the same forces acting in the opposite direction. The direct proportionality between deformation and deforming force is known as Hooke's law. This relation was initially found by Robert Hooke, a versatile person of the 17th century who is also noted for his fascinating work on microscope design and in recording the new, small aspects of the world that he was able to see with his new instruments.

The amount of distortion produced on an object (as shown in Figure 7–2) depends on many factors. These are the force applied, the cross-sectional area, the length, and the type of material. It is found in practice that if the area of a cross-section is doubled, the elongation for a given force is cut in half. If the rod is doubled in length and the same force is applied, each half elongates and the total elongation is doubled. These relations

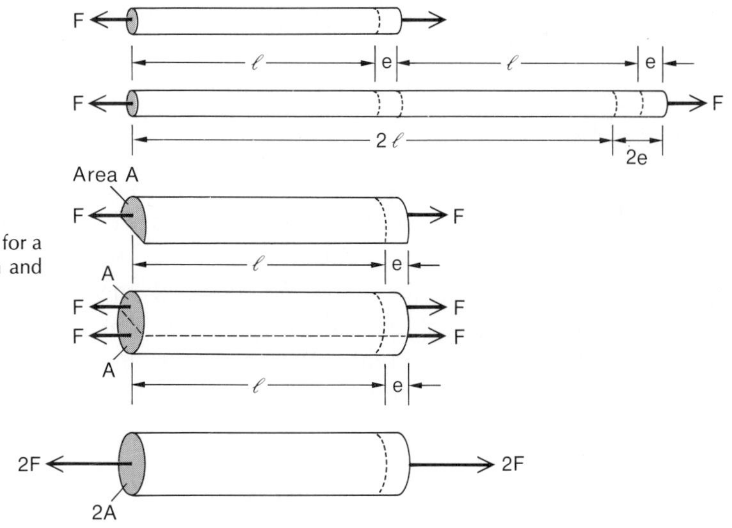

Figure 7–2   The change in length of a rod for a given force depends on the original length and the cross-sectional area.

can be summarized by letting e be the elongation, F the force, A the cross-sectional area, and l the length. Then:

$$e \text{ is proportional to } F, \text{ to } 1/A, \text{ and to } l$$

These combine to give (where the symbol $\propto$ means "is proportional to")

$$e \propto Fl/A \qquad \text{or} \qquad e/l \propto F/A$$

The last relation states that the elongation per unit length is proportional to the force per unit cross-sectional area. Writing it in the reverse order,

$$F/A \propto e/l$$

The force per unit cross-sectional area, F/A, is given the name of *stress*. The distortion per original unit distance, e/l, is called *strain*. The equation expressed in words is then that stress is proportional to strain; note that it applies only within the elastic limit of the material. The two quantities, stress and strain, are directly related: as one increases so does the other, and their ratio is constant for a given type of material. The value of the proportionality constant obtained from the ratio of stress to strain is called the **modulus of elasticity**. In the case of stretching or compressing it is called **Young's modulus,** often represented by E_y. In symbols:

$$E_y = \frac{F/A}{e/l} = \frac{Fl}{Ae}$$

Numerically, E_y is the force per unit area that would cause an elongation equal to the original length, though for elastic materials like steel or bone this would never be achieved. Long before such a stretch was achieved, the material would break or at least be permanently distorted.

If Hooke's law applies, the graph of stress (or force) against strain (or elongation) will be straight within the elastic limit, as is shown in Figure 7–3. As the force is increased, the elongation increases. As the force is reduced, the elongation will also be reduced and the points will all fall along the same line. This describes the stress-strain relation very closely for highly elastic materials such as steel or glass. The basic concepts in dealing with elasticity are these:

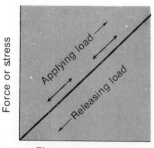

Figure 7–3 The form of the graph of stress (or load) against strain (or elongation) when Hooke's law applies.

Stress is force per unit area causing a strain. In general, stress $= F/A$.

TABLE 7–1   Moduli of elasticity for some selected materials. All are typical values; there is considerable variation from one sample to another. *

MATERIAL	YOUNG'S MODULUS, E N/m^2	BULK MODULUS, E_b N/m^2	SHEAR MODULUS, E_s N/m^2
Aluminum	7.0×10^{10}	7.46×10^{10}	2.7×10^{10}
Gold	8.0×10^{10}	16.6×10^{10}	2.8×10^{10}
Copper	12×10^{10}	14×10^{10}	4×10^{10}
Steel	20×10^{10}	17×10^{10}	8×10^{10}
Glass	7×10^{10}	5×10^{10}	3×10^{10}
Rubber	4×10^6	—	2×10^6
Oak	1.3×10^{10}	—	—
Bone	2.0×10^{10}	—	—
Rocks	—	3×10^{10}	1.5×10^{10}
Water	N.A.	2.17×10^9	N.A.

*The dash indicates that the author could not decide on a typical value, and N.A. indicates that the quantity described does not apply to that type of material.

Strain is a change in a dimension divided by an initial dimension. For stretching or compressing, strain $= e/l$.

Modulus of elasticity is the ratio stress/strain.

Some typical values of Young's modulus for a few materials are included in Table 7–1.

One of the applications of Hooke's law is in the measurement of forces (stresses) that occur in the beams and columns of buildings, bridges, aircraft, and other structures. The measurements are made by applying what is called a "strain gauge" or a "strain meter" to the point being investigated. The strain meter measures the change in length in a known initial length (i.e., the strain, e/l) at that position. The modulus of elasticity will be known for the material and the stress F/A can be calculated. In this way the forces in complex structures under different types of loading can be measured, and even the forces in hundreds of places in an aircraft in flight while performing various maneuvers can be found.

EXAMPLE 1

A certain steel wire 0.5 mm in diameter and 61 cm long is seen to stretch by 0.071 mm when the load on the wire is increased by 0.5 kg. Find Young's modulus for that wire.

Use

$$E_y = \frac{Fl}{Ae}$$

where

$F = 0.5 \times 9.8 \text{ N} = 4.9 \text{ N}$
$l = 61 \text{ cm} = 0.61 \text{ m}$
$e = 0.071 \text{ mm} = 7.1 \times 10^{-5} \text{ m}$

$A = \pi d^2/4 = \pi \times (0.5 \times 10^{-3} \text{ m})^2/4$
$\qquad = 1.96 \times 10^{-7} \text{ m}^2$

Note how all the quantities were changed to SI units. Then

$$E_y = \frac{4.9 \text{ N} \times 0.61 \text{ m}}{1.96 \times 10^{-7} \text{ m}^2 \times 7.1 \times 10^{-5} \text{ m}}$$

$$= 2.1 \times 10^{11} \text{ N/m}^2$$

Young's modulus for that wire was 2.1×10^{11} N/m².

7-1-1 THE ELASTIC CONSTANT

For a given object (rather than the material of which it is made), a constant can be found to describe the relation between force and distortion. In the case of stretching or compressing, this constant is related to the dimensions of the object as well as to Young's modulus. For a cylindrical object, for example, the relation among the various quantities is described by:

$$E_y = \frac{Fl}{Ae}$$

Solving for F:

$$F = (AE_y/l)\,e$$

Letting AE_y/l be represented by K, this expression becomes:

$$F = Ke \qquad \text{(The units for } K \text{ are N/m)}$$

The quantity K is referred to as the **elastic constant** for that particular object. It is sometimes easier to use K rather than the more fundamental expression involving Young's modulus. If the object is of a complicated shape or structure, such as an irregular bone or a honeycombed or porous object, then a force F can be applied to that object and a measurement of e will yield a value of the elastic constant K. In such a case it may not be possible to derive the expression in terms of the modulus of elasticity of the material because the cross-sectional area is not well defined.

A helical spring is also an example of a situation in which the elastic constant is used, and in this case it is referred to as the "spring constant." The elongation of the spring varies directly with the applied force, as shown in Figure 7-4. A spring scale is used to determine the force of gravity on a mass by measuring the elongation of the spring. The elongation depends on the value of g at that place, so a spring scale cannot be used for precise measurement of mass at two different places on earth.

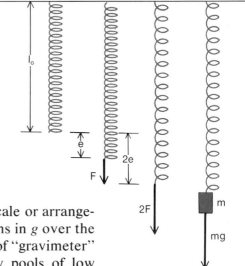

Figure 7–4 A spring used to demonstrate Hooke's law, that strain is proportional to stress. The force being exerted by gravity acting on a mass is shown in the last diagram.

On the other hand, a very delicate spring scale or arrangement of springs can be used to measure variations in g over the surface of the earth. This is the basis of one type of "gravimeter" used to detect local variations in g caused by pools of low density oil or concentrations of high density ores.

EXAMPLE 2

   I have a spring scale that is used to find the mass of objects like fish, as in Figure 3–2. The distance between the 0 mark and the 10 kg mark is 5.0 cm. What is the spring constant in SI units?
   It will be assumed that the scale is to be used where g is 9.8 N/kg.
   When a 10 kg object hangs on the scale, the force pulling on the spring is 98 N. The spring is stretched 5.0 cm or 0.050 m by 98 N. The spring constant K is then given by

$$K = \frac{98 \text{ N}}{0.050 \text{ m}} = 1960 \text{ N/m}$$

Answer: The spring constant is 1960 N/m.

7–2 RIGIDITY OR SHEAR

   Changes in the length of an object, either stretching or compressing, are described by Young's modulus; but there are other forms of distortion described by other elastic moduli. One of these types of distortion is called **shear**.
   The phenomenon called shearing is illustrated in Figure 7–5(a) by opposing forces on the two sides of a book. The originally rectangular end of the book is distorted to a parallelogram by the forces shown. This phenomenon can also occur in a solid, though the distortion is usually very small; it is often important but not so vivid as an illustration.
   The definition of stress in this instance is the force F per unit area A, as shown in Figure 7–5(b). Strain is measured by the

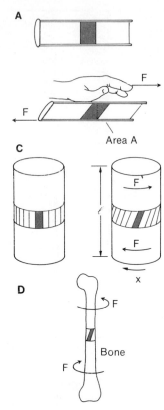

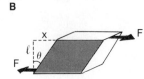

Figure 7–5 The phenomenon called shearing.

angle θ in the figure; since the angles are ordinarily small, θ can be replaced by x/l. This is a ratio of lengths and is what is called strain for the process of shearing. The modulus of elasticity in this case is also described by stress divided by strain. It is referred to as either the **shear modulus** or the **modulus of rigidity.** With the symbols shown in Figure 7–5, the shear modulus is described by

$$E_s = \frac{F/A}{x/l} = \frac{Fl}{Ax}$$

Some typical values of the shear modulus for common substances are listed in Table 7–1. Liquids do not show a shearing effect: they have no shear modulus. A similar quantity in flowing liquids is called viscosity.

Another example of shearing is the distortion of crustal plates of the earth across an old fault line. The plates move slowly and cause a shearing of the rocks near the old fault line, as in Figure 7–6. When the elastic limit of the rocks is reached, slippage and thus an earthquake occurs.

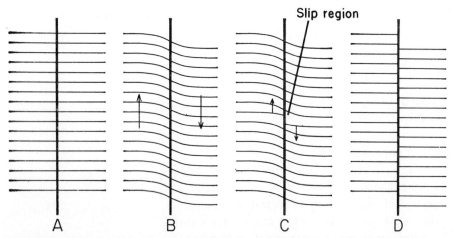

Figure 7–6 Shearing of the rocks near and across an old fault line in the crust of the earth. As the rock plates on either side of the fault slowly move, a shearing stress is built up in the rocks across the fault line. When the elastic limit is reached, slippage—and an earthquake—occurs. (From H. Benioff, *Science,* Vol. 143, pp. 1399–1406, March 1964. Copyright 1964 by the American Association for the Advancement of Science.)

7–2–1   TWISTING

The shear modulus is also important in the phenomenon of torsion or twisting. As in Figure 7–7, parallel lines marking a rectangle on the side of an untwisted cylinder are distorted to a parallelogram when a torque is applied. This phenomenon of the twisting of a rod, whether it is a thick rod or a rod thin enough to be called a fiber, has many uses. Torsion bar suspensions are used in some automobiles, where large torques are involved; and some of the most delicate of scientific instruments measure small forces by the twisting of a fiber. These include the instruments used to measure the gravitational attraction of masses in the lab and also some for the detection of very weakly magnetic objects.

The twisting of a thin-walled cylinder will be analyzed in detail, even though the instruments will usually use a solid rod or fiber. A solid cylinder can be considered to be made up of a large number of concentric thin-walled cylinders. The extension of the result to a solid cylinder requires use of the branch of mathematics called calculus. The analysis of the thin cylinder will demonstrate much about the process.

Consider, as in Figure 7–7, a cylinder of length l and radius r, with a wall thickness of Δr. The cylinder is twisted by a force F at the circumference. The other end is either clamped solidly or has an opposing force F. The strain in the cylinder wall is the angle shown by θ, and for small angles $\theta = x/l$. The angle

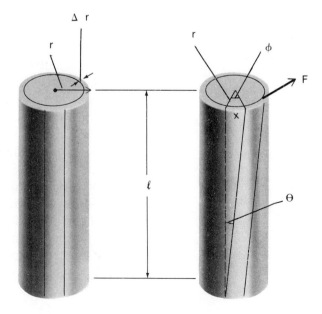

Figure 7–7   The phenomenon of shearing in the twisting of a thin-walled cylinder. Two parallel lines on the side of the untwisted cylinder form a rectangle, but they are distorted to form a parallelogram when the cylinder is twisted.

turned through by the end of the rod is the angle shown as ϕ (Greek phi); ϕ is x/r, which is much larger than θ. This is, in fact, the source of the sensitivity — the angle ϕ can be so much larger than the angle of shear θ.

The cross-sectional area of the sheared material is shown as A in Figure 7–7 and is the distance around the cylinder times the wall thickness Δr. That is, $A = 2\pi r\Delta r$.

Use

$$E_s = \frac{Fl}{Ax} \qquad \text{where} \qquad A = 2\pi r\Delta r$$

$$E_s = \frac{Fl}{2\pi rx\Delta r}$$

Multiply the numerator and denominator by r to obtain Fr, which is the twisting torque, L, and solve for L.

$$L = \frac{2\pi E_s r^2 \Delta r}{l} x$$

Multiply the right side again by r/r to obtain the quantity x/r, which is the angle ϕ. This gives

$$L = \frac{2\pi E_s r^3 \Delta r}{l} \phi$$

The relation between the torque and the angle of twisting of the thin cylinder is a direct relation, which can be written

$$L = k\phi \qquad \text{where} \qquad k = \frac{2\pi E_s r^3 \Delta r}{l}$$

The quantity k is called the torsion constant for that particular object. The smaller k is, the smaller the torque required for a given angular displacement and hence the more sensitive the device. To obtain a small k, r must be small and l must be as large as convenient.

The apparatus used to measure the force of gravity between objects in the laboratory, called the Cavendish apparatus, is shown in Figure 7–8. Two small balls are on the ends of a rod hanging on a thin torsion fiber. The large balls are brought near the small ones, and the gravitational force produces a torque to twist the fiber. A very small angle, ϕ, of twisting is detected by reflecting a light beam from a mirror attached to the bottom of the rod. The total displacement can be increased by moving the large balls to pull the small ones in the opposite direction, as in Figure 7–8.

Figure 7–8 The Cavendish apparatus, used to measure the gravitational force between two metal balls. The force twists a fine fiber.

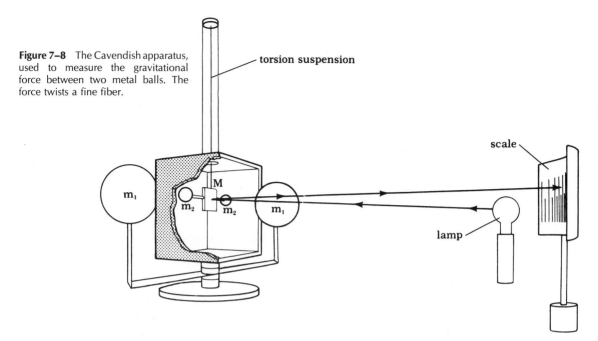

EXAMPLE 3

Calculate the torque that would twist a copper pipe through 10° (0.175 radian). The length is 2.0 meters, the radius is 0.8 cm, and the wall thickness is 1.0 mm. The shear modulus is 4×10^{10} N/m².

Use $$L = k\phi \qquad \text{where} \qquad \phi = 0.175 \text{ radians}$$

Then $$k = \frac{2\pi E_s r^3 \Delta r}{l}$$

in which

$$E_s = 4 \times 10^{10} \text{ N/m}^2$$
$$r = 0.008 \text{ m} = 8 \times 10^{-3} \text{ m}$$
$$r^3 = 5.12 \times 10^{-7} \text{ m}^3$$
$$\Delta r = 1 \text{ mm} = 10^{-3} \text{ m}$$
$$l = 2.0 \text{ m}$$

$$k = \frac{2\pi \times 4 \times 10^{10} \text{ N} \times 5.12 \times 10^{-7} \text{ m}^3 \times 10^{-3} \text{ m}}{2.0 \text{ m} \qquad \text{m}^2}$$

$$= 64 \text{ N m}$$

The torque constant is 64 N m/radian; then

$$L = k\phi$$

$$= 64 \frac{\text{N m}}{\text{rad}} \times 0.175 \text{ rad}$$

$$= 11 \text{ N m}$$

The torque required is 11 N m.

7–3   CHANGES IN BULK

Changes in volume or in "bulk" are described by another modulus of elasticity, often called the bulk modulus. Again the modulus is given by the ratio of stress to strain. The change in volume is caused by a change in forces on all sides of the object; in this case, the change in pressure ΔP is the stress. The strain is the ratio of change in volume $-\Delta V$ divided by the original volume V. The bulk modulus is described by

$$E_b = \frac{\Delta P}{-\Delta V/V}$$

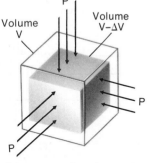

or

$$E_b = \frac{V\Delta P}{-\Delta V}$$

Figure 7–9 The change in volume of an object resulting from an increased pressure.

(see Figure 7–9). The negative sign is usually omitted, but it occurs because a positive pressure increase results in a decrease in volume. With this understanding, the bulk modulus will be considered as a positive quantity.

For most liquids and solids, the bulk modulus is very high; it is said that these materials are almost incompressible. Some typical bulk moduli are shown in Table 7–1.

EXAMPLE 4

Find by what fraction rocks are decreased in volume at a depth of 5 km below the surface of the earth compared to the surface volume. Though the bulk modulus for rocks varies from one type to another, and also rock is not homogeneous, a typical value of E_b is about 3×10^{10} N/m^2.

The pressure at 5 km is about 1500 times normal atmospheric pressure, or 1.5×10^8 N/m^2, and this is the change in pressure, ΔP.

Use

$$E_b = \frac{V\Delta P}{\Delta V}$$

from which

$$\frac{\Delta V}{V} = \frac{\Delta P}{E_b}$$

$$= \frac{1.5 \times 10^8 \text{ N/m}^2}{3 \times 10^{10} \text{ N/m}^2}$$

$$= 5 \times 10^{-3}$$

$$= 0.005$$

The rocks are compressed by a factor of 0.005, or the volume is decreased by 0.5 per cent.

This answer does not consider the whole problem because at 5 km depth the temperature is about 180°C above surface temperature. The added temperature causes expansion, and the net result will have to take this into account.

The bulk modulus of a gas, even though a gas is easily compressed, is never listed in tables. An analysis will show the reason.

If the original gas pressure is P_1 and it is increased to a value P_2, then $\Delta P = P_2 - P_1$. The corresponding volumes are V_1 and V_2, so $-\Delta V = V_1 - V_2$. Then the bulk modulus is described by

$$E_b = \frac{V_1(P_2 - P_1)}{V_1 - V_2} = \frac{P_2 V_1 - P_1 V_1}{V_1 - V_2}$$

For a gas, if the change in volume occurs at a constant temperature, then Boyle's law describes the situation and $P_1 V_1 = P_2 V_2$. Substituting, this results in

$$E_b = \frac{P_2 V_1 - P_2 V_2}{V_1 - V_2} = \frac{P_2(V_1 - V_2)}{(V_1 - V_2)} = P_2$$

The **bulk modulus of a gas** reduces merely to the **pressure**. Whereas the result above is actually P_2, the changes in practical situations are usually small and it is considered, oddly enough, that the bulk modulus of a gas is equal to the pressure P_1. This is the reason why the bulk moduli of gases are not found listed in tables. It must be remembered that this result is only for isothermal processes, for which the product of pressure and volume is a constant. If the gas is allowed to heat as it is compressed (or cool as it expands) by having no heat exchange with its surroundings, the process is called adiabatic (see Section 6–4). A given compression will require a greater force because of the heating, and the bulk modulus is increased by an amount γ, the ratio of the specific heats of the gas (Table 7–2). The adiabatic bulk modulus is then given by

$$E_b = \gamma P$$

TABLE 7–2 Gamma, γ, for some gases and vapors

Argon	A	1.667
Helium	He	1.66
Carbon monoxide	CO	1.404
Hydrogen	H_2	1.41
Hydrogen chloride	HCl	1.40
Nitrogen	N_2	1.404
Carbon dioxide	CO_2	1.304
Hydrogen sulfide	H_2S	1.32
Nitrous oxide	N_2O	1.303
Steam (100°C)	H_2O	1.324

7–4   ENERGY IN A DISTORTION

When a particular object is distorted, work is required. In stretching, for example, an applied force is moved through a certain distance. The work required gives potential energy to the distorted object. A spring may be stretched downward and hooked onto a weight; when it is released, it could possibly lift the weight. That is, the potential energy of the spring can be used to do work.

The potential energy of a stretched object is equal to the work done in stretching it. If Hooke's law applies, then for a stretched distance x the force is described by $F = kx$ where k is the elastic constant. The force changes as the object is stretched, and the work cannot simply be described by force times distance.

In finding the work done by a force which is not constant, the trick used is to consider very small displacements over which the force is almost constant. Then the bit of work done is the force times that small displacement. In Figure 7–10 is a graph of a force changing with distance in an arbitrary way. Over the small displacement shown as Δx, the force is the value shown as F. The work done in moving the force the distance Δx is $F\Delta x$. That is the height of the shaded strip in Figure 7–10(a) times its width, or the area. If the whole displacement x of Figure 7–10(b) is divided into strips, the area of each being the bit of work for a small displacement, then the total work in the displacement shown as X is the area under the curve.

The conclusion of this argument is that in the case of a force which changes with position, the work done in moving the force is the area under the curve of force against position, between the appropriate initial and final positions.

This is an instance in which it is useful to consider the work as being given by the area under the curve of force against displacement: Referring to Figure 7–11, the work done in stretching an object from $x = 0$ to an amount $x = X$ is given by the area of the shaded triangle. The height of the triangle is the force at $x = X$, and is given by kX. The base of the triangle is X, so the area is $\frac{1}{2}kX^2$.

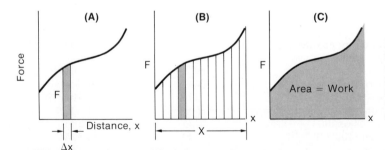

(A)   **(B)**   **(C)**

Force

F

F

F

Area = Work

Distance, x

X

x

x

Δx

Figure 7–10 Finding the work done if the force varies with position. (a) In a small movement Δx, the work is $F\Delta x$ or the area of the strip shown. (b) The total distance is divided into strips, and the total work is the sum of the areas of the strips, or the total area under the curve, as in (c).

The conclusion is that the work done to stretch an object by an amount X, if the force is described by $F = kx$, is given by:

$$\text{work} = \text{PE} = \tfrac{1}{2}kX^2$$

The work required to stretch an object can be expressed as simply $\tfrac{1}{2}kX^2$ in the region in which Hooke's law applies.

If the force is applied to the object being stretched, work is done on it and the object acquires potential energy. If, after a stretching force is applied, that force is slowly reduced, the stretched object pulls back on the object applying the force and does work on it. If the material is perfectly elastic, it will return to its original size and it will do as much work on whatever applied the force as was done on the material in producing the stretch.

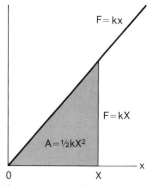

Figure 7–11 In the case of a force that is described by $F = kx$, the work done in moving a distance X is the area of the triangle shown.

EXAMPLE 5

A 1000 kg object moving horizontally at 5 m/s hits a spring-loaded bumper, which compresses 0.50 meter in bringing the object to rest. What is the spring constant of the bumper?

Solution: Find the kinetic energy of the object, and the potential energy of the compressed spring must be the same as that value.

$$KE = mv^2/2 = 1000 \text{ kg} \times 5^2 \text{ m}^2/2 \text{ s}^2$$
$$= 1.25 \times 10^4 \text{ N m}$$
$$PE = kX^2/2 = KE$$
$$k = \frac{2 \text{ KE}}{X^2} = \frac{2 \times 1.25 \times 10^4 \text{ N m}}{0.5^2 \text{ m}^2}$$

$$= 1.0 \times 10^5 \text{ N/m}$$

Answer: The spring constant must be 1.0×10^5 N/m.

This analysis leads to the subject of what happens when the stress goes beyond the limit for which Hooke's law applies, or when the type of material is not perfectly elastic. The phenomena actually become very complex, but an introduction to some of them will be given.

In Figure 7–12 are two curves for the extension of a human hair. With a small load on that particular specimen the elongation occurs quite suddenly and remains constant. With a higher load the specimen continues to stretch, and when the load is removed a permanent elongation remains. This phenomenon of a permanent change after a high stress for a long time occurs not only in biological materials but also in rocks. It is common to

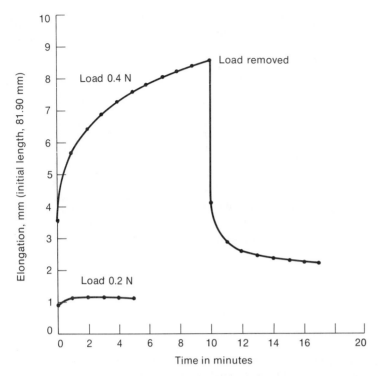

Figure 7–12 The stretching of a biological material with time. A hair has been used as an example.

see layers of sedimentary rock that have been bent or folded after the application of a large stress over a long period of time. Figure 7–13 shows an example.

The energy considerations are also important outside the region of application of Hooke's law. If an object is stretched

Figure 7–13 Rocks folded by being stressed beyond the elastic limit at their temperature at the time. (Photo by the author.)

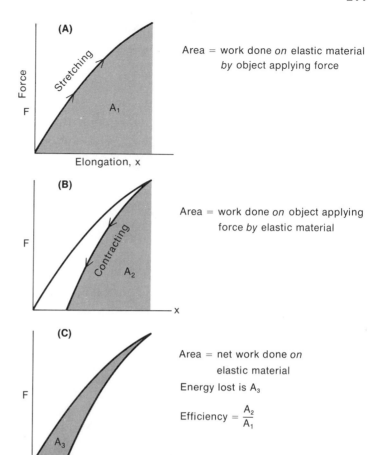

(A)

Force
F

Stretching

A_1

Elongation, x

Area = work done *on* elastic material
by object applying force

(B)

F

Contracting

A_2

x

Area = work done *on* object applying
force *by* elastic material

(C)

F

A_3

x

Area = net work done *on*
elastic material

Energy lost is A_3

Efficiency $= \dfrac{A_2}{A_1}$

Figure 7–14 As an object is stretched, the work done on it is the area shown in (a); as the object pulls back, it does work on the surroundings of an amount shown by the shaded area in (b). There is a net loss of energy, which is the area between the two curves in (c).

with a stress-strain curve as in Figure 7–14(a), the work done in stretching it is the shaded area under the curve. When the object is allowed to contract, the work done by the material on whatever applied the force is the shaded area in Figure 7–14(b). This work is not as much as was done on the material. It is possible to store energy in an elastic material for later use. For example, in the operating mechanism of the wings of some insects, the wings are not beat entirely by muscular energy; rather, the energy is stored in an elastic material for part of the cycle of the beat and then released. A simplified drawing of the system is in Figure 7–15. With each stretch and release of the elastic material there is some energy loss, given by the area between the curves in Figure 7–14(c). The efficiency of the material is the ratio of the work that it could do, A_2, divided by the energy stored in it, A_1. In this way elastic materials can be rated by their efficiency. The elastic material in the insect wing mechanism, called *resilin*, is an extremely efficient material. Resilin has an efficiency of 97 per cent, compared with 91 per cent for rubber.

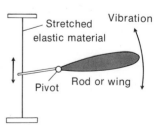

Stretched
elastic material

Vibration

Pivot

Rod or wing

Figure 7–15 A much simplified picture of how energy for the beating of an insect's wing is stored in an elastic material.

Elastic   Inelastic

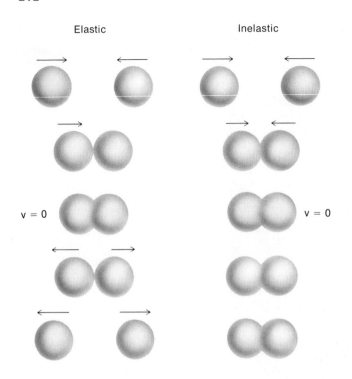

v = 0   v = 0

Figure 7–16 Elastic and inelastic collisions. In an elastic collision, the objects are distorted but return to their original shapes, pushing each other apart. In an inelastic collision, the objects remain together; there is no tendency for them to return to their original shapes after distortion.

7–5   COLLISIONS

When two objects collide, the force between them builds up and each of them is distorted by the force. If the objects are elastic, after reaching the maximum force and distortion they push apart as they return to their original, undistorted shape, as in Figure 7–16. If the objects are completely inelastic, after the distortions have occurred there is no tendency for them to be removed; the objects do not push apart but remain in contact.

7–5–1   PERFECTLY INELASTIC COLLISIONS

In a perfectly inelastic collision the objects stay together and move as one after the collision. The forces of each on the other are at all instances the same size, though opposite in direction, so the impulses are of the same size and momentum is conserved. However, work is done by each in distorting the other, and the energy for this comes from the kinetic energy of the objects. This amount of kinetic energy is changed to heat, the heat of percussion. So in an inelastic collision, momentum is conserved but some kinetic energy is lost.

In any collision there are the two objects moving toward each other — the "before" picture. After the collision each object has a different velocity, giving an "after" picture. A positive velocity may be defined as being to the right for all objects, or a diagram with vectors shown may be used to define positive directions for each velocity.

Consider the case of an object m_1 moving to the right with a velocity v_1 while m_2 is moving toward it (to the left) with a velocity v_2 as in Figure 7–17. Let us describe the situation after the collision if the collision is perfectly inelastic.

The total momentum before the collision is $m_1v_1 - m_2v_2$ (the momentum to the left will be negative if the momentum to the right is positive). After the collision the objects move together with a common velocity v_3. The momentum after the collision is $(m_1 + m_2)v_3$. Note how any collision situation has two pictures, a "before" and an "after."

By conservation of momentum

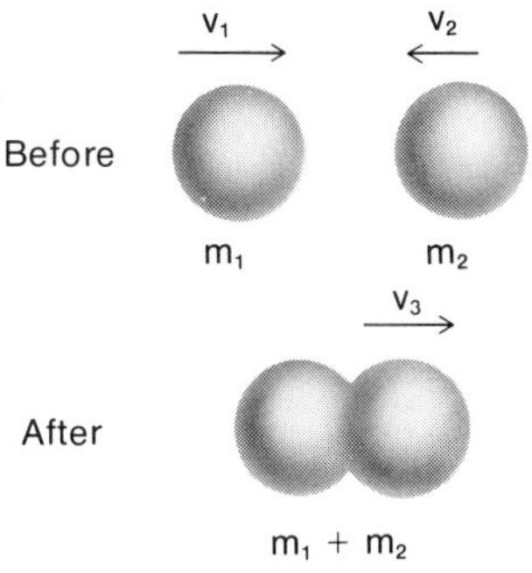

Figure 7–17 An inelastic collision. The velocity vectors show the positive direction for their numerical values.

$$m_1v_1 - m_2v_2 = (m_1 + m_2)v_3$$

EXAMPLE 6

An object of mass 0.2 kg moving to the right at 5 m/s strikes a mass of 1.5 kg that was moving to the left at 2 m/s. The collision is perfectly inelastic. Find the common velocity after the collision and also find the total kinetic energy before and after the event.

Figure 7–17 will serve as a diagram with the appropriate numerical values; v_3 is the unknown.

Use $$(m_1 + m_2)v_3 = m_1v_1 - m_2v_2$$

Solve for v_3 $$v_3 = \frac{m_1v_1 - m_2v_2}{m_1 + m_2}$$

where $$m_1 = 0.2 \text{ kg}$$

$$v_1 = 5 \text{ m/s}$$
$$m_2 = 1.5 \text{ kg}$$

$$v_2 = 2 \text{ m/s}$$

Then $$v_3 = \frac{(0.2 \text{ kg})(5 \text{ m/s}) - (1.5 \text{ kg})(2 \text{ m/s})}{1.7 \text{ kg}}$$

$$= -1.18 \text{ m/s}$$

The velocity is −1.18 m/s in the direction shown by v_3 in Figure 7–17. So the velocity after the collision is 1.18 m/s to the left.

The kinetic energy before was

$$\text{KE (before)} = \tfrac{1}{2}m_1v_1^2 + \tfrac{1}{2}m_2v_2^2$$
$$= 2.5 \text{ J} + 3 \text{ J} = 5.5 \text{ J}$$

The kinetic energy after the collision is

$$\text{KE (after)} = \tfrac{1}{2}(m_1 + m_2)v_3^2$$
$$= \tfrac{1}{2}(1.7 \text{ kg})(-1.18 \text{ m/s})^2$$
$$= +1.18 \text{ J}$$

This is less than the kinetic energy before the collision. The kinetic energy is reduced by the amount that was required to cause the distortion of the objects. It appears as heat.

7-5-2 ELASTIC COLLISIONS

The elastic collision, or perfectly elastic collision, is one in which there is conservation of kinetic energy as well as conservation of momentum. These two conditions allow two equations to be written, and hence two unknowns can be found. These may be the velocities of the two objects after the collision. The momentum involves the first power of the velocity and the kinetic energy involves the square of the velocity. As a result, the unknowns are a little awkward to find and a technique has been developed to simplify this. The technique is also aplicable to collisions that are neither perfectly elastic nor perfectly inelastic. So the équations will be set down and manipulated to show the new approach.

In this discussion the diagram will be made with all velocities to the right, so that will be the positive direction, and negative numerical values will be to the left. The elastic collision is shown in Figure 7–18. Writing "momentum before equals momentum after."

$$m_1 v_1 + m_2 v_2 = m_1 v_3 + m_2 v_4$$

or
$$m_1 (v_1 - v_3) = m_2 (v_4 - v_2)$$

The total kinetic energy is the same before as after, so

$$\tfrac{1}{2} m_1 v_1^2 + \tfrac{1}{2} m_2 v_2^2 = \tfrac{1}{2} m_1 v_3^2 + \tfrac{1}{2} m_2 v_4^2$$

Multiply this equation by 2, and gather the terms in m_1 and in m_2 to get

$$m_1 (v_1^2 - v_3^2) = m_2 (v_4^2 - v_2^2)$$

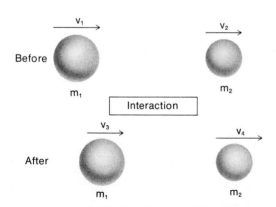

Figure 7–18 Diagrams illustrating an elastic collision.

Use the general relation, $(a^2 - b^2) = (a - b)(a + b)$, to write

$$m_1(v_1 - v_3)(v_1 + v_3) = m_2(v_4 - v_2)(v_4 + v_2)$$

but we had

$$m_1(v_1 - v_3) = m_2(v_4 - v_2)$$

These two equations will apply only if

$$(v_1 + v_3) = (v_4 + v_2)$$

and thus

$$v_1 - v_2 = v_4 - v_3$$

The quantity $v_1 - v_2$ is the speed at which the two objects are getting closer together. If v_1 was 3 m/s to the right and v_2 was −1 m/s to the right or 1 m/s to the left, then they are approaching at 4 m/s and $v_1 - v_2 =$ 3 m/s − (−1) m/s = 4 m/s. Similarly, $v_4 - v_3$ is the speed at which they are moving apart or the speed of separation after the collision. That last equation then says that in an elastic collision,

$$\text{speed of approach} = \text{speed of separation}$$

This relation can be used in place of the conservation of kinetic energy, along with conservation of momentum, to solve problems in perfectly elastic collisions.

EXAMPLE 7

A mass of 5 kg moves to the right at 10 m/s while a mass of 0.2 kg moves toward it, to the left at 10 m/s. Find the velocities of the masses after the collision.

Referring to Figure 7–18, $m_1 = 5$ kg and $m_2 = 0.2$ kg; v_1 is 10 m/s and v_2 is −10 m/s. By conservation of momentum,

$$m_1v_1 + m_2v_2 = m_1v_3 + m_2v_4$$

Putting in the numerical values and cancelling the kilogram units gives

$$5 v_3 + 0.2 v_4 = 48 \text{ m/s}$$

The speed of approach is 20 m/s, and this is the same as the speed of separation, so

$$v_4 - v_3 = 20 \text{ m/s}$$

Solve this for v_4, put it in the equation above, and solve for v_3:

$$v_4 = 20 \text{ m/s} + v_3$$

$$5 v_3 + 0.2(20 \text{ m/s} + v_3) = 48 \text{ m/s}$$

Solving for v_3,

$$5.2 v_3 = 44 \text{ m/s}$$
$$v_3 = 8.5 \text{ m/s}$$

and it follows that

$$v_4 = 28.5 \text{ m/s}$$

After the collision the 5 kg mass will have lost only a small amount of speed and will be moving at 8.5 m/s, while the small mass will be moving to the right at 28.5 m/s.

This situation is somewhat like a heavy bat striking a baseball, though the figures are not claimed to apply accurately.

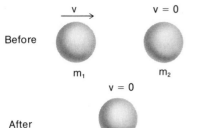

Before

v = 0

m₁   m₂

v = 0

After

m₁   m₂

v

Figure 7–19 A perfectly elastic collision between two equal masses, one of them initially at rest.

An interesting special case of an elastic collision is one in which the masses are equal and one of the bodies is initially at rest; that is, $m_1 = m_2$ (call it m), and $v_2 = 0$. Then

$$m_1 v_1 + m_2 v_2 = m_3 v_3 + m_2 v_4$$

becomes, since the masses cancel and $v_2 = 0$,

$$v_1 = v_3 + v_4$$

Since speed of approach equals speed of separation,

$$v_1 = v_4 - v_3$$

Then $v_4 - v_3$ must be the same as $v_4 + v_3$, and this will occur only if v_3 is zero and $v_1 = v_4$. The striking object will stop; the object struck will move off with the speed with which it was struck. This is shown in Figure 7–19. Have you ever seen this phenomenon?

7–5–3 INELASTIC COLLISIONS

In a perfectly inelastic collision the speed of separation is zero, and in a perfectly elastic collision the speed of separation is equal to the speed of approach. Between these two limiting types of collision is the inelastic collision for which the speed of separation can be anything from zero to nearly the speed of approach. The degree of elasticity of a particular collision is measured by the ratio of those speeds and is called the **coefficient of restitution,** for which the symbol e is often used.

$$e = \frac{\text{speed of separation}}{\text{speed of approach}}$$

If $e = 0$, the collision is perfectly inelastic.

If $e = 1$, the collision is perfectly elastic.

Using the velocities shown in Figure 7–18,

$$v_4 - v_3 = e\,(v_1 - v_2)$$

EXAMPLE 8

If a ball is dropped from a height of 1.00 meter and on the first bounce rises to 0.50 meter, what is the coefficient of restitution between the ball and the surface on which it fell?

After falling 1 meter the speed, from $v = \sqrt{2gh}$, is 4.4 m/s. In order for it to rise 0.50 meter, the speed of rebound, also from $v = \sqrt{2gh}$, is 3.1 m/s.

The ball hit at 4.4 m/s and rebounded at 3.1 m/s. The coefficient of restitution is

$$e = \frac{3.1 \text{ m/s}}{4.4 \text{ m/s}} = 0.70$$

7–6 MEMBRANE TENSION

The phenomenon of tension in films or curved membranes occurs in a variety of situations. These include toy balloons, bladders, the heart wall, pipes or rubber tubing, arteries, the cell membrane, and even soap bubbles. The pressure inside a curved membrane must be greater than that outside, and the pressure difference is related to the tension in the membrane. The tension may be caused by pressure, as in a container holding a liquid under high pressure; or the pressure may be caused by the tension, as in the case of the heart producing the pressure in the blood.

If a membrane is under tension, at any place on the membrane a line could be drawn and forces imagined pulling in opposite directions on the two sides of that line. If the membrane were cut it would separate; but otherwise, molecular forces act across that line, as in Figure 7–20, to hold it together. The membrane tension would be described by the force per unit length T along the line. In the case of a liquid this is called surface tension. For water, the surface tension at room temperature is about 0.072 N/m. A film of water, as in a soap bubble, has two

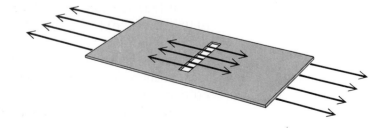

Figure 7–20 There are tension forces pulling both ways across any line on a stretched membrane.

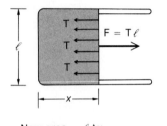

New area = $\ell \Delta x$

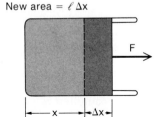

Figure 7–21 A film over a frame can be stretched to produce a new area. Work is done to produce that new area.

surfaces (inside and outside), so the tension is double the surface tension.

There are actually two basic ways to interpret membrane or surface tensions. One, already mentioned, is the force per unit length of the edge of the surface; the other is the energy stored in the surface. Referring to Figure 7–21, consider the membrane or film stretched across a U-shaped wire, with one end held in position by a sliding bar. The force needed to hold the bar in position, F, is the force per unit length along it times the length, l. That is, $F = Tl$. If the bar is moved a short distance Δx (the Δ symbol again means "a small change in"), the amount of work done is $F \Delta x$. In terms of the membrane tension this work is given by

$$W = F \Delta x = Tl \Delta x$$

but the quantity $l\Delta x$ is the increased area ΔA, and the result is that the work W (or energy) used to produce the area ΔA is $W = T \Delta A$. The energy needed to produce a unit area is $W/\Delta A$, and thus the result is:

$$W/\Delta A = T$$

This indicates that the surface tension is also given by the energy per unit area of this surface.

7–6–1 TYPES OF MEMBRANES

Membranes can be classified according to the way in which the tension varies with the amount that the membrane is stretched. One type is that which has a constant tension (there is no change in tension with change in area). An example of this type of membrane is the liquid surface. The surface tension is due to the forces between molecules on the surface, and these forces are not dependent on the total surface area. Bubbles formed from water have a constant tension in the surface, irrespective of size.

Some liquids do have a variable surface tension. The material in the lungs is one of these. As the area increases, so does the surface tension. It may vary from a low of 5×10^{-5} N/m (0.05 dynes/cm) to 5×10^{-2} N/m (50 dynes/cm) with a five-fold increase in area. This variation is illustrated in Figure 7–22.

Some membranes are elastic, and as they are stretched the tension increases. The change in tension is in many cases proportional to the elongation of the membrane; that is, Hooke's law will often describe the variation in tension. In Figure 7–23

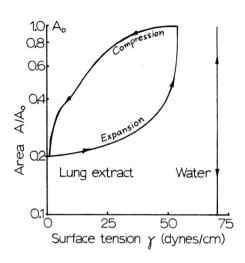

Figure 7–22 The surface tension of a film of lung material changes as the area changes. This curve is from the work of Hildebrandt and Young (1961). The surface tension of water is constant, independent of area.

is shown the manner in which the tension in the wall of the bladder of a dog depends on the circumference. The tensions were calculated from measured pressure-volume data considering a spherical surface. At small radii, when the bladder is probably adjusting to its eventual almost spherical shape, the calculated tension is not proportional to size; but then over a considerable region there is a linear relation between tension and circumference. The elastic constant for this particular membrane is 18 N/m. The curve of Figure 7–23 was obtained with the animal under general anesthesia and also with local anesthesia to prevent muscle tension. Without anesthesia the muscles of the bladder wall will at some stage be stimulated, and the tension will be greatly increased. The heart wall was cited also as an example of the type of membrane in which a large tension is produced on stimulation of the muscles in the membrane.

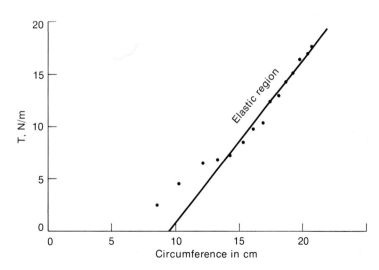

Figure 7–23 The tension in the wall of the bladder of a dog changes with the bladder circumference. The points are based on data from T. C. Ruch and H. D. Patton (eds.), *Physiology and Biophysics,* 19th Edition, W. B. Saunders Company, Philadelphia, 1965.

7–6–2 SPHERICAL MEMBRANES

Anyone who has blown a bubble or a balloon is aware that a pressure is required inside and that the amount of pressure depends on the "stiffness" of the membrane (or, rather, on the membrane tension). To find how membrane tension and pressure are related, consider a spherically curved surface with a small portion sliced off, as in Figure 7–24(a). This "cap" is held on by the membrane forces shown as T, while a force F, caused by the inside pressure, tends to push it off. F pushes only against the downward component of the tension forces.

The sideways tension forces on opposite sides of the cap just pull against each other. These components are shown in part (b) of the figure. The force F pulls only against the component $T \sin \theta$, where T is the force per unit length around the cap. The total force around the edge of the cap, working against F, is the force per unit length, $T \sin \theta$, times the distance around the cap, $2\pi r \sin \theta$, as in part (c) of the figure. So

$$F = T \sin \theta \times 2\pi r \sin \theta$$
$$= T \times 2\pi r \sin^2 \theta$$

To hold the membrane in a spherical shape, there must always be such an outward force F. For a toy balloon to be spherical there must be a higher pressure inside the balloon than outside it. The spherical shape of a membrane, in fact, is maintained by this pressure. If the pressure outside is P_o, the pressure inside must be greater; say it is $P_o + P$. The net outward

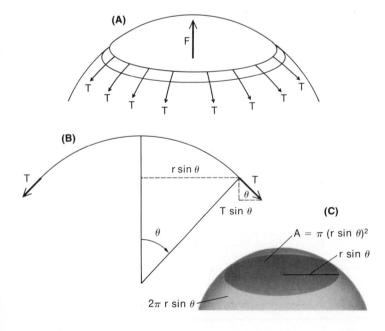

Figure 7–24 A "cap" from part of a spherical membrane is held on by tension forces around the edge, and pushed out by pressure forces.

force on each unit area of the cap is P. The total outward force is P times the area of the cap. If a small "cap" is considered to be almost (but not quite) flat, the area is $\pi(r \sin \theta)^2 = \pi r^2 \sin^2 \theta$. (See Figure 7–24.) The outward pressure force is then given by $F = P \times \pi r^2 \sin^2 \theta$. This balances the inward tension force on the cap, which was given by $F = T \times 2\pi r \sin^2 \theta$. Equating these,

$$P\pi r^2 \sin^2 \theta = T \times 2\pi r \sin^2 \theta$$

which reduces to

$$Pr = 2T \quad \text{or} \quad P = 2T/r$$

This equation, relating the pressure inside a spherical membrane, the tension, and radius of curvature, is known as the **Law of Laplace as applied to a spherical surface.** Some examples of the application of this law to a variety of situations follow.

First consider a soap or water bubble in air. This is the type of film for which the tension is constant no matter what the size, and it will be just double the listed surface tension of water because the film has two surfaces. For bubbles of different sizes, it is shown by the equation above that the excess pressure in the bubble can be expected to vary inversely as the radius. Large bubbles will have lower pressure in them than small ones. Very small bubbles can, in fact, have fairly large pressures. This seems to be nonsense, but a little reflection shows that it could be valid. A large bubble may have a lot of air in it, but it could be at only a small pressure excess above atmospheric.

One way to check this is to produce two bubbles, a large one and a smaller one, in contact with each other as in Figure 7–25. The dividing membrane will bulge into the volume that is at the lowest pressure. This is left for observation.

A second example to consider concerns the tension forces in the walls of the heart required to produce a given pressure P on the blood enclosed. Solving for T yields

$$T = rP/2$$

To produce a given pressure P, the tension in a wall varies directly as the radius. An enlarged heart will require greater muscle tension to produce the same pressure than would a smaller heart. Hence, the walls of the large heart would be more liable to damage from stress or excess tension than would the walls of a small heart.

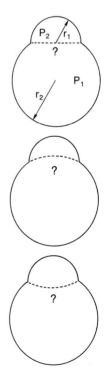

Figure 7–25 Bubbles of different sizes have different pressures. In what direction does the membrane separating two bubbles bulge?

EXAMPLE 9

Find the tension in the wall of a spherical chamber if the radius is 0.5 cm (5×10^{-3} m) and the pressure inside is 1.33×10^4 N/m².
Using the law of Laplace,

$$T = rP/2$$
$$P = 1.33 \times 10^4 \text{ N/m}^2$$
$$r = 5 \times 10^{-3} \text{ m}$$
$$T = \tfrac{1}{2} \times 5 \times 10^{-3} \text{ m} \times 1.33 \times 10^4 \text{ N/m}^2$$
$$= 0.334 \times 10^2 \text{ N/m}$$
$$= 33.4 \text{ N/m}$$

The tension is 33.4 newtons per meter.
This situation is similar to a chamber of the heart producing the blood pressure.

The red cell is also an example of another aspect of this subject. If red cells are put into distilled water, the water moves by osmotic pressure through the membrane into the cell. The resulting pressure may cause the tension in the cell wall to exceed its limit, and it ruptures, spilling its contents into the water and leaving behind an empty shell or "ghost" cell. If the cell is not to be ruptured, physiological saline solution must be used as a medium for suspension.

A further situation concerns the data that are presented graphically in Figure 7–23 about the tension in a bladder wall. This tension is not measured directly; the measurable quantities are the pressure P and the volume V. The radius can be calculated from the volume ($V = 4\pi r^3/3$). Then P and r can be substituted into the law of Laplace to find the tension in the wall.

EXAMPLE 10

Find the tension in the wall of a circular chamber, given that when the volume is 100 cm³ the pressure is 980 N/m². (This is the same as the pressure at the bottom of a column of water 10 cm high, and is sometimes referred to as a pressure of 10 cm of water.)
From the volume calculate the radius:

$$V = 4\pi r^3/3$$

$$r = \sqrt[3]{\frac{3V}{4\pi}}$$

$$V = 100 \text{ cm}^3$$

$$r = \sqrt[3]{\frac{300 \text{ cm}^3}{4\pi}}$$

$$= 2.88 \text{ cm} = 2.88 \times 10^{-2} \text{ m}$$

$$P = 9.8 \times 10^2 \text{ N/m}^2$$

Using the law of Laplace,

$$T = rP/2$$
$$T = \tfrac{1}{2} \times 2.88 \times 10^{-2} \text{ m} \times 9.8 \times 10^{2} \text{ N/m}^2$$
$$= 14 \text{ N/m}$$

The tension in the wall is 14 newtons per meter.
These data apply to the bladder of the dog as illustrated in Figure 7–23.

7–6–3   LIQUID SURFACES

Liquid drops give the appearance of being held together by a membrane. Liquid surfaces exhibit a force across a line or along an edge, and the force per unit length is given the name **surface tension.** Soap bubbles have already been mentioned. Another demonstration of the surface tension forces is the rise of a liquid such as water in a narrow tube, a capillary tube for example. The phenomenon is illustrated in Figure 7–26.

If the liquid in a glass tube wets the surface of the tube, the liquid surface is parallel to the glass at the point of contact. The surface tension forces, T in Figure 7–26, pull upward around the edge and it is that upward force that supports the weight of the liquid in the tube. The upward force is the distance around the tube $(2\pi r)$ times the force per unit length, T. The total upward force is then $2\pi r T$. The mass of the liquid of density d in a

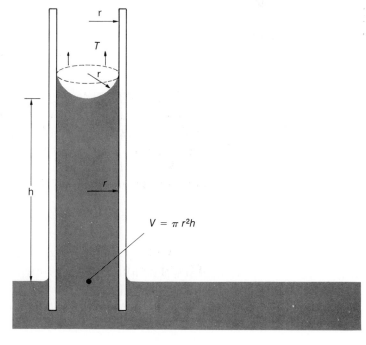

Figure 7–26   The rise of fluid in a capillary tube.

column of length h and radius r is the volume $\pi r^2 h$ times d. The weight is mg or $\pi r^2 hdg$. Equating the upward and downward forces gives

$$2\pi rT = \pi r^2 hdg$$

or
$$T = rhdg/2$$

From this equation, the height to which the liquid would rise is given by

$$h = 2T/rdg$$

This shows that the smaller the radius, the greater the height to which the liquid rises. Capillary tubes are often used to obtain blood samples. Capillary rise is one of the phenomena that cause sap or other fluids to rise in trees and plants.

EXAMPLE 11

Find the height to which water would rise in a tube of radius 0.05 mm or 5×10^{-5} m. (The diameter is a tenth of a millimeter.) The surface tension of water is about 0.072 N/m.

Use $\quad\quad h = 2T/rdg$

where
$$T = 0.072 \text{ N/m} = 0.072 \text{ kg/s}^2$$
$$r = 5 \times 10^{-5} \text{ m}$$
$$d = 1000 \text{ kg/m}^3$$
$$g = 9.8 \text{ m/s}^2$$

$$h = \frac{2 \times 0.072 \text{ kg/s}^2}{5 \times 10^{-5} \text{ m} \times 10^3 \text{kg/m}^3 \times 9.8 \text{ m/s}^2}$$

$$= 2.9 \times 10^{-1} \text{ m} = 29 \text{ cm}$$

The water will rise 29 cm.

The capillary rise phenomenon is more complicated if the inside of the tube is not wetted by the liquid so that the surface tension forces do not pull vertically but at an angle θ, as in Figure 7–27. The upward component is $T \cos \theta$ and the total upward force is $(T \cos \theta) \times 2\pi r$. This balances the weight of the liquid in the tube, and in this case the height is given by

$$h = 2T \cos \theta/rdg.$$

Capillary rise is often used to measure surface tension. One of the difficulties with the method is that the liquid may not wet the inside surface of the tube so there will be an angle of contact θ. With water in a glass tube, for example, the glass must be extremely clean and the tube must be moved up and down in

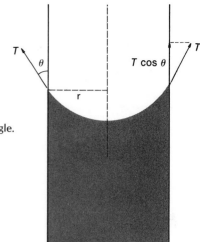

Figure 7–27   Capillary rise in the case in which the fluid meets the wall at an angle.

the water to be assured that the inside surface is wetted above the line of contact.

7–6–4 MEASUREMENT OF SURFACE TENSION

If surface tension effects are important, it may be necessary to measure them sometime. How is this done?

In a pinch, the capillary rise method can be used. This requires only dipping a capillary tube into the solution and measuring the height to which the liquid rises. The relation between surface tension and height has already been derived.

A more precise method to measure surface tension is based on actual measurement of the tension along a length of film. The basic part of the apparatus is shown in Figure 7–28. The force needed to just break the film may be measured by any delicate force-measuring apparatus. Often a torsion balance arrangement is used. This makes use of the force required to twist a fine wire. Each piece of commercial apparatus for measuring surface tension may have different details, but the principles will be similar. Sometimes, for instance, in place of a flat plate, the film is formed around a wire loop that is pulled out of the solution. If l is the length of the plate or distance around the loop, the downward force just before the film breaks is $2lT$. The 2 enters the formula because the film has two surfaces. If the force needed to pull the film out is F, then

$$F = 2lT$$
or $$T = F/2l$$

In using this type of apparatus, the dish must be extremely

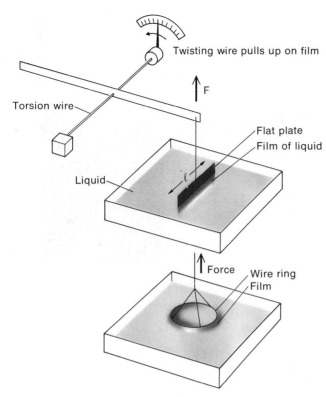

Twisting wire pulls up on film

Torsion wire

Flat plate
Film of liquid

Liquid

F

l

Force   Wire ring
          Film

Figure 7–28 Apparatus used to measure surface tension. The force needed to break a linear film (a) or a circular film (b) is measured.

clean, and the ring or plate will have to be cleaned in either a cleaning solution or a gas flame.

7–6–5 LIQUID DROPS

The basic shape for a liquid drop is spherical. One of the reasons for this is that the sphere has a lower ratio of surface to volume than does any other shape. Thus, the surface energy is lowest for a spherical shape. When many small drops coalesce to form a large one, the surface area of the large drop is less than the total surface area of all the small ones. The loss of surface energy is converted to heat energy to warm the drops and the surroundings slightly. It is similar to the heat of fusion (or condensation) and its origin is the same — the molecular forces.

In practice, drops often deviate from spherical shapes. A drop on a table is flattened, and its shape depends on whether or not it wets the surface. Raindrops are distorted by the air rushing past them, and they take on a variety of shapes as they fall.

7–6–6 CYLINDRICAL MEMBRANES

Rubber tubes and blood vessels are another shape of membrane in which film tension forces confine a fluid under pressure.

The analysis used to relate the pressure, radius, and film tension is similar to that for spherical surfaces, and is carried out below.

Figure 7–29(a) shows a tube with a small portion cut from the side, and part (b) of the figure shows the end of such a section. The tension forces *along* the tube have no components downward and contribute nothing to balance the pressure in the tube. The downward components holding that section onto the tube are $T \sin \theta$ per unit length. The total downward force on both sides of the section amounts to $T \sin \theta \times 2l$.

The outward force is the pressure times the area of the slab. The width of the slab is $2 \times r\theta$, where θ is the angle in radians. The length is l, so the area becomes $2r\theta l$. Equating inward and outward forces,

$$P \times 2r\theta l = T \sin \theta \times 2l$$

If a small section is considered so that θ is small, then $\sin \theta$ is equal to θ and will cancel with it. The 2 and the l on each side cancels to leave

$$Pr = T \quad \text{or} \quad P = T/r$$

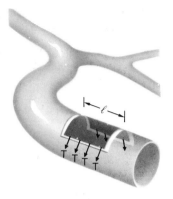
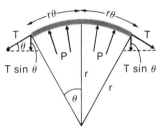

Figure 7–29 A cylindrical tube (such as an artery) with a small portion cut from it to show the tension forces that balance the pressure forces.

This is the law of Laplace applied to a cylindrical surface. For a spherical surface the relation was that $P = 2T/r$ or $T = Pr/2$. For a given pressure and radius, the tension in a cylindrical object is double that for a spherical object. This can readily be seen with a long rubber balloon (Fig. 7–30). The side of such a balloon is much more rigid than the end because of the higher tension in the cylindrical portion.

Some of the implications of this relation, $T = Pr$, are of interest. To contain high pressure fluids in a rubber or plastic tube, the tension in the wall depends not only on the pressure but directly on the radius. For example, a tube with a bore of 1 mm will, for a certain wall thickness, and therefore possible tension, withstand 10 times as much pressure as would a 10 mm (1 cm) diameter tubing of the same wall thickness. Laboratory apparatus working with high pressures may sometimes make use of small bore tubing that has a relatively thin wall. Tubing of large bore would require a much thicker wall to contain the same pressure.

The same relation applies to arteries in the body. For a large artery with blood at a high pressure, the tension around the walls can be fairly high. For the smaller branch arteries the necessary wall strength decreases directly with the radius, until the capillaries can have very high pressures with very thin walls.

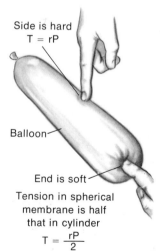

Side is hard
$T = rP$

Balloon

End is soft

Tension in spherical membrane is half that in cylinder

$$T = \frac{rP}{2}$$

Figure 7–30 For a given pressure, the tension forces around a cylinder are double those in a sphere; this is demonstrated by the rigidity of the long side of a balloon and the comparative "softness" of the round end.

EXAMPLE 12

A calculation is of interest here. Consider a tube of radius 1 cm (10^{-2} meter) and a pressure of 1.33×10^4 newtons per square meter. These figures are approximate for a large artery. Then

$$T = Pr = 1.33 \times 10^4 \text{ N/m}^2 \times 10^{-2} \text{ m}$$
$$= 133 \text{ N/m}$$

The tension in the wall is 133 N/m or 1.33 N/cm. A piece of this membrane one inch wide would be under this tension if it was supporting a weight of three quarters of a pound.

EXAMPLE 13

A cylindrical structure 1.5 m in radius and 12 m high is to hold a fluid, or even a pelleted solid or a powder that in such a situation would behave like a fluid. A typical pressure at the bottom of such a structure would be 1.0×10^5 N/m². Find the tension around the wall at the bottom.

The tension is again given by the law of Laplace for a cylindrical membrane.

$$P = Tr$$
$$= 1.0 \times 10^5 \text{ N/m} \times 1.5 \text{ m}$$
$$= 1.5 \times 10^6 \text{ N/m}$$

The tension around the wall is 1.5×10^6 newtons per meter across a vertical line on the wall. Such structures often have extra strengthening around the wall.

An infinitely long tube would require no forces along its length, but if the tube terminates in any way there are outward forces on each end, which amount to the pressure times the cross-sectional area of the tube. A longitudinal tension in the walls is required to hold them together. The size of this longitudinal tension T_L may be easily calculated. The total outward force on an end is pressure times area, $P \times \pi r^2$. If, at some position along the tube, a line is imagined around it as in Figure 7–31, the force tending to separate the tube along the line is T_L per unit length. The distance around the tube is $2\pi r$, so the total force is $T_L \times 2\pi r$. This balances the pressure forces on the ends, so

$$T_L \times 2\pi r = P \times \pi r^2$$

from which $\qquad T_L = \dfrac{Pr}{2}$ (tension along the tube)

The tension around the circumference was

$$T = Pr \text{ (tension around the tube)}$$

The conclusion is that the longitudinal tension in the wall of a cylindrical tube is just half of the tension around the circumference. Tubes are more likely to rupture in a line along their length because the forces across that line are twice as great as those along it.

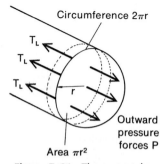

Circumference $2\pi r$

T_L
T_L
T_L
r

Outward pressure forces P

Area πr^2

Figure 7–31 There must be forces across a line drawn around a tube, because it must terminate somewhere and the forces on the ends must be balanced. This longitudinal force is only half of the force around the tube.

DISCUSSION PROBLEMS

7-1   1.   Make a list of at least twenty materials, with a short comment on the elasticity of each. Include various metals, woods, even tissue. (Poke your finger into a fleshy part of yourself and remove it. Is it elastic? In some pathological conditions the tissue is not.) How is a test for elasticity used to tell whether or not a cake is done? Use your imagination in completing the list and, whenever possible, make a test of the material.

7-1-1   2.   Discuss the use of a spring scale to determine the masses of objects at various places on earth. Could it be graduated in mass units, or would it be necessary that standard masses be used at each location?

7-2-1   3.   When a helical spring is stretched, does the wire of the spring elongate or twist? Which elastic modulus is involved?

7-5-2   4.   (a) Show analytically that if two equal masses moving in a straight line are in a perfectly elastic collision, they merely interchange velocities.

(b) Four equal masses of perfectly elastic material are in a line, at rest and in contact, and are struck by a fifth similar mass moving with a velocity, v. Describe the result.

7-6   5.   Soap films assume the surface of least energy considering their enforced boundaries. It is instructive and enjoyable to make a series of shapes from wire (squares, cubes, cylinders, and so forth) and dip them into a soap solution to produce films. A detergent solution with glycerine or sugar added will make long lasting films.

7-6-6   6.   Discuss the relative merits of cylindrical and spherical shapes for vessels designed to contain high pressure.

PROBLEMS

7-1   1.   What is the stress on a wire that is 0.5 mm in diameter when it supports an object having mass of 2 kg? Express the answer in N/m^2.

7-1   2.   If a 20 cm object is stretched 0.3 mm, what is the strain?

7-1   3.   If a wire 20 cm long and 0.50 mm in diameter, in supporting a 2 kg mass, elongates by 0.3 mm, what is Young's modulus for the material of which the wire is made?

7-1   4.   A metal rod 0.30 meter long and 3 mm in diameter stretches by 0.5 mm when pulled with a force of 200 newtons. Find (a) the stress, (b) the strain, and (c) the modulus of elasticity.

7-1   5.   A bone with a cross-sectional area of 1.5 cm^2 is 10 cm long and is subject to a compressional force of 2000 newtons. Young's modulus is 2×10^{10} N/m^2. Find (a) the stress, (b) the strain, and (c) the change in length.

7-1   6.   The diameter of a very fine glass fiber is to be found by measuring the elongation produced by a known force and making use of Young's modulus for glass. The data are: the length is 0.15 m, and when supporting 2.0 grams the fiber stretches 0.5 mm. Young's modulus is 7×10^{10} N/m^2. Express the diameter in μm.

7-1-1   7.   Find the elastic constant of a spring that stretches 2.9 cm when it supports a mass of 0.15 kg.

7-1-1   8.   Calculate the elastic constant of a steel wire 0.15 cm in diameter and 60 cm long. Young's modulus is 2.0×10^{11} N/m^2.

7-2   9.   A block of steel 2.54 cm on a side is subjected to opposing forces of 6.00×10^5 N on opposite faces. Through what angle is the block sheared? Express the angle in radians and in degrees.

7-2   10.   What is the shearing stress when over a distance of 10 km rocks are sheared through a distance of 5 m? The shear modulus is 1.5×10^{10} N/m^2.

7-2-1   11.   A tube of aluminum 1.27 cm in radius, of wall thickness 1.0 mm and length 90 cm, is twisted so that one end moves through 10° while the other end is clamped.
(a) Find the strain.
(b) Find the stress.
(c) Find the torque being applied.
(d) What is the torsion constant of the tube?

7-2-1   12.   A thin horizontal wire, as in the apparatus shown in Figure 7-28, is to be designed to give a twist of 10° when the force at a distance of 0.15 m is 1.5×10^{-3} N. What must be the torsion constant in N m/rad?

7-3   13.   By what fraction would the volume of a block of aluminum be decreased if it was sunk to a depth of 1000 meters in the sea? At that depth the pressure is 1.00×10^7 N/m^2.

7-3   14.   At a depth of 100 km in the earth, where the pressure is about 1.0×10^9 N/m^2, by what fraction would rocks be compressed?

7-4   15.   Find the work needed to stretch a spring by 0.1 meter (about 4 inches) if the force required to hold it at that elongation is 25 N and the force is known to be proportional to the amount by which the spring is stretched.

7-4   16.   What work is done in stretching a spring by 5 cm if the spring constant is 50 N/m?

7-4   17.   If the pull on a bow increases from zero to 120 N (about 30 lb) as the arrow is pulled back a distance of 0.5 meter, find (a) the work done, (b) the potential energy of the system, (c) the speed of the arrow when it is released, if it is given all the energy. The mass of the arrow is 40 grams.

7-5-1   18.   A 14.1 kg object moving to the left at 7.09 m/s collides and sticks to a 30 kg object moving to the right at 3.33 m/s. Find the velocity of the system after the collision and comment on the energy of the system.

7-5-1   19.   A small mass m moving to the left at 66 m/s hits and sticks to a mass of 155 kg moving to the left at 21 m/s. After the collision the velocity is 23 m/s to the left. Find m.

7-5-2   20.   A mass of 0.55 kg moving to the left at 25 m/s is struck by a mass of 6.6 kg moving to the right at 4.4 m/s. The collision is elastic. Find the velocity of each mass after the collision.

7-5-2   21.   Two equal masses are moving to the left, one at 30 m/s and one at 5 m/s. They collide elastically. Find the velocities after the collision.

7-5-3   22.   A ball is dropped from a height of 1.00 meter, and bounces with a coefficient of restitution of 0.70. It continues to bounce up and down until it stops. For how long will it continue to bounce? (Hint: The answer involves an infinite geometric series with a finite sum.)

7-5-3   23.   A ball falls from a height h onto a hard surface, and then continues bouncing up and down. The third bounce is to just half of the original height. What is the coefficient of restitution?

7-5-3   24.   A mass m moves at 2.5 m/s toward an equal mass that is stationary. The resulting collision is inelastic, with a coefficient of restitution of 0.9. Find the velocity of each mass after the collision.

7-6   25.   Find the energy in the surface of a bubble 5.0 cm in radius. The bubble actually has two surfaces, and the surface tension of water (the material of the bubble) is 0.072 N/m.

7-6   26.   (a) Consider a droplet of water 1 μm in radius. Find the surface area and the energy in the surface. The surface tension of water is 0.072 N/m.
(b) Find the volume and the mass of the drop in (a).
(c) How many droplets 1 μm in radius can be made from 1 kg of water?
(d) What is the total surface energy in water drops of 1 μm radius with a total mass of 1 kg?
(e) When 1 kg of drops of water 1μm in radius coalesce to form 1 kg of water, by how much will the temperature rise?

7-6-2   27.   Find the pressure inside soap bubbles of the following radii. Use $T = 0.072$ N/m; there are two surfaces. Express the result in pascals.
(a) $r = 0.5$ mm, (b) $r = 1.0$ cm, (c) $r = 5.0$ cm, (d) $r = 7.0$ cm.

7-6-2   28.   Find the radius of curvature of the surface dividing two "soap" bubbles, one of which has a radius of 5.0 cm and the other a radius of 7.0 cm. Make a diagram to show the bubbles and the surface between.

7-6-2   29.   Consider two spherical chambers; the wall tension of each produces a pressure inside. One is small, of radius 0.5 cm, and the other somewhat larger, of radius 0.8 cm. The pressure produced in each is 1.33×10^{-4} N/m². How do the tensions compare?

7-6-2   30.   A spherical container 2.5 m in radius is to contain a gas under a pressure of 1.50 megapascals. What will be the tension in the walls?

7-6-2   31.   A spherical craft is to descend 10,000 m into the ocean, where the pressure is 1.00×10^8 N/m². The radius is 1.50 m. What is the compressional force in the wall?

7-6-3   32.   What must be the diameter of the bore in a tube in order that fluid with a surface tension of 0.085 N/m will rise to a height of 5.0 cm? The density is 1035 kg/m³.

7-6-3   33.   A fluid rises 1.8 cm in a tube with a bore of 1 mm diameter. The density of the fluid is 1034 kg/m³. What is the surface tension? The fluid is seen to meet the wall of the tube at an angle of 0°.

7-6-6   34.   A pipe has walls that will withstand a tension of 5×10^5 N/m. What is the limiting pressure that can be put into the pipe? The radius is 0.44 m. Express the pressure in terms of standard atmospheric pressure (101.3 kPa).

CHAPTER EIGHT

FLUIDS

A fluid is a material that will flow. Gases and liquids are fluids and, in general, the solids are not. Some solids are difficult to categorize properly. The rocks in the mantle of the earth are under such high pressure that they flow very slowly, though they transmit shock waves like a solid rather than like a fluid. Glass that has been in windows for hundreds of years is now found to be thicker at the bottom because it actually flows, although slowly to be sure. Glass behaves like a very viscous fluid in some situations, and like a solid in others. Viscosity is just one of the properties of fluids that will be studied. Density, pressure, buoyancy, and the description of flow are among the topics also included in this chapter.

8–1 FLUID VOLUME

The unit for measurement of the volume of a fluid in SI units is the same as for any other volume; it is the cubic meter (m^3). This is an inconveniently large volume for many applications, and a thousandth of a cubic meter is frequently used. It is known as a liter, symbol ℓ. On a smaller scale still is the cubic centimeter, which is just a thousandth of a liter, one milliliter, ml. On the basis of the most recently adopted definitions, one ml is identical to one cm^3.

The units used for fluid volume have been varied and complex. The U.S. gallon is equal to 231.00 cubic inches, which in SI units is 3.785412×10^{-3} m^3. The Imperial gallon is the volume occupied by 10 pounds of water, which is equivalent to 4.546085×10^{-3} m^3. A U.S. fluid ounce is 1/128 of a U.S. gallon, whereas an Imperial fluid ounce is 1/160 of an Imperial gallon. These two systems, the U.S. and the Imperial, are the major non-metric systems in the world. Unfortunately, they use the same words for volume units but the definitions differ. The resulting potential source of confusion will be eliminated with the adoption of SI units.

8–2   DENSITY

The density of a fluid is mass per unit of volume, which for a volume V and a mass M is

$$d = M/V$$

Relative density is the density compared to the density of water for liquids and also frequently for gases, although sometimes air is used as a standard. The relative density of a liquid is obtained with high precision by using what is called a specific gravity bottle. The relative density of liquids is important in some types of analysis (for example, in urinalysis), in determining the concentration of a solute, and also in home wine-making as an indicator of alcohol content.

The specific gravity bottle shown in Figure 8–1 has a stopper with a narrow capillary tube. It is filled with distilled water, and when the ground glass stopper is inserted the outside of the bottle can then be wiped dry and the liquid level made to just reach the top of the capillary tube. The procedure is to weigh in turn the dry clean bottle, the bottle filled with distilled water, and then the bottle filled with the liquid whose relative density is to be found. The ratio of the mass of the

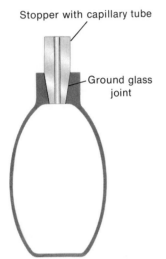

Stopper with capillary tube

Ground glass joint

Figure 8–1   A specific gravity bottle. Note the capillary tube in the stopper.

TABLE 8–1   The density and relative density of some common fluids

Fluid	Density, kg/m^3 (at s.t.p. for gases)	Relative Density, Water as Standard* (at 4°C)	Relative Density, Air as Standard (at s.t.p.)
Ethyl alcohol	792	0.792	
Mercury	13,600	13.60	
Olive oil	920	0.92	
Water	999.8 at 0°C	0.9998	
	1000.0 at 4°C	1.000	
	998.2 at 20°C	0.9982	
Sea water	1025	1.025	
Heavy water	1105	1.105	
Air	1.293	0.001293	1.000
Helium	0.179	0.000179	0.138
Chlorine	3.21	0.00321	2.48

*The relative density with water as a standard is numerically the same as the density in g/cm^3.

liquid to the mass of water can be found very precisely in this manner. Precision may warrant temperature corrections, and often such bottles have a thermometer attached or "built in."

The density and the relative density of some fluids are shown in Table 8-1.

8-3 PRESSURE

Pressure is defined as the force per unit area between two surfaces in contact. The materials in contact may be two solids, a liquid and a solid, or a gas and a liquid or a solid. To find the pressure, the total force F is divided by the area A over which it acts. Thus, $P = F/A$. Also, total force is pressure times area, or $F = PA$. The force contributing to the pressure is that which is normal to the surface, not parallel to it.

8-3-1 A VARIETY OF PRESSURE UNITS

A large variety of units have been used for the description of the magnitude of a pressure. Some of these will be described and related to the basic SI unit, the N/m^3 or pascal, which was described in Chapter 7. The pascal was named in honor of Blaise Pascal, who in 1643 demonstrated the functioning of a mercury-type barometer. He persuaded his brother-in-law to climb a mountain with a mercury barometer, and it was shown that the height of the mercury in the barometer decreased with increasing altitude. That fact vividly demonstrated that the barometer indicated air pressure. Normal sea level atmospheric pressure is just over 100,000 pascals, or 100 kilopascals (kPa).

At sea level the normal height of a column of mercury in a barometer is 760 mm, and the pressure can be described by giving the height of such a mercury column. For example, as a low pressure weather system moves in, the barometer may fall to 750 mm. Mercury columns are frequently used to measure blood pressure. The blood pressure will support a column of mercury about 100 mm high. Any pressure can be expressed in the height of such a mercury column, so the unit is **mm of mercury.** The mm of mercury as a pressure unit is also given the name **torr** in honor of Evariste Torricelli, who in about 1640 invented the barometer. In terms of the SI unit, 1 mm of mercury is 133 N/m^2 or pascals.

1 mm of mercury = 1 torr = 133 Pa

Another pressure unit sometimes encountered is based on

234

normal barometric pressure. Normal pressure is 1.013×10^5 N/m², and 1.000×10^5 N/m² is called one **bar**. A thousandth of a bar is a **millibar,** mb. Normal atmospheric pressure is 1013 mb, just over 1 bar.

Pressure is also expressed in a unit called an **atmosphere.** One atmosphere is the same as normal atmospheric pressure, 101.3 kPa.

8-3-2 CALCULATION OF PRESSURE

If a vertical cylinder of cross-sectional area A is filled to a depth h with a mass m of liquid, the weight supported by the base is mg. This force is exerted over the area A, so the pressure is mg/A. The mass m can be expressed in terms of density d and volume V, being

$$m = dV$$

The volume is given by $V = Ah$, so $m = dAh$ (see Figure 8-2). The pressure is then $P = mg/A = dAhG/A$. The A's cancel, leaving $P = hdg$. It is very significant that the area cancelled, for it means that the **pressure is a function only of depth.**

In subsequent work, the expression generally used for pressure will be the one that gives the result in fundamental units. At a depth h in a liquid of density d, the pressure due to the fluid will be described by

$$P = hdg$$

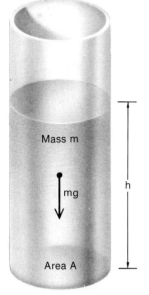

Figure 8-2 The pressure at the bottom of a container.

EXAMPLE 1

What would be the height of the atmosphere if its density were constant at sea level density and the pressure was 101,300 Pa? The density of the atmosphere at sea level is 1.293 kg/m³.

Use $P = hdg$ or $h = P/dg$

$$h = \frac{1.013 \times 10^5 \text{ N/m}^2}{1.293 \frac{\text{kg}}{\text{m}^3} \times 9.8 \frac{\text{m}}{\text{s}^2}}$$

Write N = kg/ms², and cancel some units to leave just meters, m. Then

$$h = 8000 \text{ m} = 8 \text{ km}$$

If the atmosphere was of uniform density, it would be just 8 km thick. Of course, as the pressure decreases with altitude, so does the density; there is no definite end to it. A height is often assigned to it by phenomena that begin at some altitude. For example, meteors begin to glow at about 100 km, while artificial satellites at 160 km slowly lose their energy because of friction with the atmosphere at that altitude.

The analysis just discussed considered the pressure due only to a column of liquid. If an open column on earth is considered, the pressure at the top surface where the depth $h = 0$ is not zero, but is atmospheric pressure. The pressure due to the liquid alone is often what is of interest, though, and this is what is called **gauge pressure.** The total pressure is gauge pressure plus atmospheric pressure, and is called **absolute pressure.** If the pressure due to the atmosphere is represented by P_A, the absolute pressure is described by

$$absolute\ pressure = hdg + P_A$$

In referring to blood pressure, to water pressure, or to automobile tire pressure, it is almost invariably the gauge pressure that is quoted without specifically saying so.

Pressure is often measured by using a U-tube with liquid in it. Such a device is called a **manometer.** The difference in pressure between the two ends of the U is shown as a difference in level of the fluid. If this difference is a height h, as in Figure 8–3(a), and the fluid density is d, then the pressure difference is hdg in basic units. The pressure difference can also be expressed as the distance h. If the fluid is mercury and the distance h is expressed in millimeters, the pressure difference is in **mm of Hg** or **torr.** If the fluid is water, the pressure difference can be expressed in **mm of water.**

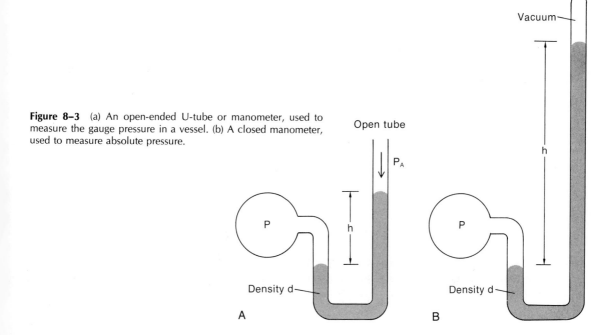

Figure 8–3 (a) An open-ended U-tube or manometer, used to measure the gauge pressure in a vessel. (b) A closed manometer, used to measure absolute pressure.

An open-ended manometer attached to a pressure vessel as in Figure 8–3(a) reads the difference between the atmospheric pressure and the pressure in the vessel. That is, it measures gauge pressure. If the end of the manometer is closed and there is a vacuum above the fluid (Figure 8–3(b)), the difference in levels is a measure of the absolute pressure. The difference in the fluid levels in the arms of the manometer is greater than that in (a) by an amount necessary to balance the atmospheric pressure, shown as P_A in (a).

To illustrate the difference between absolute and gauge pressure, let us find the total force on a person's chest due to the absolute pressure. The absolute pressure in this case is only normal atmospheric pressure, which is about 10^5 N/m². The area of a chest is about 0.1 m², so the force is 10,000 newtons. That is about the force of gravity on 1000 kg, a tonne. Yet you do not feel it. This is because the pressure inside your body is the same as that outside. The situation is like a paper bag closed at the top. There is no net pressure to collapse it. If the bag is immersed in water (after waterproofing), the extra pressure outside will make it smaller. Similarly, if a person dove deeply into water his chest cavity would collapse like the bag. The ear drums would also give way. In diving below about 3 meters, it is necessary to pressurize the lungs to prevent such collapse. Similarly, in going to high altitudes there must be a pressure equalization, especially of the ears through the Eustachian tubes. Unequal pressure on the two sides of the eardrum can be extremely painful. Clearing of the Eustachian tubes may sometimes be accomplished by swallowing, chewing, or blowing the nose.

EXAMPLE 2

Find the gauge pressure and the absolute pressure at a depth of 10.33 meters in water at sea level.

First find the gauge pressure, that is, due to water alone:

Use $P = hdg$, where

$$h = 10.33 \text{ m}$$
$$d = 1000 \text{ kg/m}^3$$
$$g = 9.81 \text{ m/s}^2$$
$$P = 10.33 \text{ m} \times 100 \text{ kg/m}^3 \times 9.81 \text{ m/s}^2$$
$$\text{(cancel only one m)}$$

$$= 101{,}300 \frac{\text{kg m}}{\text{s}^2 \text{ m}^2}$$

$$= 101{,}300 \text{ N/m}^2$$
$$= 101.3 \text{ kPa}$$

The gauge pressure is 101.3 kilopascals, which is the same as normal atmospheric pressure. This amount of pressure is also called one atmosphere. Water pressure is approximately 1 atmosphere for each 10 meters in depth.

The absolute pressure is $hdg + P_A$, which is 202.6 kPa or two atmospheres.

8–3–3   THE BAROMETER

Though there are mechanical devices that give a direct reading of air pressure, such instruments must be calibrated against a standard—a mercury barometer. The barometer is, as in Figure 8–4, basically a closed tube inverted in a cup of mercury that is open to the atmosphere. At the level of the mercury in the cup, the pressure is atmospheric. Above the mercury in the closed tube there is a vacuum. If the height of the mercury in the tube is h, the pressure at the level of the surface of the cup is given by hdg. The relative density of mercury varies only slightly with temperature, and at 0°C is 13.595. The normal height of a mercury barometer at sea level is 0.76 meter. The value of g at sea level and 45° latitude is 9.806 m/s². Then atmospheric pressure is

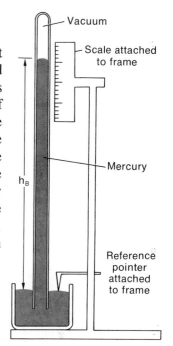

Figure 8–4   A mercury barometer.

$$P = hdg$$

$$= 0.76 \text{ m} \times 13.595 \times 1000 \, \frac{\text{kg}}{\text{m}^3} \times 9.806 \, \frac{\text{m}}{\text{s}^2}$$

$$= 1.0132 \times 10^5 \text{ N/m}^2$$

$$= 101.32 \text{ kPa}$$

(This value is slightly more precise than previously given).

The normal pressure at sea level is 101.32 kilopascals. Pressure decreases with altitude in a manner shown in Figure 8–5. You may get an approximate idea of the normal pressure

Figure 8–5   The change in normal pressure with altitude.

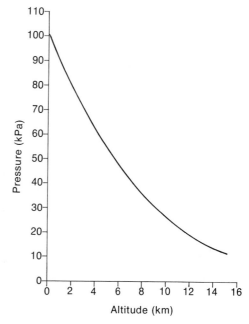

in your locality from the figure if you know your altitude. The pressure ordinarily given on a weather broadcast is corrected to sea level, in order to compare pressures across the country to see where it is higher or lower than normal.

8-4   BUOYANCY

When objects are immersed in a fluid, either a gas or a liquid, a buoyant force is exerted on the object. If the buoyant force exceeds the "gravitational" force, the object will rise; if the "gravitational" force exceeds the buoyant force, the object sinks. If the two forces are equal, the object will not tend to rise or fall, but will, in that system, be effectively weightless. The origin of the buoyant force is in the pressure forces. The downward pressure on the top of the object will be less than the upward pressure on the bottom because of the difference in depth, leaving a net upward force. A detailed analysis for a cylindrical object immersed in a liquid, as in Figure 8-6, will be carried through. With an object as shown, the sideways pressures cancel. Even with irregularly shaped objects, the sideways pressures must still cancel or the object would spontaneously move sideways in the liquid. Such a phenomenon is not observed. If the density of the fluid is d_f, the pressure on the top is $hd_f g$ and the force on the top, of area A, is $hd_f gA$. The pressure on the bottom, which is at a depth $h + l$, is $(h + l)d_f g$. The upward force is this quantity times the area A. The difference between the upward and downward forces is the net upward force, or the buoyant force, B. It will be given by:

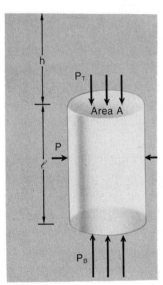

Figure 8-6 A diagram to assist in the calculation of buoyant forces.

$$B = (h + l)d_f gA - hd_f gA$$
$$= hd_f gA + ld_f gA - hd_f gA$$
$$= ld_f gA$$

The quantity lA is the volume V of the object. The product Vd_f is the mass m_f of the fluid that could occupy the volume V. This can be expressed as the volume of fluid displaced by the object. Then $m_f g$ would be the weight of that displaced fluid. The buoyant force, $ld_f gA = m_f g$, is then equal to the weight of the fluid displaced. This was first found by Archimedes in about 250 B.C. He even managed to carry out an analysis for objects of different shapes and reached the conclusion that the buoyant force did not depend at all on the shape. Furthermore, he carried through the analysis with a body of water having a spherical surface, the radius being the radius of the earth.

In *Archimedes on Floating Bodies,* Book 1, his "Proposition 2" is "The surface of any fluid is the surface of a sphere whose center is the same as that of the earth." This was about 250 B.C.! When did the concept of a round earth originate? One of Archimedes' drawings to calculate buoyancy is reproduced as Figure 8–7. The center of the earth is at O and the object is lettered $EFGH$.

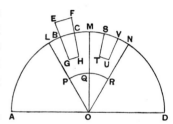

Figure 8–7 One of Archimedes' drawings for the calculation of buoyant forces. The center of the earth is at O. (From *The Works of Archimedes*, ed. by T. L. Heath. New York, Cambridge University Press, 1897, republished by Dover Publications, Inc., 1953.)

One interesting point is that the depth h cancelled from the equation. From this it can be concluded that the buoyant force is independent of depth unless the fluid density depends on depth. Water and most other liquids are almost incompressible (i.e., their densities are independent of pressure), and objects that sink will usually sink all the way to the bottom.

To carry the analysis further, the weight of the object when not immersed in the fluid is given by its mass m_o times g. The mass m_o is found from the density of the object d_o times the volume V. The symbol $\mathscr{G}$ will describe this "gravitational" force. The downward force due to "gravity" is given by

$$\mathscr{G} = d_o V g$$

The buoyant force is $B = d_f V g$. The net downward force is $\mathscr{G} - B$, described by

$$\begin{aligned} \mathscr{G} - B &= d_o V g - d_f V g \\ &= (d_o - d_f) V g \end{aligned}$$

These forces are illustrated in Figure 8–8(a). The net downward force is $\mathscr{G} - B$, so in order to keep the object at rest in the fluid a supporting force S would be required, as in Figure 8–8(b). **The quantity $\mathscr{G} - B$ could be called the weight, W, in the medium.** If

Figure 8–8 The weight of an object immersed in a fluid. In (a) there is a net downward force. In (b) the supporting force needed to keep the object at rest is S. The weight is the negative of S, which in this case is $\mathscr{G} - B$.

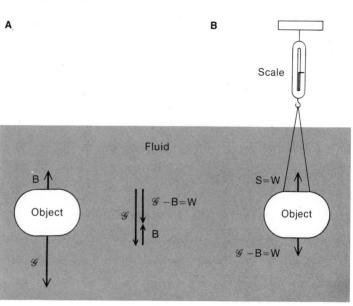

$\mathscr{G}$ is greater than B, or if $d_o > d_f$, the object will sink; if $\mathscr{G} = B$, the weight in the fluid is zero and the object will remain at whatever level it is put. If B is greater than $\mathscr{G}$, or $d_f > d_o$, the object will rise to the surface. The relative densities of some common materials are listed in Table 3–1. Materials with a relative density greater than 1 sink, those with a relative density less than 1 float in fresh water.

The case of a water animal, such as a fish, is of interest. The density of the flesh and bones of a fish is slightly greater than that of water. By means of an internal gas bag, the fish can adjust its average density so that the buoyant force will be equal to its weight, and it will neither sink to the bottom nor float to the top but will remain at a constant level. The fish can go upward by allowing the bag to expand or downward by contracting it, or by propelling itself in an upward or downward direction. The fish in water is weightless because no external force need be applied to keep it at rest in the medium.

There are some creatures of the sea that move about by walking on the bottom. The jellyfish and the octopus are among these; an interesting thing to note is that their "legs" are not rigid to support their weight as are the legs of a land animal. Out of the water these creatures cannot stand; underwater their bodies are supported by the buoyant force of the water, while the legs are used principally for locomotion across the bottom.

A brain is another example of the use of buoyant forces in nature. The brain is such a fragile structure that it cannot support its own weight in air. However, inside the skull it is immersed in a fluid having a relative density of 1.007; the relative density of the material of the brain is 1.040, which is only slightly higher than that of the fluid. The brain, then, has a weight in fluid that is only about a thirtieth of the weight it would have in air. If the supporting fluid is removed, as in some x-ray procedures or surgery, the brain is easily damaged and intense pain can occur because the nerves and blood vessels are put under strain when they take over part of the task of supporting the brain. The weight in the fluid, $(d_o - d_f)Vg$, also depends on the local value of g, which in turn depends on the acceleration of the reference system. In other words, inertial forces are to be included. If the head is subject to a high acceleration, as in an accident, or if the head is the target of a blow, the excess weight may cause damage to the fragile structure.

8–4–1 FLOATING OBJECTS

Objects float in a fluid with enough of their volume submerged that the buoyant force will just balance the force of

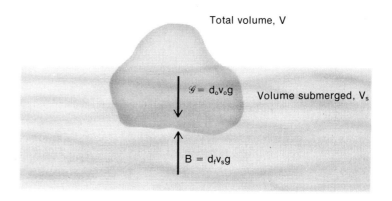

Figure 8–9 A floating object. When there is equilibrium, the gravitational force is equal to the buoyant force.

gravity on them. The buoyant force that results from the upward components of the pressure forces is equal to the weight of fluid displaced.

As in Figure 8–9, when an object of volume V_o and density d_o is in a fluid of density d_f, a volume V_s is submerged. The gravitational force is

$$\mathcal{G} = V_o d_o g$$

The buoyant force is

$$B = V_s d_f g$$

When the object is in equilibrium, $\mathcal{G} = B$.

The fraction submerged, V_s/V_o, is found from the above equation to be

$$\frac{V_s}{V_o} = \frac{d_o}{d_f}$$

EXAMPLE 3

An example of this principle is an iceberg in sea water. The ice has a density $d_o = 920$ kg/m³. Sea water has a density $d_f = 1025$ kg/m³. The fraction of the iceberg that is under the water is

$$\frac{V_s}{V_o} = \frac{920}{1025} \quad \text{(units cancel)}$$

$$= 0.90$$

Nine tenths of an iceberg will be under water.

There is a common and quite precise type of hydrometer (device to measure relative density of fluids) that is based on the concepts of buoyant forces and floating objects. That

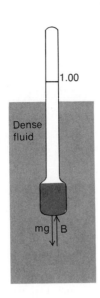

Figure 8–10 One type of hydrometer. The instrument floats at a level that depends on the density of the fluid. Density markings are made along the stem.

hydrometer, as shown in Figure 8–10, consists of a hollow glass tube with a relative density scale and a weighted base. The device is floated in distilled water and the level of the fluid on the rod is marked as relative density 1.000. If the hydrometer is floated in a fluid less dense than water, it will sink more deeply — until the weight of the fluid displaced is equal to the weight (*mg*) of the hydrometer. In a fluid more dense than water it will float higher. Such hydrometers, depending on the care in construction, may read relative density to an accuracy of 0.1 per cent or even better.

8–5   FALLING IN A FLUID

If the weight of an object in a fluid is not zero, the object will either sink or rise at a rate that depends not only on the weight but also on other factors such as its size and shape, and the viscosity of the fluid. For example, red blood cells have a relative density of 1.098. The rate at which the cells settle, called the sedimentation rate, is of clinical significance. The rate at which they fall in the plasma can be increased by increasing g, as occurs in a centrifuge. The value of g experienced by the cell is the centrifugal acceleration, $r\omega^2$, when it is rotating, where r is the radius of rotation and ω is the angular velocity.

An analysis of the rate of falling in terms of weight in the medium is worthwhile because it can, for example, indicate the way to change centrifuging time to compensate for a change in centrifugal force.

When an object moves in a fluid it experiences a resisting force R, shown in Figure 8–11. The resistance force increases

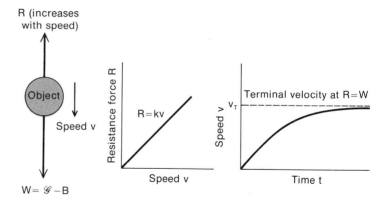

Figure 8–11 The resistance force on an object falling in a fluid.

with speed; for very low speeds the resistance is proportional to the speed (but opposite in direction, of course). The size of the resistance force will then be described by

$$R = kv$$

where k is a constant for a given object in a given fluid. An object falling in a medium will increase its speed until the resistance force balances the weight. This limiting speed is called the **terminal velocity,** v_t. At the terminal velocity

$$R = \mathscr{G} - B$$

or

$$kv_t = \mathscr{G} - B = W$$

and the terminal velocity is given by

$$v_t = W/k$$

where
$$\begin{aligned} W &= \text{weight in medium} \\ &= (d_o - d_f)Vg \end{aligned}$$

According to this equation, the terminal velocity or the speed at which an object falls in a medium is proportional to the weight in the medium. This may sound a bit like Aristotle; in fact, it is just what he said. It applies to the terminal velocity for objects falling in a resisting medium, but only if the resistance constant k is the same for each object.

Aristotle did not specify all these conditions rigorously but he did, in his writings, say that he referred to fall in a resisting medium. He also said that in a vacuum all objects would fall at the same speed; but he said that since a vacuum was impossible, it was also impossible for different objects to fall at the same rate.

FLUIDS

EXAMPLE 4

A person falling in air has a terminal velocity of about 50 m/s. If the mass of that person is 70 kg, what is the resistance force constant? The buoyant force is negligible.

The gravitational force is mg, or 686 N. At the terminal velocity this is equal to the resistance force R.

$$R = kv$$
$$k = R/v$$
$$= 686 \text{ N}/(50 \text{ m/s})$$
$$= 13.7 \text{ N s/m}$$

The resistance force constant is 13.7 N s/m, which could be read as 13.7 newtons per meter per second or 13.7 newton seconds per meter.

EXAMPLE 5

A certain glass sphere 1.5 cm in diameter, relative density 2.5, has a terminal velocity of 0.3 m/s downward in water. Assume that $R = kv$.

Find the terminal velocity of a wooden sphere, of the same size but relative density 0.85, in water.

To do this type of problem, work with ratios and there is a lot of cancellation. For the glass sphere,

$$v_t(\text{glass}) = \frac{W(\text{glass in water})}{k}$$

For the wooden sphere

$$v_t(\text{wood}) = \frac{W(\text{wood in water})}{k}$$

The ratio is

$$\frac{v_t(\text{wood})}{v_t(\text{glass})} = \frac{W(\text{wood in water})}{W(\text{glass in water})}$$

The weight in a fluid, $\mathcal{G} - B$, is:

$$W(\text{glass in water}) = (d_{\text{glass}} - d_{\text{water}})Vg$$

$$W(\text{wood in water}) = (d_{\text{wood}} - d_{\text{water}})vg$$

Then

$$\frac{v_t(\text{wood})}{v_t(\text{glass})} = \frac{d_{\text{wood}} - d_{\text{water}}}{d_{\text{glass}} - d_{\text{water}}} \quad (V \text{ and } g \text{ cancel})$$

$$= \frac{0.85 - 1}{2.5 - 1}$$

$$= -\frac{0.15}{1.5}$$

$$= -0.1$$

Therefore, the ratio v_t(wood)/v_t(glass) $= -0.1$, from which

$$v_t \text{(wood)} = -0.1 \ v_t \text{(glass)}$$
$$= -0.1 \times 0.3 \text{ m/s downward}$$
$$= -0.03 \text{ m/s downward}$$
$$= 0.03 \text{ m/s upward}$$

The wooden sphere will rise at 0.03 m/s or 3 cm/s.

In centrifuging, the particles being separated must move a certain distance s in the tube; the required time is given by

$$t = \frac{s}{v_t}$$

where v_t is the terminal velocity. The higher the terminal velocity, the shorter the time. Substituting for the weight, the terminal velocity is

$$v_t = (d_o - d_f) V g / k$$

From this, the time for a particle to move a distance s is

$$t = \frac{sk}{(d_o - d_f) V g}$$

In a centrifuge the value of g experienced by a particle is $r\omega^2$ where ω is the angular velocity. Then

$$t = \frac{sk}{(d_o - d_f) V r \omega^2}$$

It is interesting to relate the time, distance and centrifuging speed ω. For given particles and solution, all the other quantities are constant; lump them into a constant c. This leaves

$$t = \frac{cs}{\omega^2}$$

The time for the particles to move a distance s varies inversely with the square of the speed of the centrifuge. If ω is doubled, for example, the centrifuging time may be cut to a quarter. If the centrifuge speed is cut in half, then t must be increased by a factor of 4 to move the particles the same distance s. This adjustment of centrifuging time inversely as the *square* of a change in centrifuge speed should be known to any laboratory worker.

8-6 PRESSURE AND SPEED

Oddly enough, the pressure in a fluid depends on its state of motion. As the speed increases, the pressure decreases. This effect will be analyzed as it applies to incompressible fluids moving in rigid tubes. The general features of the results will apply to more general situations, though the equations will not be exact if the conditions specified are not met.

8-6-1 SPEED AND AREA OF TUBE

The first aspect to consider is how the speed of the fluid changes with the size of the conducting tube. If the fluid is not compressible, the same volume of fluid must pass each point per unit time. Figure 8–12 illustrates a tube that varies in size. At the point marked 1 the cross-sectional area is A_1, and at the point marked 2 the area is A_2. If the fluid in a length l_1 (volume V_1) flows past the point 1 in the time t, the fluid in the length l_2 will flow past point 2 in the same time, and l_2 must be such that the volume $A_2 l_2$ is the same as the volume $A_1 l_1$. Then

$$A_1 l_1 = A_2 l_2$$
or
$$A_1/A_2 = l_2/l_1$$

The length l of the fluid that flows past any point in a given time depends on its speed. Consider a time t; from the relation that time is distance over speed,

$$t = l_1/v_1 = l_2/v_2$$

where v_1 and v_2 are the speeds. From this it follows that $l_2/l_1 = v_2/v_1$. Combining this with the relation for the area gives

$$A_1/A_2 = v_2/v_1$$

or
$$A_1 v_1 = A_2 v_2$$

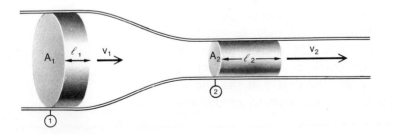

Figure 8–12 A fluid flowing in a tube with a changing cross-sectional area.

This says that the product of the area of a cross-section and the speed of the fluid at that point is a constant everywhere along the tube. Where the tube is small, the velocity is higher than where the tube is large.

EXAMPLE 6

Find the speed of the fluid flowing in a pipe at the rate of 0.1 liter per second. One part of the pipe has a diameter of 2 cm, and the pipe then narrows to a diameter of 1 cm. Find the speed in each part.

The cross-sectional area of the large pipe, $\pi d^2/4$, where $d = 0.02$ m, is 3.14×10^{-4} m^2. In one second, 0.1 ℓ or 10^{-4} m^3 flows.

The fluid in a length l flows past any point per second, where l is such that

$$Al = 10^{-4} \text{ m}^3 \text{ and } A = 3.14 \times 10^{-4} \text{ m}^2$$
$$l = 10^{-4} \text{ m}^3/3.14 \times 10^{-4} \text{ m}^2$$
$$= 0.318 \text{ m}$$

The speed in the 2 cm diameter pipe must be 0.318 m/s.

The speed in the smaller pipe must be such that the product of area times speed is the same as in the larger pipe.

$$A_1 v_1 = A_2 v_2$$
$$v_2 = v_1 A_1/A_2$$

where

$$v_1 = 0.318 \text{ m/s}$$
$$A_1 = \pi \times 2^2 \text{ cm}^2/4$$
$$A_2 = \pi \times 1 \text{ cm}^2/4$$
$$A_1/A_2 = 4$$
$$v_2 = 4v_1$$
$$= 1.27 \text{ m/s}$$

The smaller pipe has a quarter the area, so the speed is 4 times as great or 1.27 m/s.

8-6-2 KINETIC ENERGY AND PRESSURE CHANGE

A change in the speed of the fluid implies a change in its kinetic energy. For the speed to increase, there must be work done on the fluid, and this work is done by the pressure forces. The pressure must be highest in the regions of low speed so that the extra pressure can accelerate the fluid to its higher speed. The tube may also vary in height, so the potential energy may also vary along the tube. The general case is illustrated in Figure 8-13, in which the tube varies in cross section and in elevation. The area, the velocity, and the pressure all vary along the tube. At the two positions in the tube marked 1 and 2, equal volumes of the liquid, containing the same mass m, have been indicated. In each position the mass m has some kinetic energy and some potential energy; the sum of these two forms of energy is not necessarily the same at the two points because of work being done by pressure forces.

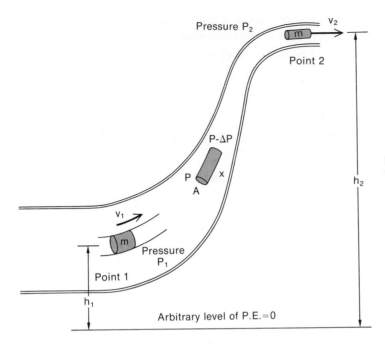

Figure 8–13 Fluid flowing in a tube of varying height and area.

To investigate the pressure effects, consider the small element of the fluid shown in the central portion of the tube in Figure 8–13. There is a net force of ΔP times A moving this fluid along, where ΔP is the difference in pressure between the ends of that element. The speed of the small element of fluid changes, and this implies that there must be a force (pressure change). When the small element of fluid moves by its length x, the work done on it is the force $\Delta P \times A$ times the distance x or $\Delta P \times Ax$. The quantity Ax is the volume V; so the work done on the fluid element, which is also the change in energy of that element, is ΔPV. This can be written as $\Delta PV = \Delta E$. This is the work done in moving the small bit of fluid just a small distance. The total work done on the volume V of fluid in moving it from the point 1 to the point 2 is the sum of the bits of work that moved it each short bit of the way. This is the sum of all the changes in pressure ΔP times V. V is a constant, so the sum of all the small changes in P (or ΔP's) times V is the total change in pressure, $(P_1 - P_2)$, times V. The total work on the volume V is $(P_1 - P_2)V$, and this is the total change in energy of that volume, $E_2 - E_1$. The energy of a mass m of the fluid is the sum of its kinetic energy and its potential energy. At the points marked 2 and 1, the energies E_2 and E_1 are the sums of the kinetic and the potential energy at each point,

$$E_2 = (1/2)\,mv_2^2 + mgh_2$$
$$E_1 = (1/2)\,mv_1^2 + mgh_1$$

The difference is given by

$$(P_1 - P_2)V = E_2 - E_1$$
$$= (1/2)mv_2^2 + mgh_2 - (1/2)mv_1^2 - mgh_1$$

In the expressions for the energies, the mass m can be written as the product of the density and the volume, $m = dV$. The volume V will then conveniently cancel to leave

$$(P_1 - P_2) = (1/2)dv_2^2 + dgh_2 - (1/2)dv_1^2 - dgh_1$$

The terms can be rearranged to give

$$P_1 + (1/2)dv_1^2 + dgh_1 = P_2 + (1/2)dv_2^2 + dgh_2$$

The positions 1 and 2 were arbitrary, so what this equation implies is that at all points in a tube **the sum of the pressure, the kinetic energy per unit volume, and the potential energy per unit volume is the same.** This is known as **Bernoulli's law,** and in the mathematical form it is often called **Bernoulli's equation.** It describes fluid motion in which there are no energy losses owing to what could be called fluid friction.

Bernoulli's equation can also be called upon to find the velocity of fluid in a tube. A constriction is put into the tube, and in this constriction the velocity is increased and consequently the pressure is reduced. The drop in pressure depends on the velocity in the large tube and the relative cross-sectional areas. Some flow-meters work on this principle. Paint sprayers, perfume atomizers, and aspirators also depend on Bernoulli's principle. The canvas top on a convertible bulges out as it travels down the road because the high velocity air outside has lower pressure than the stationary air inside, resulting in a net outward pressure.

Balls in golf, baseball, or ping-pong are made to move in a curved path by making them spin; because of viscosity or friction, the spinning ball drags a layer of air around with it. On one side this partly cancels the air flow due to motion of the ball, and the air speed on that side is reduced. On the other side the spinning ball adds to the air velocity due to the motion of the ball, and the speed is increased. On the side with high speed there is lower pressure than on the side with the lower air speed. This creates a net pressure or **Bernoulli force** that causes the ball to curve (see Figure 8–14).

One of the most efficient pumps for high vacuums consists of a high-speed jet of either vaporized oil or mercury. The air molecules are drawn into the low pressure area associated with the high-speed molecules. Such a pump is called a *diffusion pump.*

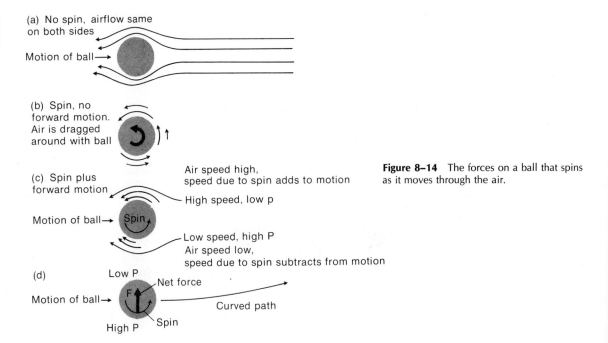

(a) No spin, airflow same on both sides

Motion of ball→

(b) Spin, no forward motion. Air is dragged around with ball

(c) Spin plus forward motion

Air speed high, speed due to spin adds to motion

High speed, low p

Motion of ball→ Spin

Low speed, high P

Air speed low, speed due to spin subtracts from motion

(d)

Low P

Net force

Motion of ball→

Curved path

High P   Spin

Figure 8–14   The forces on a ball that spins as it moves through the air.

EXAMPLE 7

The statement was just made that the speed of flow of fluid in a tube could be measured by putting a constriction in the tube and measuring the change in pressure. As shown in Figure 8–15, this device, called a **venturi flowmeter,** has a constriction of area A_2 put into the main tube of area A_1 in which the speed v_1 is to be measured. To measure the pressure difference, a U tube with mercury in it is connected as in the figure. The pressure difference $P_2 - P_1$ is found from the difference in mercury levels. The problem is to find how v_1 is related to $P_2 - P_1$, or the difference in height, h.

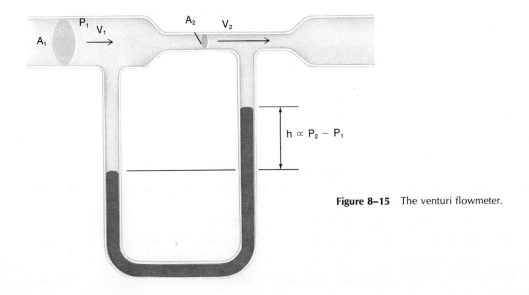

P_1 V_1

A_2 V_2

A_1

$h \propto P_2 - P_1$

Figure 8–15   The venturi flowmeter.

The tube is level so that the potential energy is constant, and Bernoulli's equation can be written

$$P_1 + (1/2)\,dv_1^2 = P_2 + (1/2)\,dv_2^2$$

where d is the density of the flowing fluid. Also,

$$A_1 v_1 = A_2 v_2$$

or

$$v_2 = v_1 A_1 / A_2$$

Square this, substitute it into Bernoulli's equation, and gather the terms in v_1 and the pressure terms together to get

$$(1/2)\,dv_1^2 \left(\frac{A_1^2}{A_2^2}\right) - (1/2)\,dv_1^2 = P_1 - P_2$$

from which

$$v_1 = \sqrt{\frac{2\,A_2^2}{d\,(A_1^2 - A_2^2)}}\ \sqrt{P_1 - P_2}$$

For a given meter and fluid the quantities under the first root sign are all constant, so they can be combined into one constant C_1, giving

$$v_1 = C_1 \sqrt{P_1 - P_2}$$

The speed is directly related to the square root of the difference in pressure between the wide and narrow parts of the tube.

8-6-3   THE PITOT-STATIC TUBE

Bernoulli's principle is used in the design of an instrument to measure air speed. Its principal use is, not unexpectedly, in aircraft, where it serves the same function as the speedometer in an automobile. As the speed of the aircraft nears the speed of sound the resistance force, or drag, rises rapidly. This is the *sound barrier,* and it is important that the pilot know how his airspeed is related to the speed of sound. The speed of sound varies with temperature, being highest in the warm air near the ground and lowest at high altitudes. The variation may be from 300 m/s (670 mi/hr) to 350 m/s (780 mi/hr). The ratio of the speed of the aircraft to the speed of sound is called the **Mach number.** A Mach number less than one indicates a speed below the speed of sound. Mach 1 is the speed of sound, and speeds faster than sound are indicated by Mach numbers greater than 1. The pitot-static tube, or Prandtl's tube as it is sometimes called, can be adapted to indicate either air speed or Mach number no matter what the speed of sound may be.

The pitot-static tube is shown in cross-section in Figure 8-16. The pressure measuring part can be a mechanical or

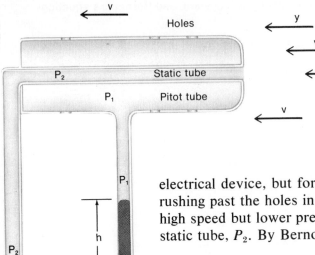

Figure 8–16   The pitot-static tube.

electrical device, but for simplicity a U tube is shown. The air rushing past the holes in the outer casing, the pitot tube, is at a high speed but lower pressure (P_1) than the stationary air in the static tube, P_2. By Bernouilli's principle,

$$P_1 + (1/2)\,dv^2 = P_2$$

From this equation the speed v can be obtained, with corrections to be applied for the appropriate air density. The method for measuring Mach number will be clearer after further analysis.

The density of the air is d, and this can be expressed in terms of temperature using the general gas law in the form

$$PV = \frac{M}{\mathscr{M}}\,RT$$

or $\qquad\qquad P = \frac{M}{V}\frac{RT}{\mathscr{M}} \qquad$ and $\qquad \frac{M}{V} = $ density d

Substitute this to get, in the pitot tube,

$$d_1 = P_1\,\frac{\mathscr{M}}{RT}$$

where $\mathscr{M}$ is the molar mass, R is the gas constant and T is the kelvin temperature.

In the chapter on waves it will be shown that the speed of sound in a gas is given by

$$v_s = \sqrt{\frac{\gamma RT}{\mathscr{M}}}$$

from which

$$\frac{\mathscr{M}}{RT} = \frac{\gamma}{v_s^2}$$

Then

$$d_1 = P_1 \frac{\gamma}{v_s^2}$$

so $P_1 + (1/2)P_1 \frac{\gamma}{v_s^2} v^2 = P_2$

or $\frac{v^2}{v_s^2} = \frac{2}{\gamma} \frac{(P_2 - P_1)}{P_1}$

and the Mach number is given by

$$M = \frac{v}{v_s} = \sqrt{\frac{2}{\gamma}} \sqrt{\frac{P_2 - P_1}{P_1}}$$

In order to measure the Mach number, the pressure difference as well as the absolute pitot pressure must be measured by the instrument. The instrument is then calibrated in terms of Mach number according to the square root of the ratio of those quantities.

8–7 VISCOSITY

"Blood is thicker than water." That, in more scientific language, says, "The coefficient of viscosity of blood is higher than the coefficient of viscosity of water." Fortunately, all writing is not in scientific terms.

The property called viscosity is of importance in many phenomena: lubrication, fluid flow in pipes or blood vessels, sedimentation rates, and many others. What is viscosity? How is it measured?

When a fluid moves across a surface, the fluid that is directly in contact with the surface is held by molecular forces and does not move at all. That stationary layer of fluid retards the layer next to it, and the fluid velocity gradually increases away from the surface as in Figure 8–17. Within a thin layer of fluid between two surfaces, one of which is moving, the velocity will increase gradually between them as in Figure 8–18. This situation can be illustrated by a "puddle" of syrup on a table with a flat plate placed on the surface of the fluid. A force will be required to move the plate across the surface. Some "thought analysis" will show how the various factors may affect the force required to move the plate. It might be expected that the force would depend on the velocity v, on the area A, and inversely on the fluid layer thickness x. This can be written as

$$F \propto Av/x$$

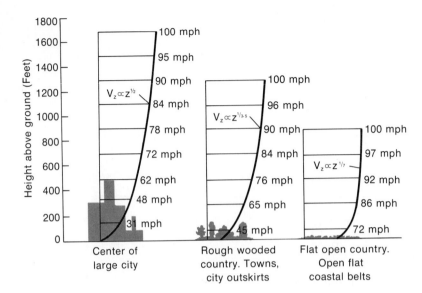

Figure 8–17 The manner in which air velocity varies above a surface. (After J. E. Allen, *Aerodynamics: A Space Age Survey*, Harper & Row, New York, 1963.)

This is a proportionality relation. The proportionality constant that can be put in to make it an equation depends on the fluid used. The more viscous the fluid, the greater the force, all other quantities held constant; and the constant is called the **coefficient of viscosity.** It is usually represented by the Greek letter eta, η. That is,

$$F = \eta A v/x$$

or

$$F/A = \eta v/x$$

The quantity F/A is not a pressure, for the force is tangential to the surface. The quantity v/x is the change in velocity per unit thickness of the fluid layer and is called the **velocity gradient.** Solving for η.

$$\eta = Fx/Av$$

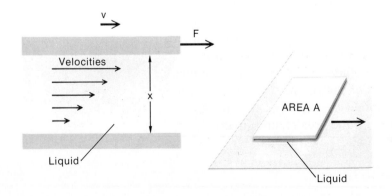

Figure 8–18 The velocity in a thin layer of fluid between two plates. This situation can be obtained with a "puddle" of syrup and a flat plate.

TABLE 8–2   The viscosities of a few substances

MATERIAL	TEMPERATURE, °C	VISCOSITY, Pa s
Water	0	0.179
Water	20	0.101
Water	40	0.066
Ethyl alcohol	20	0.119
Methyl alcohol	20	0.059
Glycerine	20	83
Olive oil		8.4
Whole blood	37	0.27
Air	23	1.8×10^{-4}

The units for description of viscosity are found from that formula, putting in the SI unit for each quantity:

F is in newtons, N
x is in meters, m
A is in square meters, m^2
v is in meters per second, m/s
and $1/v$ is in s/m

Showing units only

$$\eta = \frac{N \cancel{m} s}{m^2 \cancel{m}}$$

Cancel an m and write $N/m^2 =$ Pa (pascals). Then the units for viscosity, η, are pascal seconds, Pa s.

Much literature will still list viscosity in an older unit, the **poise,** which was based on the cgs system. The **centipoise,** a hundredth of a poise, is also common. There are 10 poise or 1000 centipoise in one Pa s. In reverse, to change from poise to the modern unit Pa s, divide by 10.

Some typical viscosities are shown in Table 8–2.

EXAMPLE 8

Find the torque required to turn a cylinder that is inside another but separated from it by a layer of oil. The inside cylinder is 2.5 cm in radius and 0.10 m long; the outside cylinder is 2.6 cm in radius. The situation is shown in Figure 8–19. The viscosity of the oil is 8.3 Pa s. Neglect the effect at the end of the cylinder. The torque will depend on rotational speed: use $\omega = 1$ rad/s and 2 rad/s.

The area of the sheet of material is the distance around the cylinder times the length. But do you use the area of the inner or the outer cylinder? There is not much difference; use the average, so $r = 2.55$ cm or 0.0255 m.

$$A = 2\pi \times 0.0255 \text{ m} \times 0.10 \text{ m}$$
$$= 0.0160 \text{ m}^2$$

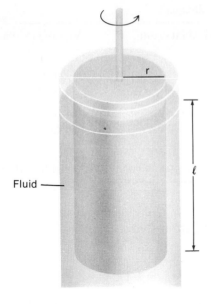

Figure 8–19 Concentric cylinders with fluid between them, as in Example 8.

Fluid ——

The thickness is $x = 1\ mm = 1 \times 10^{-3}\ m$

The speed v is $r\omega = 0.0255\ m/s$ for 1 rad/s
and $r\omega = 0.051\ m/s$ for 2 rad/s

Solve for F and then find Fr:

$$\eta = Fx/Av$$

$$F = \frac{\eta Av}{x}$$

$$Fr = \frac{\eta Avr}{x}$$

Insert the values for $\omega = 1$ rad/s:

$$Fr = 8.3\ \frac{N}{m^2}\ s \times \frac{0.0160\ m^2 \times 0.0255\ m/s}{10^{-3}m}\ 0.0255\ m$$

$$= \frac{8.3 \times 1.6 \times (2.55)^2 \times 10^{-6}}{10^{-3}}\ N\ m$$

$$= 0.086\ N\ m$$

The torque needed to turn the cylinder at 1 rad/s is 0.086 N m.

At 2 rad/s the numbers would be the same except for v, which would be doubled, so the torque is 0.172 N m.

Note that if a viscous fluid separates two solids surfaces, then the force required to move them, which corresponds to a frictional force, depends on speed. With solid surfaces in contact the frictional force does not depend on speed.

DISCUSSION PROBLEMS

8–3–2   1. Discuss the pressure differences at a depth of 2 meters between the following locations. Consider both absolute and gauge pressure.
(i) In a lake at sea level.
(ii) In a lake at 10,000 feet above sea level.

(iii) At the bottom of a 2-meter glass tube filled with water.

(iv) In the Pacific ocean.

8-4   2.   (a) What is the average density of a human being? How can you deduce it to a high degree of precision? (b) A person normally has air in the lungs; based on part (a), what would you conclude about the average density of bone and tissue?

8-6-2   3.   When a car with a canvas top is traveling along the road, the top "balloons" out. Explain why.

8-6-3   4.   Describe how a ball should be thrown in order that it would drop faster than normally.

8-6-3   5.   How could the Bernoulli effect be used to lift water out of a container?

PROBLEMS

8-1   1.   Convert an automobile gasoline consumption of 10 km/ℓ to the consumption in mi/U.S. gal.

8-1   2.   If an automobile will give 35 mi/U.S. gal, how many kilometers will it go on one liter?

8-2   3.   Calculate the total mass of air in a room that is 3.0 m high, 4.0 m wide, and 5.0 m long. Use the density of air given in Table 8-1.

8-2   4.   Calculate the mass of air in a small cylindrical container 1.00 cm in diameter and 2.00 cm long. The density of air at 0°C and 101.3 kPa is 1.293 kg/m³. The air in the container is at 20°C and 97.6 kPa. Express the answer in kilograms, in grams, and in milligrams.

8-3   5.   (a) Calculate the mass of mercury in a container 0.5 cm in diameter and 760 mm high.

(b) Calculate the weight of the mercury in part (a); this force will act on the bottom of the column.

(c) What is the force per unit area on the bottom (F/A)?

8-3-2   6.   Calculate the water pressure at a depth of 1 meter, at 2 meters, and at 3 meters.

8-3-2   7.   At what depth in water is the water pressure the same as normal atmospheric pressure?

8-3-2   8.   Find the pressure at a depth of 30 km in effectively fluid rock of density 3000 kg/m³.

8-3-2   9.   Two tubes are connected as in Figure 8-20, and fluid is drawn up into both tubes by producing a low pressure at A. The pressure at the bottom of each column is the same. In one arm is water at 20°C; the water column rises 31.3 cm. In the other arm is another fluid, which rises 28.3 cm. Find the density of the other fluid.

8-3-2   10.   What height of a column of water would produce the same pressure as a column of mercury 100 mm high? If a giraffe produced a blood pressure of only 100 mm of mercury but had a neck 2 m long, what would happen to his head when it was held up? The density of blood is fairly close to that of water (about 1060 kg/m³).

8-3-2   11.   A collection of grains of solid material is in many ways like fluid. Wheat and corn are in this category. Find the pressure at the bottom of a structure filled to a depth of 25 m with wheat. The density of wheat is 750 kg/m³.

8-3-3   12.   By how much would the height of the column in a barometer fall if it were taken from sea level to an altitude of 100 meters? Assume a constant sea level air density for that 100 meters.

8-4   13.   (a) Calculate the downward force due to pressure on the top of an object in water and the upward force due to the pressure on the bottom. The object is a rectangular solid 0.319 m on a side and 0.20 m high. The top of the object is 1.0 m below the surface of the water. Find also the net upward force.

(b) Calculate the weight of water that would occupy a rectangular container 0.319 m × 0.319 m × 0.20 m.

8-4   14.   Find the force needed to support a block of aluminum when it is immersed in water. The volume of the block is 1.00×10^{-6} m³. The relative density of aluminum is 2.70.

8-4   15.   Find the mass of a killer whale for which measurements on a series of photographs showed that its volume was 4.60 m³. Express the mass in kg, in tonnes, and in pounds. A killer whale is weightless when immersed in sea water.

8-4   16.   A piece of rock is balanced by a mass of 141 g in air (in the older terminology, it "weighs" 141 g), and by 101 g when the rock is immersed in water. Find the density of the rock.

8-4   17.   A block of aluminum 10^{-6} m³ in volume is balanced by a mass of 0.270 kg when the block is in air. If that block, when it is immersed in a fluid, is balanced by a mass of 0.180 kg in air, what is the density of the fluid?

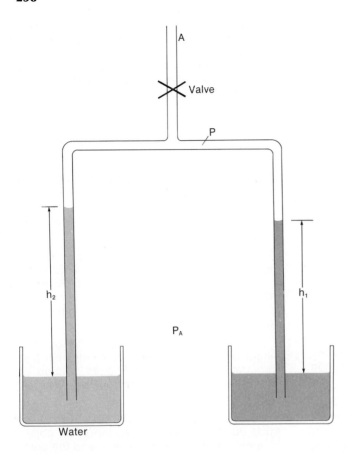

Figure 8–20 A device for comparing the densities of two liquids, as in Problem 9.

8–4   18. Find the difference between the supporting force on the brain, relative density 1.040, when it is in air and the supporting force when it is immersed in a fluid of relative density 1.007. The mass of the brain is 1.2 kg.

8–4   19. An empty cylindrical container (scientific language for a bucket) with a diameter of 0.25 m is pushed downward to a depth of 0.20 m in water, as in Figure 8–21.
(a) Find the force required to do this by finding the pressure on the bottom, the area, and the upward force.
(b) Find the weight of the water when the container in (a) is filled to a depth of 0.2 m.
(c) Discuss the effect of air pressure in part (a) pushing the bucket downward.

8–4–1   20. How deeply will a cylindrical container sink into water if it is 0.25 m in diameter, has a mass of 1.50 kg, and is filled with a fluid, density 750 kg/m^3, to a depth of 0.10 m?

8–5   21. Find the constant k in the resistance force equation ($R = kv$) for the case in which a mass of 0.0013 kg falls at 0.30 m/s in air. Neglect the buoyancy of the air.

8–5   22. A spherical object, 0.0010 m (1 mm) in diameter, falls through water at a con-

stant speed of 1.1 cm/s. The density of the object is 3000 kg/m^3. Find the resistance force constant, k.

8–5   23. If certain particles are just moved to the bottom of a tube in a centrifuge spinning at 500 rev/min in 3.2 minutes, how long will it take at a speed of 2000 rev/min?

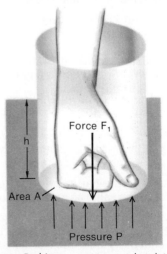

Figure 8–21 Pushing an empty container into water. See Problem 19.

8-6-1   24.   (a) If fluid flows at 0.25 liters per second in a tube 1 cm in diameter, what is the average speed along the tube?

(b) If the tube in (a) is connected to a tube only 0.8 cm in diameter, what will be the speed in that smaller tube?

8-6-1   25.   (a) Find the kinetic energy per kilogram of the fluid moving in a tube if at one place it moves at 3.16 m/s. (b) In a narrower part of the tube, where it moves at 4.9 m/s, what is the kinetic energy per kilogram? In each case the fluid is water.

8-6-2   26.   A fluid moves in a tube that forms a loop in a vertical plane. The density of the fluid is 1060 kg/m³. The top of the loop is 1.8 m above the bottom, and at a point 1.3 m above the bottom the pressure is 13,000 N/m² (13 kPa). The speed is the same everywhere in the system.

(a) Find the fluid pressure at the top of the loop.

(b) Find the fluid pressure at the bottom.

The system is a much simplified model of blood flow; in (a) the blood pressure is found in the head, while in (b) it is at the feet.

8-6-2   27.   If a fluid moves in a tube of area A_1 at a speed of 1.5 m/s and under a pressure of 13 kPa, what area (as a fraction of A_1) would cause a pressure reduction to zero in the fluid as it goes through? The tube is level and the fluid is water.

8-6-2   28.   When air of density 1.3 kg/m³ is moving at 40 m/s past a window, what is the pressure produced according to Bernoulli's equation? In what direction is the resulting force on the window? The speed of the air inside the window is zero.

8-6-2   29.   In an aircraft flying at the speed of sound, what will the ratio between pitot and static pressure be?

8-7   30.   A block of material is pulled along a surface separated from it by a layer of material 1.0 mm thick. The block is 0.30 m × 0.15 m. The force required to move the block at a constant speed of 0.50 m/s is 190 N.

(a) What is the velocity gradient in the material?

(b) What is the force per unit surface area?

(c) What is the ratio of the force per unit area to the velocity gradient?

(d) What is the coefficient of viscosity?

CHAPTER NINE

VIBRATIONS
AND WAVES

Regular periodic vibrations of some sort control the speed of any clock. It may be a pendulum, a balance wheel, a tiny tuning fork, or a quartz crystal. They are all mechanical vibrations and are described by the laws of force and motion.

A vibrating part of any instrument transmits its vibrations to the air, and they travel away as sound waves. An explosion or an earthquake sends waves through the earth. Light waves tell us most of what we know of stars and of galaxies beyond ours. Microwaves are the carrier for our telephone conversations and our T.V. programs.

All vibrations and waves have common characteristics, and this chapter contains the basic material for understanding them. Mechanical vibrations and mechanical waves will be considered here. The waves of sound and of light will be described in later chapters.

9-1 SIMPLE HARMONIC MOTION, S.H.M.

Harmonic motion is illustrated by a child on a swing. The speed is greatest through the central equilibrium position, and it slowly drops to zero at the maximum displacement (Figure 9-1). There, where the velocity changes direction, the acceleration is actually a maximum. Through the central position, as the speed changes from increasing to decreasing, the acceleration is momentarily zero.

Some of the characteristics of s.h.m. are these: at zero displacement the speed is maximum and acceleration is zero, while at maximum displacement the speed is zero and acceleration is maximum. As the object moves away from the equilibrium position it slows down, so the acceleration is in the

Figure 9–1 The motion of a swing is a demonstration of harmonic motion.

Displacement maximum
Speed zero
Acceleration maximum
  as velocity changes
  direction

Displacement zero
Speed maximum
Acceleration zero

direction opposite to the motion. The acceleration is always back toward the center: if the displacement is in a direction chosen as positive, the acceleration is in the negative direction, but when the object swings to the negative direction the acceleration back toward the center is in the positive direction.

This type of motion can also be seen by watching a point on a rotating wheel from a direction in the plane of the wheel, as in Figure 9–2. The displacement of the point goes to one side and then the other from the central position. The observer sees a linear, back-and-forth motion of the point. The speed is maximum at the center; at the maximum displacement the speed is seen as zero as the point moves away from (or toward) the eye. **Simple harmonic motion is described by one component of motion in a circle at constant speed.**

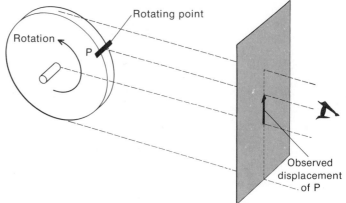

Rotating point

Rotation

P

Figure 9–2 The edge view of a point on a rotating wheel shows the type of oscillation called simple harmonic motion.

Observed
displacement
of P

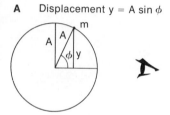

A   Displacement $y = A \sin \phi$

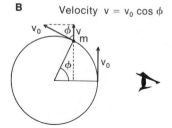

B   Velocity $v = v_0 \cos \phi$

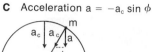
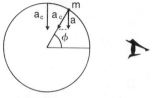

C   Acceleration $a = -a_c \sin \phi$

Figure 9–3 (a) One component of circular motion describes the displacement in s.h.m. (b) One component of the speed in a circle describes the speed of an object moving with s.h.m. (c) One component of centripetal acceleration describes the acceleration in s.h.m.

The maximum displacement from the equilibrium position in s.h.m. is called the **amplitude, A,** of the vibration; this is also the radius of the reference circle, as in Figure 9–3(a). The angle ϕ from the equilibrium angle is referred to as the **phase angle** in the vibration. (Note that the greek lower case phi, ϕ, will be used for the phase angle. In an angular vibration, for instance, θ will be reserved for angular displacement.)

9–1–1   DISPLACEMENT, VELOCITY, AND ACCELERATION

In terms of ϕ, the displacement observed is given by

$$y = A \sin \phi$$

The speed in the reference circle is the same as the maximum speed observed, which is at $y = 0$ and which is referred to as v_0. In Figure 9–3(b) the speed at any time during a vibration, in terms of the angle, is shown to be

$$v = v_0 \cos \phi$$

The observed acceleration is the component of the centripetal acceleration shown in Figure 9–3(c) and is described by

$$a = -a_c \sin \phi.$$

Other aspects of the vibratory motion are as follows:

The period T is the time required for a complete vibration, or the time taken to go around the reference circle. The frequency f is the number of vibrations per unit time and is given by $f = 1/T$. A frequency of **one cycle per second is called one hertz.**

The angular speed in the reference circle, in radians per second if T is in seconds, is given by

$$\omega = 2\pi/T \quad \text{or} \quad \omega = 2\pi f$$

The angle ϕ, if $t = 0$ at the center line as in Figure 9–3(a), is given by

$$\phi = \omega t \quad \text{or} \quad \phi = 2\pi \frac{t}{T} \quad \text{or} \quad \phi = 2\pi f t$$

The speed v_0 is given by the distance around the reference circle divided by the time required, or

$$v_0 = \frac{2\pi A}{T} \quad \text{or} \quad v_0 = 2\pi f A = \omega A$$

The centripetal acceleration, which is also the maximum acceleration observed in the vibratory motion, is given by:

$$a_c = -\frac{v_0^2}{A} \quad \text{or} \quad a_c = -4\pi^2 f^2 A = -\omega^2 A$$

The negative sign has been included because of the direction, which is apparent from examination of Figure 9–2(c).

These relations can be combined in various ways to obtain equations describing the displacement, the velocity, or the acceleration in simple harmonic motion. These equations are summarized below.

Phase angle at time t:

$$\phi = \omega t \quad \text{where } \omega = 2\pi/T = 2\pi f$$
so $\quad \phi = 2\pi t/T$
or $\quad \phi = 2\pi f t$

displacement: $\quad y = A \sin \phi \quad$ where A is the amplitude
velocity: $\quad\quad\; v = v_0 \cos \phi \quad$ where $v_0 = \omega A$
acceleration: $\quad a = a_c \sin \phi \quad$ where $a_c = -\omega^2 A$

EXAMPLE 1

A child on a swing makes a complete oscillation in 3 seconds ($T = 3$ s) and swings 2.0 meters on each side of the center ($A = 2.0$ m). Find the maximum speed of the swing.

The motion of the swing is harmonic (not exactly simple harmonic, but for this problem assume that the simple harmonic motion equations apply with sufficient accuracy).

Use $\quad\quad\quad\quad\quad\quad\quad\quad v_0 = \omega A$
where $\quad\quad\quad\quad\quad\quad\;\; \omega = 2\pi/T$
$\quad\quad\quad\quad\quad\quad\quad\quad\quad = 2\pi/3$ s
$\quad\quad\quad\quad\quad\quad\quad\quad\quad = 2.1/$s

The angular speed in the reference circle is 2.1 rad/s. Then

$$v_0 = \frac{2.1}{s} \times 2.0 \text{ m}$$

$$= 4.2 \ m/s$$

The maximum speed is 4.2 m/s. This can also be expressed as 15 km/h, which is a considerable speed.

EXAMPLE 2

Plot the position for every tenth of a second of an object oscillating with a period of one second and a maximum displacement of 5 cm to each side of the equilibrium position. The object starts ($t = 0$) at the equilibrium position, going upward.

Use $y = A \sin \phi$ with $\phi = 2\pi t/T$ to fit the given data.

$$y = A \sin 2\pi t/T$$
$$A = 5 \text{ cm}$$
$$T = 1 \text{ second}$$
$$2\pi/T = 6.28 \text{ (radians/sec)}$$

TABLE 9–1   Data of example 2

t	$\phi = 2\pi\dfrac{t}{T}$	$\sin \phi$	$y = A \sin \phi$
0.0	0°	0.000	0.00
0.1	36°	0.588	2.94
0.2	72°	0.951	4.75
0.3	108°	0.951	4.75
0.4	144°	0.588	2.94
0.5	180°	0.000	0.00
0.6	216°	−0.588	−2.94
0.7	252°	−0.951	−4.75
0.8	288°	−0.951	−4.75
0.9	324°	−0.588	−2.94
1.0	360°	0.000	0.00

The quantity $2\pi t/T$ is in radians. Change it to degrees if necessary in order to look up the sines in a table of trig functions. Tabulate the data, because repetitive calculations are involved. A sample calculation will be carried through.

For $t = 0.1$ second, $2\pi t/T = 0.628$ radian; 1 radian is 57.3°, so 0.628 radian = 36°. The sin of 36° is 0.588. Multiply this by the amplitude A (5 cm), so the displacement at 0.1 sec is 2.94 cm.

The other data are shown in Table 9–1, and the graph showing the position with time is in Figure 9–4.

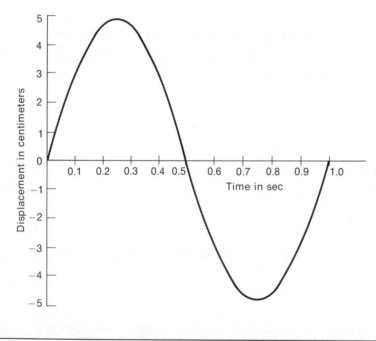

Figure 9–4 The graph of position against time for an object vibrating with a period of 1 second and an amplitude of 5 cm.

9-1-2   ANGULAR VIBRATIONS

Some objects twist back and forth with angular harmonic motion. The balance wheel in a clock or a watch is one example; an object suspended on a thin rod or fiber, twisted and then released, is another (Figure 9–5).

Angular vibrations are described similarly to linear vibrations. The angular symbols are used rather than those for linear motion. With θ_0 referring to the maximum angular amplitude, these are:

Hair spring

Quantity	Linear Vibration	Angular Vibration
Phase angle	$\phi = 2\pi ft$	$\phi = 2\pi ft$
Displacement	$y = A \sin \phi$	$\theta = \theta_0 \sin \phi$
Velocity	$v = 2\pi fA \cos \phi$	$\omega = 2\pi f\theta_0 \cos \phi$
Acceleration	$a = -4\pi^2 f^2 A \sin \phi$	$\alpha = -4\pi^2 f^2 \theta_0 \sin \phi$

9–2 FORCE AND S.H.M.

If a mass is oscillating back and forth with simple harmonic motion, there must be a force acting. For example, you hang a mass carefully on a spring, letting it hang at rest. If the mass is pulled down as in Figure 9–6, the spring pulls it back toward its equilibrium position. If the mass is lifted slightly, gravity tends to pull it down. Whenever the mass is displaced, there is

Figure 9–5 The balance wheel in a watch and an object at the end of a rod or fiber are examples of angular vibrations.

Figure 9–6 A mass on a spring hangs at an equilibrium position (a); if it is pulled downward as in (b), there is an upward force. If it is lifted as in (c), there is a downward force. Thus, if it is displaced in either direction, it will tend to return to the equilibrium position.

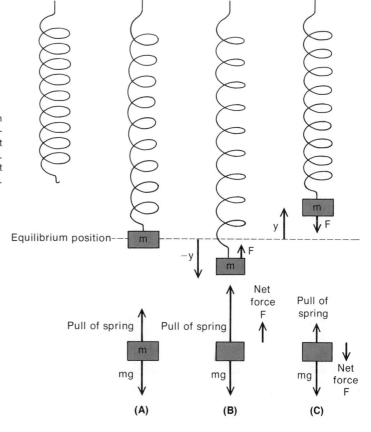

Equilibrium position

Pull of spring

Pull of spring

Net force F

Pull of spring

mg

mg

mg

Net force F

(A)   (B)   (C)

a force pulling it back to the equilibrium position. If the displacement is up (positive), the restoring force is down (negative). If the displacement is down (negative), the restoring force is upward (positive). The restoring force is opposite in sign to the displacement.

The force is related to the acceleration by Newton's second law, $F = ma$. In s.h.m. the acceleration is described by $a = a_c \sin \phi$, where $a_c = -\omega^2 A$. The displacement y is described by $y = A \sin \phi$, and using this in the equation for acceleration gives

$$a = -\omega^2 y$$

Applying Newton's second law,

$$F = ma = -m\omega^2 y$$

This is of the form $F = -ky$

where $$k = m\omega^2 = 4\pi^2 m/T^2$$

These last equations form the basis for the definition of motion of the type called simple harmonic. *If an object moves under the influence of a force described by $F = -ky$, then simple harmonic motion will result and the period will be found from the relations*

$$k = \frac{4\pi^2 m}{T^2}$$

or

$$T = 2\pi \sqrt{\frac{m}{k}}$$

Forces of the form $F = -ky$ occur frequently in natural situations because the force exerted by elastic materials is described by such an equation, with k being called the elastic constant and y the deformation. The minus sign occurs because the force needed to deform an elastic material by an amount y is described by $F = ky$, but the material pulls back on whatever is producing the distortion by $F = -ky$. This type of force is illustrated in Figure 9–7, and it is, in fact, the force described by Hooke's law.

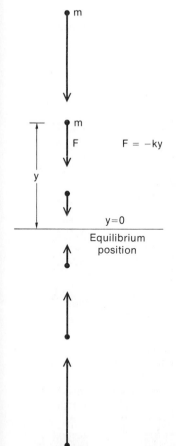

Figure 9–7 A representation of a force described by $F = -ky$, the type that results in simple harmonic motion.

EXAMPLE 3

Consider a spring scale in which each newton of force applied stretches the spring by 2 cm (0.02 m). The spring constant k, given by $k = F/y$, is therefore 50 N/m. Find the period of oscillation, T, if a mass of 2 kg is suspended on that spring and made to move up and down.

The period is related to the spring constant by

$$T = 2\pi \sqrt{m/k}$$

$$= 2\pi \sqrt{\frac{2 \text{ kg}}{50 \text{ N/m}}}$$

$$= 2\pi \sqrt{\frac{2 \text{ kg m}}{50 \text{ kg m/s}^2}}$$

$$= 1.26 \text{ s}$$

The period of oscillation is 1.26 seconds.

9-2-1 THE NATURAL PERIOD IN S.H.M.

In situations in which the force on a mass is described by $F = -ky$, the mass will vibrate with a period described by

$$T = 2\pi \sqrt{\frac{m}{k}}$$

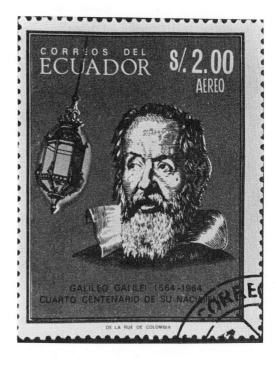

Figure 9-8 Galileo, depicted in one of his great moments of discovery.

The amplitude does not affect the period; it does not matter whether the vibrations are small or large. As long as the governing force is given by $F = -ky$, the period will be the same, the natural period (or frequency, where $f = 1/T$) for that situation. Stories have it that Galileo noted this happening with a swinging chandelier in church one day (Figure 9–8). The period seemed to be constant no matter what the amplitude. After making this simple observation, he went home and began to design a clock governed by a swinging pendulum. The pendulum clock became a timing device of quite amazing precision and remained the standard type of timing device for several hundred years.

The quartz crystal used in many watches today is made to vibrate by an electric field, and it too has a natural frequency governed by its mass and its elastic constant. But the pendulum phenomenom is still of value and will be analyzed in more detail.

9–2–2 THE PENDULUM

The simplest form of pendulum consists of a small heavy mass on a light string. It is often referred to as a point mass on a weightless string—an idealized situation.

The pendulum is illustrated in Figure 9–9. When the mass is drawn aside and released, it will move back toward the center along the arc s. The forces on the mass are shown in Figure 9–9(b). The vector sum of T and mg must be in the direction that it moves, and this sum is shown in (c). The net force F is given by

$$\frac{F}{mg} = \sin \theta = \frac{x}{l} \text{ in magnitude}$$

The restoring force is opposite in direction to the displacement x; taking this into account,

$$\frac{F}{mg} = -\frac{x}{l}$$

or
$$F = -\frac{mg}{l} x$$

The pendulum is displaced by an amount s, not x. However, if the angle of the swing (θ) is small, then the straight line distance x and the curved distance s are very close. At $\theta = 5°$ the ratio x/s is 0.999, and at 10° the ratio is 0.995, so using x as the displacement of the mass instead of s introduces only a small dis-

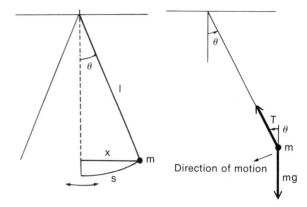

Figure 9–9 Diagrams to assist in the analysis of the simple pendulum.

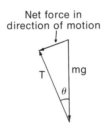

Net force in direction of motion

crepancy. With this approximation the restoring force is related to the displacement x by

$$F = -\frac{mg}{l} x$$

which is of the form

$$F = -kx$$

where

$$k = \frac{mg}{l}$$

The period of oscillation is described, as for any s.h.m., by

$$T = 2\pi \sqrt{\frac{m}{k}}$$

$$= 2\pi \sqrt{\frac{ml}{mg}}$$

$$= 2\pi \sqrt{\frac{l}{g}}$$

The period of the pendulum is independent of the mass and almost independent of the angle of swing. The period does depend on the length, and it does depend on the value of g. The first definite evidence that the earth bulges out at the equator (and hence g is less there) was that pendulum clocks transported from Europe to tropical countries ran more slowly—the period of swing of the pendulum was increased. In fact, a pendulum can be used to measure g with high precision.

EXAMPLE 4

Find the period of a pendulum with a length of 1.000 meter where $g = 9.80$ m/s^2. Use

$$T = 2\pi \sqrt{\frac{l}{g}}$$

$$= 2\pi \sqrt{\frac{1.000 \text{ m s}^2}{9.80 \text{ m}}}$$

$$= 2.007 \text{ s}$$

The period will be 2.007 seconds. During one complete period a pendulum goes past the center line twice, once each way. That pendulum with $l = 1.000$ meter would pass the center every 1.0035 seconds. That this is so close to one second is not an accident, for the original idea for the meter was the length of a pendulum that would pass the center every second. Later there were many changes in the definition of the meter, and now the "seconds pendulum" is not a meter long. In Table 9–2 is shown the length of a "seconds pendulum" at various places on earth.

There are many examples of harmonic motion; some are shown in Figure 9–10. Clamp a bar or a steel scale at one end, pull it aside a bit, and let go. It vibrates with a natural frequency. Float a block of wood or a weighted rod in water, push it down a bit, and let go; it bounces up and down with s.h.m. An equal arm balance set swinging will move with s.h.m., and the period may be several seconds. The more sensitive the balance, the longer will be the period, for that will imply a small restoring force constant, k. The period, T, is inversely proportional to the square root of k, so a small value of k means a large value for T.

Example (f) in Figure 9–10 is of special interest. It is a pivoted rod with two elastic bands holding one end. The rod

TABLE 9–2 The length, l, of a seconds pendulum at various places on earth

Location	l, m
Equator	0.9909
Pole	0.9962
Paris	0.9939
New York	0.9932
Quebec	0.9937
Washington	0.9930
Tokyo	0.9927
Melbourne, Australia	0.9929
Cambridge, England	0.9942
Moscow	0.9945

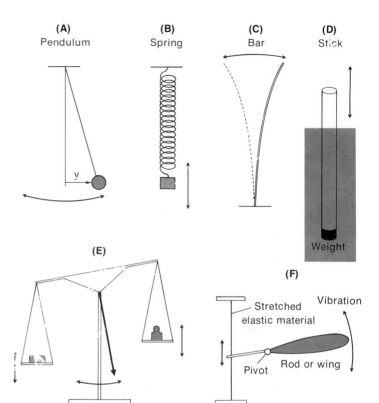

Figure 9–10 Some examples of motion that is simple harmonic and for which there will be a natural period of vibration: (a) pendulum; (b) weight on spring; (c) rod clamped at one end; (d) floating object; (e) equal-arm balance; (f) rod with a pivot and elastics.

can be made to vibrate up and down against the force of the elastic bands; the vibration will slow down mainly because elastic materials are not perfect and energy is lost as they are stretched and released. In the beating of a wing, such as occurs in an insect, energy would have to be used in every stroke to speed it up and slow it down. But in many insects an elastic medium is used much as in Figure 9–10(f), though in a more complex arrangement that gives the wing a rotary pattern of motion as well as up and down and back and forth slightly. Once the wing is started in harmonic motion, energy is stored in the elastic material during each beat and then used to accelerate the wing for the next beat; the muscles need supply energy only to make up the small losses in the elastic.

9–2–3 TORQUE AND ANGULAR VIBRATIONS

The torque required to twist a thin rod or a fiber through an angle θ depends directly on the angle. When it is twisted, the rod or fiber pulls back on whatever is twisting it with a torque, L, described by:

$$L = -k\theta$$

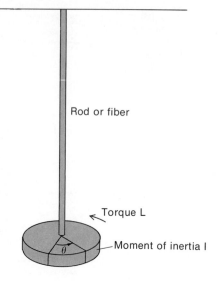

Rod or fiber

Figure 9–11 A system with angular vibration.

Torque L

Moment of inertia I

The torsion constant is k, as was described in Section 7–2–1.

If a mass on a rod is twisted as in Figure 9–11 and then released, angular s.h.m. will result. By analogy with the derivations for linear motion, the period can be shown to be described by

$$T = 2\pi \sqrt{\frac{I}{k}}$$

where I is the moment of inertia of the object on the fiber.

In the case of a balance wheel in a watch, the constant k will be the torque needed to twist the hair spring through one radian.

The apparatus used to measure the force of gravity between two masses (Figure 7–8) made use of a twisting fiber for which the torque constant would have to be known. The forces and torques involved are very small, and the method used to measure the torque is to allow the small lead balls to oscillate back and forth to measure the period T. The moment of inertia of the balls can be calculated, and k can be found from $T = 2\pi \sqrt{I/k}$:

$$k = \frac{4\pi^2 I}{T^2}$$

This trick is often used in working with very sensitive apparatus.

9–3 ENERGY IN A VIBRATION

When an object is to be set in vibration, it is first displaced from its equilibrium position and then released. Work is done in the displacement, so with respect to the equilibrium position the object has potential energy. As the vibrating object passes the center point, all the energy is kinetic; during the vibration there is a continuous change from kinetic to potential energy, but the total is constant. The total can be found either from the potential energy at the maximum displacement or from the kinetic energy at zero displacement. With a displacing force given by $F = ky$, the work required for a total displacement A (as described in Section 7–4) is

$$\text{work} = \text{P.E.} = \frac{1}{2} kA^2$$

The force constant k is given by $m\omega^2$ or $4\pi^2 f^2 m$. Then the potential energy of the mass at its maximum displacement, when the velocity is momentarily zero, is

$$\text{P.E.} = 2\pi^2 f^2 A^2 m$$

At the center the kinetic energy is

$$\text{K.E.} = \frac{1}{2} mv^2$$

and v_0 is given by ωA so

$$\text{K.E.} = \frac{1}{2} m\omega^2 A^2$$

$$= 2\pi^2 f^2 A^2 m$$

This is the same as the potential energy at maximum displacement, which is as expected. Each is an expression of the total energy. The result is that if a mass m vibrates with a frequency f and amplitude A, the energy is given by

$$\text{energy} = 2\pi^2 f^2 A^2 m$$

An interesting aspect of this relation is that the energy depends on the square of the product of frequency and amplitude. Sound is produced by a vibrating object such as a string, vocal chord, or speaker cone, and is carried through the air as a vibration. The pitch of a sound is related to the frequency, and the intensity (or loudness; these terms are slightly different but here the difference is minor) is related to the energy in the vibration. For two sounds of different frequency but of the same intensity, the product fA must be the same.

EXAMPLE 5

Consider a very low 100 hertz note and a very high 10,000 hertz note that are of equal intensity. How will their amplitudes compare?

The product fA must be the same for both, so

$$f_1A_1 = f_2A_2$$
or
$$A_2/A_1 = f_1/f_2$$

Now
$$f_1 = 100/sec$$

and
$$f_2 = 10,000/sec$$

Then the ratio of amplitudes is 100/10,000 or 1/100. The amplitude of the high note is only 1/100 of that of the low note for the same intensity. In the case of a recording, the sound is inscribed on the disk as a "wiggle" in the grove. The low notes require very high amplitude "wiggles" and the high notes require very low amplitude "wiggles." Figure 9–12 is a photomicrograph of the grooves of a record. In making a recording, the low notes must be suppressed so the grooves do not overlap, and the high notes must be enhanced so the first play does not erase the very small irregularities or "wiggles." In playing a record, the reverse must be done; the low notes must be amplified by the amount by which they were suppressed, and the high notes must be cut down to the original comparative intensity. Each recording company may have its own pattern of adjustment, and a record player must be able to compensate correctly if it is to reproduce the original sound.

Figure 9–12 A photomicrograph of the grooves of a record.

9–4 SOME GENERAL CHARACTERISTICS OF WAVES

The material waves that are familiar to us, water waves and sound waves in particular, are disturbances being carried along through a medium. A wave starts as a disturbance in an elastic medium—something dropped into water, a string vibrating in air, a moving surface such as a drum, or even an earthquake. Because of the elasticity of the medium, a disturbance or deformation in one spot distorts the medium next to it, and then the portion of the medium next to that, and so on, so the disturbance propogates. If the disturbance is continuous and harmonic, back and forth, a continuous wave is propogated away from the source of disturbance. It may be a water wave, a sound wave in air, or a seismic wave through the earth.

Another example of a wave is that which occurs in a long stretched spring or even in a rope, as in Figure 9–13. A wave of this type, in which each part of the rope or spring moves perpendicularly to the direction of travel of the wave, is called a *transverse wave* or a *shear wave*.

One of the most impressive demonstrations of the propagation of a disturbance is done with apparatus such as that in Figure 9–14. If you can actually see this, it is worthwhile; not tremendously exciting, just a bit weird or fascinating. It consists of a glass tube about 2.5 cm in diameter and a meter long. A thin rubber membrane is stretched and tied across each end, and ping-pong balls are hung to just touch each membrane. One ball is pulled aside as in Figure 9–14(a) and released. Almost as soon as the ball hits the membrane, the other ball is "kicked" away, as in Figure 9–14(c). A pressure pulse has traveled through the tube as in Figure 9–14(b) and pushed the other

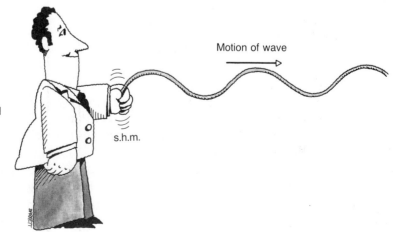

Motion of wave

Figure 9–13 A wave in a stretched spring or rope.

s.h.m.

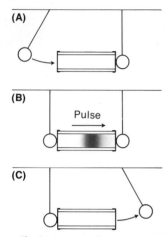

Figure 9–14 A demonstration of a pressure pulse in the air in a tube.

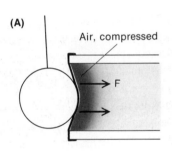

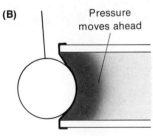

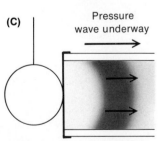

Figure 9–15 The process of formation of a pressure pulse.

membrane. There is a time delay as the pulse has traveled through the tube, but with a one-meter length it is only 1/330 of a second.

The formation of the pressure pulse is due to one membrane being pushed in, compressing the air beside it as in Figure 9–15. This compression pushes on the air beside it; the compression moves ahead, and this proceeds down the tube. The membrane, however, moves back, and the air it pushed into the compression also moves back. This also goes on down the tube; each bit of air moves forward as it is pushed by the compressional wave and then moves back again. Each part of the air makes only a small forward-and-back motion while the pulse travels along.

If the membrane is made to oscillate back and forth, then a series of pulses or waves travels along the tube. If the membrane moves with simple harmonic motion, that same harmonic motion is performed by each bit of air, as in Figure 9–16, though there is a time delay as the impulse travels along the tube.

In such a wave in air (it could be a sound wave), there are places of compression and rarefaction produced by the vibrating bits of air. These condensations and rarefactions are illustrated in Figure 9–17(a). The pressure along the tube goes up and down with time as in Figure 9–17(c). The amplitude of the vibration can be expressed in terms of the actual distance by which individual bits of air move from their equilibrium position, or it can be expressed as what is called pressure amplitude. The pressure amplitude is the maximum change in pressure from the pressure when there is no wave passing. Since the motions of the bits of air are back and forth in the direction that the wave travels, or along the direction of the wave motion, it is called a **longitudinal wave** or a **pressure wave**.

9–4–1 TYPES OF WAVES

Two types of waves have been introduced, the *transverse wave* in which the direction of vibration in the medium is perpendicular to the direction of travel of the wave, and the *longitudinal* or *pressure wave* in which the vibration of each part of the medium is along the direction of motion of the wave. Sound in air is a longitudinal wave, and a guitar string vibrates with a transverse wave. A wave on the surface of water is a combination of these, in that each particle on the surface of the water moves back and forth horizontally as well as up and down, the result being a circular motion. We will avoid analysis of this complex type of wave but you are encouraged, when the oppor-

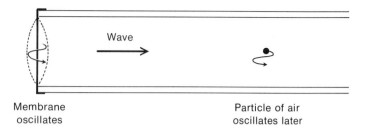

Figure 9–16 The motion of a bit of air in a tube follows the motion of the membrane at the end, but delayed by the pulse travel time.

Membrane oscillates

Particle of air oscillates later

tunity arises, to watch the motion of a small object floating on water waves.

Most of the wave properties apply equally to longitudinal and transverse waves. There are few differences; for instance, transverse waves do not propogate through a liquid. In an earthquake, both transverse waves and longitudinal waves are produced and spread out through the earth. When produced by an earthquake, the transverse waves are called S waves and the longitudinal waves are called P waves.

The S waves do not travel through a central part of the earth about 1400 km in radius, and it is therefore suggested that this "inner core" is liquid. The inner core will transmit P waves, for P waves can go through a solid or liquid.

Why is the core liquid? Was the earth formed hot, and if so, why has it not cooled down in the four to five thousand million years of its existence? Does the moon also have a liquid core? Does Mars?

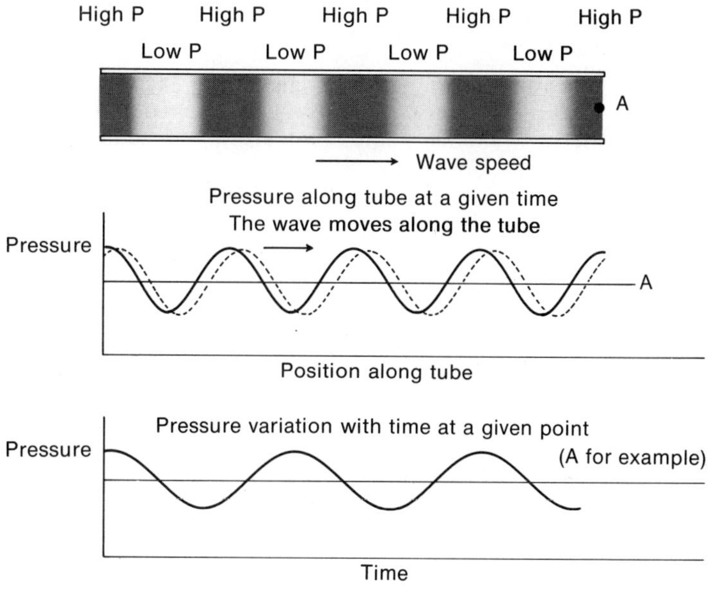

High P   High P   High P   High P   High P

Low P   Low P   Low P   Low P

A

Wave speed

Pressure along tube at a given time
The wave moves along the tube

Pressure

A

Figure 9–17 A pressure wave in a tube.

Position along tube

Pressure variation with time at a given point

Pressure

(A for example)

Time

9–5   WAVELENGTH, FREQUENCY, AND SPEED

A source of waves goes through a complete cycle of its vibration a certain number of times per second and produces the same number of waves per second. This is called the frequency, f. The unit for frequency, cycles per second, is also called hertz (abbreviated Hz) after Gustav Hertz, the discoverer of radio waves. The note A on the common musical scale is 440 hertz, which means that when it is played, 440 waves strike my ear each second. If 440 waves come per second, the time between waves is 1/440 second and this is the period, T. T is $1/f$.

The source sends out f waves each second and they travel away from the source at a wave speed V. In a time t, ft waves are produced and these are spread over a total distance Vt in the medium. Each wave covers a distance given by the total distance divided by the number of waves, Vt/ft or V/f (see Figure 9–18). The distance covered by one wave is the wavelength, often designated by the lower case Greek lamda, λ. Hence, the relation between wavelength, frequency, and speed is

$$\lambda = V/f \quad \text{or} \quad V = f\lambda$$

These equations are equivalent and also are probably among the most basic and most used of all equations describing waves.

EXAMPLE 6

A certain radio station broadcasts at 540 kHz. Find the wavelength of those waves, knowing that radio waves travel at the speed of light, 3×10^8 m/s. The term kHz means kilohertz, where kilo- again indicates a thousand. Then

$$f = 540,000/s$$
$$= 5.40 \times 10^5/s$$
$$V = 3 \times 10^8 \text{ m/s}$$

From
$$v = f\lambda$$
$$\lambda = V/f$$

Then
$$\lambda = \frac{3 \times 10^8 \text{ m/s}}{5.4 \times 10^5/s}$$

$$= 556 \text{ m}$$

The wavelength is 556 meters.
What are the wavelengths of the radio waves broadcast by your favorite radio or T.V. station?

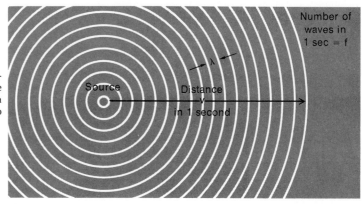

Figure 9-18 The relation between wavelength, frequency, and wave speed. In one second, f waves are produced. Each has a length λ, and they cover a distance v, so $v = f\lambda$.

9-6 THE SPEED OF A WAVE

The speed with which a wave propagates through a medium depends on two main factors, its density and its elastic modulus. For a high elastic modulus, a given displacement produces a large force to set the adjoining matter in motion. With a low modulus of elasticity, the force available to propagate the wave would be less and the wave would move more slowly. If the material is dense, or has a large inertia per unit volume, the rate at which the wave moves would be smaller than if the medium had a lower density. The precise way in which the speed depends on these factors, modulus of elasticity and density, will be deduced by a method referred to as **dimensional analysis.** The result will be a method for calculating the speed of a wave (sound, for instance) in any gas, liquid, or solid.

9-6-1 DIMENSIONAL ANALYSIS

The units on the two sides of an equation must be the same, and the idea of dimensional analysis is that you find the form of an equation by adjusting it until the units are consistent. This type of analysis will not show the existence or the value of a dimensionless constant, and also it will work only if you have put in the quantities that are important. The suggested results must be in accord with experiment. The method is powerful enough to warrant an introduction.

For a simple example, you may remember that an equation exists relating distance, acceleration, and time but you do not remember the form. You know that

$$s \text{ is a function of } a \text{ and } t$$
$$s = f(a, t)$$

The units of s are length, L. The units of a are length over time squared, or L/T^2. The question is, "How do you combine acceleration and time to get a unit of length?" It is obvious that

$$\frac{L}{T^2} \times T^2 = L$$

That is, acceleration must be multiplied by the square of time to give a unit of length. The equation for s must have at^2 in it. You must find in some other way (or remember) that it is

$$s = \frac{1}{2} at^2$$

The problem of finding the speed of a wave in a medium is a little more complex than the above example. If part of the medium is distorted, the force on an adjacent part will depend in some way on a modulus of elasticity. The *force* depends on this, but the *rate* at which the adjacent part moves depends on its inertia, or on its density. So make the assumption that perhaps the wave speed is a function of a modulus of elasticity of the medium and of the density. Further, try the idea that the speed depends on some power of each of these quantities.

Speed has the dimensions of a length divided by a time (m/s) or, in general terms, L/T. Then the problem is to find a combination of modulus of elasticity and density that has the resulting units of L/T or LT^{-1}. Let

$$v = \text{constant} \times E^a \times d^b$$

where a and b are the unknown powers that are to be found.

The units of density d are mass/volume. In general terms, using M for a mass unit and L for length, the units of density are M/L^3 or ML^{-3}.

A modulus of elasticity is stress/strain. Strain is a ratio of dimensions which cancel, so it is dimensionless. The elastic modulus E has the units of stress, which are force/area. The units of a force, which is mass times acceleration, are ML/T^2 or MLT^{-2}, and force/area is $MLT^{-2}L^{-2}$. To summarize,

> units of v are LT^{-1}
> units of E are $MLT^{-2}L^{-2}$ or $ML^{-1}T^{-2}$
> units of d are ML^{-3}

Writing the equation $v = \text{constant} \times E^a \times d^b$ showing units only,

$$LT^{-1} = \text{const} \times (ML^{-1}T^{-2})^a \times (ML^{-3})^b$$
$$LT^{-1} = \text{const} \times M^a L^{-a} T^{-2a} M^b L^{-3b}$$

Combining similar units on the right-hand side,

$$LT^{-1} = M^{(a+b)} L^{-a-3b} T^{-2a}$$

The mass unit does not occur on the left side. Therefore, from $M^{(a+b)}$ on the right, it is apparent that $(a + b) = 0$ or

$$a = -b$$

Looking at the units of T, on the left there is T^{-1} and on the right there is T^{-2a}. Then $2a = 1$ or

$$a = \frac{1}{2}$$

and it follows from $a = -b$ that $b = -\frac{1}{2}$.

The unknown powers a and b have been found: a is $\frac{1}{2}$ and b is $-\frac{1}{2}$.

It follows that the speed of a wave is described by

$$v = \text{constant} \times E^{1/2}\, d^{-1/2}$$

A power 1/2 signifies a square root; thus, the equation can be written

$$v = \text{constant} \times \sqrt{E/d}$$

The constant can be evaluated by other types of analysis, and it often turns out to be equal to 1 if *SI units are used. The speed of a wave in a medium is therefore given by*

$$v = \sqrt{E/d}$$

where E is an appropriate modulus of elasticity and d is the density. The modulus of elasticity involved will depend on the type of wave and to some extent on the shape of the medium.

9–6–2 A LONGITUDINAL WAVE IN A THIN ROD

If a thin rod is struck on the end with a hammer as in Figure 9–19(a), a momentary compression will travel along the rod. The speed depends on Young's modulus and is given by

$$v = \sqrt{E_y/d}$$

where d is the density of the material. Note that the measurement of a wave speed can be a method for determining a modulus of elasticity.

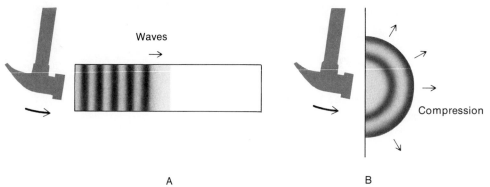

Figure 9–19   Compressional waves in a thin rod and in an extended medium.

EXAMPLE 7

Find Young's modulus for the material of a thin rod for which the speed of a longitudinal wave is measured to be 4800 m/s. The density is 7700 kg/m³.

Use $\qquad\qquad v = \sqrt{E_y/d}$

from which $\qquad E_y = v^2 d$

and $\qquad\qquad v = 4.8 \times 10^3$ m/s
$\qquad\qquad\qquad v^2 = 2.3 \times 10^7 \text{m}^2/\text{s}^2$
$\qquad\qquad\qquad d = 7.7 \times 10^3 \text{ kg/m}^3$

$$E_y = 2.30 \times 10^7 \frac{\text{m}^2}{\text{s}^2} \times 7.7 \times 10^3 \text{ kg/m}^3$$

$$= 1.77 \times 10^{11} \text{ kg} \frac{\text{m}}{\text{s}^2} \text{m}^2$$

$$= 1.77 \times 10^{11} \text{ N/m}^2$$

Young's modulus for that material is 1.77×10^{11} N/m².

9–6–3   A LONGITUDINAL WAVE IN AN EXTENDED SOLID OR A LIQUID

A longitudinal wave in an extended solid is not quite the same as that in a thin rod, because the rod can change dimension laterally while the extended solid cannot. This type of wave is, for example, the P wave produced in earthquakes. Analysis that is beyond our scope shows that the speed is given by

$$v = \sqrt{\frac{E_b + (4E_s/3)}{d}}$$

where E_b is the bulk modulus of elasticity and E_s is the shear modulus.

A liquid has no shear modulus, so the speed of a longitudinal or pressure wave in a liquid is given simply by

$$v = \sqrt{E_b/d}$$

EXAMPLE 8

Calculate the speed of a compressional wave in water, using the value of the bulk modulus given in Table 7–1. Compare this value with the values measured for water.

Use

$$v = \sqrt{E_b/d}$$
$$E_b = 2.17 \times 10^9 \text{ N/m}^2 \text{ at } 25°C$$
$$d = 997.1 \text{ kg/m}^3 \text{ at } 25°C$$

$$v = \sqrt{\frac{2.17 \times 10^9 \text{ N m}^3}{0.9971 \times 10^3 \text{ m}^2 \text{ kg}}}$$

$$= 1475 \text{ m/s}$$

The speed of sound in water is therefore expected, on the basis of this theory, to be 1475 m/s. The speed of sound in water at 25°C is listed in one handbook as 1498 m/s at 25°C. The discrepancy is small, so it can be assumed that the theory is valid.

9–6–4 TRANSVERSE WAVES IN A SOLID

Transverse waves in a solid produce distortions as in Figure 9–20. If the solid is initially marked out into blocks as in Figure 9–20(a), the transverse wave distorts these blocks as in (b) as it passes, and this is a shearing effect. The wave speed is governed by the shear modulus, and is described by

A

$$v = \sqrt{\frac{E_s}{d}}$$

where, again, d is the density.

One example of the production of shear waves is in earthquakes, where they are referred to as S waves.

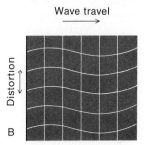

B

Figure 9–20 Transverse or shear waves in a solid.

DISCUSSION PROBLEMS

9–2–1   1. Relax a leg, let it swing freely, and measure or estimate the period. Walk with an easy gait and find the average time per step. Relate this, with discussion, to the natural swinging period.

9–2–2   2. Why was the motion of a child on a swing referred to in Section 9–1 as harmonic motion, not simple harmonic motion?

9–2–2   3. The period of a simple pendulum of a given length does not depend on the mass of the bob. The period of a mass oscillating on a spring does depend on the mass. Why is there a difference?

9–5   4. When you hear a distant band, the notes are heard in the same order that they are played. From this, what do you deduce about the speed of notes of different frequency? What would be the effect if the speed increased with the frequency?

9–6–2   5. Discuss the use of sound measurements in solid materials like steel for the determination of elastic constants, and the type of equipment needed to make direct measurements of those constants.

9–6–4   6. During an earthquake both S and P type waves are produced. The speeds of these waves in the earth's crust are known. Discuss how an earthquake position can be found by timing the arrival of the two types of waves at more than one seismic station.

PROBLEMS

9–1   1. An object moves with simple harmonic motion; the period is 3.14 s, and the amplitude is 0.50 m. Find:
(a) the radius of the reference circle,
(b) the angular velocity in the reference circle,
(c) the maximum velocity in the s.h.m.,
(d) the maximum acceleration in the s.h.m.

9–1   2. A wheel 0.8 m in diameter rolls along the ground at a speed of 2.5 m/s and, as it does, a spot on the rim moves up and down with respect to the ground.
(a) Sketch a graph showing the distance, h, of the spot above the ground during the first five seconds. (Make a few calculations of the height to do this). Let it start at the level of the axle, moving upward.
(b) Find the equation expressing the height of the spot as a function of time.

9–1–1   3. A child on a swing is in motion with a period of 4 seconds and an amplitude of 2.5 meters.
(a) Find the maximum speed.
(b) Find the maximum value of the acceleration (neglect direction).
(c) At what points does the maximum acceleration occur?
(d) In what direction is the acceleration at the positions where it is maximum?
(e) If you were on the swing, at the points of maximum acceleration, in what direction would the inertial force be?

9–1–1   4. Find the maximum speed of a mass that is vibrating with a period of 2.4 s and an amplitude of 0.60 m.

9–1–1   5. Find the maximum speed and the maximum acceleration of an object that vibrates with a frequency of 5000 Hz and an amplitude of 0.10 mm.

9–1–1   6. For a vibration of amplitude 0.10 m to have a maximum acceleration of g, what would the period have to be?

9–1–1   7. (a) Find the phase angle when an object making a vibration with an amplitude of 2.5 meters and a period of 4 seconds is 1.0 meter from the central position, moving outward in a positive direction.
(b) What are the four possible phase angles for the system in (a) but for a displacement of ±1.0 meter in the first cycle of the motion? The velocity may also be positive or negative.

9–2–1   8. Find the period of vibration of a mass on the end of a thin steel bar if a force of 25 N displaced the bar by 0.015 m. The mass is 2.5 kg.

9–2–1   9. Find the force constant for which a mass of 0.015 kg would oscillate with a period of 11 seconds. (This is a method sometimes used to find force constants of laboratory apparatus.) The units must be manipulated to the form N/m.

9–2–1   10. A certain system vibrates with a period of 0.950 s when the mass is 0.500 kg. What mass would have to be added to increase the period to 1.000 s?

9–2–2   11. (a) Calculate the length of a pendulum that has a period of 1.000 s where $g = 9.806$ m/s^2.
(b) Repeat (a) but for a period of 2.000 s.

9–2–2   12. What would be the period of a simple pendulum that has a length of 60.0 m?

9–2–2   13. You are asked to find the height of a room and, being without a tape measure but in possession of a watch, you hang a string from the ceiling, put a weight on the other end, and adjust it so that it just about touches the floor. Then you swing it as a pendulum and find that the period is 3.56 seconds. How high is the room?

9–2–2   14. A pendulum with a massive bob is often used to deliver a blow or impulse of known size to an object. A massive steel ball on a cable is used to wreck a building, and in the laboratory a similar device on a smaller scale can test the impact strength of materials. Find the momentum at the lowest part of its swing of a pendulum bob of 2.5 kg. The length of the pendulum is 2.45 m and it is drawn aside 0.75 m before it is released.

9–2–2   15. The speed of a bullet is to be found by firing it into a block of wood, mass 1.7 kg, which forms the bob of a pendulum of length 2.45 m. Find the speed of the bullet if the block swings 0.18 m aside when a 2 g bullet is fired into it.

9–2–3   16.   Find the period of oscillation of a disk, radius 0.12 m and mass 0.36 kg, suspended at its center from a thin wire. A torque of 0.0032 N m twists the system through 28.65°.

9–2–3   17.   As in Figure 7–8, two small lead balls, mass 15 grams each, are on a thin rod suspended at its center by a fiber. Each ball is 5.0 cm from the fiber. The system oscillates with a period of 11 minutes. Find the torsion constant.

9–3   18.   Compare the energy in two vibrations, one at 3/s and one at 30/s. The vibrating masses and amplitudes are the same.

9–3   19.   How must the amplitudes compare in two vibrations that are of the same period and energy if one vibrating mass has a thousand times the mass of the other? A crude example of this is sound vibrations in air and in water.

9–3   20.   How do the amplitudes compare in two vibrations if one has eight times the frequency of the other but both have the same energy and mass?

9–5   21.   Find the speed of sound in air, using the fact that a note of 1000 Hz is found to have a wavelength of 0.343 m.

9–5   22.   Find the wavelength of a sound of 15,000 Hz in air where the speed of sound is 340 m/s, and then in water where the speed of sound is 1410 m/s.

9–6–1   23.   Use dimensional analysis to find the form of the formula relating the speed of a wave in a stretched string to the tension in it.

From thought and possibly experience you expect that the speed depends on the tension in the string (the force on it). It also depends on the inertia of the string, or its mass per unit length; a massive string will move more slowly under a given force than will a light one. This property, mass per unit length, can be called linear density. It will be total mass over total length.

Let the speed be given by the product of a power of the force and a power of the linear density:

$$v = (F)^a (M/L)^b$$

Find a and b and write the equation. In SI units the constant, which this analysis does not determine, is unity.

9–6–1   24.   Use dimensional analysis to find the form of the relation that gives the energy of a vibrating mass. Consider that the energy will depend on a power of the mass, times a power of the frequency, times a power of the amplitude. Solve for those powers. Find the constant in the equation by referring back to the text.

9–6–2   25.   Find Young's modulus for the material of a thin rod, given that a wave travels 6.00 meters through the rod in 0.0011 second and that the density of the rod is 2700 kg/m³.

9–6–2   26.   A continuous railroad track 3.0 km long is struck with a hammer, and a longitudinal wave travels through the track. How long will it take to get to the other end? Young's modulus is 2.0×10^{11} N/m² and the density of the track is 7700 kg/m³.

9–6–3   27.   Find the distance from a seismic station to the point of occurrence of an earthquake. The P wave traveling at 5.5 km/s arrives 340 seconds before the S wave, which moves at 3.2 km/s.

9–6–3   28.   A compressional wave produced deep in the ocean is seen to travel at 1450 m/s. What is the bulk modulus of the water? (Is this not perhaps an easier way to measure a bulk modulus than to exert an extremely high pressure on a sample of the water and measure the change in volume?)

9–6–4   29.   Near the surface of the earth, S waves travel 3.2 km/s. What is the shear modulus of the surface rocks? The density is 3300 kg/m³.

CHAPTER TEN

SOUND

10–1 WHAT IS SOUND?

In this chapter the topic of waves will be continued, but with special reference to what we call sound. This topic is set apart because it is an area of such importance. Also, many of the topics will tell more of the basic properties of all waves.

By sound we usually mean the pressure waves in air that are in the frequency range that is detected by a normal human ear, from about 20 to 20,000 hertz. Waves of higher frequency, such as are used by bats but undetected by humans, or such as those used in special equipment for medical investigations or even for cleaning laboratory apparatus, are referred to as ultra-sound. In most ways ultra-sound waves behave the same as sound.

We commonly associate sound with air, but sound can pass through solids or liquids, too; in those materials the waves are still sound waves. "Sonar" apparatus is used in water to measure depth and to locate schools of fish, sunken boats, or unwelcome undersea vessels. The name "sonar" indicates that these waves in water are considered also as sound waves. There is no need to give a restrictive definition to the word "sound."

The understanding that sound is carried by a wave in air seems quite ancient, for Diogenes of Sinope, in a lecture in the Portico in Athens about 300 B.C., introduced his topic with these words:

> "According to the promises I have made you, gentle men of Athens, nothing might be more reasonably expected from me, than to give, before I proceed farther, such a definition of what I understand by the man in the moon, as that each of you, **as often as a vibration of the air, caused by the sounds, of which this name is composed,** reaches his ear, ..."
> (Boldface type added by the author.)

The topic of the lecture, the man in the moon, was chosen to ridicule those learned men of the day who would at times be inclined to talk long and loudly about something of which little or nothing was known. Of course, no one does such things today. Or do we, too, have to learn to judge what is known and what is not? Perhaps that is even one of the functions of education.

10-1-1 THE TERMINOLOGY FOR SOUND

There are commonly used words for which there are special scientific meanings. Some of these will be described.

The **pitch** of a sound is related to the frequency with which the waves encounter the ear or other detector. The higher the frequency, the higher the pitch. The human ear detects frequencies in the range from 10 or 20 hertz to 15,000 or 20,000 hertz. Definite limits cannot be put on either end. Very low frequencies can be detected as individual vibrations, and the change from sound to vibration is not at a definite frequency. At the high frequency end of the scale, different people have different limits. In general, the limit is highest for young people and decreases with age. The ear is normally most sensitive at aout 3500 hertz.

A **pure tone** is one in which the vibration is simple harmonic and of just one frequency. The notes on a musical scale are pure tones and, moreover, the frequencies of the notes of the Basic Scale bear simple numerical relations to each other. Notes one **octave** apart, for example, have frequencies related by a factor of two. A note on the scale may be heard as such even if other frequencies that are harmonics (multiples of the frequency of the basic note) are present. Different instruments have harmonics of different frequencies and of different intensities. It is the harmonics that give a note its quality.

The **intensity** of a sound is related to the energy being transported in the wave. Loudness is related to intensity.

Sound was described in Chapter 9 as a longitudinal wave, usually occurring in a gas. The topic will be introduced by investigating the speed of sound in a gas and looking at some of the interesting aspects of the results of the analysis.

10-2 THE SPEED OF SOUND IN A GAS

The equation for the speed of a wave is

$$v = \sqrt{E/d}$$

For a gas, the modulus of elasticity E is equal to the pressure P if the process occurs at a constant temperature, and to γP if the compressions are adiabatic (page 190). The equations for the wave speed would be

$$v = \sqrt{P/d} \quad \text{(isothermal process)}$$

$$v = \sqrt{\gamma P/d} \quad \text{(adiabatic process)}$$

For a wave in a gas, which of these applies depends on other factors. For normal sound the energy does not have time to move from the places where there is compression, with a resulting temperature rise, to the cooler low pressure areas. So for normal sound the process is adiabatic. At extremely short wavelengths, the process is closer to isothermal.

At first sight it would appear that the speed of a wave (like sound) in a gas would vary with the pressure; but the pressure and density are related: increase the pressure and the density increases. To take this effect into account, the general gas law can be used.

The general gas law is often written in the form

$$PV = \frac{MRT}{\mathcal{M}}$$

where P is the pressure
   V is the volume
   M is the mass of the gas
   T is the absolute temperature
   R is the universal gas constant; in SI units, the value of
       R is 8.31 joules/mole K
   $\mathcal{M}$ is the molar mass

Dividing by the volume V leaves M/V on the right-hand side, and this is density, d, so

$$P = \frac{M}{V}\frac{RT}{\mathcal{M}}$$

$$= d\frac{RT}{\mathcal{M}}$$

from which $P/d = RT/\mathcal{M}$

The ratio P/d, which occurs in the equation for the wave speed, can be replaced by $RT/\mathcal{M}$. Then

$$v = \sqrt{\frac{\gamma RT}{\mathcal{M}}}$$

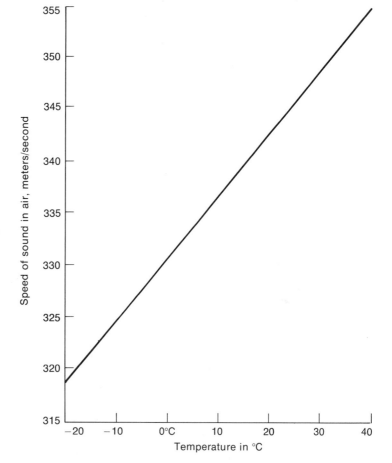

Figure 10–1 The speed of sound in air as a function of temperature.

where γ is the constant given in Table 7–2
  R is the universal gas constant, 8.31 joules/mole K
  T is the absolute temperature
  $\mathcal{M}$ is the molar mass

This equation can be used to calculate the speed of sound in different gases at different temperatures. It also shows that the speed of sound in a gas increases as the temperature increases. The speed of sound in air is shown as a function of Celsius temperature in Figure 10–1. An example of a calculation is given below.

EXAMPLE 1

Calculate the speed of sound in helium at 20°C (293 K).

Use

$$v = \sqrt{\frac{\gamma RT}{\mathcal{M}}}$$

γ for He = 1.66 (see Table 7–2)
R = 8.31 joules/mole K
T = 293 K
$\mathcal{M}$ = 4.00 × 10^{-3} kg/mole

$$v = \sqrt{\frac{1.66 \times 8.31 \text{ joules} \times 293 \text{ K}}{4.00 \times 10^{-3} \dfrac{\text{kg}}{\text{mole}} \text{ mole K}}}$$

$$= \sqrt{\frac{1.66 \times 8.31 \times 2.93 \times 10^5 \text{ kg m}^2}{4.00 \text{ kg s}^2}}$$

$$= \sqrt{10.11 \times 10^5 \frac{\text{m}^2}{\text{s}^2}}$$

$$= \sqrt{1.011 \times 10^6} \text{ m/s}$$

$$= 1005 \text{ m/s}$$

The speed of sound in helium gas at 20°C is therefore 1005 m/s.

The equation for the speed of sound, $v = \sqrt{\gamma R T / \mathcal{M}}$, shows also that the speed depends on the molar mass of the gas. The smaller the molar mass, the higher the speed. In hydrogen or helium, for instance, the speed is higher than in air. In some high pressure undersea chambers, helium is mixed with the oxygen rather than nitrogen, which dissolves in the blood. If normal air were used and the pressure were released (by return to the surface) too quickly, bubbles of nitrogen would form in the blood, causing the often fatal "bends." It has been found that this does not occur with helium. However, when a person in an atmosphere of helium talks, the sound traveling faster in the throat and mouth resonates at a higher frequency than normal. The pitch of the voice is therefore raised. A person talking in a helium-diluted atmosphere speaks at almost an octave above the normal pitch.

The speed of sound in various solids, liquids, and gases is shown in Table 10–1.

10–3   WAVE PROPAGATION: HUYGENS'S PRINCIPLE

The propagation of a wave through a medium can be analyzed by a process devised by Christian Huygens (1629–1695). The idea behind it is that a distortion in a medium affects the adjacent areas and the distortion propagates away at the wave speed. If a point is disturbed as in Figure 10–2(a), a wave spreads out in all directions. In a time T, the period, the wave travels a distance equal to the wavelength and a wavefront is at a distance λ from the source as in Figure 10–2(b). The wave will appear as an expanding circle around the source. In an

TABLE 10–1   The speed of sound in various substances

SUBSTANCE	SPEED OF SOUND, METERS/SECOND
Aluminum	5100
Iron	4900
Gold	2080
Lead	1230
Mahogany	4300
Cork	500
Rubber	70
Water, 25°C	1498
Sea water, 25°C	1531
Air, 0°C	331.4
Air, 20°C	343.3
Air, −42°C	304.9
Hydrogen, 20°C	1303
Helium, 20°C	1005

illustration, a line is drawn along points that are all at the same phase in the vibration. Such a line shows a **wavefront.**

Each point on the wavefront, as at P in Figure 10–2(b), is itself a disturbance; if that disturbance alone occurred, a wave would travel in a circular pattern away from that point. If a small wave is considered to emanate from each point on the wavefront as in Figure 10–2(b), the lateral effects from adjacent points cancel and the new wavefront is the envelope of the

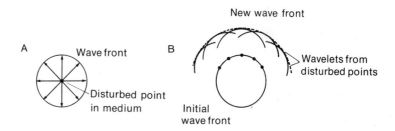

Figure 10–2   In (a), a disturbance at a point leads to a circular wavefront. Each point on the wavefront in (a) results in wavelets that together form the new wavefront as in (b). The disturbed points on a plane wave (c) send out wavelets that together form a new plane wave. These are examples of the use of Huygens's principle.

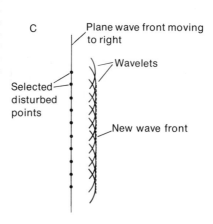

wavelets from all points on the old wavefront. In this way it is possible to see how the circular wave pattern is formed about a point source.

A plane wave in cross-section, as in Figure 10–2(c), can be considered as a long line of sources, and it continues to propagate as a plane wave. From this we also see that a plane wave results from a long linear source.

If that was all we learned from Huygens's principle, it would be of little value. However, it will be applied to a few other situations to show its power.

10–3–1   DIFFRACTION

Consider a plane wave that encounters an obstacle, as in Figure 10–3(a). Marking several points along the wave just at the obstacle shows that the wave bends around the obstacle, spreading into the shadow region. This phenomenon of a wave bending around an obstacle is called **diffraction.**

If there are two obstacles, as in Figure 10–3(b), forming just a small opening, the wavefront bends both ways; in fact, the wave spreads out from that opening just as though the opening was a point source. To show how this actually happens, Figure 10–4 is a photograph of water waves passing through a narrow opening between two islands.

The amount of energy in the part of the wave that bends around the obstacle depends on the wavelength because the energy going "around the corner" comes only from near the obstacle. For a short wavelength, it will come from only a short distance along the wavefront; for a longer wave it will come from a correspondingly longer portion of the wavefront at the obstacle. Consequently, the low notes (long waves) of music are carried better around corners than are the high notes. You may have noticed that if you hear noise coming from a room down the hall and around a corner in a building, it is a low rumble—the high notes do not diffract as well.

The phenomenon of diffraction occurs with any wave phenomenon; the longer the wave, the greater the effect. Another example involves radio waves. The long waves, especially those at the lower end of the A.M. band, bend around the curved earth and can be detected at long distances. The very short waves used for F.M. broadcasting or television bend very little over the horizon and have a short range. The waves in the "shortwave" bands can often be heard at long distances because they are reflected from electrically charged layers in the upper atmosphere; that is a different phenomenon from diffraction.

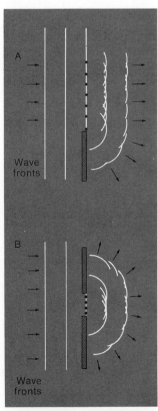

Figure 10–3 Using Huygens's principle to follow a wave pattern past an obstacle or through a narrow slit.

Figure 10-4 Diffraction of water waves (a) passing the end of a barrier and (b) passing through a narrow opening.

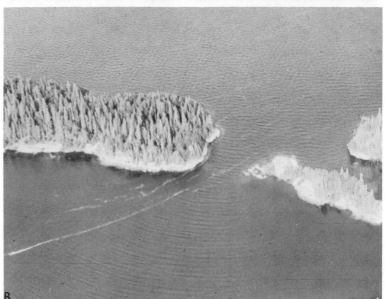

10-3-2 WAVES IN INHOMOGENEOUS MEDIA

If there is a gradual change in wave speed through a medium, it can be taken into account if the wave propagation is analyzed using Huygens's principle. After a wavefront is established, the next one occurs at a distance from it that is determined by the wave speed, V. Considering a time of just one period, the distance will be one wavelength. The wavelength is given by $\lambda = VT$. The wavefronts will be close together where the speed is low and far apart where it is high. Successive wavefronts are constructed, and at each point the distance between the wavefronts is proportional to the speed of the wave at that point.

For example, consider the propagation of sound in the atmosphere. If the atmosphere were the same temperature at all altitudes, the waves would propagate in hemispherical wavefronts from a point source, as in Figure 10–5(a). Usually, however, the air temperature decreases with altitude and the wavefronts get closer together at higher altitudes. The result, as in Figure 10–5(b), is that the sound is focused upward. Sailplane pilots and those who hang-glide often comment on how clearly voices on the ground can be heard in the air above. If there is a "temperature inversion," the temperature increasing with altitude, then the sound waves are as in Figure 10–5(c). This results in the sound being carried along the ground, even focused back to the ground. Explosions can often be heard at very distant points even 200 km away, where the sound is focused back down. At these times the explosion will not be heard at intermediate points, e.g., at 100 km. In 1951 in Nevada, an atomic

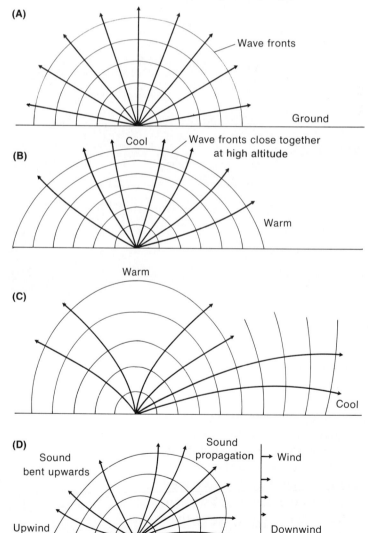

Figure 10–5 Propagation of sound in the atmosphere. (a) The wave pattern that would appear if the temperature were constant. If the air cools with altitude, the wave pattern is as in (b); if there is a temperature inversion, the pattern is as in (c). A wind distorts the wave pattern as in (d).

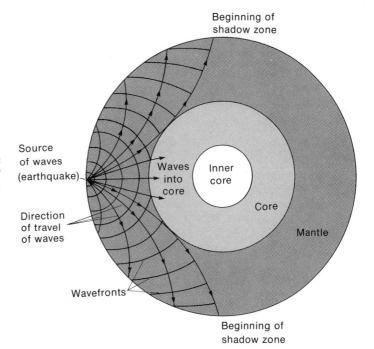

Figure 10-6 The curving path of seismic waves in the mantle of the earth, caused by the increase of speed with increasing depth.

bomb was detonated at the time of a temperature inversion and windows in Las Vegas, Nevada, 130 km away, were broken.

If sound is being propagated in a wind, the speed used for construction of the wavefronts is the speed of sound added to the horizontal wind speed. The result is that the wave fronts "lean" as in Figure 10–5(d). The sound is focused up into the air "upwind" and along the ground "downwind."

Watch for all these phenomena!

A further example of a wave in an inhomogeneous medium is the propagation of seismic waves through the earth. In Table 10–2 are listed the approximate speeds of the P waves and S waves at various depths. The speed increases with depth in the mantle, so the wavefronts become further apart with depth. The direction of propagation is curved as in Figure 10–6. The waves change direction sharply at the core boundary, although that is not shown in the figure.

TABLE 10-2 The speed of seismic waves at various depths in the mantle of the earth

Depth, km	P wave, km/s	S wave, km/s
100	8.0	4.5
500	9.1	5.2
1000	11.4	6.4
1500	12.1	6.7
2000	12.8	6.9
2500	13.4	7.2

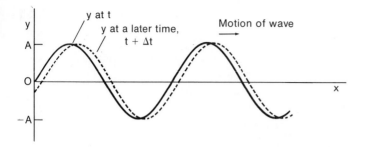

Figure 10–7   A traveling wave.

10–4   THE EQUATION OF A WAVE

Physics does not consist of developing equations for the sake of equations only, but the mathematical manipulation of some equations can tell us much about natural phenomena: what is their process, and how can we make use of them?

A vibration, either s.h.m. or a combination of such motions, occurs in the source to produce a wave in the surrounding medium. Then, as illustrated in Figure 10–7, the various parts of the medium perform a similar vibration but at a later time. As in Figure 9–16, let there be s.h.m. at the origin; then at the point marked x the vibration will be the same but delayed by the time needed to travel the distance x. This time is given by x/v.

Let the s.h.m. at the origin be described by

$$y = A \sin \omega t$$

The phase angle in the vibration is ωt. At the distance x the phase angle is that which occurred at the origin at an earlier time, earlier by x/v. The s.h.m. at the distance x is then

$$y = A \sin \omega \left(t - \frac{x}{v} \right) \qquad \text{where } \omega = 2\pi f$$

This equation describes the vibration in time of the medium at any distance x from the origin. Also, at any given time (t constant), it describes the vibration as a function of position x; that is, the equation describes the shape of the wave. In Figure 10–7 the wave is shown at a time t and then at a slightly later time. The wave has moved along. The equation describes a wave moving in the positive x direction, to the right in the diagram.

If the wave is traveling toward the origin from the right, or in the negative x direction, then the displacement at point x occurs before the corresponding displacement at $x = 0$. That is, for a wave in the negative direction use $(t + x/v)$ in place of t. Such a wave is described by

$$y = A \sin \omega \left(t + \frac{x}{v} \right)$$

Now consider what would happen if two waves of equal amplitude were moving at the same time, one in the positive x direction and one in the negative x direction. You might expect a complex mess, but analyze it mathematically to see what really happens.

Let one wave cause a displacement y_1 and the other a displacement y_2:

$$y_1 = A \sin\left(\omega t - \frac{\omega x}{v}\right)$$

$$y_2 = A \sin\left(\omega t + \frac{\omega x}{v}\right)$$

Use the relation $\sin(a + b) = \sin a \cos b + \cos a \sin b$ to rewrite these equations as

$$y_1 = A\,(\sin \omega t \cos \omega x/v - \cos \omega t \sin \omega x/v)$$

$$y_2 = A\,(\sin \omega t \cos \omega x/v + \cos \omega t \sin \omega x/v)$$

Add these two equations, and the last terms in each cancel to leave

$$y_1 + y_2 = 2A \sin \omega t \cos \omega x/v$$

Let $y_1 + y_2$ be the net displacement y. Also interchange terms to get

$$y = (2A \cos \omega x/v)\sin \omega t$$

This is no longer the equation of a traveling wave, but merely that of simple harmonic motion. The equation for s.h.m. of amplitude A' is

$$y = A' \sin \omega t$$

Thus, the amplitude of the s.h.m. resulting from the two waves depends on the position x and is

$$A' = 2A \cos \omega x/v$$

This will have more meaning if some substitutions are made. Write

$$\omega = 2\pi f \quad \text{and} \quad v = f\lambda$$

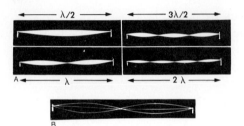

Figure 10–8 A time exposure photograph of a standing wave on a string. (From F. W. Sears and M. W. Zemansky, *College Physics,* 4th ed., Addison-Wesley Publishing Co., 1974.)

so $\omega x/v$ is just $2\pi x/\lambda$ (λ is wavelength). The amplitude as a function of x is then:

$$A' = 2A \cos 2\pi x/\lambda$$

Remember that $\cos \theta = 0$ if $\theta = \pi/2,\ 3\pi/2,\ 5\pi/2$, etc., and these values occur when $x = \lambda/4,\ 3\lambda/4,\ 5\lambda/4$, etc. There is no vibration at all at these values of x. (These are **nodes**.)

Also, $\cos \theta = 1$ and the amplitude A' is maximum if $\theta = 0$, $\pi,\ 2\pi$, etc. or $x = 0, \dfrac{\lambda}{2}, \lambda$, etc. This is called an **antinode**.

The result of all this analysis is that the two waves in opposite directions produce what is called a **standing wave**. A time exposure of a standing wave in a string is shown in Figure 10–8, along with distances marked in terms of wavelengths. In our theory, the position $x = 0$ occurs at an amplitude maximum. The nodes are separated by half a wavelength.

10–4–1 PRODUCTION OF A STANDING WAVE

It has just been shown that a standing wave results from two waves traveling in opposite directions—a situation not so unlikely as you might suspect. If a wave is reflected, it goes back and then you have the two waves. There can even be multiple reflections, back and forth in a restricted space. In such a case the standing waves will result only if the distance between reflecting points is such that the waves traveling in each direction are in phase. For example, when the string of a guitar is plucked, a wave reflects back and forth in it setting up a standing wave. There will be a node at each end where the string is fixed, so the length is just half a wavelength. The string can be made to resonate with two loops if, as it is plucked, a finger is held lightly in the center to produce a node at that point. The note will then be an octave above that for which it is tuned.

An air column can also have a standing wave in it, but only of a frequency for which the wavelength is related simply to the length of the column. There must always be a node at a closed end and an antinode at an open end. It is said that the air column will *resonate* to certain frequencies.

Some patterns of resonances in vibrating strings and air columns are shown in Figure 10–9.

10–4–2   RESONANT FREQUENCY

The phenomenon of resonance may occur when a sound is confined in a limited space and is reflected back and forth from the boundaries. If the size of the space is such that the sound waves moving in each direction are always in phase, a resonance occurs. For instance, a sound of low intensity may be fed into a closed tube. It is reflected from the ends to go back and forth. If the length is such that the waves going along the tube are always in step with those fed in and in step with the previous reflections, then the intensity of the vibration in the tube in-

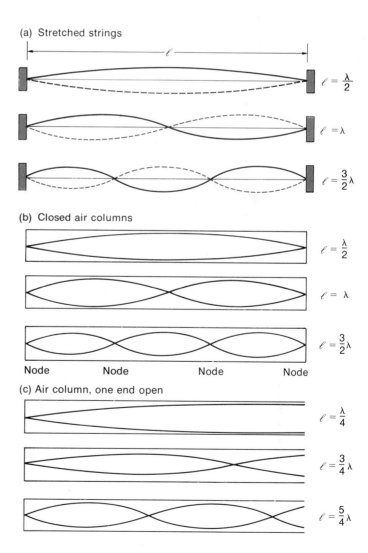

(a) Stretched strings

$\ell = \dfrac{\lambda}{2}$

$\ell = \lambda$

$\ell = \dfrac{3}{2}\lambda$

(b) Closed air columns

$\ell = \dfrac{\lambda}{2}$

$\ell = \lambda$

$\ell = \dfrac{3}{2}\lambda$

Node   Node   Node   Node

(c) Air column, one end open

$\ell = \dfrac{\lambda}{4}$

$\ell = \dfrac{3}{4}\lambda$

$\ell = \dfrac{5}{4}\lambda$

Figure 10–9 Standing waves on strings and in air columns.

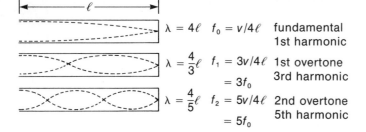

$\lambda = 4\ell \quad f_0 = v/4\ell \quad$ fundamental
1st harmonic

$\lambda = \dfrac{4}{3}\ell \quad f_1 = 3v/4\ell \quad$ 1st overtone
$\qquad\qquad = 3f_0 \quad$ 3rd harmonic

$\lambda = \dfrac{4}{5}\ell \quad f_2 = 5v/4\ell \quad$ 2nd overtone
$\qquad\qquad = 5f_0 \quad$ 5th harmonic

Figure 10–10 Overtones and harmonics for an air column that is open at one end.

creases. A limit occurs when loss of energy at the reflections just equals the energy fed in.

The topic of resonant frequencies is very common in science, and the basic concepts will be examined by analyzing one resonant system, an air column open only at one end. The standing wave produced by feeding in even a low intensity sound will have a node at the closed end and an antinode at the open end. There can be any number of nodes in between, or none. The possible shapes of the standing waves are shown in Figure 10–10, where the curves indicate the amplitude of the vibration along the tube.

To find the resonant frequencies, we use the ideas that a wavelength is two complete loops of the standing wave, and that $v = f\lambda$ (or $f = v/\lambda$) where v is the speed of the wave. In Figure 10–10(a) there is half a loop in the column of length l so $\lambda = 4l$. It follows that the frequence, f, is $v/4l$. This is the lowest resonant frequency, referred to as the **fundamental.** There will be other resonant frequencies, referred to as **overtones.** The overtones will be integral multiples of the fundamental, referred to as **harmonics.** The number given to a harmonic is the number by which the fundamental is multiplied to obtain the frequency of that harmonic.

From Figure 10–10 you will see that with an open-ended air column only the odd-numbered harmonics occur. With some other resonant systems all harmonics will occur.

The resonant frequencies are related to the length of the column and the speed of sound in the column. It is possible to find the speed of sound by using a known frequency and finding the length of column that produces resonance.

EXAMPLE 2

Find the fundamental frequency and the first overtone of sound that would resonate in a room (with solid walls) that is 2.5 meters long. Treat it as a resonant air column and use 330 m/s as the speed of sound.

With both ends closed there must be a node at each end, so the length of the room is half a wavelength. For the fundamental frequency, then,

$$\lambda = 2l = 5.0 \text{ meters}$$

Use $f = v/\lambda$

$$f_0 = \frac{330 \text{ m/s}}{5.0 \text{ m}} = 66/s$$

The fundamental frequency is 66 per second or 66 hertz.
    The first overtone would have a node at each end and one in the middle, so $l = \lambda = 2.5$ m

$$f_1 = \frac{330 \text{ m/s}}{2.5 \text{ m}} = 132/s$$

The first overtone is 132 Hz.
    Middle C is 256 Hz, and low C is half of this or 128 Hz.
    Our representative figures would correspond to those for a small room, and such a room would resonate to low notes. This is why men sing so well in the bathroom.

10–5 POWER IN A WAVE

In creating a sound wave in air, the source pushes and pulls on the air, doing work on it at some rate. The energy is transferred to the air as it is made to vibrate; then the vibration, and hence the energy, travels away from the source. If the energy flows always into a hemisphere as in Figure 10–11, the energy becomes spread over a larger and larger area and the intensity of the wave decreases. By the **intensity of a wave** is meant **the rate of flow of energy through a unit area parallel to a wavefront.** But rate of flow of energy, energy per unit time, is power. The intensity of a wave is measured basically in units of power per unit area. For example, in the SI system it is in **watts per square meter.**

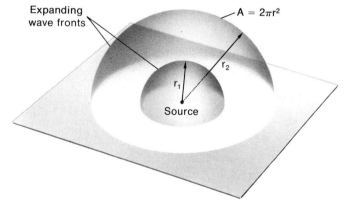

Figure 10–11 The power output of the sound source flows out through the hemispheres shown.

EXAMPLE 3

Find the sound intensity at 3 meters from a speaker that is placed on the ground outdoors, feeding sound evenly into a hemisphere above it. Let the sound power of the source be 2 watts. The situation is shown in Figure 10–11.

The area of a sphere is given by $A = 4\pi r^2$, and that of a hemisphere is then $2\pi r^2$. The area of the hemisphere of 3 meter radius is then

$$A = 2\pi \times 3^2 \, \text{m}^2$$
$$= 56.5 \, \text{m}^2$$

The total power of 2 watts flows through this area, so the power per unit area, the sound intensity, is

$$I = \frac{2 \, \text{watts}}{56.5 \, \text{m}^2} = 3.54 \times 10^{-2} \, \text{watts/m}^2$$

The sound intensity at 3 meters from that source is therefore 3.5×10^{-2} watts/m².

Sound intensities detectable by the ear may range from a low of about 10^{-12} watts/m² up to about 1 watt/m². This is a tremendous range; in order to simplify the description of sound intensities, a system that in effect deals only with exponents of 10 has been devised. This is what is behind the decibel system.

10–5–1   SOUND INTENSITY LEVEL

Whereas the basic description of a sound **intensity** is in watts/m², the common designation of a sound **intensity level** is in a unit called a decibel or db. This unit is related to watts/m² in the following way.

The intensity I of a sound is compared to a standard zero level of sound intensity. This zero level is often chosen to be the intensity at the threshold of hearing. There is by no means a universally agreed-upon zero intensity level, although frequently the zero level I_0 is taken to be 10^{-12} watts/m². The comparison of a sound of intensity I to the zero level I_0 is I/I_0. For example, if I is 10^{-4} watts/m² and $I_0 = 10^{-12}$ watts/m², then

$$\frac{I}{I_0} = \frac{10^{-4} \, \text{watts/m}^2}{10^{-12} \, \text{watts/m}^2} = 10^8$$

That sound intensity is 10^8 times the chosen zero level. The units cancel in the ratio; measurements of sound levels have no units. The power of 10 occurring in the ratio I/I_0 is the sound level and is referred to as the sound intensity level in **bels.**

Remembering that $\log_{10} 10^x = x$, the sound intensity level in bels is given by

$$\text{sound level in bels} = \log I/I_0$$

For convenience, one tenth of a bel is often used as a unit and given the name **decibel,** abbreviated db, where

$$1 \text{ bel} = 10 \text{ decibels}$$

From this, the sound level in decibels is seen to be given by

$$\text{sound level in db} = 10 \log (I/I_0)$$

EXAMPLE 4

Find the intensity level in db of a sound of intensity 3.5×10^{-2} watts/m². Use a zero level of 10^{-12} watts/m².

$$\text{sound level in db} = 10 \log_{10} I/I_0$$
$$I = 3.5 \times 10^{-2} \text{ watts/m}^2$$
$$I_0 = 10^{-12} \text{ watts/m}^2$$
$$\text{sound level in db} = 10 \log_{10} (3.5 \times 10^{-2}/10^{-12})$$
$$= 10 \log_{10} 3.5 \times 10^{10}$$

and
$$\log 3.5 = 0.54$$
$$\log 10^{10} = 10$$

so
$$\text{sound level in db} = 10 \times (10 + 0.54)$$
$$= 105.4$$

The sound intensity level will be 105.4 db. This is actually a very high intensity of sound.

Although the level I_0 is often chosen to be 10^{-12} watts/m², the reference level should always be stated.

Approximate sound intensity levels of some common sources are shown in Table 10–3.

TABLE 10–3 Some typical sound intensity levels

SOURCE AND LOCATION	INTENSITY LEVEL, db
Jet plane at 300 feet	140
Near industrial furnace	110
Maximum recommended for 8 hour day	90
Among heavy traffic	85
Inside car at 60 mi/h	65–75
Conversational voice at 3 feet	70
English sparrow at 20 feet	65
Quiet office, some air conditioning noise	45

EXAMPLE 5

If the intensity of a sound in watts/m^2 is doubled, by how much is the db level changed?

Let the first intensity be I_1 and the intensity level be db_1. The intensity is then increased to I_2, where $I_2 = 2\,I_1$ and the intensity level is db_2. The problem is to find the change $db_2 - db_1$.

$$db_1 = 10\ \log_{10} \frac{I_1}{I_0}$$

$$db_2 = 10\ log_{10} \frac{2I_1}{I_0}$$

Since $\log ab = \log a + \log b$,

$$\log 2I_1/I_0 = \log 2 + \log (I_1/I_0)$$
$$db_2 = 10(\log 2 + \log I/I_0)$$
$$= 10 \log 2 + 10 \log I/I_0$$
$$= 10 \log 2 + db_1$$

So
$$db_2 - db_1 = 10 \log 2$$
$$= 3.01$$

The sound intensity level is increased by only 3 db when the intensity is doubled.

EXAMPLE 6

What is the resultant sound level when a 70 db sound is added to an 85 db sound?

It is the energy flow per unit area that adds, so to do this problem it is necessary to find the intensity associated with each sound level, add the intensities, and then convert back to intensity level in db.

Let the intensities be I_1 and I_2. For the first sound

$$70 = 10\ \log_{10} I_1/I_0$$

$$\log_{10} I_1/I_0 = 7$$
$$I_1/I_0 = 10^7$$
$$I_1 = 10^7\ I_0$$

For the second sound
$$85 = 10\ \log_{10} I_2/I_0$$
$$\log_{10} I_2/I_0 = 8.5$$
$$I_2/I_0 = 3.16 \times 10^8$$
$$= 31.6 \times 10^7$$
$$I_2 = 31.6 \times 10^7\ I_0$$

Adding the two sounds,

$$I = I_1 + I_2 = 10^7\ I_0 + 31.6 \times 10^7\ I_0$$
$$= 32.6 \times 10^7\ I_0$$
and
$$I/I_0 = 3.26 \times 10^8$$

For the sum, the intensity level in db is

$$db = 10\ \log_{10} 3.26 \times 10^8$$
$$\log 3.26 = 0.513$$
$$\log 10^8 = 8$$
$$\log 3.26 \times 10^8 = 8.513$$
$$db = 85.13$$

The 70 db sound added to the much louder 85 db sound results in a level of only 85.13 db. This would hardly be a noticeable increase.

10-5-2 SOUND PRESSURE LEVEL

The power associated with a sound is usually extremely small, so devices that respond to sound make use of another effect; when a sound wave hits something, a pressure is exerted, so sound measuring devices actually respond to sound pressure. Corresponding to any given intensity there will be a certain sound pressure. There is not a direct relation; rather, the intensity and pressure are related by an equation of the form

$$I = kP^2$$

It is a square law relation. A doubling of pressure is associated with a fourfold increase in intensity. Sound intensity levels in decibels can be expressed in terms of the sound pressures P. If the zero level is called P_0, where

$$I_0 = kP_0^2$$

and an intensity I corresponds to a pressure P, where

$$I = kP^2$$

then the ratio $I/I_0 = P^2/P_0^2 = (P/P_0)^2$

The constant k has cancelled. The level in decibels is given by:

$$\begin{aligned} db &= 10 \log_{10} I/I_0 \\ &= 10 \log_{10} (P/P_0)^2 \\ &= 20 \log_{10} P/P_0 \end{aligned}$$

The sound level in terms of pressures is given in terms of that last equation. The zero pressure level is usually taken as

$$\begin{aligned} P_0 &= 2 \times 10^{-5} \text{ N/m}^2 \\ &= 2 \times 10^{-4} \text{ dynes/cm}^2 \\ &= 2 \times 10^{-4} \text{ microbar} \end{aligned}$$

The sound levels in terms of pressures are used in the same manner as are those in terms of intensities.

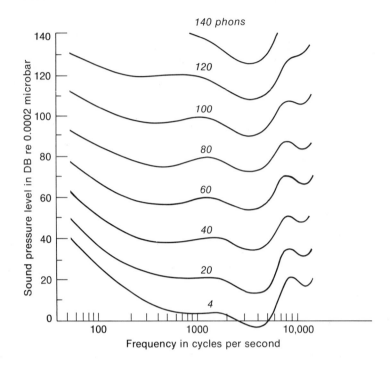

Sound pressure level in DB re 0.0002 microbar

140 phons

Frequency in cycles per second

Figure 10–12 Loudness levels at different frequencies shown as a function of sound pressure level. (After W. A. Munson in *Handbook of Noise Control*, C. M. Harris (ed.), McGraw-Hill Book Company, New York, 1957.)

10–5–3 LOUDNESS LEVELS

The ear does not have the same sensitivity for sounds of all pitches or frequencies. It is most sensitive to low intensity sounds at about 3500 cycles/s (3500 Hz). That is, the intensity at the threshold of hearing is lower at this frequency than at any other. The intensity level in db as outlined in the previous section is not then a measure of loudness. For example, a sound of 1000 cycles/s (Hertz) may be created at an intensity level (or pressure level) of 20 db. A sound of a lower frequency, say 100 Hz, would have to reach about 36 db to be judged to be of the same loudness. At 3000 cycles/s (Hz) the pressure level would have to be only about 16 db to be judged to be of the same loudness as the 20 db sound at 1000 Hertz. Actually, these measurements cannot be made with precision, for loudness is not measurable with a meter. The loudness tests are based on the judgments of a large number of people. Usually a note with a frequency of 1000 cycles/s (Hertz) is used as a standard. A sound of known intensity level is created at this frequency, and then a sound at some other frequency is created and adjusted in intensity until the *observer* judges it to be of the same loudness as the 1000 cycle note. In this way the actual intensity level (or pressure level) of a sound at any frequency can be found that has the same loudness level as the standard 1000 cycle sound.

Loudness level is expressed in a unit called a **phon.** At 1000 Hz, loudness level and sound intensity level units are chosen to be equal. A sound of intensity level I db at 1000 Hertz has a loudness level of A phons, and $I = A$. At any other frequency, a sound of the same loudness has a loudness level of A phons even though it probably is of different intensity. Figure 10–12 is a chart showing the relationship between intensity level and loudness level in phons.

10-5-4   INTENSITY OF A WAVE IN A MEDIUM

The intensity of a sound wave can be expressed in terms of the properties of the medium, density and wave speed, and in terms of the factors that describe the vibration of the parts of the medium, the amplitude and frequency. The expression for intensity in terms of these quantities is necessary for understanding the problem of transmission of sound from one medium to another, and that of the way in which the middle ear has been constructed to transmit sound from the air to the fluid in the inner ear. The design of the ear is nothing short of amazing. This will be apparent even though all of the aspects will not be considered.

If a medium is vibrating, each part of it doing simple harmonic motion as the wave goes by, then there is a certain amount of energy in the vibration. For example, if a source feeds sound into a tube for a short time t and then stops, as in Figure 10–13(a), the medium in the length l is set vibrating. These vibrations that make up the wave travel past the point of observation at Q as in Figure 10–13(b), and after a time t the wave has completely passed Q. The energy in the wave in (a) has been carried in the wave past the point Q as in (c). The

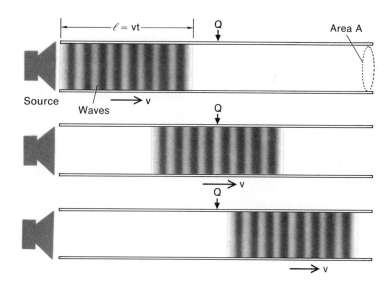

Figure 10–13 A sound wave being fed into a tube. The energy in the vibration to the left of Q in part (a) flows past Q in part (b), until all of the energy has moved to the right as in (c).

energy per unit area of the tube passing Q per unit time is the intensity. This can be expressed in terms of the properties of the medium, as will be derived below.

Consider that the sound source in Figure 10–13(a) operates for a time t feeding a wave into the tube. This wave will occupy a length given by $l = vt$, where v is the wave velocity. All of the mass in the length l has been set in vibration. The amplitude of the vibration of each portion of the medium can be represented by A and the frequency of the source by f. It was shown in Section 9–3 that the energy (kinetic plus potential) of a vibrating mass m is described by

$$E = 2\pi^2 m f^2 A^2$$

The mass vibrating in the tube is that of the medium in a length l. That will be the volume la (where a is the cross-sectional area) times the density d. Also $l = vt$, so

$$m = vtad$$

The expression for the energy is then

$$E = 2\pi^2 vtad f^2 A^2$$

This is the energy that flows past the observation point Q in a time t. The intensity is the power (E/t) per unit area a, so

$$I = \frac{E}{at} = 2\pi^2 vd f^2 A^2$$

This expression describes the intensity of the wave in terms of frequency f, amplitude A, density of the medium d, and wave speed v. The product of wave speed times density, vd, occurs very frequently in equations for waves, so it has been given a name, the impedance of the medium. The name arises from an analogy with equations in electricity; the equations used in working with electricity and with wave propagation in a medium often have similar forms. The quantity vd occurs in the wave equations in the same places as does the impedance (or resistance) in the electrical equations.

EXAMPLE 7

Find the amplitude of a sound that has a frequency of 1000 Hz and an intensity level of 10 db in air. The reference intensity is $I_0 = 10^{-12}$ watts/m^2.
We start by finding the intensity in watts/m^2:

$$10 = 10 \log_{10} I/I_0$$
$$\log I/I_0 = 1$$
$$I/I_0 = 10$$
$$I = 10^{-11} \text{ watts/m}^2$$

Solve $I = 2\pi^2 vdf^2A^2$ for A, to get

$$A = \frac{1}{\pi f} \sqrt{\frac{I}{2vd}}$$

$f = 1000/s = 10^3/s$

$I = 10^{-11}$ watts/m^2 = 10^{-11} kg/s^{3*}

$v = 331$ m/s (speed of sound in air)

$d = 1.29$ kg/m^3 (density of air)

so

$$A = \frac{1}{\pi \times 10^3/s} \sqrt{\frac{10^{-11} \text{ kg/s}^3}{2 \times 3.31 \times 10^2 \text{ m/s} \times 1.29 \text{ kg/m}^3}}$$

$$= 3.18 \times 10^{-4} \text{ s } \sqrt{1.17 \times 10^{-14} \text{ m}^2/\text{s}^2}$$

$$= 3.44 \times 10^{-11} \text{ m}$$

$$= 0.034 \text{ nm}$$

The amplitude of vibration of the air for a sound of 1000 Hz and 10 db (very low intensity, audible only to good ears) is only 0.034 nm, less than the radius of an atom!

---

*Note that 1 watt = 1 J/s = 1 (kg m^2/s^2)/s, so 1 watt/m^2 = 1 kg/s^3.

10–5–5 SOUND WAVES IN DIFFERENT MEDIA

The intensity of a wave was shown above to be described by

$$I = 2\pi^2 vdf^2A^2$$

where

v is the wave speed
d is the density of the medium
f is the frequency
A is the amplitude

Consider a wave in air and a wave in water, both with the same intensity. The frequency is to be the same but wave speed, density, and amplitude will all be different. Using subscripts a for air and w for water, then

$$I_a = 2\pi^2 (v_a d_a) f^2 A_a^2$$
$$I_w = 2\pi^2 (v_w d_w) f^2 A_w^2$$

If $I_a = I_w$, these two expressions can be equated, and then we can compare the amplitudes needed in order for the intensities to be the same. The quantities $2\pi^2 f^2$ cancel to leave

$$v_a d_a A_a^2 = v_w d_w A_w^2$$

and the ratio of amplitudes is

$$\frac{A_a}{A_w} = \sqrt{\frac{v_w d_w}{v_a d_a}}$$

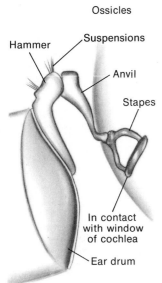

Ossicles

Hammer

Suspensions

Anvil

Stapes

In contact
with window
of cochlea

Ear drum

Figure 10-14 The three tiny bones (ossicles) that connect the ear drum to the oval window of the cochlea act as a double level system to decrease the amplitude of the vibration and hence to increase the pressure on the fluid in the cochlea.

Sound waves in air
Large amplitude
vibrations, small
force

Ear drum

Lever

Small amplitude
vibrations,
large force

Pivot

Cochlea

"Window" Fluid
(thin membrane)

Figure 10-15 A simplified diagram to illustrate the function of the ossicles of the middle ear.

This example has been chosen using air and water because this is not unlike the situation of a sound wave impinging on the ear and being transferred to the fluid in the cochlea. If the intensity in the fluid is to be the same as that in the air, the amplitudes of the vibrations will be given by the above expression. Putting in typical values for air and water,

$$v_w = 1530 \text{ m/s (at 37°C)}$$
$$d_w = 1000 \text{ kg/m}^3$$
$$v_a = 331 \text{ m/s}$$
$$d_a = 1.29 \text{ kg/m}^3 \text{ (at stp)}$$

then

$$\frac{A_a}{A_w} = \sqrt{\frac{1530 \times 1000}{331 \times 1.29}} \quad \text{(all units cancel)}$$

$$= 60$$

The amplitude of the vibration in the air would have to be 60 times the amplitude in water for the intensities to be the same.

One of the functions of the set of tiny bones in the middle ear (Figure 10-14), which connect the ear drum to the thin window in the cochlea, is to change the amplitude of the vibration. The large amplitude but low pressure vibration of the ear drum is changed to a low amplitude but more forceful vibration on the window of the cochlea. The action is not unlike that of a simple lever system, as in Figure 10-15. The vibrations in the air make the ear drum vibrate quite easily, so they are not reflected as would be the case if they fell directly onto the liquid surface.

The calculation above showed that the lever arm should give a decrease in amplitude of 60. The bones of the middle ear actually give a reduction in amplitude of only about 22. It is quite amazing that this phenomenon is taken into account in the design of the ear. The figures differ, but remember that our calculations were approximate and did not take into account the stiffness of the membranes and other factors.

10-6 REFLECTION OF WAVES

Whenever a wave impinges on a boundary between two media in which the wave speeds are different, only a part of the energy will enter the second medium and the remainder will be reflected.

The direction of the reflected wave is the subject of this section.

Consider the plane wave in Figure 10-16. The wavefront *AB* is moving toward the boundary, and reflection begins when

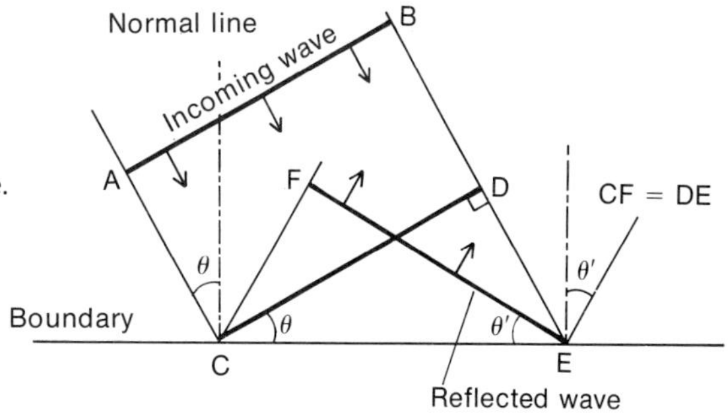

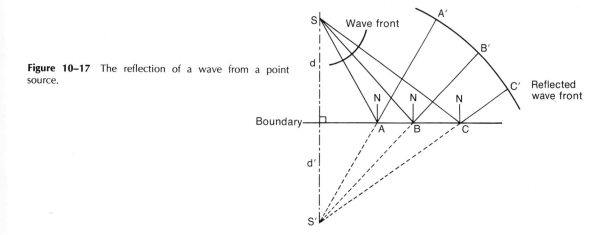

Figure 10-16 The reflection of a plane wave.

Normal line

Incoming wave

Boundary

CF = DE

Reflected wave

the edge at C meets the boundary. The angle used to describe the wave is usually the angle between the direction of motion of the wave and a line normal (perpendicular) to the surface — the angle θ in Figure 10-16. When the edge of the front at D reaches E, the whole wave has been reflected. The edge at C will have moved to F, a distance such that $CF = DE$. Also, the direction of travel of the wave is normal to the wavefront, so the angle at F is a right angle. The triangles CFE and CED are congruent, so the angle θ is equal to θ' and the wave moves away from the surface at the same angle at which it hit. Calling θ the angle of incidence and θ' the angle of reflection, the conclusion is that the angle of incidence is equal to the angle of reflection. This is called the **first law of reflection.**

If a spherical wave is reflected as in Figure 10-17, the analysis uses a series of lines normal to the wavefronts to indicate the direction of motion of the waves. Such *"rays"* are indicated by SA, SB, and SC. For each ray the angle of incidence is made equal to the angle of reflection. A given wavefront is at all places the same distance from the source; then $SAA' = SBB' = SCC'$. Then the new reflected wavefront is $A'B'C'$ and the center of this wavefront is at S'. The distances $S'A'$, $S'B'$, and $S'C'$ are equal to SAA', SBB', and SCC'. The new wave-

Figure 10-17 The reflection of a wave from a point source.

front will appear to be spreading out from a source that is the same distance behind the reflecting surface as the source was in front of it. Also, the line joining S and S' is normal to the surface. The geometrical proofs of the above statements are left as exercises.

Figure 10–17 is a cross-section of the spherical wave in a plane perpendicular to the reflecting boundary, and it could be in any direction obtained by rotation around the line SS'. In all cases an incident ray, the normal line (N in the diagram), and the reflected ray would lie in a plane. This is the **second law of reflection.**

An interesting extension is the situation of multiple reflections from two parallel boundaries, as in Figure 10–18. S_0 is the original source; after reflection at A, the reflected wave appears to have originated at S_1. The ray leaving A is reflected at B, and the wave leaving B appears to have originated at S_2. Another reflection occurs at C, and the wave leaving C appears to have come from S_3. This continues along the channel formed by the reflectors. Down the channel at Z, the wave will appear to have come from a very large number of sources all in a line. In Section 10–3 it had been shown that a long line of sources leads to a plane wave in the forward direction. If a wave is confined to a narrow channel between reflecting boundaries, it develops into a plane wave.

Sound will travel in almost undiminished intensity in a smooth metal pipe. This was the principal of the speaking tubes once used principally aboard ships but now usually replaced

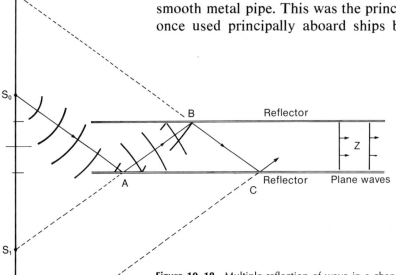

Figure 10–18 Multiple reflection of wave in a channel has the same effect as emission from a long line of point sources. The result is a plane wave in the channel.

with electronic apparatus. Water waves in long narrow channels develop into plane waves moving along the channel perpendicular to the shores. Very short radio-waves, called microwaves, are made to travel in hollow channels called waveguides. Even light will travel as a plane wave along a sufficiently narrow channel or transparent fiber. Fan noise from an air conditioner can also travel along air ducts as an almost undiminished plane wave in this way.

10-7   ACOUSTICS

The topic of acoustics has to do with the ability to hear sound, and the quality of the sound in a room, be it a small room or an auditorium. A room with hard, bare walls, no furniture and no drapes, "echoes." A room, especially a large room with an excess of sound-absorbing material such as large soft drapes and soft, stuffed furniture, may have a quality referred to as "dead." The sound does not carry. Both types of rooms would have what are called "poor acoustics." The quality of a room in these cases is measured by what is called reverberation time. Too long a reverberation time caused by the sound bouncing back and forth almost undiminished from the surfaces leads to echoes. Too much absorption leads to too short a reverberation time and a "dead room."

One of the problems in the design of rooms dedicated to good sound is to obtain the proper reverberation time. The appropriate reverberation may be from a fraction of a second for a small room to over a second for a large auditorium.

The analysis for reflection of sound in a room will be carried out with a one-dimensional room—a long, narrow room with the sound traveling only back and forth in one dimension. This will indicate some of the basic ideas, and the extension to a three-dimensional room will be discussed. Some of the ideas apply also to lighting. That is the case in much of physics; the general ideas apply in many areas.

10-7-1   SOUND INTENSITY IN A ROOM

If sound is fed into a long, narrow room as in Figure 10-19, the sound is reflected back and forth. The sound level in what is shown as the detector volume is not only the original sound level but also that of all the reflected waves. The waves in this

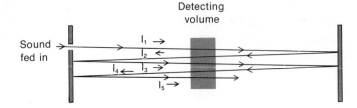

Figure 10–19 The intensity of sound in a closed space includes all of the reflected waves.

example are assumed to travel only in one dimension. The analysis is much simpler than for a three-dimensional room, but the ideas to be found will be general.

At each reflection only a part of the incident sound is reflected; some is absorbed, and the fraction reflected depends on the material. Hard plaster walls reflect most of the sound, drapes very little, and an open window, none. Let the fraction reflected be f; then, if the primary wave is of intensity I_0, the first reflected wave is of intensity $I_1 = fI_0$. Each reflected wave is f times the previous one.

$$
\begin{aligned}
\text{primary wave} \quad & I_0 \\
\text{on one reflection} \quad & I_1 = fI_0 \\
\text{and also} \quad & I_2 = fI_1 = f^2 I_0 \\
& I_3 = fI_2 = f^3 I_0 \\
& I_4 = fI_3 = f^4 I_0 \\
& \text{etc.}
\end{aligned}
$$

The sound intensity in the detecting volume shown is the sum of all the reflected waves

$$
\begin{aligned}
I &= I_0 + I_1 + I_2 + \ldots \\
&= I_0 \left(1 + f + f^2 + f^3 \ldots \right)
\end{aligned}
$$

When the sound source is first turned on, the intensity is I_0; then on reflection, I_1 is added, then I_2, and so on. The intensity slowly builds up. (By slowly is usually meant in a second or less in a case like this.) The ultimate intensity, with an infinite number of reflections, is the sum of the series for I carried out to an infinite number of terms.

$$
I = I_0 \left(1 + f + f^2 + f^3 + \ldots \text{ to infinity} \right)
$$

This is a geometric series; the sum is finite and is given by

$$
I = \frac{I_0}{1 - f}
$$

EXAMPLE 8

In the situation being described above, compare the steady state sound intensities in two rooms. In the first room the walls reflect 95 per cent of the sound ($f = 0.95$), and in the second room they reflect 10 per cent of the sound ($f = 0.10$).

For $f = 0.95$, $I = \dfrac{I_0}{1 - 0.95} = 20\,I_0$

For $f = 0.10$, $I = \dfrac{I_0}{1 - 0.10} = 1.11\,I_0$

These compare as 20/1.11 or 18.

The intensity in the room with walls with low absorption will be 18 times the intensity in the room with high absorption.

The less the absorption, the higher will be the steady state intensity of the sound. To hear well in a room or auditorium it may seem that the walls should absorb little of the sound. This would lead to echoing, however. Too much absorption results in too low an intensity. A balance must be obtained.

It should be pointed out, too, that this analysis applies also to light. The ultimate light intensity in a room depends not only on the total intensity of the sources but also on the multiply reflected light. Dark walls and furnishings can result in a low level of illumination even with intense sources.

10–7–2 REVERBERATION TIME

When a sound is first introduced to a room as in Figure 10–19, the intensity is first I_0; then I_1 is added, then I_2, and so on. The intensity slowly builds up to a steady state value; let it be I_∞. This is

$$I_\infty = I_0 (1 + f + f^2 + f^3 + \ldots \text{ to infinity})$$
$$= I_0/(1 - f)$$

If the sound is suddenly cut off, then I_0 disappears from the sum. This cuts out the first term in the series to leave an intensity I',

$$I' = I_0 (f + f^2 + f^3 + \ldots \text{ to infinity})$$
$$= I_0 f(1 - f) = f I_\infty$$

The next reflected wave is cut out in a short time to leave

$$I'' = I_0 (f^2 + f^3 + \ldots \text{ to infinity})$$
$$= I_0 f^2/(1 - f) = f^2 I_\infty$$

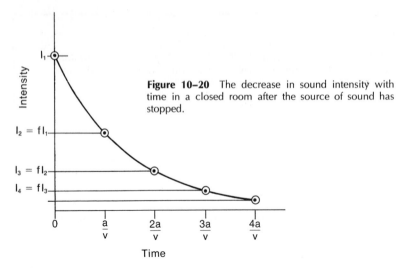

Figure 10–20 The decrease in sound intensity with time in a closed room after the source of sound has stopped.

The time between the cutting out of successive waves depends on the size of the room. In our one-dimensional case, let the detector be in the center of the room, which is of length a. The time between the cutting out of successive waves is the time required to travel a length a or $t = a/v$, where v is the wave speed. The intensity drops by a factor f in each of these time intervals:

$$
\begin{array}{ll}
\text{at } t = 0 & I = I_\infty \\
\text{at } t = a/v & I' = fI_\infty \\
\text{at } t = 2a/v & I'' = f^2 I_\infty \\
\text{at } t = na/v & I = f^n I_\infty
\end{array}
$$

In each interval of time, a/v, the intensity drops by the fraction f; as shown in Figure 10–20, it never reaches zero but does approach it.

The reverberation time for a room is agreed to be the time needed for the intensity to drop to one millionth (10^{-6}) of its initial value.

EXAMPLE 9

Find the reverberation time in our idealized one-dimensional room if the length $a = 10$ m, the fraction reflected, f, is 0.5. The speed of sound is 330 m/s.

The quantity a/v is $(10 \text{ m})/(330 \text{ m/s}) = 0.030$ s. The intensity after n times a/v is given by $I = I_\infty f^n$. We want $I/I_\infty = 10^{-6}$, or $f^n = 10^{-6}$. To solve this, use logarithms:

$$\log f^n = \log 10^{-6}$$

where
$$n \log f = -6$$
$$f = 0.5$$
$$\log f = -0.30$$

Thus,
$$n \times (-0.30) = -6$$
$$n = (-6)/(-0.30)$$
$$= 20$$

The time na/v is $20 \times 0.030 \text{ s} = 0.6$ second. The reverberation time of that room would be 0.6 second.

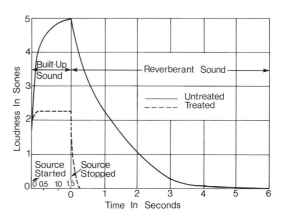

Figure 10–21 The build-up and decay of sound in a room. (From H. J. Sabine in *Handbook of Noise Control*, C. M. Harris (ed.), McGraw-Hill Book Company, New York, 1957.)

Figure 10–21 shows the buildup of sound in a room, and the decay of the sound when the source is turned off. The data are first for a room with highly reflective walls and then for the same room with absorbing material added. The addition of absorber decreases the reverberation time as well as the steady-state intensity.

Reverberation time can be expressed in terms of sound level (in db) rather than intensity. If the initial intensity is I_i, and the final intensity I_f is $10^{-6} I_i$, then in db

$$db_i = 10 \log I_i / I (\text{standard})$$
$$db_f = 10 \log I_f / I (\text{standard})$$
$$db_i - db_f = 10 \log(I_i / I_f)$$

This results because subtracting logs implies division and then the standard intensity cancels. Also, $I_i / I_f = 10^6$ and log $10^6 = 6$. Then

$$db_i - db_f = 10 \times 6 = 60$$

The reverberation time is the time required for the intensity to drop by 60 decibels.

Extension should be made to a real three-dimensional room. The reverberation time in one dimension is proportional to the time needed for the sound to travel the length of the room, a. In a three-dimensional room this time depends on all three dimensions, a, b, and c. It therefore depends on the volume, V, of the room.

The reverberation time depends on the absorption of all parts of the room, of the walls, furniture, people, and so forth. Some parts may be highly absorbent, and some absorb only a little. The reverberation time will depend on the sum of the products of the absorption coefficient and the area of each ab-

TABLE 10–4 Sound absorption coefficients of some materials. These are representative values for the type indicated and are for 500 Hz.

MATERIAL	ABSORPTION COEFFICIENT, c
plaster	0.02
gypsum board	0.05
concrete block, coarse	0.31
concrete block, painted	0.06
linoleum	0.03
wood	0.10
carpet	0.40
acoustic tile	0.80

sorber. The greater the absorption, the shorter the reverberation time, so it is an inverse relation when referring to absorption. Reverberation time is proportional to (volume)/(sum of absorption coefficients times areas). In SI units the proportionality constant is 0.16 s/m. The reverberation time, T_r, in seconds of a room of volume V is given by

$$T_r = 0.16 \frac{\text{s}}{\text{m}} \frac{V}{\Sigma cA}$$

In the denominator, c is the absorption coefficient of each object of area A, and Σ indicates the sum of the products of absorption coefficients and areas. The absorption coefficient is $(1 - f)$, wher f is the fraction reflected. A few representative values of c are shown in Table 10–4. These are for a frequency of 500 Hz; there is variation with frequency and also with the precise type of material.

In English units the reverberation time is

$$T_r = 0.049 \frac{\text{s}}{\text{ft}} \frac{V}{\Sigma cA}$$

In this relation V is in ft^3 and A is in ft^2. The product cA for each area in ft^2 is called the absorption in *sabines* for that area. To convert from sabines to the equivalent cA value in SI units, multiply by 0.0929. The absorption in sabines per unit area is the same in SI or in English units. For example, each square meter of a room occupied by upholstered chairs contributes 0.8 to the cA sum, and each square foot contributes 0.8 if English units are used.

EXAMPLE 10

First find the reverberation time in a room with linoleum on the floor and with gypsum board on walls and ceiling. Then find the reverberation time when the ceiling is covered with acoustic tile and the floor with a carpet. The room is 2.4 m high, 3.6 m wide, and 4.6 m long. The volume is therefore 39.7 m³. The absorption coefficients are in Table 10–4.

In the first instance the cA values are:

walls	39.4 m² × 0.05	cA = 1.97 m²
floor	16.5 m² × 0.03	cA = 0.50 m²
ceiling	16.5 m² × 0.05	cA = 0.83 m²
	sum	cA = 3.30 m²

$$T_r = 0.16 \frac{s}{m} \times \frac{39.7 \text{ m}^3}{3.30 \text{ m}^2} = 1.9 \text{ s}$$

With the carpet and acoustic tile, the values are:

walls	39.4 m² × 0.05	cA = 1.97 m²
floor	16.5 m² × 0.40	cA = 6.60 m²
ceiling	16.5 m² × 0.80	cA = 13.20 m²
	sum	cA = 21.8 m²

$$T_r = 0.16 \frac{s}{m} \times \frac{39.7 \text{ m}^3}{21.8 \text{ m}^2} = 0.3 \text{ s}$$

With the addition of carpet and acoustic tile, the reverberation time of that room drops from 1.9 s to 0.3 s. A reverberation time of 1.9 s would be perceived as an echoing in such a small room. In a large auditorium it would probably not be objectionable. The 0.3 s reverberation time in the small room would be satisfactory.

10–7–3 REFLECTION AT CURVED SURFACES

When sound is reflected from a curved surface, as in Figure 10–22(a), it may be focused. The curved surface shown could be the ceiling or the wall of a room. The sound waves spreading out from the source at A reflect from the curved surface with the angle of reflection, r, equal to the angle of incidence, i, at each point. The sound is then focused to point B. If a person even whispers at point A, a person at point B may hear the whisper very distinctly. It is usually advisable to avoid such focusing effects. The location of a point of focus may be found by using such a ray drawing, or it may be calculated by using $1/y_1 + 1/y_2 = 2/R \cos^2 \theta$, for which the symbols are as in Figure 10–22(a).

Curved sound reflectors can be used to project a sound in the direction of an audience or to focus a distant sound source into a microphone. If the sound source is halfway between the center of curvature and the reflector, the waves reflected in the direction of the center are in an almost parallel beam as in Figure 10–22(b). Reversing the direction of travel of the sound

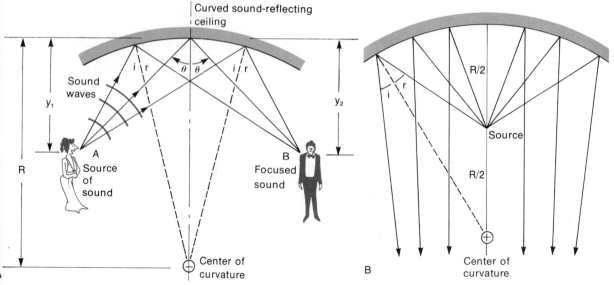

Figure 10–22 Some effects of sound reflected from curved surfaces.

in Figure 10–22(b), it can be seen that sound from a distant source will be focused to a point halfway between the center of curvature and the reflector. Such a system is used with a microphone at the point of focusing to record bird calls or other distant sounds.

More on the reflection of waves from curved surfaces is given in the section on light, for much use is made of curved mirrors for image formation. The same theories apply to light as to sound.

DISCUSSION PROBLEMS

10–1   1. Use reference books or the library to find the ranges of sound frequencies used by bats, various insects, and toothed whales (including dolphins). Relate those ranges to the 10 Hz to 20 kHz detected by humans.

10–1–1   2. Make an investigation of some of the musical scales. Find the frequencies of the various notes and the ratios of the frequencies of the notes of the scale for each.

10–2   3. If you want to go as fast as possible in an aircraft but are limited to Mach 0.95, would you chose a cool or a warm day? High altitude or low?

10–4   4. Read the opening paragraph of Section 10–4 and elaborate on it with a specific example from the section. Also, compare the

effectiveness in science of reasoning by discussion and by using mathematics.

10–5–1   5. Investigate the types of noise laws in your community.

PROBLEMS

10–1   1. What is the range of the wavelengths of sound detected by the human ear?

10–1   2. Some bats detect sounds in the range from 10 kHz to 100 kHz. What is this range expressed in terms of the wavelengths in air rather than in terms of frequency?

10–2   3. The speed of sound in air is measured to be 331.4 m/s at 0°C (273 K); in order to compare this with a theoretical value, the average molar mass of air must be known. Air is a mixture of mainly O_2 and N_2 with some

CO$_2$ and A. Use the theoretical relation for the speed of sound to calculate an effective molar mass, $\mathcal{M}$, for air. Both O$_2$ and N$_2$ are diatomic, so $\gamma = 1.40$.

10-2   4.   Calculate the speed of sound in air at 0°C as if it were an isothermal process. Use $\mathcal{M} = 0.0289$ kg/mole.

10-2   5.   Calculate the speed of sound in krypton gas at 27°C. The atomic weight of krypton is 0.0837 kg/mole. Krypton gas is monatomic, so $\gamma = 1.67$.

10-2   6.   Express the equation for the speed of sound in air in terms of Celsius temperature, t, where absolute temperature $T = 273 + t$. Use a binomial expansion of the surd in the equation and show that in the region close to 0°C the speed of sound is given by

$$v_{t°C} = v_{0°C}\left(1 + \frac{t}{2 \times 273}\right)$$

Extend this to show that the speed of sound near 0°C increases by 0.60 m/s for each 1°C rise in temperature.

10-3-2   7.   Use Huygens's principle to follow a plane wave that goes from a medium in which it has a speed V to a medium in which the speed is reduced to $V/2$. The boundary between the media is a plane and the wave hits the surface at 45°.

10-4   8.   Show that the equation for a wave traveling in the positive x direction, which is described in the text by $y = A \sin \omega\left(t - \frac{x}{v}\right)$, can also be described by $y = \sin 2\pi \left(\frac{t}{T} - \frac{x}{\lambda}\right)$.

10-4   9.   Make a diagram of the wave resulting from two waves of equal amplitudes and frequencies traveling in opposite directions. Plot the wave according to the following data and put the waves of various times on the same sketch, labeling the different curves. To make the drawings use only the points calculated at the maximum, minimum, and zero values of the cosine function. The use of a table to list the y values for each wave and the sum is recommended.

Use $A = 1$ cm
$\lambda = 4$ cm
$0 \leqslant x \leqslant 8$ cm

Draw the waves for the values of t for which:
(a) $\omega t = 0$
(b) $\omega t = \pi/2$
(c) $\omega t = \pi$
(d) $\omega t = 3\pi/2$
(e) $\omega t = 2\pi$

10-4   10.   Two waves are traveling in opposite directions. Each has an amplitude of 4.56 cm and a frequency of 0.00974 Hz. The wavelength is 8.00 cm. Plot the displacement as a function of time at $x = 2.00$ cm. To do this well, calculate the displacement, y, every fifth of a second for the first 10 seconds.

10-4-2   11.   A cavity 20 cm long is filled with material in which the speed of sound is 1500 m/s, and it is made to resonate with a node at each end. What is the resonant frequency?

10-4-2   12.   Find the resonant frequency of a narrow air channel open at one end only, if its length is 2.5 cm. It resonates with a node at the closed end and an antinode at the open end, so $l = \lambda/4$ as in Figure 10-10. The speed of sound is about 330 m/s.

The channel described is not unlike that in the outer ear, between the outside world and the ear drum. Would this perhaps have something to do with the human ear being most sensitive to sounds have a frequency of about 3000 Hz?

10-4-2   13.   Find the fundamental frequency of the vibration in an air column open at one end, and also find the first two overtones. To what harmonics do those overtones correspond? The length of the column is 0.330 m and the speed of sound in the column is 338 m/s.

10-4-2   14.   If an organ pipe (which is a resonant air column) gradually warms during a concert, what would happen to the pitch produced by the pipes? This phenomenon can be discussed, but a calculation shows how significant it is. If a pipe resonates at 440 Hz (A on the musical scale) when the temperature is 20°C, what would be the resonant frequency at 25°C?

10-5-1   15.   If two watts of sound power are allowed to spread out evenly in all directions, what is the intensity level at a distance of 10 meters from the source?

10-5-1   16.   If two watts of sound power are fed into a cone of one steradian, what is the intensity level 10 meters from the source? The area on the surface of a sphere of radius r included within a solid angle of one steradian is equal to r^2.

10-5-1   17.   What is the power per square meter where there is a sound level of 85 db? Use 10^{-12} watts per square meter as a zero level.

10-5-1   18.   What intensity level would result when two sounds that separately give levels of 60 and 65 db are combined?

10-5-1   19.   If a single source gives a sound level of 50 db, what sound level would result when 20 such sources are operated together?

10-5-1   20.   Derive a general expression for the increase in sound intensity level when n sources of equal intensity are combined.

10-5-1   21.   An estimated one million mosquitoes give a sound level of 60 db. What would be the sound level from one mosquito?

10-5-1   22.   Show that when the distance from a sound source in the open air is doubled, the intensity level will drop by 6 db. Consider the conditions to be such that an inverse square law describes the change in intensity with distance.

10-5-1   23.   Find the power in watts entering the outer ear when a sound of 10 db is heard. The detecting area of the ear is 1.0 cm in diameter. Write the answer in decimal notation and comment on the sensitivity of the human ear.

10-5-4   24.   Find the power per square meter carried in a sound in air if the speed of sound is 330 m/s, the density of air is 1.3 kg/m³, the frequency is 440/s (hertz), and the amplitude is 0.1 mm.

10-5-4   25.   Compare the amplitudes of two sounds of equal intensity in air if the frequencies are 100 Hz and 1000 Hz.

10-5-4   26.   Compare the amplitudes of two sounds of the same intensity and same frequency in helium and in air.

10-6   27.   Make a careful diagram similar to Figure 10-17 but continue the wave past A', B', and C' to be reflected from another boundary perpendicular to the one shown. This will allow the location of a second image s''. Describe the location of this image with respect to s'.

10-7-1   28.   (a) Find the intensity to which the sound level will rise in a long, narrow room as discussed in the text, if the fraction reflected at each end is 0.33. The initial intensity is I_0 with no reflection.

(b) Does the maximum intensity depend on the length?

(c) What would the effect of variation in length be?

10-7-1   29.   Calculate points to make curves similar to those in Figure 10-21. Use a long, narrow room, 33 m in length (the speed of sound is 330 m/s). Calculate the intensity as each reflection is added for 1 second while the source steadily emits an intensity I_0. Then at 1 second the source is stopped. Plot the decay for one second after this. Use a reflection coefficient of 0.5 at each end.

10-7-1   30.   (a) Find the difference between the intensity level of a reflected sound and the intensity level of the incident sound when the intensity, I, reflected is f times the incident intensity, I_0. Use $f = 0.9$.

(b) How many reflections would be required to reduce the level by 60 db?

10-7-2   31.   Find the reverberation time in a corridor that is 2.0 m wide, 2.5 m high, and 30.0 m long. The walls and ceiling are of gypsum board and the floor is carpeted. The absorption coefficients are 0.05 and 0.4, respectively.

10-7-2   32.   A room 4.5 m wide, 2.5 m high, and 6.0 m long has walls of plaster and a wooden floor ($c = 0.02$ for plaster and 0.10 for wood). The ceiling is of plaster also, but enough acoustic tile ($c = 0.8$) is to be put on to reduce the reverberation time to 0.7 s. What must be the area of acoustic tile?

10-7-2   33.   A room 2.4 m high, 3.5 m wide, and 6.0 m long is measured to have a reverberation time of 1.5 s. What is the absorption coefficient of the material of the walls, ceiling, and floor if they are all of the same material?

10-7-3   34.   A plane wave approaches a curved mirror of radius R. Make a diagram showing rays approaching parallel to an axis through the center of curvature. Show radius lines (normal lines) to the places at which the rays hit the surface, and draw the reflected rays. What do you deduce about the position of the point to which the rays are focused?

CHAPTER ELEVEN

LIGHT WAVES

There are many phenomena that occur with light and are due to its wave nature. Light waves exhibit the properties of other waves, properties already discussed; but there are other things that occur with waves, things not yet discussed because they are particularly impressive with light.

When the opportunity arises, take a good look at soap bubbles or soap suds. Note the colors of the reflected light, sometimes brilliant, continually changing colors. These result from the wave nature of light. The wave properties of light are more diverse and more important than the colors in soap bubbles, yet that is a simple, familiar phenomenon that demonstrates the existence of light waves.

The waves of light are very short and they cover a broad range. The wavelength is associated with color. The longest waves, those at the red end of the spectrum, are about 0.00007 cm in length in air (0.7 μm or 700 nm) while at the violet end the wavelength is 0.00004 cm (0.4 μm or 400 nm). Because of this small wavelength, is it any wonder that the establishment of the wave nature of light was a long and difficult process?

Light is not a material vibration like sound, but it is an oscillating electric and magnetic field. Because of this, light can pass through a vacuum whereas sound, which is a material wave, cannot. The speed of light is 3.00×10^8 m/s (300,000 km/s) in a vacuum, almost exactly the same in air, but somewhat less in a denser medium.

11–1 DEVELOPMENT OF THEORIES OF LIGHT

One of the early formal studies of light was the work of Sir Isaac Newton, as he published it in his book *Opticks* in 1708. This is still a fascinating book for any scientist to read. It is the product of a great mind, sometimes called the greatest analytical mind of all time. Fortunately, it has been reprinted

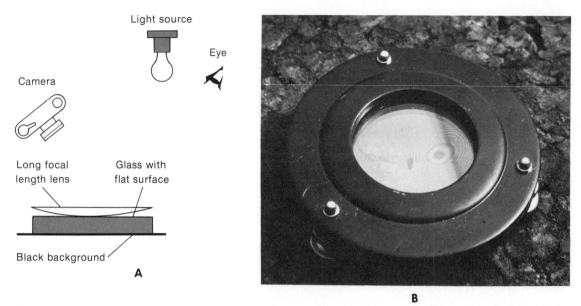

recently and is available readily and inexpensively. In his book Newton described the circles of color that he saw when he put together two pieces of glass (one flat and one slightly curved) with a thin film of air between them (see Figure 11–1). Today the production of these colors is explained on the basis of the wave properties of light. Newton tried to explain them, and it seems that he wanted to treat light as a wave. From these experiments on thin air films he even measured a distance associated with each color, which is related to what is now considered the wavelength. He did not have the wave theory available to allow interpretation of his experiments; the distances he associated with the colors were not what are now known as the wavelengths, but they were just half of the wavelengths. With yellow light, for instance, he gave a size of 1/89,000 of an inch. Doubling this value and converting it to the metric system, it is 570 nm, which is within the range of the wavelengths associated with yellow light. When you measure wavelengths in the laboratory you might see how well your value for yellow light corresponds with Newton's.

Actually, Newton also saw the phenomenon of light bending around corners, though not to the extent that he expected. He did see phenomena that are now interpreted as interference between light waves that are sometimes in step (in phase) and sometimes out of step (out of phase). However, he could not visualize a wave traveling through a vacuum. A material wave requires an elastic medium for propagation; if light was a wave, he reasoned, there would have to be "something" in which the wave occurred. The planets that circle the sun then would have to move through this medium, and since they apparently showed no frictional drag associated with a medium, he could not accept that there was one. As one reads of his arguments for and against a wave theory, one gets the feeling that he reluctantly, though firmly, rejected the concept of light being a wave motion, at least as it passed through a vacuum, for as he says:

> . . . motions of planets and comets cannot be explained by means of a dense fluid and are better explained without it — there is no evidence for such a dense fluid medium and it should be rejected — this means rejecting the wave theory of light. . .

But there are other parts to the story. Francesco Grimaldi (1618–1663) had noted the bending of light around obstacles, which is now referred to as diffraction, and described it in a book published in 1665, more than 40 years before Newton's *Opticks*. Some of Newton's contemporaries, principally among them Christian Huygens, staunchly supported the wave theory. Huygens presented ample evidence for the wave nature of light in his book *Traité de Lumière* in 1690, yet the authority of Newton was such that the theory did not become accepted for almost another hundred years. You must always watch yourself lest you believe on the basis of authority rather than on evidence.

In 1803, Thomas Young published papers that described experiments overwhelmingly in support of the wave theory. It then became generally accepted that light is a wave. The wave properties (specifically interference, diffraction, and polarization) were demonstrated so convincingly that the wave theory had to be accepted, even though it seemed that the conducting medium, in order to give light its known high velocity of propagation, would have to have unusual elastic properties.

Michael Faraday in 1846 showed that the direction of polarization of light (that is, the direction of the vibrations that are perpendicular to the direction the light travels) rotates in a magnetic field. This hinted at a connection between magnetism and light, and magnetism was already associated with electricity! James Clerk Maxwell in 1865 published his electromagnetic theories and equations, in which light was considered as an electromagnetic wave. The equations described practically everything there was to describe about the behavior of light. It was still considered that an "aethereal medium" or "aether" was necessary for the propagation of the wave, even though the aether could not otherwise be detected. The suggested medium would have to have fantastic properties; for instance, it would have to be stronger than steel. Scientists were learning to live and work with incomplete theories. Yet the wave theory worked so well that they accepted it and left the understanding of the medium of propagation for later. In fact, Michelson and Morley in 1887 performed experiments that showed almost conclusively that there was no aether.

A hundred years after Young introduced the wave theory, a hundred years of finding evidence in favor of the wave-like nature of light, the particle theory came back. It was shown by Max Planck in 1901 and by Albert Einstein in 1905 that some of the properties of light, properties that will be discussed in later chapters, were in contradiction to wave theories but were satisfactorily explained by a particle theory. The particles are now referred to as photons or quanta of light.

It is hard to imagine a worse situation from the scientific viewpoint. Light sometimes behaved like a wave, but there was no medium to carry it. It also behaved at times like a particle, and yet wave and particle phenomena are very different. Perhaps the best analogy or model is what we call a wave packet. This is a short burst of waves that travels in a little "packet" and has both wave and particle properties. The wave theory can then be used to explain the phenomena that have been shown to depend primarily on the wave properties of the light. Without qualms, the particle theory can also be used to analyze other phenomena in which the particle properties predominate.

We do not say that light is a wave; neither do we say that light is a flow of particles. We do attribute both wave properties and particle properties to light. That is, at some times a wave model is best, while at other times a particle model is best. The analogy of a flower is appropriate here. A plastic model of a flower exhibits one property of the flower, its shape. A perfume exhibits another property, its smell. Neither the plastic model nor the perfume model is the flower, just as neither the wave model nor the particle model is light. The models only assist us in describing some of the properties of the real thing.

The major changes in thought about light took place just after the turn of each century:

1708	Newton's particle theory
1803	Young's wave theory
1901–1905	Planck and Einstein's particle theory

Is there a regularity here that will continue? Such speculation has no basis except in fun, but it is useful as a memory aid.

11–2 WAVELENGTH, SPEED, AND FREQUENCY

The speed of light in a vacuum is one of the fundamental constants of nature. To four significant figures it is 2.998×10^8

m/s, and to three figures it is 3.00×10^8 m/s. This is an example of the case in which the zeros are significant. In a medium the speed is less than in a vacuum. The ratio of the speed in a vacuum to the speed in a medium is referred to as the *index of refraction* of the medium. The symbols commonly used are:

speed of light in a vacuum: c
speed of light in a medium: v
index of refraction: μ (lower case Greek mu)

By definition,

$$\mu = c/v$$

Some typical indices of refraction are listed in Table 11–1. The indices are so specific that the measurement of an index can be used to identify a mineral or to measure a solution concentration.

The index of air is so close to one that only in special circumstances is the speed of light in air considered to be different from c, the speed in a vacuum.

The wavelength, frequency, and wave speed are related by the general relations

$$v = f\lambda \quad \text{or} \quad f = v/\lambda$$

As a wave travels past any point there will be a certain

TABLE 11–1   Indices of refraction. The index listed is for yellow light, $\lambda = 589$ nm.

Material	Index
Air	1.00029
Carbon dioxide	1.00045
Helium	1.000034
Water (20°C)	1.3330
Ethyl alcohol	1.3617
Methyl alcohol	1.3292
Benzene	1.5014
Carbon disulfide	1.6279
Sugar solution, 25%	1.3723
Sugar solution, 50%	1.4200
Sugar solution, 75%	1.4774
Glass, light crown	1.517
Glass, dense crown	1.588
Glass, light flint	1.579
Glass, heavy flint	1.647
Canada Balsam	1.530
Fluorite	1.434
Diamond	2.417

number of vibrations per second; as it goes along, that same number of vibrations must pass any other point, too, even if the medium changes. That is, the frequency must be the same at all points along the wave. In a medium where the speed is v_1, the frequency is given by $f = v_1/\lambda_1$; and in a medium where the speed is v_2, the frequency is given by $f = v_2/\lambda_2$. These two frequencies are the same, so if the velocities are different it follows that the wavelength is different, but is such that

$$\frac{v_1}{\lambda_1} = \frac{v_2}{\lambda_2}$$

The speeds are related to the indices of refraction μ_1 and μ_2 by the general relation $\mu = c/v$. From this:

$$v_1 = c/\mu_1 \quad \text{and} \quad v_2 = c/\mu_2$$

The wavelengths in terms of these indices are then related by

$$\frac{c}{\lambda_1 \mu_1} = \frac{c}{\lambda_2 \mu_2}$$

from which $\lambda_1 \mu_1 = \lambda_2 \mu_2$.

For the product $\lambda\mu$ to be constant, if the index goes up the wavelength must go down.

EXAMPLE 1

Find the wavelengths in the various media when blue light of 450 nm in a vacuum goes from air to glass, back to air, and then into the eye. The index of the glass in question is 1.550 and that of the material in the eye is 1.336. The index of air is considered to be 1.000.

In the glass, $1.550 \lambda = 1.000 \times 450$ nm
                     $\lambda = 290$ nm

In the eye, $1.336 \lambda = 1.000 \times 450$ nm
                     $\lambda = 337$ nm

The answer is that in air the wavelength is 450 nm. It changes to 290 nm in the glass, back to 450 nm in air, and then to 337 nm in the eye.

11-2-1 COLOR, FREQUENCY, AND WAVELENGTH

The sensation of color is in our brain. We do know that light that has certain wavelengths in air will, when it enters the eye and strikes the sensitive cells on the retina at the back of the eye, give the sensation of a certain color. There are probably three types of color sensitive cells, each type being sensitive to a certain range of wavelengths or colors. The many colors that our brain distinguishes arise from the relative stimulation of the different color cells. Some people have one or even two sets of the color-sensitive cells missing, or perhaps not present in the normal number. These people will have some degree of defective color vision.

The question is often asked whether it is the frequency or the wavelength that determines the color seen. The frequency determines the wavelength in the eye, so the question has no answer. Also, the wavelength in air determines the wavelength in the eye, so it is usual to designate various colors by the wavelength in air. In the visible spectrum, a wavelength of about 700 nm correspond to the red end; in going to shorter wavelengths, the colors passed through are red and then orange. At around 580 nm there is a sensation of yellow, which slowly merges into green, then blue, and finally violet; the cells become insensitive at about 400 nm. The colors merge from one to another without sharp boundaries. Even the color names are almost arbitrary. One could argue that yellow-green should be labeled separately. Perhaps one called cyan should be there, or chartreuse. This can never be settled, but the order of colors listed above is often used in scientific work.

When you have an opportunity in a lab you should measure for yourself the wavelengths that you associate with various colors, and also measure the wavelengths at the ends of the spectrum as seen by your eyes.

11-2-2 OPTICAL PATH LENGTH

In optics it has been found to be convenient to use a quantity called optical path length. The idea of an optical path length will be described with an example.

Consider a piece of glass 1.0 cm thick with an index of refraction of 1.5. The waves in the glass will be shorter than the light waves in air by the factor 1.5, and there will be more waves in that 1.0 cm of glass than there would be in 1.0 cm of air. In fact, the distance in air that would have the same number of waves is 1.5 cm. Considering the number of waves, that 1.0 cm of glass is equivalent to 1.5 cm of air.

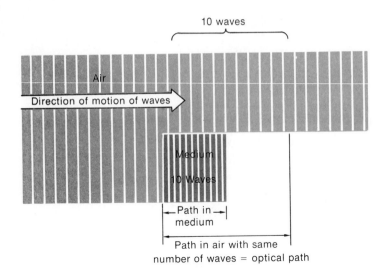

Figure 11-2 Illustration of optical path.

The equivalent path in air is called the **optical path length.** The term **optical thickness** is also used. The actual length of the light path through 1.0 cm of glass is, of course, 1.0 cm. The equivalent air path is 1.5 cm, and it is said that the optical path length through the glass is 1.5 cm. Expressing it another way, the *actual thickness* of the glass is 1.0 cm; the *optical thickness* is 1.5 cm.

optical path = index of refraction × actual path

The concept of optical path length is illustrated in Figure 11-2.

EXAMPLE 2

If a light beam goes through 10 cm of air, 2 cm of glass ($\mu = 1.60$), and 3 cm of water ($\mu = 1.33$), what is the actual path length and what is the optical path length?

actual path length = 10 cm + 2 cm + 3 cm
　　　　　　　　 = 15 cm
optical path length = 10 cm + 1.60 × 2 cm + 1.33 × 3 cm
　　　　　　　　　 = 17.2 cm

The actual path is 15 cm long, whereas the optical path is 17.2 cm.

11–3   INTERFERENCE OF LIGHT WAVES

An important phenomenon that occurs with waves is called interference. It results when two (or more) waves arrive at the same place at the same time. If the two waves arrive in step, or in phase, the resulting disturbance is greater than would be caused by either wave alone. If the two waves are exactly out of step, the vibrations will cancel, leaving nothing. It might seem odd that two light waves could arrive at the same point and cancel each other, leaving darkness. This does not happen with two flashlights because the waves from one are not all in step with the waves from the other. The light coming from each flashlight originates at various places over the surface of each filament. The waves from one light may be in step (in phase) with the waves from the other light at one moment, and out of step (out of phase) the next. It is said that the waves are not coherent.

The vibration caused by one wave at a given point in a medium is described by $y_1 = A_1 \sin \phi_1$, where ϕ_1, the phase in the vibration, depends on position and time. A second wave at the same position would be described by $y_2 = A_2 \sin \phi_2$. The phase ϕ in each case is given by $\phi = 2\pi f\left(t - \dfrac{x}{v}\right)$. If each wave originates at different places from moment to moment (as in the flashlights), the distance x will keep changing and so will the phase. However, if the two waves are of the same frequency and the distance x is constant for each, the phase ϕ_1 will not necessarily be the same as the phase ϕ_2 but at least the difference will be constant. The waves are said to be **coherent.** If the phase difference is always zero, the two waves are said to be in phase. In non-scientific terms, they are **in step.** If the phase difference $\phi_2 - \phi_1$ is always 180°, or π radians, the waves are said to be out of phase (or out of step). The phase difference can be any amount, but as long as it is a constant for the two waves they are said to be coherent. If two sources produce coherent waves, the sources are said to be coherent.

One of the common methods of achieving coherence of two waves is to separate a light beam into two, make each beam travel a different path, and then bring the two back together to produce interference.

11–3–1   CONSTRUCTIVE AND DESTRUCTIVE INTERFERENCE

To analyze this interference phenomenon, it will be admitted that two or more coherent sources are obtainable. Then a few ways in which such sources are obtained, the effects produced, and the uses made of them will be investigated.

Consider the two coherent light sources S_1 and S_2 of Figure 11–3(a). If each source is the same distance from the point P on the screen, the two waves will arrive in phase, reinforce each other, and result in more intense light than would arrive from either one alone. The interference, it is said, is **constructive** when the path S_1P is equal to the path S_2P.

If the source S_2 is moved by half a wavelength, as in Figure 11–3(b), so that the difference between the paths S_1P and S_2P is $\lambda/2$, the waves arrive exactly out of step at the point P on the screen. No light at all will be seen at P; this interference is said to be **destructive.** At points P' and P'' of Figure 11–3(b), the waves will not be exactly out of step, so there will be some

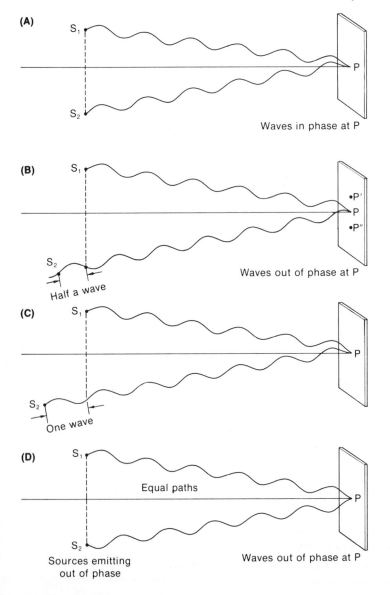

(A)

S_1

S_2

Waves in phase at P

(B)

S_1

•P'

P

•P''

S_2

Half a wave

Waves out of phase at P

(C)

S_1

P

S_2

One wave

(D)

S_1

Equal paths

P

S_2

Sources emitting
out of phase

Waves out of phase at P

Figure 11–3 S_1 and S_2 are coherent sources. In (a) the distances S_1P and S_2P are equal, so there is constructive interference at P. In (b) the difference between the two paths is half a wavelength, so there is destructive interference at P. In (c) the path difference is one wavelength, and again the waves arrive in phase at P to give constructive interference. In (d) the path lengths are equal as in (a), but the waves are emitted out of phase and the result at P is destructive interference.

light there. The distribution of light and dark on the screen illuminated by two coherent sources will be a pattern of light and dark bands.

If the source S_2 is moved farther back, so that the path to P is increased by a whole wavelength as in Figure 11–3(c), the waves again arrive at P in phase, and bright light will occur.

A little thought will show that bright light due to constructive interference will occur if the path difference for light from two coherent sources is 0, one wavelength λ, 2λ, 3λ, ...; that is, $n\lambda$ where n is zero or an integer. Destructive interference occurs when the path difference is $\lambda/2$, $3\lambda/2$, $5\lambda/2$, ... The general expression for destructive interference is that the path difference is $(2n + 1)\lambda/2$, where $n = 0, 1, 2, \ldots$.

If the path difference occurs in a medium, the path difference for constructive interference must be an integral number of the wavelengths *in that medium*. The equivalent distance in air is μ times the geometrical path. It is this, the optical path, that is important in interference phenomena.

For destructive interference in a medium, the optical path difference must be $(2n + 1)\lambda/2$, where $n = 0, 1, 2, \ldots$.

If the actual path difference is represented by P.D., then the optical path difference in a medium of index μ is μP.D. Then for constructive interference:

$$\mu\text{P.D.} = n\lambda \qquad n = 0, 1, 2 \ldots$$

and for destructive interference:

$$\mu\text{P.D.} = (2n + 1)\lambda/2 \qquad n = 0, 1, 2 \ldots$$

There is one more situation to consider; that is when the light from the two sources is not emitted in phase but rather completely out of phase. In Figure 11–3(d) is shown the same case as in Figure 11–3(a), but the wave from S_2 is opposite in phase to that from S_1. Then for equal paths the light at P is out of phase and destructive interference results. Such a "phase inversion" in one of the waves is equivalent to adding an extra half wavelength to the optical path. This type of phase inversion often occurs when one of the beams is reflected from the surface of a higher index material. If both paths have such a reflection, the effect of the inversion cancels.

11–4   THIN FILM PHENOMENA

It has been stated that coherent sources can be obtained by dividing one beam into two, but it has not been explained how

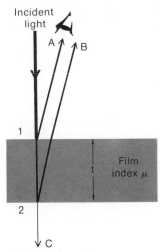

Figure 11–4 The way in which reflection from a thin film produces two waves, A and B, that will interfere with one another. Whether the interference is constructive or destructive depends on the optical thickness of the film.

this is achieved. One of the important ways is to use reflection from two surfaces. Consider a thin film of material, as in Figure 11–4. There is always some reflection when light encounters a surface at which there is a change of index. For air-glass surfaces and normal incidence of light, about 4 per cent is reflected. A film has two surfaces, and some light is reflected from each. This results in the two rays shown as A and B in Figure 11–4. For a film thickness t and normal incidence, ray B traveled an extra geometrical distance of $2t$. The optical path difference between the rays A and B is $2\mu t$ if the index of refraction of the material of the film is μ. The reflected light from the film may range from bright to nothing at all, depending on the size of that path difference.

11–4–1 COATED LENSES

One application of interference of light reflected from a thin film is in the reduction of the amount of light reflected from a surface. An air-glass (or glass-air) surface reflects about 4 per cent of the light incident on it. In an instrument, such as a good microscope, there may be 10 lenses, and thus 20 surfaces. Each surface reflects 4 per cent of the light arriving at it, so the reflections result in a large light loss through the system. Also, an appreciable amount of light will undergo more than one reflection, eventually putting either "ghosts" or haze on the image. If each surface could be coated with a thin film such that the light reflected from its two surfaces canceled, this reflection problem would be eliminated. In practice the solution is not perfect, but coated lenses are of such value that high quality optical instruments have this kind of coating on all air-glass surfaces. To understand the effects associated with such coated surfaces, the process will be examined in more detail.

It is interesting to note that, about 1660, Newton saw the cancellation of reflected light from a thin film. Anti-reflection coatings could have been made any time after that; yet it was not until 1935* that coated lenses were produced. When we look at it now, the idea seems so obvious. Why did almost 300 years elapse before someone started doing it? How many such ideas are now waiting for someone to put them to use?

To obtain complete cancellation, the amplitude or intensity of the light reflected from each surface must be the same. The fraction of the incident light reflected from any surface depends on the change of index across the boundary. A large change

*Patented by Carl Zeiss, 1935.

results in a greater amount of reflection than does a small change of index. The actual relation depends in a fairly complex manner on the indices of refraction on each side of the boundary. The index of refraction of the thin film on a coated lens must be approximately halfway between those for air and for glass, so that the amplitude reflected at each surface will be approximately the same. The two reflected rays will very nearly cancel.

The thickness of the coating must be precisely chosen. The optical path, $2\mu t$, for destructive interference must be $\lambda/2$, $3\lambda/2$, ... or $(2n + 1)\lambda/2$, where $n = 0, 1, 2, \ldots$. The path $2\mu t$ is for exactly normal incidence. If the ray is not normal to the surface, the path difference will be different from this. Approximately destructive interference will occur over the widest angle if the film is chosen at its thinnest possible value (that is, for which the path difference is $\lambda/2$), rather than any of the other possible thicknesses. So the anti-reflection coating must be of a thickness t such that

$$2\mu t = \lambda/2$$

or

$$\mu t = \lambda/4$$

The quantity μt is the optical thickness. This must be a quarter of a wavelength, and such a coating is referred to as a **quarter wave film.** The geometrical thickness is given by $t = \lambda/4\mu$.

To calculate the thickness of an anti-reflection film to be put on the surface of a lens, what should be used for the wavelength λ? The wavelength varies across the spectrum. The film can, therefore, be made anti-reflecting for only one wavelength or color. This is one reason why the idea does not work perfectly. To get maximum cancellation, the wavelength on which the thickness calculations are based should be near the center of the spectrum or, for instruments for visual use, at the wavelength of maximum sensitivity of the eye. This is in the yellow-green at about 550 nm or 0.55 μm. The variation of wavelengths toward either end of the spectrum is not so large that constructive interference occurs for other colors. It means only that the least reflection occurs for yellow-green light, and the most occurs for red and for violet. Because of this, coated lenses designed for visual use will often have a purple or red hue in reflected light.

The actual thickness of an anti-reflection coating may be calculated on the basis of $\lambda = 0.55$ μm and the index $\mu = 1.25$. The result is that the thickness in micrometers is $t = 0.55/(4 \times 1.25) = 0.11$. Expressed in centimeters this value is more impressive; it amounts to 0.000011 cm. That is, the film is just 11 millionths of a centimeter thick. It is apparent that coating a lens is a very delicate and precise process. A variation of only 11 millionths of a centimeter in thickness would result in increased rather than decreased reflection. It follows that a coated lens surface must be cleaned with care so that the thin coating is not rubbed off.

The extent of the reduction of reflection actually obtained with a thin film on a glass surface is illustrated in Figure 11–5. Reflection is shown to be reduced by a single film from just over 4 per cent to slightly more than 1 per cent. To reduce the reflection even further, another anti-reflection film can be placed on the first anti-reflection film. The curve in Figure 11–5 for a double film shows a reduction to less than 1/4 per cent over a narrow range of wavelengths. Three films can be designed to give low reflection over a broader part of the spectrum. Lenses ordinarily have a single film; special optical equipment may be provided with two or three, though the multiple coatings are becoming more and more common as techniques are refined and costs are reduced.

The idea of the anti-reflection coating has uses outside the field of optics. In large halls, such as cathedrals or auditoria, it is sometimes necessary to absorb selectively some range of frequencies of sound to improve the acoustics. Slots or cavities that are a quarter wavelength deep can be put into the walls. The ancient Greek and Roman architects, in fact, sunk appropriately sized clay pots into the walls. Sound waves entering such cavities cannot escape, because after traveling in and out of the cavity they are out of phase with waves reflected from the wall surrounding the cavity. The sound intensity builds up inside the cavity and is eventually absorbed.

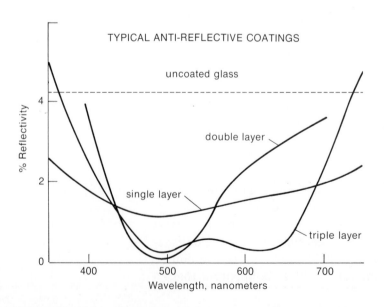

TYPICAL ANTI-REFLECTIVE COATINGS

Figure 11–5 The effectiveness of one or more anti-reflective coatings on glass. The data are from an Oriel Optics catalogue.

11-4-2 COLORS OF THIN FILMS

The colors of thin oil or gasoline films on water are inter-ference phenomena. The colors of soap films or bubbles, the areas of unwanted color sometimes seen in photographic trans-parencies mounted between glass, and color patches seen under microscope cover glasses are all due to the interference of the light reflected from the two surfaces of a thin film. There are others, too; watch for them in your everyday living. They are demonstrations of the wave properties of light. A color appears at a position where the film thickness is such that constructive interference occurs for the waves of that color reflected from the two surfaces of the thin film.

11-4-3 INTERFEROMETERS

A thin film produces two waves that come together after a very short path difference and interfere with each other. It is quite possible to have very long path differences between two beams and still see the interference phenomena. In fact, it is in this way that very high precision measurements are made. The meter, which is now defined in terms of wavelengths of light, was set up by "counting" the number of light waves along the metal bar that was previously the standard.

The instrument to be described is just one of a number of devices called interferometers, but it has been one of the more important in science. It is called the **Michelson interferometer.**

The Michelson interferometer is shown in Figure 11-6. Light from the source S is directed to the beam splitter, which is a piece of glass with a very thin coating of silver on one side — just enough so that half of the light is reflected and half passes through. The light that goes straight through is reflected back from the mirror M_1, and then half of it is reflected by the beam splitter to the eye. The light reflected toward M_2 is also re-flected back toward the eye, and interference occurs between the light beams from M_1 and M_2. The light to M_1 passed through the glass beam splitter in its travels to M_1 and back; to equalize the beams, a piece of the same glass from which the beam splitter is made is placed in the path to M_2. That is called the compensating plate.

If the distances l_1 and l_2 are equal, there will be constructive interference between the two rays going toward the eye. The mirrors will then appear bright.

If mirror M_1 is moved a distance Δl, a path difference of $2\Delta l$ is introduced between the two beams. If this path difference is half a wave, there will be destructive interference and the mirrors will appear dark. To illustrate the sensitivity of such an

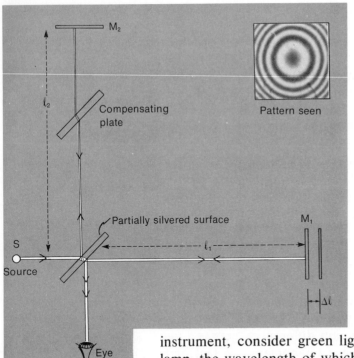

Compensating plate

Pattern seen

M_2

l_2

Partially silvered surface

M_1

S

Source

l_1

Δl

Eye

Figure 11–6 The Michelson interferometer.

instrument, consider green light such as that from a mercury lamp, the wavelength of which is 546 nm. For a change in the field from bright to dark,

$$2\Delta l = 546 \text{ nm}/2 = 273 \text{ nm}$$
$$\Delta l = 137 \text{ nm}$$
$$= 0.000014 \text{ cm (to two figures)}$$

A movement of the mirror of only 14 millionths of a centimeter changes the light seen from bright to dark. The statement that fine measurement can be made with an interferometer is adequately illustrated!

The oblique rays do not have the same path difference as the central rays, so the pattern of light seen in the interferometer is a series of rings (inset in Figure 11–6).

To measure a long distance with the interferometer, the number of changes of the center from bright to dark is counted and for each change the motion of the mirror is just a quarter of a wave. Light of only a single wavelength must be used, or else the point will be reached at which one wavelength will be reinforced while the other will not and the pattern will be lost. This was the basic reason for the choice of the element krypton for the definition of the meter. The light from it is extremely "pure." It was by using a Michelson interferometer that the standard meter bar was found to contain 1,670,763.73 wavelengths of a certain spectral line of krypton, and that is now the length standard.

With white light, which is a mixture of all colors or wavelengths, the interferometer shows bright white light only when

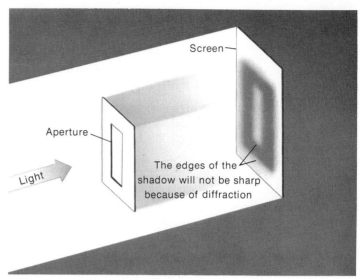

Figure 11–7 A light source, aperture, and screen. Because of the wave nature of light, the shadow on the screen cannot have perfectly sharp edges.

the paths l_1 and l_2 are the same within a small fraction of a wave. When a path difference is introduced, some colors will cancel while others won't and a pattern of colored fringes appears.

11–5 DIFFRACTION OF LIGHT

Diffraction, the bending of a wave into a shadow zone, occurs with any wave. Sound is an example, and certainly sound does travel around corners. To explain why the effect is not so obvious with light waves, the phenomenon will have to be analyzed in detail. Though the effect is not ordinarily obvious with light, it is of great importance in some scientific work and in what we learn about the natural world from diffraction effects.

Because of diffraction, shadows do not have perfectly sharp edges; it would be expected that a shadow would be as in Figure 11–7. This is not ordinarily obvious because the extent of the spreading into the shadow region is small if wide apertures are used. But with narrow apertures the effect becomes large. In Figure 11–8 are actual photographs obtained by putting light through narrow slits onto a photographic film. The narrower the slit, the wider the pattern of light on the film!

Figure 11–8 The pattern produced on a film when the light is passed through slits of different sizes. The numbers give the slit widths; the narrower the slit, the wider the pattern. Note also the secondary bright fringes.

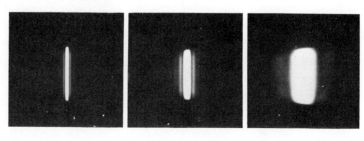

To analyze this phenomenon, Huygens's principle will be used to show the direction of motion of the wave. In any given direction there will be interference between the waves that, using Huygens's concept, originate at various parts of the wavefronts.

11–5–1 LIGHT THROUGH AN APERTURE

When a light wave passes through an aperture, the disturbance that makes up the wave in the aperture sends a wave forward and contributes to the light ahead. To apply Huygens's principle, consider the wave at the slit or aperture to be broken into a series of narrow strips; as in Figure 11–9, each contributes to the light on a screen or film. The edge view of the aperture is shown in the figure. The light at any point P on the screen actually comes from all parts of the aperture, and the phase difference from each part must be taken into account. There is bright light straight ahead, because the light from all portions of the slit is in phase; but as P moves into the geometrical shadow (outside the projection of the slit on the screen), the intensity does not drop immediately to zero. When the position is reached at which the light from the edge of the slit is out of phase with the light from the center, these two rays cancel. The light from the second strip from the edge cancels the light from the strip just below the center, and so on across the aperture. This condition is shown in Figure 11–10. The angle θ at which the light on the screen reaches zero is described by

$$\sin\,\theta = \frac{\lambda/2}{a/2} = \lambda/a$$

The diagram is correct only if the distance to the screen is large compared to the width of the aperture, so that rays A and B are effectively parallel. If a lens is put in so that the parallel rays are focused on the screen, the analysis is correct for any aperture size.

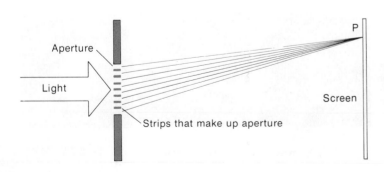

Figure 11–9 An aperture considered as a series of strips. The wave disturbance in each strip contributes to the light at P on the screen. The resulting intensity at P depends on the interference between rays from all parts of the aperture.

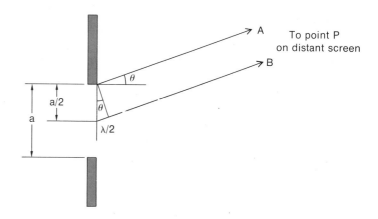

Figure 11-10 The light intensity on the screen drops to zero in the direction for which the ray from the center travels half a wavelength farther than does the ray from the edge.

Examination of the formula explains some phenomena that are otherwise obscure. If the aperture width, a, is large compared to the wavelength so that λ/a is a small number, then θ will be small and the edge of the shadow of the slit will appear sharp. But if the width a becomes small so that the fraction λ/a increases, then θ increases and the apparent spread of the wave past the edge of the aperture becomes large. Thus, for light going through large openings like doorways, there is very little spreading into the shadows. If the aperture is made very small, like a narrow slit, the light may spread out well into the geometric shadow. With sound waves, whose wavelengths are about the same size as doorways, the diffraction effects are large and we hear the sound well into the geometrical shadow. Because wavelengths of light are so much shorter than sound waves, we do not ordinarily notice diffraction effects with light as we do with sound. This analysis also explains why the high notes (short wavelengths, much shorter than the widths of doors) are not transmitted out of rooms and down halls as well as are the low notes (long waves); in a building, distant noises are heard as a low pitched rumble. It also explains why Sir Isaac Newton was reluctant to accept a wave model of light: he did not see the diffraction of the light to the extent that he had expected.

Beyond the angle θ that has just been discussed, there is a region in which there is again incomplete cancellation of the light, and a bright band results. Still further out, the intensity again drops to zero, and beyond that there is again light. This results in a series of light and dark bands in the geometrical shadow. The bands can be seen clearly in Figure 11-8.

Figure 11-8 shows also the extent to which the light spreads out through apertures of different sizes. For the relatively wide aperture, there is only a little spreading. For the very narrow slit, the waves spread out almost as though that slit were a source sending the waves out in all directions. This actually occurs when the slit width is less than half a wavelength.

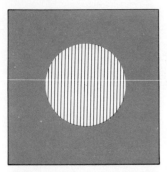

Figure 11–11 To analyze for light intensity on a screen after the beam passes through a circular aperture, the aperture is considered as a series of narrow strips as shown.

Here we have stumbled on another means of obtaining coherent light sources: use two narrow slits close together. But before analyzing this, let us mention what happens with circular holes rather than rectangular openings.

If the aperture is circular, the above analysis does not quite explain the situation because the total light from a strip at the edge of the hole, as in Figure 11–11, is not enough to cancel the light from the middle. When the analysis is carried out taking this into account, the result is similar to that for rectangular slits except for a constant of 1.22; the equations for finding the angle from the edge of the opening to the first position at which the light intensity is zero are:

for a rectangular aperture: $\sin \theta = \lambda/a$
for a circular aperture: $\sin \theta = 1.22\lambda/a$

For a circular aperture, the light intensity does not drop to zero until the path difference between the rays from the center and those from the edge is $1.22\lambda/2$, rather than just $\lambda/2$.

To show the extent of this effect, the angles to the first minimum for circular apertures of various diameters are shown in Table 11–2. The angles are in degrees and are calculated for light of different colors.

11–5–2   DOUBLE SLITS

Light passing through two slits results in one of the most unexpected and fascinating phenomena in optics. With a single slit the diffraction pattern is as in Figure 11–8; however, when another similar slit is placed near the first, the central bright region breaks into a large number of bright and dark bands.

TABLE 11–2   The angles to which light is diffracted into the shadow region by circular apertures of different sizes. The angle is that to the first zero-intensity position and is shown for different colors.

APERTURE DIAMETER, a		ANGLE TO FIRST MINIMUM		
μm	mm	Violet	Green	Red
2000	2	0.016°	0.019°	0.023°
1000	1	0.032°	0.039°	0.045°
500	0.5	0.063°	0.077°	0.091°
100	0.1	0.32°	0.38°	0.45°
50	0.05	0.63°	0.77°	0.91°
35	0.035	0.90°	1.10°	1.30°
10	0.01	3.15°	3.85°	4.55°

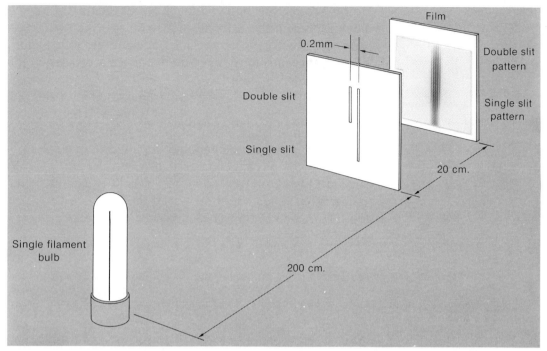

Figure 11–12 The diffraction pattern produced by light passed through a double slit (top) compared with the single slit pattern (bottom). The secondary maxima did not register in this exposure.

Actually, the same thing happens with the bands on both sides, but they are not so striking. The unusual thing is that when two broad, uniform patterns from two slits overlap, the result is not a brighter and uniform pattern, but a highly contrasting array of bright and dark bands, as shown in Figure 11–12.

This double slit phenomenon demonstrates the wave character of light. At the centers of the bright bands, the waves from both slits are in phase and reinforce each other. At the centers of the dark bands, the light from the two slits completely cancels. The analysis of the situation is shown in Figure 11–13. This figure shows an enlarged view of the ends of the slits. The distance to the screen in the diagram is not at all to scale. In prac-

Figure 11–13 A diagram to assist in the analysis of double slit interference. There is no light on the screen in directions for which the path difference (P.D.) is $(2n + 1)\lambda/2$.

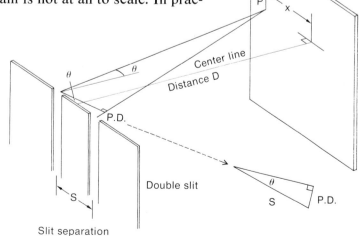

tice, the screen will be far enough away (compared to the slit spacing s) that the rays from the two slits to the point P on the screen are effectively parallel. To achieve the effect, the light falling on the slits must itself be from a very narrow source, such as another slit or a laser.

If the point P were at the center line, the light from the two slits would be in phase at the screen. Rays from both slits would travel the same distance to that center point. As the point P is moved away from the center line, a path difference P.D. is introduced. When P.D. becomes half a wavelength, $\lambda/2$, the intensity at P is zero. At P.D. $= \lambda$, the light is again bright. The zero-intensity bands occur as before for P.D. $= \lambda/2, 3\lambda/2, \ldots$, $(2n + 1)\lambda/2$, and the bright fringes occur for P.D. $= n\lambda$.

EXAMPLE 4

In Figure 11–12, the two slits are 0.2 mm apart and the screen is 20 cm away. Find the angles from the center of the film to the first two minima on one side, and also the corresponding distances on the film from the center line to those minima. Also find the separation of the minima on the film. Use a green light of $\lambda = 550$ nm.

For the first minimum, the path difference between the rays from the two slits is $\lambda/2$. From the small triangle shown in Figure 11–13, P.D./$s = \sin \theta$. Call the angle to the first minimum θ_1; then

$$\sin \theta_1 = \frac{\lambda}{2s}$$

Since $\lambda = 550$ nm $= 550 \times 10^{-6}$ mm and $s = 0.2$ mm, we find

$$\sin \theta_1 = 550 \times 10^{-6} \text{ mm}/0.4 \text{ mm}$$
$$= 0.00138$$

This is a small angle, and for small angles the sine and the radian measure are the same. Thus, using $1' = 0.0003$ radians

$$0.00138 \text{ radians} = 4.7'$$

In calculating the angle to the second minimum, we recognize that it will be small, and we let $\sin \theta_2 = \theta_2$ in radians. For this second minimum, P.D. $= 3\lambda/2$ and

$$\theta_2 = 3 \times 550 \times 10^{-6} \text{ mm}/2 \times 0.2 \text{ mm}$$

This is just $3 \times \theta_1$, and it will be 0.00414 radians or 14'.

If the slit spacing is small compared to the distance to the film, then the angle θ is given by x/D, where x is the distance on the film from the center to the minimum and D is the distance from the slits to the film. If x_1 is for the first minimum:

$$\frac{x_1}{D} = \theta_1$$

$$x_1 = D\theta_1$$
$$= 20 \text{ cm} \times 0.00138$$
$$= 0.0276 \text{ cm}$$
$$= 0.276 \text{ mm}$$

For the second minimum, the angle is three times as great, so x_2 is three times as big as for x_1:

$$x_2 = 3 \times 0.276 \text{ mm}$$
$$= 0.828 \text{ mm}$$

The angles are 4.7' to the first minimum and 14' to the second minimum. These correspond to distances on the film of 0.276 mm and 0.828 mm. The separation of these dark bands is 0.552 mm. Since the wavelength was not so precise, it would be better to say that the distances are 0.28 and 0.83 mm, and the separation is 0.55 mm.

Because of the finite width of each slit, the double slit pattern occurs only over the area defined by the pattern from one slit alone.

An added feature of the phenomenon is that a pattern produced by red light spreads out more than one produced by violet or blue light. This is because waves of red light are longer than those of blue light. If white light is used, the observed pattern is produced by the overlapping systems from the different colors. The result is a multicolored series of bands.

The double slit pattern can be used to measure wavelengths; but there is a better way, involving more slits, which will now be discussed.

11–5–3 THE DIFFRACTION GRATING

The diffraction grating consists basically of a large number of parallel lines ruled on a transparent or an opaque base. The transparent gratings are usually used for transmitted light, and the opaque or metal ones are used with reflected light. Probably no other single piece of equipment has contributed as much to our present knowledge of atoms and molecules as has the diffraction grating. A grating is also so beautiful that some jewelry now marketed is decorated with reflection gratings to give sparkling, pure spectral colors.

It has been said that the production of diffraction gratings is as much an art as a science; certainly, the utmost skill of the scientist must go into their production. When you consider that a grating for scientific work may have 6000 to 12,000 lines engraved per centimeter of its length, and that these lines must be properly shaped and precisely spaced, you may realize that the task of grating production dwarfs even the art of diamond cutting and polishing. Some fairly high quality plastic replica gratings are now made from precisely ruled masters. Some poor quality ones are also made; one cannot do good scientific work with inferior equipment.

The principle of operation of a grating is just an extension of that of the double slit. The transparent slits between the engraved lines are so narrow that the light passing through is diffracted to spread out into a semicircle centered on the slit; thus, each line acts as a source sending waves out in all directions. In certain directions the waves from all slits are in phase, and highly reinforced light occurs in these directions. In Figure 11–14 is shown a greatly enlarged end view of some of the slits of a grating. The grating is shown perpendicular to the direction of the incident light. In Figure 11–14(a) the waves from each slit are shown spreading out, and the directions in which all the crests fall together are indicated. In Figure 11–14(b), only the rays in a particular direction θ from the direction of the incident light are shown. There is a path difference between the rays from adjacent slits. If the path difference is one wavelength or an integral number of wavelengths, the waves from all slits will be in phase in that direction. If a lens is used to bring these waves together, the amplitude will be greatly enhanced. If, say, a 1 centimeter grating of 6000 lines per centimeter is used, the amplitude will be 6000 times the amplitude of the incident wave. In the directions for reinforcement, the intensity (which depends on the square of the amplitude) would be 6000^2 or 36,000,000 times what it would be from one slit. The slits are very narrow, and each transmits very little light, but in the directions for reinforcement the intensity is high.

The path difference for light from adjacent slits increases as the angle increases. Since different colors have different

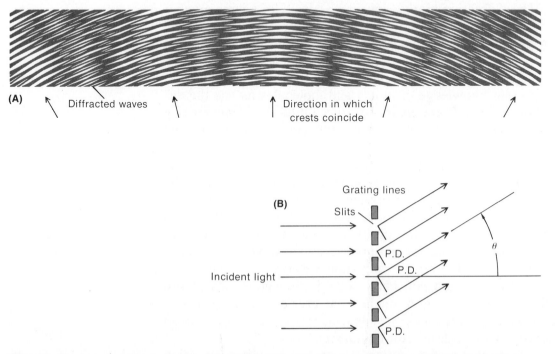

(A) Diffracted waves   Direction in which crests coincide

(B) Grating lines
Slits
P.D.
P.D.
Incident light
θ
P.D.

Figure 11–14 A much enlarged end view of the slits of a diffraction grating. In (a) it is shown that the crests of the waves spreading out from each slit will coincide in certain directions. In (b) the rays from adjacent slits are shown to have a path difference in a direction θ.

wavelengths, each color is reinforced in a different direction. If light of many colors is incident on the grating, the different colors are reinforced and sent off in different directions. The grating therefore separates light into its component colors just as a prism does. However, a grating has several advantages over a prism. One is that the grating spectrum can be spread out over a much wider angle than can a spectrum from a prism. Another advantage is that the wavelengths can be measured with the grating. This allows precise identification of any part of the spectrum. For instance, light with a wavelength of 500 nm appears green, but so does light of 510 nm. The grating could separate and measure these two colors, whereas even the human eye could not.

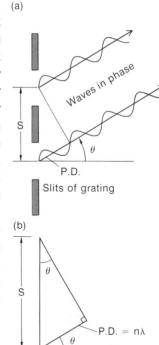

(a)
Waves in phase
S
θ
P.D.
Slits of grating

(b)
θ
S
P.D. = nλ
θ

Figure 11–15 The light from two adjacent slits of a grating. In (a) the path difference is shown as one wavelength. In (b) is a triangle showing the relation between line spacing, angle for constructive interference, and wavelength. The path difference shown in (a) could be any number of wavelengths.

The principle of the measurement of wavelengths can be derived from Figure 11–15. The distance between lines on the grating is shown as s. If the grating has 6000 lines per centimeter, as is common, the spacing s is 1/6000 centimeter. This amounts to $s = 0.0001667$ cm or 1667 nm. This spacing is in the same size range as the length of light waves.

From the triangle of Figure 11–15(b), the relation P.D./$s = \sin \theta$ is apparent. For bright light, the path difference P.D. is equal to $n\lambda$. Then we get $n\lambda/s = \sin \theta$, or

$$\lambda = \frac{s \sin \theta}{n}$$

This relation is used to calculate wavelengths from measurements with a grating. The angle of deviation θ for reinforcement of the light can be measured. Knowledge of the number of lines per unit distance allows calculation of s. The number n is called the **order** of the spectrum. For the first direction for reinforcement, the path difference is one wavelength and $n = 1$. As θ increases further, there is again reinforcement for the same color when the path difference is 2λ. For this angle, $n = 2$ and the result is called a second order spectrum. Diffraction gratings can create two, three, or more spectra on either side of the center, depending on the number of lines per centimeter.

EXAMPLE 5

Consider light of 450 nm, which falls somewhere in the blue part of the spectrum. In what directions would this blue light be reinforced when it is put through a grating of 6000 lines per centimeter? To find these angles, solve the grating equation for the angle θ;

$$\sin \theta = n\lambda/s$$

λ is given as 450 nm; s for a 6000 lines/centimeter grating we have already worked out to be 1667 nm. To find the direction of the first order light, put $n = 1$:

$$\sin \theta_1 = \frac{1 \times 450 \text{ nm}}{1667 \text{ nm}} = 0.2699$$

$$\theta_1 = \pm 15.66°$$

For $n = 2$:

$$\sin \theta_2 = \frac{2 \times 450 \text{ nm}}{1667 \text{ nm}} = 0.5399$$

$$\theta_2 = \pm 32.68°$$

For $n = 3$:

$$\sin \theta_3 = \frac{3 \times 450 \text{ nm}}{1667 \text{ nm}} = 0.8098$$

$$\theta_3 = \pm 54.08°$$

For $n = 4$:

$$\sin \theta_4 = \frac{4 \times 450 \text{ nm}}{1667 \text{ nm}} = 1.080$$

$$\theta_4 = ?$$

Since the sine of an angle cannot exceed 1, the case for $n = 4$ does not describe a real situation. Fourth order interference does not occur with that grating and that light.

It should be noted that in the direction straight ahead, for $\theta = 0°$ in our examples, all colors reinforce.

If light is put through any object that consists of a series of parallel slits, a series of diffraction patterns will be produced just as with a diffraction grating. The spacing s of the lines or slits can be found if the angle θ is measured to the bright line of known wavelength λ in the nth spectrum from the center. The grating equation $\lambda = s \sin \theta/n$ is merely solved for s to obtain this. The smaller the slit spacing, the wider will be the pattern of multiple spectra produced. Passing a beam of light through a feather produces two series of spectra, as in Figure 11–16(a), which occur in different directions. A narrow pattern is produced by the widely spaced barbs shown in Figure 11–16(b), and a much broader pattern is produced by the more closely spaced barbules.

The existence of such diffraction patterns shows that the material has a linear structure like a diffraction grating. It is not necessary that the slits even be visible; the existence of the diffraction pattern tells of the structure.

If two gratings are crossed with the slits perpendicular to each other, the effect is that of a rectangular array of points. When a narrow beam is put through such a system, the diffraction pattern consists of a rectangular array of spots as in Figure 11–17(a). If three gratings are crossed at 120° to each other,

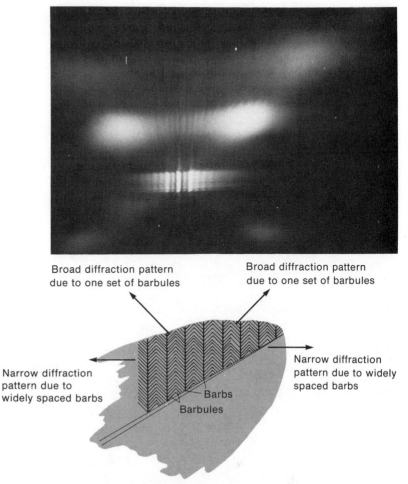

Broad diffraction pattern
due to one set of barbules

Broad diffraction pattern
due to one set of barbules

Narrow diffraction
pattern due to widely
spaced barbs

Narrow diffraction
pattern due to
widely spaced barbs

Barbs

Barbules

Figure 11–16 Light passing through a feather onto a film shows a diffraction pattern of the type resulting from a series of slits. There are actually two sets of patterns: the narrow pattern results from the widely spaced barbs, and the broad pattern is due to the closely spaced barbules. A diffraction pattern can tell us details about an unseen structure.

the diffraction pattern is a hexagonal array of points as in Figure 11–17(b). It is apparent that the structure of the diffracting material can be deduced and measured by using the diffraction pattern. These examples have been given, but actually any array of diffracting spots gives a unique diffraction pattern.

The above phenomena occur with light only if the spacing of the diffracting elements is greater than the wavelength of the light being used. For closer spacings shorter wavelengths can be used: ultraviolet light or even, very commonly, x-rays. The x-ray diffraction pattern of x-rays passed through some materials, lipids for example, show that they consist of an array of long parallel molecules, for the diffraction pattern is similar to that from a diffraction grating. X-ray diffraction is used to find the structure of crystals and of large molecules just in this way. The problem is complicated by the fact that the scattering centers, the electrons, are in a three-dimensional, diffuse array.

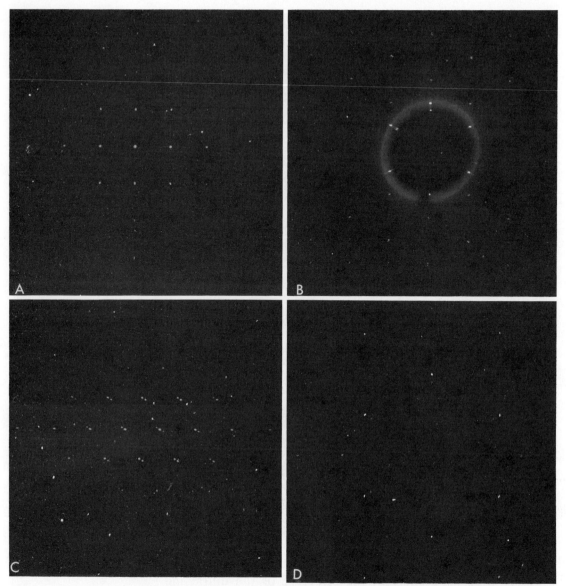

Figure 11–17   In (a) is the diffraction pattern of light passing through two crossed gratings that form, in effect, a rectangular array of slits. In (b) is the pattern produced by x-rays passing through a cubic crystal. Three gratings at 120° produce a hexagonal array of openings, and the light is diffracted in the pattern shown in (c). This is similar to the pattern in (d), which results from x-rays passing through a crystal in which the atoms are packed in a hexagonal array. (X-ray diffraction patterns courtesy of Dr. B. Robertson, Regina.)

The analysis from the diffraction pattern to the electron density pattern, and hence the crystal structure or even molecular structure, is extremely complex. It can be done only with the use of large computers plus intelligent analysis and programming.

In Figure 11–17(c) and (d) are reproductions of films obtained by putting a beam of x-rays through two different crystals. The intense primary beam in the center has been cut out by a small piece of lead. In comparing these to (a) and (b) of Figure 11–17, what would you say about the pattern of the arrangement of the atoms?

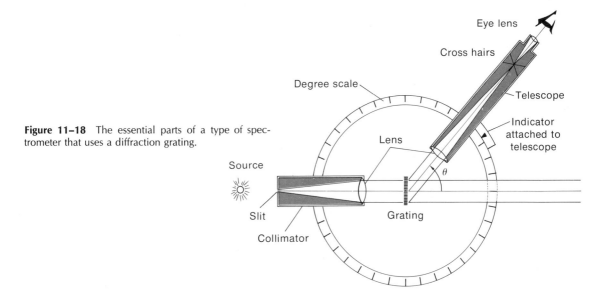

Figure 11–18 The essential parts of a type of spectrometer that uses a diffraction grating.

11–5–4 A GRATING SPECTROMETER

The essential parts of one type of spectrometer that uses a transmission grating are shown in Figure 11–18.

The collimator produces a beam of parallel light that is incident on the grating. The telescope focuses the light from the grating; it also is made to move along a protractor scale to measure the direction at which the light is reinforced. The slit is one of the keys to the spectrometer. The light to be analyzed is put through the slit; it spreads out and is focused to a parallel beam by the lens of the collimator. This parallel beam of light strikes the grating. The light emerging from the grating is focused by the objective lens of the telescope onto the cross hairs in the telescope. This light and the cross hairs are examined by means of an eyepiece, just as in a microscope. It is actually the slit that is brought to focus at the cross hairs. If light of only one wavelength comes from the source, the image of the slit will occur at a particular direction. If the source emits two wavelengths, images of the slit will be produced in two directions (in each spectral order, of course).

11–6 POLARIZED LIGHT

Light behaves like a propagating wave; but there are two basic forms of wave motion. One, in which the vibration is in the same direction as that in which the wave travels, is called a longitudinal wave. An example of a longitudinal wave is sound in air. The individual portions of the air vibrate back and forth in the same direction that the sound travels, producing alternate

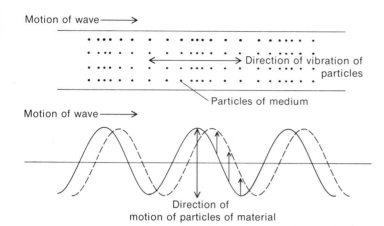

Figure 11–19 Types of waves. In (a) is a longitudinal mechanical wave, such as sound in air. The vibration of the parts of the medium is along the direction of travel of the wave. In (b) is a transverse wave, such as that in a stretched string. Each part of the string vibrates perpendicular to the direction of motion of the wave.

condensations and rarefactions that make up the sound wave. This is illustrated in Figure 11–19(a). The other type, a transverse wave, is illustrated by a wave in a rope or a string, as in Figure 11–19(b). The motion of any portion of the string is a vibration perpendicular to the direction in which the wave moves. The direction of the vibration is called the direction of polarization. Some waves have the vibrations always in the same direction, and these waves are said to be **polarized.** A longitudinal wave cannot be polarized, but light can be polarized, so it follows that light is a transverse wave. However, light is not a material wave; it is a combination of vibrating electric and magnetic fields.

In an ordinary light beam the direction of vibration of the electric field (or vector) will vary from one point to another in the beam, and from one interval of time to the next at the same point in the beam. The same is true of the magnetic vector. Such light is called unpolarized. However, a beam of light can be produced in which the vibration is always in the same direction or plane, at all places. This is called **plane polarized** light, or often just polarized light. The direction of polarization is (by convention) indicated by the direction of vibration of the electric vector. One convention to show polarized light on a diagram is shown in Figure 11–20. At some point in the beam the end view is shown with the direction of vibration indicated. In (a), we show unpolarized light; in (b) and (c), the directions of polarization are different.

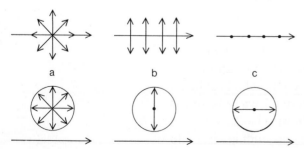

Figure 11–20 Methods of showing polarized and unpolarized light. In (a) the light is unpolarized. In (b) the direction of vibration is in the plane of the paper, and in (c) it is perpendicular to the plane of the paper.

Polarized light can be produced in many ways. The most important are:

(a) passage through certain crystals;

(b) reflection from surfaces at certain angles; and

(c) scattering of light by small particles.

An interesting phenomenon is seen with a device (called a polarizer) that will polarize light in a certain direction. If a polarized beam is put onto this device, the beam may pass through or may not, depending on its direction of polarization. If the incident beam vibrates in the direction that is passed by the polarizer, it will pass through. If the light vibrates perpendicular to that direction, it will not pass at all; for directions in between, a portion of the incident intensity will pass.

A common form of polarizer is in the form of a plastic or glass disc or sheet. These are sometimes incorporated into sunglasses. They consist of a layer of small crystals, of a type called "dichroic," all aligned in the same direction. The discs may be marked to indicate the direction in which the light (electric vector) is polarized by the device.

As indicated above, such devices that polarize light can also be used to analyze an incident beam to see if it is polarized. The word "analyze" has been used, and it is a clue that such a device has uses in analysis. In Figure 11–21, the effect of a polarizing material on a light beam is shown.

Figure 11–21 The use of a polarizer as an analyzer. In (a) and (b) the incident light is unpolarized; whatever the direction of the polarizer, the transmitted light is polarized and of the same intensity. In (c), with polarized incident light, the intensity is undiminished if the polarizer is aligned with the direction of polarization of the light; but in (d) the polarizer is perpendicular to the polarization of the incident beam and no light is transmitted.

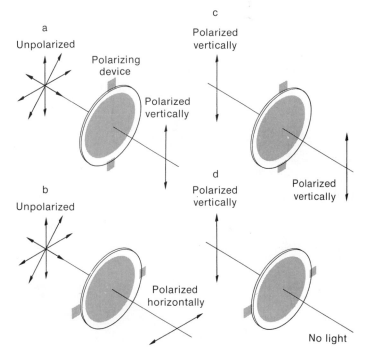

a
Unpolarized

Polarizing device

Polarized vertically

c
Polarized vertically

Polarized vertically

Polarized vertically

b
Unpolarized

Polarized horizontally

d
Polarized vertically

No light

11-6-1 CRYSTALS FOR POLARIZATION

Crystals are regular arrays of atoms. When light passes through a crystal, the vibrating electric field of the light sets the electrons of the crystal in motion. These electrons re-radiate their energy, and the beam progresses through the material. If the electrons vibrate more easily in one direction than another, the light is separated into two beams vibrating at right angles to each other and traveling at different speeds. Such a crystal has two indices of refraction, one for each direction of vibration. One of the beams may not even strictly obey the laws of refraction, and is called the extraordinary ray (*e* ray). The other ray is called the ordinary ray (*o* ray). One crystal that exhibits this phenomenon to a large degree is calcite. In Figure 11-22 is a photograph of a printed word seen through a calcite crystal. The two beams have such different indices of refraction that the image is actually double. If these two beams are observed with a polarizer (used as an analyzer), the two beams emerging from the crystal will be seen to be polarized at right angles to each other.

Many crystals show double refraction, though not to this extent. Crystals can be identified, in fact, by their double refracting properties.

Some crystals, tourmaline and quinine for instance, absorb the extraordinary ray to a far greater extent than they absorb the *o* ray. The *e* ray is absorbed in a very short distance. The crystal is opaque for light vibrating in that direction. These crystals are called dichroic, and it is these that are used in plastic polarizers. The crystals are deposited on a plastic film in a thin layer, with the optic axes of all of the crystals aligned in the same direction. Such polarizers do not work equally well at all wavelengths, often passing violet light in the *e* beam as well as the practically complete *o* beam. The intensity of the beam

Figure 11-22 A view of printed words through a doubly refracting crystal, calcite. The crystal breaks a light beam into two parts, which are polarized perpendicular to each other. (Photo by M. Velvick, Regina.)

is considerably reduced by such polarizers. Yet they can be made inexpensively and in large areas, so they have found many uses.

11-6-2 POLARIZATION BY SCATTERING

The most common example of scattered light is the blue of the sky. Sunlight is scattered by the molecules of air, but not all wavelengths are scattered with equal efficiency from such molecule-sized particles. Rayleigh found that the scattered intensity varies inversely as the fifth power of the wavelength. This means that when white light is incident on the scattering medium, the blue is scattered more than the red. The result is that the sky is blue. When the sun is near the horizon, the light passes through a long path of atmosphere and a large portion of the blue end of the spectrum is scattered from the direct sunlight. The result is that the sun appears red at sunset or at sunrise. The moon shows this effect, too, frequently being quite orange just after it rises or before it sets.

A fascinating property of blue sky light, one which human beings have no natural way of detecting, is that it is polarized.

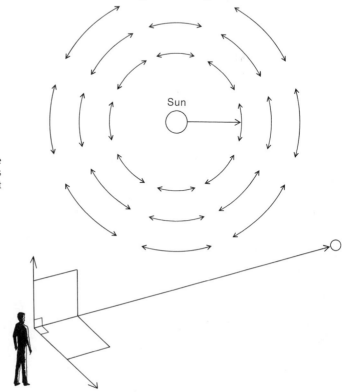

Figure 11-23 The polarization pattern of the light from the blue sky. The polarization is greatest when the viewing direction is at right angles to a line from the sun to the observer.

Light is scattered in all directions by the particles of the atmosphere, but that which is scattered at right angles is polarized most completely. The direction of polarization is at right angles to the initial beam also. Investigation of the blue sky with a polarizer used as an analyzer shows that the light is cut to a minimum when the direction of the analyzer points to the sun. This means that the light is polarized at right angles to that direction. The polarization pattern of the blue sky is shown in Figure 11–23.

Bees can undoubtedly detect this polarization of light from the blue sky and make use of it for navigation. Bees that have found a source of nectar communicate this information to the other bees at the hive by means of what is called a form of dance. The direction to the food is given in relation to the sun, but this is done even if the sun is not visible. Experimenters have found that if they change the direction of the polarization of the sunlight seen by a dancing bee, the others are sent off in the wrong direction, showing that it is the polarization that they make use of.

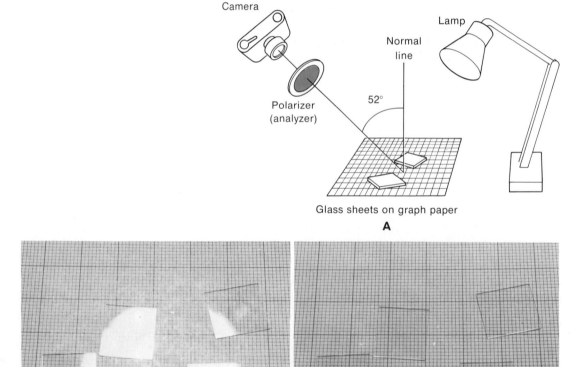

Figure 11–24 Photographs of light reflected from a shiny surface. (a) The arrangement. (b) With the polarizer direction horizontal, the reflected light is transmitted. (c) When the polarizer is used as an analyzer, with its direction vertical, the reflected light is not transmitted. (Photos by M. Velvick, Regina.)

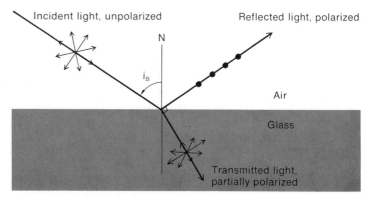

Incident light, unpolarized Reflected light, polarized

N

i_B

Air

Glass

Transmitted light, partially polarized

Figure 11–25 Polarization of reflected light. The reflected light is polarized parallel to the reflecting surface.

11–6–3 POLARIZATION OF REFLECTED LIGHT

A little investigation with some pieces of polarizing material or polarizing sunglasses will show that the light reflected from shiny surfaces is polarized to a greater or lesser degree depending on the angle at which it is viewed. Figure 11–24 is a picture of a reflecting surface viewed through a polarizing material used as an analyzer. The direction of the analyzer when the reflected light is cut out is normal to the surface, indicating that the reflected light is polarized parallel to the surface. This is shown diagrammatically in Figure 11–25.

"Sunglasses" made of polaroid material are designed to cut out reflected glare light from horizontal surfaces. They will ordinarily pass only light vibrating in a vertical direction. They will therefore not eliminate the reflected glare light from a vertical surface such as a window pane. Sunglasses may be used to polarize light, and the direction of polarization will be vertical.

11–7 INVISIBLE LIGHT: INFRARED AND ULTRAVIOLET

A continuous spectrum, what we call white light, is emitted by a hot object. The surface of the sun, which is about 6000°K, emits most of its energy in the range of visible light, with the maximum intensity in the yellow-green at the wavelength to which our eyes are most sensitive. The receptor cells in our eyes respond to a range of wavelengths around this, from about 400 nm to 700 nm. In addition to the visible light from the sun, there is also heat. The heat is in the form of radiation with wavelengths beyond the red end of the visible spectrum at 700 nm, and is known as infrared radiation, or IR. Just because our eyes do not respond to infrared does not mean that the eyes of some birds or animals do not.

Light with wavelengths shorter than that of the visible violet light of 400 nm is called ultraviolet light or UV. One effect of the ultraviolet on us is the tanning or burning of skin. Actually, only a certain band of wavelengths causes the tanning, and different wavelengths cause burning. A good suntan lotion will be transparent to the tanning wavelengths and opaque to the wavelengths that cause burning.

Radiation detectors, such as photographic film or photo-electric devices, may be made sensitive to either IR or UV. Analysis using light, such as spectral analysis, is not restricted to the visible light region. Infrared spectra, in particular, are used in the routine identification of unknown chemical compounds.

The infrared rays have some special uses related to some of their special properties. One is their connection with heat (they are emitted by sunlamps and infrared ovens); another is their ability to penetrate flesh to a greater degree than does visible light. A photograph of a person taken on film that is sensitive to infrared as well as visible light will show a network of sub-surface veins and arteries. Without the inclusion of visible light, regions that are of higher temperature than the surroundings, perhaps because of increasing metabolism caused by a lesion or a tumor, may be visible if they are close to the surface.

There is evidence that the eyes of some birds are sensitive to infrared radiation, and this may assist them in "seeing" warm-blooded prey. More work is necessary in this line.

DISCUSSION PROBLEMS

11-1   1. Make some soap bubbles, and watch them against a dark background with a light behind you or slightly to the side. Watch the color change at a particular spot and record the order of appearance of the various colors. Comment on how this order is related to the spectrum.

11-1   2. (a) Obtain some pieces of finely woven material. Hold a piece close to your eye and look toward a distant bright light. Describe what you see. This works best outdoors at night or with only a small light source in a darkened room.

(b) Repeat this with a feather from a small bird. It does not work very well with an eagle feather.

11-6-2   3. Obtain some polarizing material to investigate the polarization of sky-light for yourself.

11-6-2   4. Young children often pose the question, "Why is the sky blue?" Discuss how it could be answered. There are many questions of this type, so it is a common problem and is worthy of some thought.

11-6-3   5. Is the light from a rainbow polarized? Use polarizing sunglasses or a piece of "polaroid" material to find out.

PROBLEMS

11-2   1. Calculate the frequency of the light at the violet end of the visible spectrum ($\lambda = 400$ nm) and at the red end ($\lambda = 700$ nm), and also of "ultra high frequency" radio waves that have a wavelength of 1.0 meters.

11-2   2. If a book with 500 numbered pages is 2.00 cm thick (how many sheets of paper are there?), what is the thickness of one page? Express it in cm, in nm, in μm, and in nm. How many wavelengths of violet light ($\lambda = 400$ nm) will fit the thickness of one page?

11-2   3. How many waves of yellow light, $\lambda = 589$ nm, are there in a distance of 1.00 cm? This quantity, waves per cm, is called the wave number.

11-2   4. Find the speed and the wavelength of yellow light in water, in glass, and in the material of the vitreous humor between the lens and the retina of the eye. Let the wavelength in air be 580 nm, take the speed in air to be 3.00×10^8 m/s, and let the indices of refraction be 1.333 for water, 1.517 for glass, and 1.336 for vitreous humor.

11-2-3   5. Find the geometrical path length and the optical path length in an instrument in which the light goes through 10 cm of air, followed by 3.5 cm of glass of index 1.50, then 2.0 cm of glass of index 1.65, and finally 5 cm of air.

11-2-3   6. Find the thickness of glass of index 1.65 that would have the same optical thickness as 2.48 cm of water of index 1.33.

11-3-1   7. (a) List the first six path differences between two beams of light in air that would result in constructive interference for blue light of 450 nm in air.

(b) Repeat (a) but for yellow-green light of 540 nm.

(c) What path difference will result in constructive interference for both the blue and the yellow-green light?

11-3-1   8. Find the shortest non-zero path difference between two beams in material of index 1.25 that would result in constructive interference between light of the following wavelengths:

(a) red of 700 nm,

(b) green of 550 nm,

(c) violet of 400 nm.

Express the answers in nm, in μm, and in mm.

11-3-1   9. (a) Find the shortest path difference between two beams in material of index 1.25 that would result in destructive interference for (i) red light of 700 nm wavelength; (ii) repeat for green light of 500 nm; (iii) repeat for violet light of 400 nm. Express the answers in nm, in μm, and in mm.

(b) For any of the above thicknesses, would there be constructive interference for light of wavelength anywhere in the visible range, 700 nm to 400 nm?

11-3-1   10. An optical path difference (μt) of 900 nm can result in destructive interference for one color or wavelength in the visible range and constructive interference for another. Find the wavelengths.

11-4-1   11. At an air-glass boundary, about 4 per cent of the incident light is reflected; in an optical system with a large number of lenses, the total light loss may be appreciable.

(a) If there are 12 air-glass surfaces in a certain lens system, what fraction (express it as a percentage) of the entering light will emerge?

(b) If the lenses are coated so that the reflection is reduced to 1 per cent, what fraction of the light would get through the system that has 12 surfaces?

Measured values for an actual lens, uncoated and coated, are shown in Figure 11-26.

In doing this problem, note that the light intensity on the first surface is 100 per cent. The percentage reflected is 4 per cent, so 96 per cent hits the second surface. The amount reflected at that second surface is 4 per cent of 96 per cent; at each boundary, 4 per cent of the light incident on that boundary is reflected. Either

15 ELEMENT UNCOATED ZOOM LENS

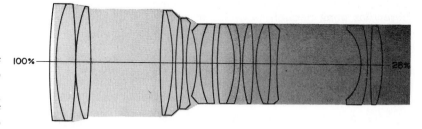

Figure 11-26 A comparison of light transmission through two telephoto zoom camera lenses, one uncoated and the other multicoated. Note that several pairs of elements are glued together to eliminate glass-air-glass transitions; each such pair eliminates two reflecting surfaces. (From B. Sherman and H. Kimata, *Modern Photography,* June, 1975.)

100% ———————————————————————— 28%

15 ELEMENT MULTICOATED ZOOM LENS

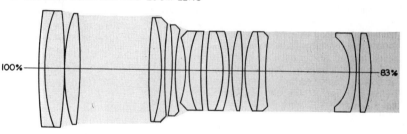

100% ———————————————————————— 83%

the surfaces must be treated one by one, or an equation may be set up to give the percentage of the light remaining after n surfaces.

11-4-1   12.   What would be the thickness of a film of index 1.30 that would give constructive interference (and enhance reflection) for infrared light of wavelength 1000 nm?

11-4-1   13.   What is the thinnest possible film for eliminating the reflection of light of $\lambda = 500$ nm if $\mu = 1.25$? What wavelengths would be reinforced by this anti-reflection film? What color would it be, or would it be in the ultraviolet or infrared?

11-4-2   14.   How thick would a film have to be so that reflected light would be reinforced in the green? Use $\lambda = 500$ nm, and disregard any possible phase changes. The index of the film is 1.25. Express the thickness of the film in terms of wavelengths and in the usual length units.

11-4-3   15.   A Michelson interferometer is to be used to find the wavelength of green light. It is adjusted with bright green in the center; then, while one mirror is moved through 1.000 mm, the center spot is seen to change through the cycle of bright to dark to bright again 3662 times. What is the wavelength in mm and in nm?

11-4-3   16.   A Michelson interferometer is balanced with a bright spot in the center using yellow light of 589 nm. In one arm there is an evacuated tube 5.00 cm long with thin plane glass ends. As air is slowly allowed into the tube, the optical path in that arm of the instrument slowly increases. The optical distance through the tube changes from 5.00 cm to the index of air times 5.00 cm. When the air in the tube reaches atmospheric pressure, the center of the mirror has gone through 49 cycles of bright-dark-bright. Find the index of refraction of the air.

Hint: Let the index be $1 + x$. The optical path in the tube changes from 5.00 cm to $5.00(1 + x)$ cm. The change in optical path is the same as though the mirror had moved that distance.

11-4-2   17.   In cold weather, ice crystals in the atmosphere result in a phenomenon somewhat like a rainbow. A ring forms around the sun, with the brightness enhanced on either side of the sun, and these bright areas are called sun dogs. The sun dogs result from light going at minimum deviation through the prism-like hexagonal ice crystals as in Figure 11-32. The crystal is like a 60° prism of ice.

(a) The inner, red edge of the dog occurs at 21.5° from the sun. What is the index of refraction of ice for red light?

(b) The blue occurs at 22.4°. What is the index for blue light?

11-4-3   18.   A rainbow effect can be demonstrated by using glass spheres rather than water drops. Find the angle to the bow formed by glass spheres of index 1.52.

11-4-4   19.   What would the index of refraction of the material of a sphere be if light from a distant object is focused by one surface of the sphere onto the other surface?

11-4-4   20.   A solid clear plastic rod has one end ground to be a portion of the surface of a sphere 5.0 cm in radius. The index of the plastic is 1.50. A lamp is put 25 cm from the curved end. How far back in the rod will the rays from the lamp be brought to a focus?

11-5-1   21.   Light passes through a circular hole 5 μm in diameter. At what angle from the center will the pattern reach a minimum? If a screen is put 10 cm from the slit, what will be the radius of the diffraction spot?

11-5-2   22.   Find the separation of two closely spaced slits if light of 550 nm is seen to be bright at 1°, 2°, and 3° on each side of the center line. Express the separation in nm and in mm.

11-5-2   23.   Two slits are carefully made with 0.100 mm between centers. Light of unknown wavelength is put through them, and the eighth bright band from the center is seen at 3.0°. What is the wavelength of the light?

11-5-3   24.   A grating has 6000 lines/cm. In what directions will green light in the first three spectra be reinforced? The wavelength is 546 nm.

11-5-3   25.   Find the spacing of the lines on a grating (and the number per centimeter) if sodium light, $\lambda = 589$ nm, appears at 44.1° from the center in the first order spectrum.

11-5-3   26.   When white light is put through a grating of 6000 lines/cm and is made to fall on a photographic film, the film is found to be affected between 10° and 23° from the center line. To what range of wavelengths is the film sensitive? Is that range in the visible spectrum?

11-5-3   27.   If a light source is viewed through a finely woven piece of cloth, the pattern seen is similar to that produced by a diffraction grating. Find the thread spacing and the number of threads per millimeter if green light ($\lambda = 0.55$ μm) occurs at 5.7° in the fifth order.

11-5-3   28.   A feather with its barbs and barbules acts like a double diffraction grating (see page 349). A certain feather gives a first order spectrum at an angle of 0.2° (0.0035 radians). The wavelength of the light is 0.55 μm. If this is a result of the barbs acting as a grating, find the spacing of the barbs and the number of barbs per millimeter.

The same feather gives a maximum for red light at 1.5° due to the barbules ($\lambda = 0.65$ μm). Calculate the spacing of the barbules and the number of barbules per millimeter.

11–5–3   29. When radiation of wavelength $\lambda = 10$ nm was passed through a thin piece of material, diffraction lines showed at 5° on each side of center. Propose a model of the structure of the material.

11–5–4   30. Find what wavelengths in the visible spectrum would reinforce at 20° from the center line of a grating that has 2000 lines per centimeter. Find also the order of interference for each of these.

11–5–4   31. Find the wavelength of the light that is reinforced in the ninth order at 51°50′ by a grating that has 1600 lines per centimeter.

11–5–4   32. Find the wavelengths of three of the lines in the spectrum of hydrogen, given that these lines are reinforced in the first order at 20°41′, 20°04′, and 16°56′ from the center line. Base the calculation on the measurement that the mercury green line of wavelength 546.1 nm is seen at 18°49′.

11–6–3   33. (a) Complete polarization occurs for reflected light when the angle between the reflected ray and the refracted ray is 90°. Show that when this situation applies, the angle of incidence, i_B, is described by

$$\tan i_B = \mu$$

The index of refraction of the reflecting material is μ and that particular angle of incidence is called Brewster's angle, hence the subscript B.

(b) At what angle of reflection is the light from water ($\mu = 1.3330$) completely polarized?

CHAPTER TWELVE

RAY OPTICS

Under this topic of ray optics we will include reflection, refraction (which refers to the change in direction of a light ray when it goes from one medium to another), and some applications of these phenomena, such as image formation and some optical instruments.

The wave property of light is dominant in image formation. The direction of travel of a wave is perpendicular to a wavefront, and lines perpendicular to the wavefronts are referred to as **rays.** The rays are a valuable aid in explaining phenomena.

We see an object because light (either emitted or reflected) diverges from the various points on it, and this light is focused by our eyes to form an image on the retina. An important concept is that we see an object at the position from which light rays diverge. At times the light rays appear to diverge from a point at which there is no object, but we say that at that place there is an image. An example is an ordinary mirror. You see objects that are apparently behind the mirror, and we say that there is an image behind the mirror. You can go behind the mirror and you will not see an image: it is not a real one, but it just appears to be there when you look from the front.

Lenses are the most common devices for image formation, and lenses use the refraction of light. Curved mirrors are similar to lenses in their image-forming properties, but they are not used as frequently as lenses for several reasons. So the phenomenon of refraction will be considered, and then we will examine the properties of lenses. This discussion will be followed by one on image formation by mirrors.

12–1 ANGLES IN OPTICS

The *normal line* is used in optics as a reference line for angles, partly because of habit and partly because it simplifies

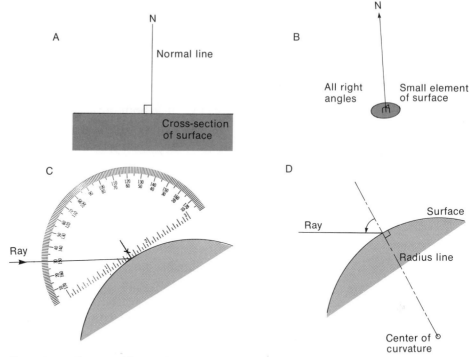

Figure 12–1 The normal line used for measurement of angles in optics.

the measurement of angles when curved surfaces are involved (Figure 12–1). The normal line forms a right angle in all directions with the small portion of the surface near the intersection. In a two-dimensional drawing, the normal line is perpendicular to the line representing a plane surface in cross-section. In the

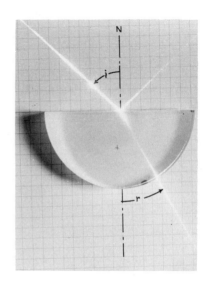

Figure 12–2 Demonstration of the refraction of a light beam as it passes from air to glass. The angles of incidence, *i*, and refraction, *r*, are indicated.

case of a curved surface, the normal line is an extension of the radius line from the center of curvature to the point of interest.

In Figure 12–2 is a photograph of a thin beam of light incident on the plane surface of a hemispherical plate of glass. The line normal to the plane surface is shown as N. The angle to the incident ray is shown as i and the angle of refraction is shown as r. As always occurs at a boundary, you will note that there is also some reflected light. You might use a protractor to see whether the angle of reflection is the same as the angle of incidence.

12–2 THE LAWS OF REFRACTION

There are two laws of refraction, the first being that the incident ray, the normal line, and the refracted ray all lie in a plane. This law was first enunciated by the Arabian investigator Alhazen, about 1000 A.D. At that time Arabia was probably the world's leader in scientific investigation. Alhazen also tried to find the law relating the angles of incidence and refraction, but he did not succeed. This relation had been sought for many centuries. Ptolemy in the second century A.D. thought he had it, claiming that the ratio of the angles i and r is constant for two given media. In Table 12–1 are some of Ptolemy's data, with the ratio i/r shown. The ratio appears to be constant up to about 30°, but beyond that there is a significant deviation. Ptolemy's law, that i/r is constant, can be used for small angles; but in practice the limit is closer to 5° than 30°, and even then you must keep in mind that it is not exact.

Johannes Kepler, who found the laws describing planetary orbits, also tried to find the law of refraction but failed.

TABLE 12–1 Ptolemy's table of refractions for light going from air to glass. The first two columns are the data attributed to Ptolemy. The last two columns, the ratios of i to r and sin i to sin r, have been added. That sin i/sin r is constant is called Snell's law.

| PTOLEMY'S DATA | | | SNELL'S LAW |
Angle of Incidence, i	Angle of Refraction, r	i/r	$\frac{\sin i}{\sin r}$
10°	7°	1.43	1.42
20°	13½°	1.48	1.47
30°	20½°	1.46	1.43
40°	25°	1.60	1.52
50°	30°	1.67	1.53
60°	34½°	1.74	1.53
70°	38½°	1.82	1.51
80°	42°	1.90	1.47

It was Willebrord Snell, working in Holland in 1621, who found the solution. He discovered that it is *the ratio of the sines of the angles of incidence and refraction* that is constant. We still know this statement as Snell's law, and it has been one of the most useful scientific relations ever found. Even Ptolemy's data (Table 12–1) show that the ratio sin *i*/sin *r* is closer to being constant than is the ratio *i*/*r*. In fact, with modern equipment the ratio sin *i*/sin *r* can be shown to be constant with a very high degree of precision.

In Snell's day there was no explanation of why it was the ratio of the sines of the angles that was constant, and it was a long time before a satisfactory theory was developed. This depended on the consideration of light as a wave motion, and on the fact that light slows down as it enters a medium. Finally, in 1850, it was shown by Foucault, using apparatus of which a spinning mirror was a central object, that light did indeed travel more slowly in water than in air.

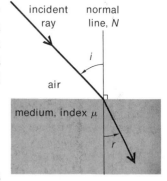

12–2–1 SNELL'S LAW

Snell discovered that when a ray of light passes from air to a medium, the ratio of the sine of the angle in air to the sine of the angle in the medium is a constant, and that the constant depends on the medium. This constant, a measure of the amount of bending or refraction, is now called the **index of refraction** of the medium. The symbol μ (lower case Greek mu) is often used for that index. Using *i* for the angle in air and *r* for the angle in the medium, **Snell's law** is

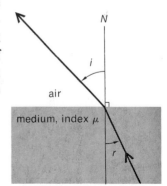

$$\mu = \frac{\sin i}{\sin r}$$

If a light ray is reversed, it still follows the same path, as shown in Figure 12–3. Then, the angle i is no longer the angle of incidence. In the form of Snell's law given above, *i* must be the angle in air and *r* the angle in the medium, no matter what the direction of the ray.

Figure 12–3 A light ray reversed will follow the same path that it came along; in Snell's law, sin *i*/sin *r* = μ, the angle *i* is measured in air no matter what the direction of the ray.

EXAMPLE 1

Find the index of refraction of water, given the experimental data that for an angle of incidence of 45° the angle of refraction is 32°.

Use sin *i*/sin *r* = μ:

$$\mu = \sin 45°/\sin 32°$$
$$= 0.7071/0.5299$$
$$= 1.334$$

The index of refraction of the water is 1.334.

EXAMPLE 2

If a ray of light is incident on a water surface at 70.5° from the normal line, what would be the angle of the ray entering the medium? Use $\mu = 1.334$.

Use sin i/sin $r = 1.334$ and $i = 70.5°$. Solving for sin r,

$$\sin r = \sin 70.5°/1.334$$
$$\sin r = 0.9426/1.334$$
$$\sin r = 0.7066$$
$$r = 45°$$

12-2-2 APPARENT DEPTH

When you look down into a medium, the depth of any object appears to be less than it really is. As you pour a clear liquid into a glass, the bottom appears to rise up. The end of a stick put obliquely into water does not appear as deep as it really is, so the stick appears to be bent. This phenomenon of the apparent depth not being the same as the real depth begs explanation. The explanation is based on Snell's law, and it will be carried out for several reasons. It illustrates an explanation of an otherwise puzzling phenomenon. The effect is made use of in some areas of science as well as in spear fishing, and the derivation of the explanation illustrates how approximations are used in physics.

The analysis of this phenomenon is most easily carried out for the case in which we look straight into the water (or other medium), so that the rays are close to the normal line. As shown in Figure 12-4(a), the rays diverging from the object in the medium appear to come from a position closer to the surface. The apparent depth of the object is not as great as the real depth.

The analysis of this phenomenon is carried out in the following way. Referring to Figure 12-4(b), the object is at O in the material of index μ. The real depth is d and the apparent depth is d'. By Snell's law, sin i/sin $r = \mu$, but we can, using the angles and distances shown in the diagram, write expressions for the tangents of these angles:

$$\tan i = x/d' \quad \text{and} \quad \tan r = x/d$$

The ratio tan i/tan r becomes (with the term x canceled):

$$\frac{\tan i}{\tan r} = \frac{d}{d'}$$

In Table 12-2 are some results of calculations based on a material of index 1.5. The ratio of tan i to tan r is, for small

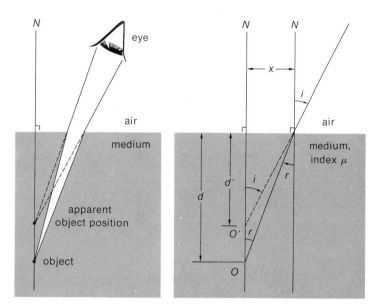

Figure 12–4 The apparent depth and real depth of an object.

angles, very close to the ratio of sin i to sin r. Therefore, for normal viewing:

$$\frac{\sin i}{\sin r} = \mu \approx \frac{d}{d'}$$

so

$$d \approx \mu d'$$

In words, this says that the real depth is greater than the apparent depth by a factor that is the index of refraction. This phenomenon is often seen in teacups and swimming pools: the depth does not seem to be as great as it really is. The relation is also sometimes used to measure an index of refraction.

The development of this expression, which does not contain the angles, shows that all rays near the normal line emanate from an apparent object position that is closer to the surface than is the real object.

TABLE 12–2 The ratios tan i/tan r and sin i/sin r for various values of i, based on a material of index 1.5.

i	tan i/tan r	sin i/sin r
1°	1.5001	1.5000
2°	1.5005	1.5000
3°	1.5011	1.5000
4°	1.5020	1.5000
5°	1.5032	1.5000
6°	1.5046	1.5000
10°	1.5130	1.5000

12–3   REFRACTION AND THE SPEED OF LIGHT

The index of refraction was historically first defined according to Snell's law. In Chapter 11 a quantity called the index of refraction was defined in terms of the ratio of the speed of light in a vacuum to the speed of light in a medium. This double definition should be justified. Also, it will be possible, using the definitions in terms of the speed of light, to obtain a very general form of Snell's law, considering light going from one medium to another, not just from air to a medium. In this work the index of refraction will be given by

$$\mu = c/v$$

from which the speed v in the medium in terms of the speed c in a vacuum is

$$v = c/\mu$$

12–3–1   THE GENERAL FORM OF SNELL'S LAW

Consider in Figure 12–5 a beam of light that is traveling in a medium of index μ_1 and is passing obliquely into another medium of index μ_2. The directions of propagation are perpendicular to the wavefronts. If the wave slows down as it enters the second medium, it will change direction as shown. This is explained by following the motion of one wavefront. In Figure 12–5 the wavefront AB is just entering the second medium at A. By the time the whole wave has entered the medium, the wavefront is in the position DE. The edge of the wavefront at B travels to E at speed v_1 in the same time that the edge at A travels to D at speed v_2 in the second medium. The

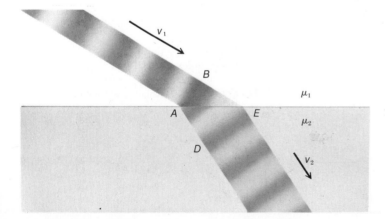

Figure 12–5 A light wave going from one medium into another.

directions of propagation are such that the angles ABE and ADE are right angles. This is because the direction of propagation in a medium is always perpendicular to the wavefronts. Figure 12–6 is a similar diagram with the angles of incidence and refraction shown.

It is necessary now to drop the use of i and r for designating the angles, not only because the rays can be reversed and i and r could be confused, but because in the form of Snell's law previously dealt with the angle i had to be measured in air. Now both angles will be in media and, in accord with the general usage of Greek letters for angles, they will be designated by the first two letters of the Greek alphabet, α and β (alpha and beta). The Greek letters are frequently used as general symbols for angles with the meaning defined in each situation, in a manner similar to the use of x and y.

The speed of light in each of the media shown in Figure 12–6 is given by $v_1 = c/\mu_1$ and $v_2 = c/\mu_2$. The time required to travel BE is given by $t = BE/v_1 = BE/(c/\mu_1)$, and the time required to go from A to D is given by $t = AD/v_2 = AD/(c/\mu_2)$. These times are the same, so

$$BE/(c/\mu_1) = AD/(c/\mu_2)$$

The c will cancel, and then we divide both sides by AE to obtain

$$\mu_1 \times BE/AE = \mu_2 \times AD/AE$$

Examination of the triangles in Figure 12–6 shows that

$$BE/AE = \sin \alpha$$

and $$AD/AE = \sin \beta$$

Figure 12–6 Diagram to assist in the analysis of the phenomenon of refraction.

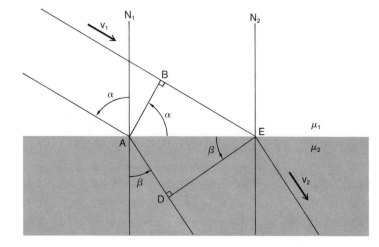

so

$$\mu_1 \sin \alpha = \mu_2 \sin \beta$$

This is the very general expression of Snell's law; the product of the index and the sine of the angle in one medium is equal to the product of the index and the sine of the angle in the other medium.

12–3–2 INDICES OF REFRACTION

In Table 11–1 (page 327) are listed some indices of refraction. From examination of the table, some interesting observations may be made. The index of air is so close to that for a vacuum that we often refer to an index measured in air rather than in vacuum.

No material has an index less than unity, from which we deduce that in no medium does light travel faster than it does in a vacuum.

In some cases the measurement of index of refraction may be used in analysis. As an example, the indices of two pure alcohols are given. Also, the index of refraction of a sugar solution depends on the concentration and may be used to measure it.

For most substances the index varies with color. In all instances the index shown in Table 11–1 is for the yellow light given off by sodium flames or arcs, the wavelength being 589 nm.

EXAMPLE 3

Consider a light ray going from water to glass and then to air, as in Figure 12–7. Let the angle in the water be 32°; the indices of refraction are:

for water, $\mu_w = 1.334$
for glass, $\mu_g = 1.517$
for air, $\mu_a = 1.000$

At each boundary, the product of the index times the sine of the angle is the same on both sides. Using this we obtain

$$\mu_w \sin \alpha = \mu_g \sin \beta$$

and
$$\mu_g \sin \gamma = \mu_a \sin \delta$$

By geometry, $\gamma = \beta$, so using this and combining the equations leaves

$$\mu_w \sin \alpha = \mu_a \sin \delta$$

It is as though the glass were not there! This realization, that the angles of the light

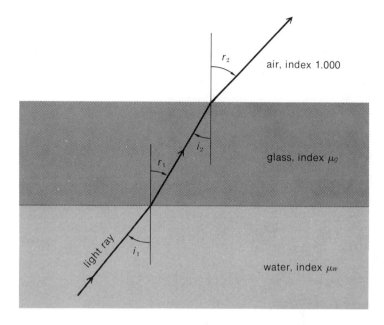

Figure 12–7 A ray going from water to glass to air.

air, index 1.000

glass, index μ_g

water, index μ_w

light ray

rays on the two sides of a parallel-sided sheet of material are the same as if that sheet was not there, can often save a lot of work.

Using the values

$$\sin \delta = 1.334 \sin 32°/1.000$$
$$= 0.7069$$
$$\delta = 45°$$

The angle in air will be 45° from the normal.

12–4 APPLICATIONS OF SNELL'S LAW

The general form of Snell's law is the starting point for the explanation of a lot of phenomena in the natural world, and it is utilized in the construction of scientific instruments. In the following subsections a few situations will be analyzed. These will show the wide importance of the law, they will be a demonstration of a scientific analysis or explanation, and they will show some phenomena that you may not have been aware of. There will be no need to go through all of them in detail.

12–4–1 TOTAL REFLECTION

You may have noticed occasionally that when you look at the bottom of an empty test tube it appears to have mercury in it — part of the bottom is reflecting light like mercury or a silvered surface. When viewed at certain angles, the side of a fish tank will be a mirror — nothing will be seen through it. An optical

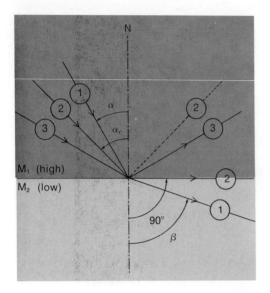

Figure 12-8 The phenomenon of total reflection. At the critical angle α_c, the emerging ray is $90°$ from the normal line. If α is greater than the critical angle, total reflection occurs.

instrument may seem to have mirrors in it, but on examination these are found to be just prisms with no silvered surfaces.

The phenomenon occurring in all of these cases is called total reflection. There is always some light reflected from a glass-air surface, but under certain conditions the reflection will be total, 100 per cent. This is even better than the best of metallic coatings.

Total reflection occurs only when a light ray strikes a boundary at which the index of refraction decreases. In that case, if α is in the higher index medium and β is in the lower index medium, β will always exceed α as shown in Figure 12-8. There will be a value of α for which β in the second medium is $90°$, and beyond that angle α the light cannot emerge into the second medium. There will be total reflection, and the laws of reflection will describe the situation.

EXAMPLE 4

Find the direction of a light ray in air if it is going from glass of index of refraction 1.50 to air. Use angles in the glass of $40°$, $41°$, $42°$, and $43°$.

Use

$$\mu_1 \sin \alpha = \mu_2 \sin \beta$$
$$\mu_1 = 1.50$$
$$\mu_2 = 1.00$$
$$\alpha = 40°, 41°, 42°, 43°$$

If $\alpha = 40°$, $\sin \beta = 1.50 \sin 40° = 0.964,$ $\beta = 74.6°$
If $\alpha = 41°$, $\sin \beta = 1.50 \sin 41° = 0.984,$ $\beta = 79.8°$
If $\alpha = 42°$, $\sin \beta = 1.50 \sin 42° = 1.004,$ β does not exist.
If $\alpha = 43°$, $\sin \beta = 1.50 \sin 43° = 1.023,$ β does not exist.

Figure 12-9 shows how the value of the sine of angles beyond $90°$ is less than one. Angles for which the sine is greater than one do not exist.

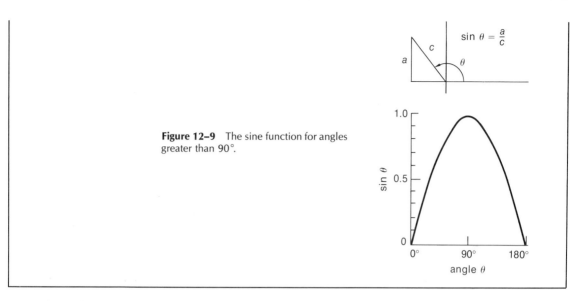

Figure 12–9 The sine function for angles greater than 90°.

The critical angle for any two media can be calculated by setting $\beta = 90°$; then $\sin \beta = 1.000$. The critical angle will be designated by α_c. Using

$$\mu_1 \sin \alpha_c = \mu_2 \sin 90°$$
$$\sin \alpha_c = \mu_2/\mu_1$$

EXAMPLE 5

Find the critical angle for a light ray going from glass of $\mu = 1.50$ to air.

Use
$$\sin \alpha_c = \mu_2/\mu_1$$
$$\sin \alpha_c = 1/1.50 = 0.667$$
$$\alpha_c = 41.8°$$

The fact that the critical angle for glass-air surfaces is less than 45° allows many practical uses of the phenomenon. If light enters a 45°-45°-90° prism as in Figure 12–10, the angle of incidence on the sloping surface exceeds the critical angle, so total reflection occurs. With the prism oriented as in Figure 12–10(a) the ray is reflected through 90°; in (b) the direction of

Figure 12–10 Prisms for total reflection.

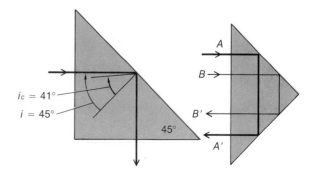

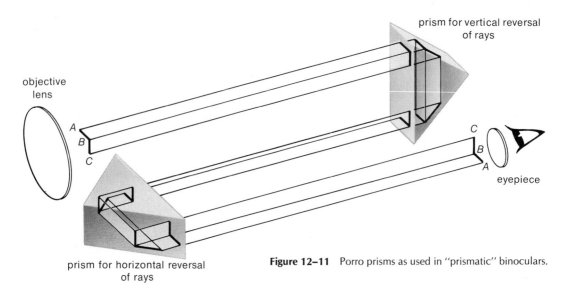

objective
lens

prism for vertical reversal
of rays

eyepiece

prism for horizontal reversal
of rays

Figure 12–11 Porro prisms as used in "prismatic" binoculars.

the ray is reversed and also shifted sideways. Optical instruments that make use of reflection will generally use this phenomenon of total reflection rather than silvered or aluminized surfaces.

Binoculars, often called **prismatic binoculars,** make use of totally reflecting prisms to obtain a long light path in a physically short space. Each half of the binoculars uses two 45°-45°-90° prisms, called Porro prisms, as shown in Figure 12–11, which also shows how the long light path is obtained. Ordinarily, an image formed by a lens is inverted and turned right to left. The two Porro prisms in the binoculars are arranged so the image is made erect and also switched right to left again, to appear through the binoculars just as it would be seen with the unaided eye.

A further example of total reflection is the **light pipe.** The term "pipe" brings to mind a long, thin object with a hole through it. Water put in one end comes out the other end, no matter how the pipe twists and turns. A light pipe is a similar device, in that light put in one end comes out the other. There is no hole, but the material must be transparent. The light pipe depends for its operation on the phenomenon of total reflection.

If light enters a polished rod of glass or transparent plastic as in Figure 12–12, total reflection will occur at the points A, B, C, and so on, provided that the curvature of the rod is not too great. The light rays that entered the rod from the source S will emerge at F, so the rod has acted like a pipe for the light.

Because of the multiple reflections, an image cannot be passed through such a pipe. But an image can be broken into a series of fine dots of various shades of light and dark, and each

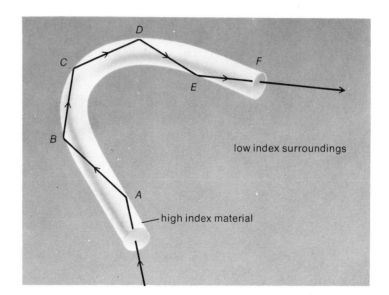

Figure 12-12 The principle of the light pipe.

portion of the image is then sent through a small light pipe (Figure 12-13). A bundle of glass fibers, each acting like a light pipe, will transmit an image if the arrangement of the emerging fibers is the same as that at the entrance end. A lens is used to cast an image on the end of the bundle, and the image is then transmitted to the other end. One necessary addition is that each fiber be coated with a material of low index so that the light in one fiber does not cross into adjacent ones at points where there is contact. Bundles in which the order of the fibers is not maintained will transmit light, though not images. They have many uses, from the illumination of places that are otherwise hard to reach to the transmission of signals or data.

The uses of this principle are so broad that a whole study area called "fiber optics" has been developed. One of the uses of such a device is in internal examination of equipment or of the human body. The fiber bundles are more flexible than systems of lenses sometimes used in instruments for similar purposes.

12-4-2 THE PRISM

A glass prism is frequently used in instruments for spectral analysis. If light consisting of several colors is put through a prism, the colors are separated and made to go in different directions.

The prism was the first device used to produce spectra. Sir Isaac Newton in 1666 wrote in a letter to the Royal Society: "I procured me a glass prism to study the celebrated phenomenon of colors." The phenomenon was well known (celebrated)

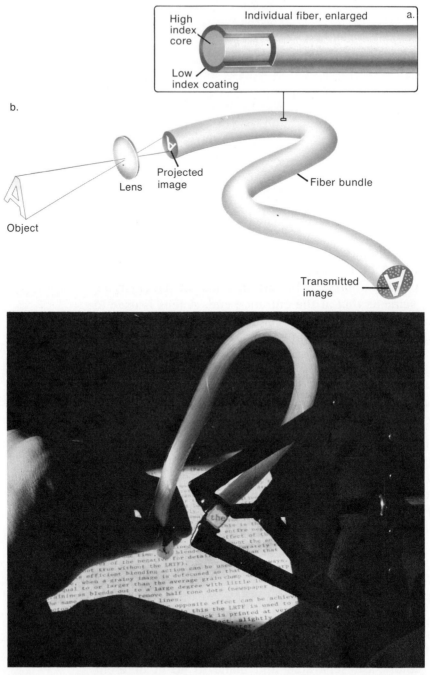

Figure 12–13 A bundle of very thin fibers, each acting as a light pipe, will transmit an image if the arrangement of the fibers is the same at both ends. (Photo by M. Velvick, Regina.)

then, though not at all understood. Prisms are used even today in many modern instruments because they are so simple. Granted, they do not spread the spectrum out as widely as do some other devices, such as the diffraction grating, but nothing else will concentrate all the incident light energy into one spectrum. The prism gives one bright though narrow spectrum, and it still has enough uses to make it worthwhile to give it further study.

To produce a spectrum, the prism is placed in the light beam oriented as in Figure 12–14. The light is bent away, or deviated, from its original direction, the amount of the deviation depending, in part, on the prism angle shown as A in Figure 12–14. Snell's law describes the amount of bending at each surface.

The total angle of deviation is the sum of the angles through which the ray bends at each surface, D_1 and D_2 of Figure 12–14. D_1 is given by $\alpha_1 - \beta_1$ and D_2 by $\alpha_2 - \beta_2$. Then the total deviation D, which is $D_1 + D_2$, is equal to $\alpha_1 + \alpha_2 - (\beta_1 + \beta_2)$. Also, by using elementary theorems of geometry, it can be shown that $\beta_1 + \beta_2$ is equal to the prism angle A, so

$$D = \alpha_1 + \alpha_2 - A$$

or

$$\alpha_1 + \alpha_2 = D + A$$

It is left as an exercise to write Snell's law for the angles at each surface and then to relate the quantities D, A, α_1, α_2, β_1, and β_2 in one equation. A further key in the analysis is the general trigonometric relation that for two angles a and b,

$$\sin a + \sin b = 2 \sin \frac{(a + b)}{2} \cos \frac{(a - b)}{2}$$

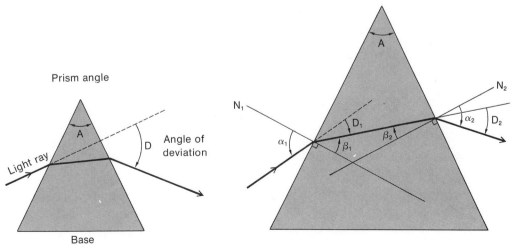

Figure 12–14 The deviation of light by a prism.

Write this equation substituting $a = \alpha_1$ and $b = \alpha_2$, and then write it again with $a = \beta_1$ and $b = \beta_2$. Apply Snell's law and a bit of algebra to get the solution

$$\sin\left[\frac{(A + D)}{2}\right] = \mu \sin (A/2) \frac{\cos\left[\frac{(\beta_1 - \beta_2)}{2}\right]}{\cos\left[\frac{(\alpha_1 - \alpha_2)}{2}\right]}$$

The quantity on the left varies with the amount of deviation, which depends on all the quantities on the right. For a given prism, A is fixed and the deviation depends only on the angle of incidence α_1, which of course determines the other angles, β_1, β_2, and α_2. The values of β are always less than the corresponding values of α, and $(\beta_1 - \beta_2)/2$ is always less than $(\alpha_1 - \alpha_2)/2$. This means that $\cos\left[(\beta_1 - \beta_2)/2\right]$ is always greater than $\cos\left[(\alpha_1 - \alpha_2)/2\right]$, because as the angle increases the cosine decreases (for angles between zero and 90°). It follows that the ratio of those cosines is always greater than one. There is one exception, for $\beta_1 = \beta_2$, and therefore $\alpha_1 = \alpha_2$. The differences are zero and the cosines are each one. For this special situation, for which the ratio of those cosines takes its least possible value, the angle of deviation D is a minimum. For this particular condition, for which the ray goes symmetrically through the prism, the relation then becomes

$$\sin \frac{(A + D)}{2} = \mu \sin A/2$$

The amount of deviation depends on the index of refraction, μ, and we know that different colors are bent by different amounts by a prism (that is why it produces a spectrum). From this we deduce that the index of refraction varies with color. Measuring the angle of deviation by a prism is a method used to determine indices of refraction very precisely.

EXAMPLE 6

Find the minimum angle of deviation of light by a prism with a prism angle of 60.00° and index of refraction of 1.615.

Use $$\sin \frac{(A + D)}{2} = \mu \sin \frac{A}{2}$$

D cannot be solved for directly, but it is done in steps.

$$A = 60.00°, \quad A/2 = 30.00°, \quad \mu = 1.615$$

$$\sin \frac{(A + D)}{2} = 1.615 \sin 30.00°$$

$$\sin \frac{(A + D)}{2} = 0.8075$$

$$\frac{(A + D)}{2} = 53.85°$$

$$A + D = 107.70°$$

$$D = 107.70° - 60.00°$$

$$= 47.70°$$

The minimum angle of deviation is 47.70°.

12-4-3   THE RAINBOW

When the sun is shining and rain has just passed over, you turn your back to the sun and ahead of you is the colored arc, the rainbow. It is more than just a bow, for you will note too that inside the bow the sky is somewhat white while immediately outside it is dark. The bow is really a white disc increasing in brightness toward the outside, and the edge of the disc is colored. The bow is as shown in Figure 12–15(a).

The rainbow results from sunlight being reflected back from water drops suspended in the air, and to explain its characteristics the reflection of light from spherical drops will be analyzed using the laws of refraction and reflection.

A ray of light is traced through a drop in Figure 12–15(b). The sunlight hits all over the drop, and a ray striking the surface at an angle α is shown. The observer sees the emerging ray at the angle D from the direction of the ray from the sun. The ray is changed in direction at each of the points A, B, and C. At B there is only partial reflection. By the law of reflection, at B, both angles shown are β. The changes in direction of the ray are:

at A, $\alpha - \beta$
at B, $180° - 2\beta$
at C, $\alpha - \beta$

The total deviation is $180° + 2\alpha - 4\beta$. This can be written as $180° - (4\beta - 2\alpha)$. The deviation is less than $180°$ by the amount shown as D in the Figure 12–15 and

$$D = 4\beta - 2\alpha$$

All values of α from $0°$ to $90°$ occur, so the corresponding

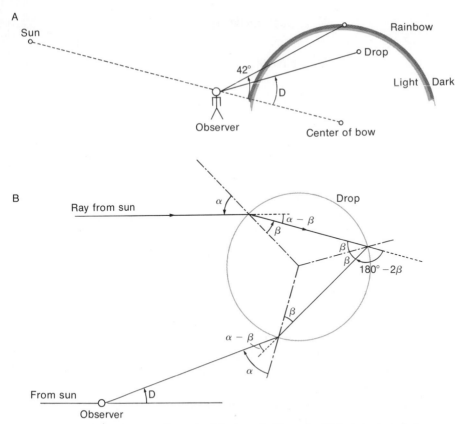

Figure 12–15 The rainbow. (a) The angle of the bow. (b) The path of light through a raindrop.

values of D can be found. α and β are related by $\sin \alpha = \mu \sin \beta$. Some calculations of D for a range of values of α are shown in Table 12–3. These were calculated for an index of refraction of 1.333, which is for yellow light in water.

From Table 12–3 two things will be noticed. First, reflected light will be seen from $D = 0°$ to a maximum of about 42°, but beyond 42° there is no light reflected back from the drops. There is also a second limit at 14.4°. The light between 14.4° and 42.1° results from two values of α. Inside 14.4° the intensity drops because only one value of α contributes. The limits of 42.1° for the main bow and 14.4° for the inner one depend on the index of refraction. The index varies with color. Red light will form a disc out to about 42.4°, and violet out to only 41.5°. The inside of the bow is white because all colors are seen there, but the different colors stretch out different distances leading to the colored edge of the disc — the rainbow.

A larger secondary bow results from the part of the light that undergoes more than one reflection inside the drop. The small 14° bow is most commonly seen from aircraft.

TABLE 12–3   The direction of light reflected from a raindrop.

α	D
0°	0°
10°	9.9°
20°	19.5°
30°	28.1°
40°	35.3°
45°	38.1°
50°	40.3°
55°	41.7°
60°	42.1°
65°	41.3°
70°	39.3°
75°	35.7°
80°	30.5°
85°	23.4°
90°	14.4°

12–4–4   FOCUSING BY A CURVED SURFACE

Curved surfaces are common in nature. The front surface of the human eye is curved outward like a portion of a sphere. Lenses have curved surfaces, and they are important in image formation.

In this analysis a curved surface in air will be dealt with. In this case the simple form of Snell's law will be used, with the angle of incidence i, and the angle of refraction r. The index of the medium will be μ and the surface will be considered to be a portion of a sphere of radius R.

Such a curved surface will be shown to have a focusing property like a lens, for rays diverging from an object point will be brought to a focus as in Figure 12–16(a).

A ray of light through such a surface is shown in Figure 12–16(b). The angles of incidence, i, and of refraction, r, are measured from a radius line that goes through the center of curvature, C. Snell's law is used in the form

$$\sin i = \mu \sin r.$$

The angles α, β, and γ as shown along the axis are used in the analysis. On examining the figure, it is seen that $i = (\alpha + \beta)$ and $r = (\beta - \gamma)$. If we assume that all the angles are small, then the sine of the angle is equal to the angle (in radians). If this is not clear, do Problem 13 of Chapter 4. Then the above form of Snell's law adapted to this situation reduces to

$$(\alpha + \beta) = \mu(\beta - \gamma)$$

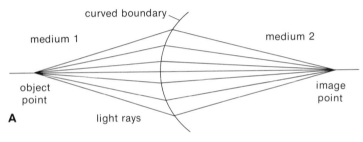

A

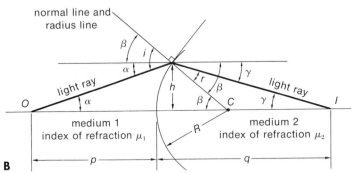

B

Figure 12–16 Focusing by a single curved surface. (a) Overall view. (b) Single ray.

The angles α, β, and γ in radian measure (the angles must again be recognized to be small) are given by h/p, h/R, and h/q respectively. The last equation then becomes, with the h canceled,

$$[(1/p) + (1/R)] = \mu[(1/R) - (1/q)]$$

This can be manipulated into the form

$$(1/p) + (\mu/q) = (\mu - 1)/R$$

The fact that h, the distance of the ray from the axis, canceled out means that all the rays emanating from the point at the distance p and hitting the surface at any distance h from the axis are focused to the same point at the distance q. Thus, an object point at O is focused at the position I.

If the object is a long distance away so that $1/p$ approaches zero, the rays approaching the surface will be parallel, and the distance q can then be called the focal length, f, of the surface. The equation then becomes

$$\mu/f = (\mu - 1)/R$$

from which the focal length is

$$f = \frac{\mu R}{(\mu - 1)}$$

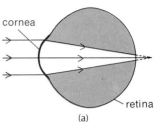

(a)

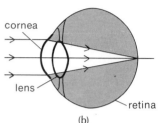

(b)

Figure 12–17 (a) Focusing by cornea alone. (b) Focusing by cornea and lens.

An example of a single surface used for focusing is the cornea of the human eye. The radius of curvature of the front surface of the cornea is about 7.7 mm or 0.0077 meter. The index of refraction of the human eye is 1.336 (μ in the above equation). The focal length of the cornea alone is given from $f = \mu R / (\mu - 1)$, where $\mu = 1.336$ and $R = 7.7$ mm. Putting these numbers into the equation yields a value of $f = 30$ mm. This is the distance from the cornea to the position at which a distant object would be focused. The normal length of the eyeball is about 24 mm. If there were no lens, the cornea alone would focus an image only 6 mm behind the retina, as illustrated in Figure 12–17(a).

The cornea alone bends the light rays just about enough to form an image on the retina. It is apparently the main image-producing surface, and the lens of the eye must be used for only minor focusing as in Figure 12–17(b). This is one of the surprising things that we learned by the application of physical principles, and numbers, to a biological situation.

The image position would vary with object distance if the cornea alone did the focusing, so there are two functions for the lens. One is to provide the extra focusing to put the image onto the retina, and the other is to adjust to focus objects at different distances sharply onto the retina. This is accomplished by change of the curvature of the lens surfaces, principally the posterior (rear) surface, and the accommodation for distance is normally such that objects from about 25 cm to infinity can be brought into focus. Unfortunately, the lens often loses its flexibility with age and this ability to accommodate is reduced.

12–5 LENSES

Without the invention of the lens and of combinations of lenses, science would have had to develop without the visual evidence of bacteria, and without the awareness of the real nature of other planets, of the Milky Way or of other galaxies. Instruments with lenses are among the most useful devices in sciences.

In this study of lenses, ray optics will be used to show image formation and the types of images formed with a single lens. Then instruments using more than one lens will be described, specifically the telescope and the compound microscope.

12–5–1 FOCAL LENGTH

You are probably familiar with the focusing properties of lenses. The common converging lens (thicker in the middle than at the edges) focuses a parallel beam of light to a point at a distance from the lens known as the **focal length,** f, as in Figure 12–18(a). This point is called the **principal focus,** and often just the **focal point.** The focal point or principal focus is also frequently indicated in diagrams by f. There is a focal point on each side of the lens. You may also, from past experience, be familiar with the phenomenon that if the source of light is at the focal point, the rays going through the lens are bent into a parallel beam; light rays have the property of following the same path when they are reversed. The bending at each surface is according to Snell's law. The focal length of such a converging lens is given a positive value.

A lens that is thinner in the middle than at the edges diverges the light as though it came from a point on the opposite side of the lens, as shown in Figure 12–18(b). Such a diverging lens is said to have a negative focal length.

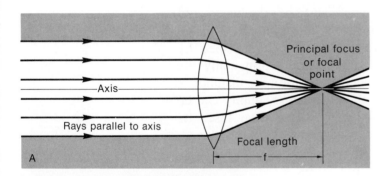

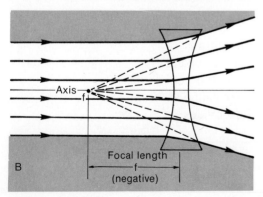

Figure 12–18 Focusing of parallel rays by a converging lens and by a diverging lens.

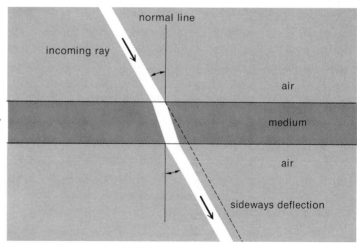

Figure 12-19 The small deflection of a ray passing through a parallel-sided medium.

12-5-2 PRINCIPAL RAYS

A lens focuses all the rays coming from any object point to a point where an image is formed. This was shown in Section 12-4-4 for a single curved surface, but the situation is similar for a double-surfaced lens.

The problem in the study of image-object relations is to be able to plot the path of some rays through a lens. One of these rays is the one that passes through the center of the lens. At the center, the two sides of the lens are effectively parallel. The case of a ray going through a parallel-sided sheet of material is illustrated in Figure 12-19. The ray is bent on entering and again on leaving. The final direction is the same as the initial direction, but the ray is shifted sideways. If the material is thin, this sideways shift is small. An example of this is that when an object is seen through a window, its direction is not changed. This phenomenon is used in considering the ray that passes through the center of a lens, at which point the two sides of the lens are parallel. Such a ray is undeviated; if we consider a thin lens, the ray through the central part may be considered merely to travel in a straight line. Our treatment of lenses will in most cases consider them to be very thin.

To assist in image location there are three rays, called principal rays, that can easily be drawn. These are shown in Figure 12-20. Of all the rays emanating from point A on the object, we choose to follow just three:

1. The ray parallel to the axis, which is deviated to go through the focal point on the far side as in Figure 12-18.
2. The ray through the center of the lens, which is undeviated.
3. The ray through the near focal point, which is deviated to emerge from the lens parallel to the axis.

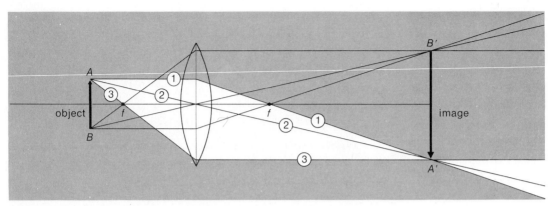

Figure 12–20   A principal ray diagram.

These rays that originate at A converge to the same point marked A' and diverge again from A'. If your eye was just beyond A', the object would be seen at the position A' in space. Only two of these principal rays would have been sufficient to locate the image position.

A similar set of rays has been shown diverging from point B and meeting at point B'. The points between A and B are focused at corresponding points between A' and B'. An image would then be formed between A' and B'. This is called a **real** image, because it can be seen on a screen placed at the position at which the light is focused.

Principal ray diagrams can be drawn to scale to locate an image position, given the object position. Even a sketch to an approximate scale will give a reasonable idea of where to expect the image.

The image shown in Figure 12–20 is referred to as a real image because the rays of light are actually focused to points in space at the image location. Such an image is produced on a movie screen or with a slide projector. A camera lens produces a real image on the film, and the effect of the light on the material of the film results in the developable picture.

EXAMPLE 7

An object is placed 3.30 cm from a lens of focal length 2.00 cm. Use a principal ray diagram to find the image position and its size compared to the object size.

The steps in making a principal ray diagram are shown in Figure 12–21. In part (a) of the figure the given data are transferred to the diagram: the object of arbitrary size is shown at 3.3 cm, and a focal point is shown on each side. In part (b) one of the principal rays is drawn, and a second principal ray has been added in part (c). The image is formed at the distance from the lens at which the rays cross. The image is shown in part (d). It is measured to be 5.3 cm from the lens, and it is 1.6 times as large as the object; this is what was asked for.

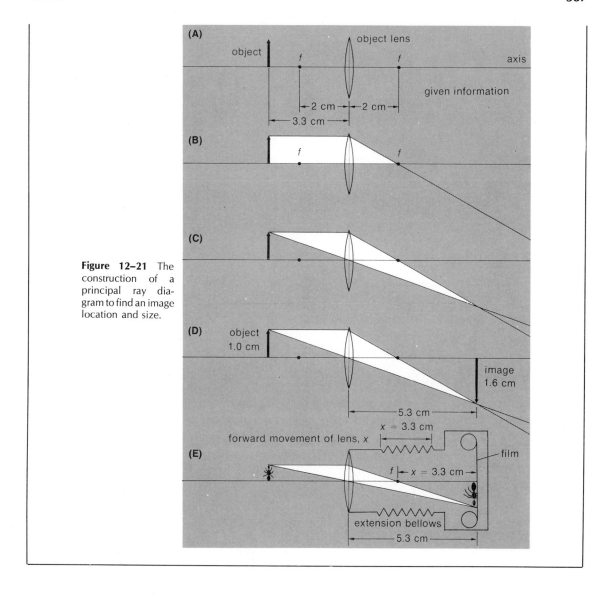

Figure 12–21 The construction of a principal ray diagram to find an image location and size.

12–5–3 IMAGE-OBJECT POSITIONS

A principal ray diagram is not always satisfactory for finding an image position. Diagrams are very limited in precision, and you might also have the problem that the object is 50 meters away and your biggest piece of paper is only 30 cm long. You might use a scale drawing, but if the focal length of the lens was only 10 mm you would be in trouble. An algebraic formula would solve such problems, and that will now be developed.

To do this, consider the ray diagrams shown in Figure 12–22. There are two pairs of similar triangles indicated by shading in opposite directions.

The triangle with base p and height O is similar to the one with base q and height I. They are both right-angle triangles

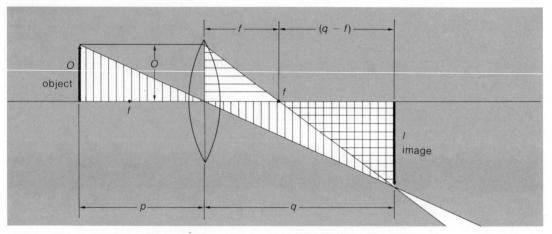

Figure 12-22 A principal ray diagram used to find the relation among image distance, object distance, and focal length.

with one of the other angles, the angle at the center of the lens, also equal. Therefore, the ratio of corresponding sides is the same so

$$I/O = q/p$$

The other pair of triangles have bases f and $(q-f)$. The equal angles occur on opposite sides of the point marked f. Again using the equality of the ratio of corresponding sides,

$$I/O = (q-f)/f$$

These two expressions for I/O can be equated to give

$$q/p = \frac{q-f}{f}$$

or $\qquad q/p = (q/f) - 1$

Divide each term by q to get

$$1/p = (1/f) - (1/q)$$

and add $1/q$ to each side to obtain

$$(1/p) + (1/q) = 1/f$$

Note that q is positive when it is measured on the side opposite the object.

This is a fundamental and useful equation relating image and object distances.

EXAMPLE 8

Solve Example 7 of Section 12–5–2 using the equation rather than a scale drawing. The given information is that $f = 2.00$ cm and the object distance $p = 3.30$ cm. Solve the expression $(1/p) + (1/q) = 1/f$ for q:

$$1/q = (1/f) - (1/p)$$

Bring the right-hand side to a low common denominator.

$$1/q = \frac{p - f}{fp}$$

Invert this:

$$q = \frac{fp}{p - f}$$

Substitute the values of f and p, including the units.

$$q = \frac{2.00 \text{ cm} \times 3.30 \text{ cm}}{(3.30 - 2.0) \text{ cm}}$$

$$= 5.08 \text{ cm}$$

The image position is 5.08 cm from the lens. The precision exceeds that obtainable with a scale drawing.

EXAMPLE 9

A camera lens of focal length 5.00 cm is used to photograph an object that is at a distance of 2 meters. Find the image distance. The relation to be used has been obtained in Example 8.

$$q = \frac{fp}{p - f}$$

where

$$f = 5.00 \text{ cm}$$

$$p = 200 \text{ cm}$$

$$q = \frac{5.00 \times 200 \text{ cm}}{200 - 5.00 \text{ cm}}$$

$$= \frac{5.00 \times 200 \text{ cm}}{195 \text{ cm}}$$

$$= 5.128 \text{ cm}$$

The image is formed at a distance of 5.128 cm from the lens or 0.128 cm outside the focal point. Note that in this case a scale diagram would have been of no value. Note also that in order to obtain a sharp photograph of an object at 2 meters using a camera with a 5 cm lens, the lens-to-film distance should be increased by 0.128 cm over that used for photographing distant objects. Cameras usually have a method to move the lens forward for photographing nearby objects. The closer the object, the farther the lens must be moved.

12–5–4 MAGNIFICATION

Magnification is defined as the ratio of image size, I, to object size, O. To show how magnification is related to object and image positions, reference can be made to Figure 12–23. This diagram is similar to Figure 12–20 but only one of the principal rays is shown, the one through the center of the lens. The two shaded triangles of Figure 12–23 can be seen by examination to be similar, so the ratio of corresponding sides is the same. That is, the ratio I/O is the same as the ratio q/p. To summarize:

By definition: $M = I/O$

For a single thin lens:

$$M = I/O = q/p$$

In words, the magnification (which is defined as the ratio of image size to object size) is, in the case of a single thin lens, also described by the ratio of image distance to object distance.

There is another expression for magnification that can be very helpful in the use of a camera as a scientific instrument for the measurement of small objects. If a camera is set to photograph distant objects, the lens-to-film distance is the same as the focal length of the lens, f. To bring close objects into focus, the lens is moved forward by an amount we call x, so the lens-to-film distance, or image distance q, is just $x + f$.

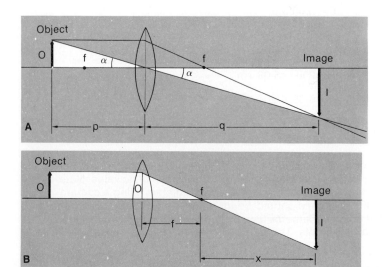

Figure 12–23 Diagrams used to analyze for magnification.

Then from Figure 12–23(b), $O/f = I/x$ by the similar triangles shown. This can be manipulated to give I/O or M as

$$M = \frac{I}{O} = \frac{x}{f}$$

In using a camera the distance x, the amount the lens is moved forward to bring the object into focus, can be easily measured and f is usually marked on the lens. Then from the measured image size on the film, the object size O can be calculated.

You will note that magnification is *defined as* I/O but in different cases other expressions can be used to *evaluate* magnification, such as q/p or x/f.

EXAMPLE 10

In photographing a small object with a camera having a focal length of 50 mm, the lens was moved forward 35 mm from the position at which very distant objects would be in focus. The size of the image on the film was 1.3 mm. What was the size of the object?

Use the relation $I/O = x/f$

and solve for O to get

$$O = \frac{f}{x} \times I$$

where $x = 35$ mm
                                            $f = 50$ mm
                                            $I = 1.3$ mm

Then $O = \dfrac{50 \text{ mm}}{35 \text{ mm}} \times 1.3 \text{ mm}$

                                            $= 1.9$ mm

You will note that this procedure does not involve measurement of an object distance. This is convenient because modern camera lenses are of such a thickness that the point from which to measure object distances is by no means clear; but the measurement of the movement of a lens, the distance x, can be obtained very precisely.

12–5–5 VIRTUAL IMAGES AND THE SIMPLE MICROSCOPE

If the object is brought closer to the lens than the focal length, then p is less than f: when you solve for q, you now find that it has a negative value. This situation is illustrated by the principal ray diagram in Figure 12–24, in which p is put equal to 2/3 of f. The light rays that go through the lens do not converge, but appear to have come from a point behind the lens, as shown by the dotted lines. The point at which these rays

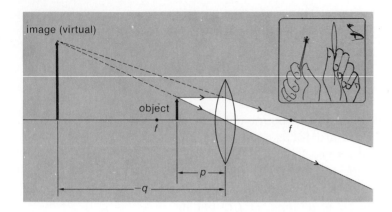

image (virtual)

object

f f

p

$-q$

Figure 12–24 The formation of a virtual image when the object is inside the focal point.

appear to have originated is the image location. It is not a real image, it cannot be put on a screen, but it can be seen by looking through the lens. Such an image is said to be **virtual.** This situation occurs when a lens is used as a magnifier to examine a small object. When the lens is used with the eye as close to the lens as possible, it is referred to as a simple microscope. It was with a lens used as a simple microscope that Antony Von Leeuwenhoek first saw protozoa and bacteria. This was quite an accomplishment!

Objects look larger when seen through a lens in this way, but the magnification as given by image size divided by object size does not describe this "enlarging" effect. The larger the image, the farther away it is, so the effect of greater magnification is canceled. In this situation we use the concept of magnifying power. This concept will be dealt with so that it can be applied to lenses and instruments.

EXAMPLE 11

Use the lens equations to find the image distance when an object is at 2/3 of the focal length from a lens.

Use $1/p + 1/q = 1/f$. Solve for q:

$$q = fp/(p - f)$$

Substitute

$$p = 2f/3$$

$$q = \frac{2f \times f}{3(\frac{2}{3}f - f)} = \frac{2f^2}{-f}$$

$$= -2f$$

The image will be at a distance of twice the focal length, and because the distance is negative the image is on the same side as the object. It will be a virtual image.

12–5–6 MAGNIFYING POWER OF A LENS

Magnifying power is an entirely different concept from magnification. Magnification, the ratio of image size to object size, is easily and precisely measurable with a scale. Magnifying power, on the other hand, is the ratio of the apparent size of an object seen with an instrument to the apparent size seen without the instrument. The problem is in the definition of **apparent size.**

Apparent size depends on distance. The closer an object is to the eye, the larger it appears. Of course, if you get the object too close your eye cannot focus on it. Apparent size could be expressed as the size of the image on the retina, but this would be a little difficult to measure. However, the size of the image on the retina depends on the angular size of the object, as shown by θ in Figure 12–25, and this can be measured easily. **The magnifying power of an instrument is defined as the ratio of the angular size of an object viewed with the instrument to the angular size of the object viewed without the instrument.** Figure 12–25 shows that the size of the image on the retina depends on the distance between object and viewer and is directly related to the angular size shown as θ.

The concept of angular size is probably new to you, but it is simply the angle between lines drawn from the eye to the two ends of the object. A small coin a meter away may have an angular size of about 1°; a person at 20 meters is about 5°, and at 15 meters is about 7°.

The radian has been introduced as a unit for the measure of an angle (Chapter 4, Section 4–2), and it is particularly convenient for the measurement of small angles such as are often encountered in optics. Small angles are frequently expressed in minutes or seconds of arc, and the relation between these and radians is worth developing. In converting between radians and

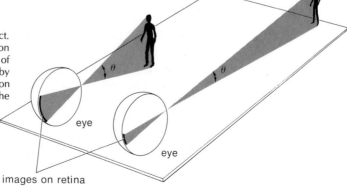

Figure 12–25 The angular size of an object. The size of the image on the retina depends on the angular size shown as θ. The angular size of the distant person is 10°, and that of the nearby person is 30°. The image of the closer person on the retina is three times as large as that of the farther person.

eye

eye

images on retina

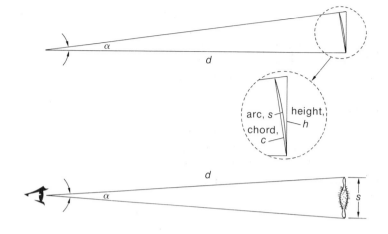

Figure 12–26 Radian measure for small angles. For a small angle, as in (a), the arc s, the chord c, and the vertical height h are almost identical. The angle α is given equally well by s/d, h/d, or c/d. In (b) the angular size of the object from the viewing position is α = s/d.

degrees, the relation was that $360° = 2\pi$ radians or $1° = 0.01745$ rad. A minute is a sixtieth of a degree or

$$1' = 0.000291 \text{ rad}$$

An arc second is a sixtieth of a minute or

$$1'' = 0.00000485 \text{ rad}$$

These conversion factors may be remembered easily and very closely (with an error of 3%) by:

> 1 minute equals point three zeros three radians
> 1 second equals point five zeros five radians

In working with lenses, small angles are commonly encountered; indeed, if the angular size of the object is not small, a lens is not needed to see it. For a small angle as in Figure 12–26(a), the arc length s, the chord c, and the vertical height h are almost identical. Though the angle in radians is defined by the arc length s divided by the radius, the error caused by using the ratio of the chord to the radius is less than a tenth of one per cent for angles up to 6°, while the ratio h/r has a tenth of one per cent error at 5°. Thus, the ratios s/r, c/r, and h/r are extremely close for small angles. If an object of size s is viewed from a distance d as in Figure 12–26(b), then the angular size θ is just s/d. That is, **for small angles the angular size of an object in radians is given by object size divided by distance.**

EXAMPLE 12

Find the angular size of the mm division on a scale when it is viewed from distances of 25 cm and 3 meters. Express the results in radians and in minutes of arc.
In radians, $\theta = s/d$

For the first part of the problem, $s = 1$ mm and $d = 250$ mm.

$$\theta = 1 \text{ mm}/250 \text{ mm}$$
$$= 0.004 \text{ radian}$$
$$= 0.004/0.0003 \text{ minutes}$$
$$= 13 \text{ minutes}$$

For the second part, $s = 1$ mm and $d = 3000$ mm.
$$\theta = 1 \text{ mm}/3000 \text{ mm}$$
$$= 0.00033 \text{ radian}$$
$$= 1.1 \text{ minutes}$$

The answers are that when viewed from 25 cm the mm marks have an angular size of 0.004 radian or 13 minutes of angle. When viewed from 3 meters the angular size is 0.00033 radian, or just over 1 minute of angle.
This example raises some questions. Is it possible to see the 1 mm marks on a scale from 3 meters away? Try it! If you can't see them at 3 meters, what is the maximum distance for you? What is the smallest angular separation of objects that you can see? What are the physical reasons for the limit to sharpness of vision?

Angular size is a function of distance, and in order to express the magnifying power of an instrument or a lens a standard distance for viewing with the unaided eye is chosen. These days 25 cm is used as the average near point or nearest distance of unstrained focusing for normal eyes.

Now we are ready to find the magnifying power of a lens held close to the eye.

Referring to Figure 12–27, which is similar to Figure 12–24 but shows only one principal ray, the magnifying power MP is the ratio of the angles β and α. The angular size α is the object

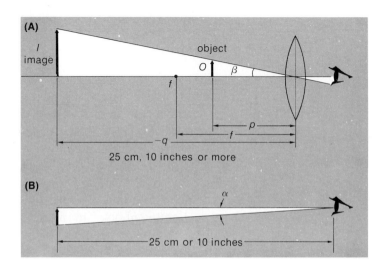

Figure 12–27 The magnifying power obtained with a lens used as a simple microscope. (a) The angular size with the lens, and (b) the angular size without the lens.

size O divided by 25 cm. The angle measured in this way is expressed in radians. The angular size of the image when seen with the lens is β, and is given by the ratio I/q (as well as by O/p). The angle in the diagram is admittedly not small; the ratio given is, in fact, the tangent. The angle shown is large only so it could be satisfactorily drawn. In practice, if the angles involved were not small, a lens would not be necessary!

The magnifying power β/α is then given by

$$\text{MP} = \frac{O/p}{O/25 \text{ cm}} = \frac{25 \text{ cm}}{p}$$

The object size O cancels. To put this in terms of focal length, find $1/p$ from the lens equation,

$$1/p = (1/f) - (1/q)$$

and then

$$\text{MP} = 25 \text{ cm}\,[(1/f) - (1/q)]$$
$$\text{MP} = (25 \text{ cm}/f) - (25 \text{ cm}/q)$$

The magnifying power does in fact depend on how far away the final image is. This distance q can range between minus infinity and the nearest distance of clear vision, which is on the average 25 cm. So q can range between $-\infty$ and -25 cm. Then

for $q = -\infty$,
$$\text{MP} = \frac{25 \text{ cm}}{f}$$

for $q = -25$ cm,
$$\text{MP} = \frac{25 \text{ cm}}{f} + 1$$

That wide range in position of the final image leads to a change of only one in the value of the magnifying power. When a lens is rated in terms of magnifying power, it is by convention that the rated MP of a lens is based on 25 cm/f. The MP obtainable can range from the rated value to the rated value plus one.

The shorter the focal length of a lens, the greater its magnifying power. A 5 cm focal length lens will have a magnifying power of 5, and a 2.5 cm lens will magnify objects 10 times.

Leeuwenhoek, in about 1676, achieved his remarkable observations of bacteria principally by using extremely short focal length lenses. Those which he presented to the Royal

Society were found to have focal lengths ranging between 1.3 mm and 5 mm. One writer, in describing some of Leeuwenhoek's microscopes, says:

> Among these lenses there are three made from so-called "Amersfoort diamond" (rock-crystal pebble) [quartz]; and of one of the microscopes it is noted that its magnifying glass is ground from a sand grain. . . .[1]

No one since Leeuwenhoek has been able, using a simple microscope only, to see the detail that he described. Unfortunately, he described his policy thus: "My method of seeing the very smallest animalcules, and the little eels, I do not impart to others; nor yet that for seeing very many animalcules all at once; but I keep that for myself alone."[2]

12-5-7 THE COMPOUND MICROSCOPE

The compound microscope as used today consists of two basic lens elements. They may each be actually a cluster of small lenses, but each element acts like a single lens. At the bottom of the microscope tube is what is called the objective lens. The objective lens is placed with the object just outside the lower focal point. A real and enlarged image is then formed above it near the top of the microscope tube, as in Figure 12–28(a). This already enlarged image is viewed with an eye lens used as a simple microscope, as in Figure 12–28(b). The complete microscope is shown in Figure 12–28(c).

The eye lens of Figure 12–28 will usually be one of two or more in a short tube at the top of the microscope. The whole combination is referred to as either an ocular or an eyepiece. A telescope also uses an ocular or eyepiece. The eye lens of the ocular is positioned to produce all or almost all of the magnifying power of the system.

In the development of the compound microscope, many varieties were tried. The two-lens system currently used was probably devised in the first decade of the seventeenth century. It is the only basic form to have survived the centuries, though improvements have been rather great. A modern compound microscope compares to a seventeenth century microscope in about the same way that a modern automobile compares to Daimler's original of 1886.

If the eye lens or ocular was used alone to examine the object, the magnifying power would be that given in Section

[1]Clifford Dobell: *Antony Von Leeuwenhoek and his 'Little Animals,'* Russell & Russell, New York, 1958, p. 328.
[2]Ibid., p. 144.

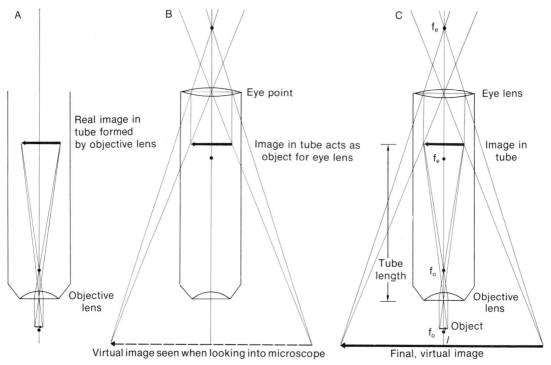

A

Real image in
tube formed
by objective lens

Objective
lens

B

Eye point

Image in tube acts as
object for eye lens

Virtual image seen when looking into microscope

C

f_e

Eye lens

Image in
tube

Tube
length

f_e

f_o

Objective
lens

Object

f_o

Final, virtual image

Figure 12–28   Construction of a ray diagram for a compound miscroscope.

12–5–6. Writing M_e for the MP of the eye lens and f_e for its focal length, the magnifying power would be just

$$M_e = (25 \text{ cm}/f_e) + 1$$

but the eye lens is used to examine an image that is larger than the object by the magnification of the objective lens. This can be described by image distance over object distance. The image distance in a microscope is usually fixed, and a measuring scale or cross hair is often put at the image position. This fixed length is often called the tube length, T, and T may be used in place of q for the image distance.

Rather than using an object distance, p, which is difficult to measure in a microscope, the magnification by the objective lens is usually expressed in terms of the focal length of the objective, f_o, which is of course a fixed quantity for a given lens. To show how this is done, use the lens equation

$$\frac{1}{p} + \frac{1}{q} = \frac{1}{f}$$

Multiply through by q to get

$$\frac{q}{p} + 1 = \frac{q}{f}$$

Set $q/p =$ the magnification of the objective lens, M_o:

$$M_o + 1 = \frac{q}{f}$$

Replace q by the tube length, T, and f by the focal length of the objective lens, f_o, and solve for M_o:

$$M_o = \frac{T}{f_o} - 1$$

This is the expression for the magnification by a microscope objective lens. A standard tube length has been adopted, 16 cm, but it is by no means universal. For a standard microscope the magnification by the objective lens is

$$M_o = \frac{16 \text{ cm}}{f_o} - 1$$

The magnifying power, MP, of the complete microscope is then the product $MP_e M_o$:

$$MP = \left(\frac{25 \text{ cm}}{f_e} + 1\right)\left(\frac{16 \text{ cm}}{f_o} - 1\right)$$

For example, if the eyepiece or ocular alone would magnify an object by 10 times, but the "object" being examined is really an image already magnified 30 times, then each part of the final image would appear 300 times larger than if the object was examined with the naked eye.

The formula for the magnifying power is usually expressed without the "ones," that is, by convention

$$MP = (16 \text{ cm}/f_o)(25 \text{ cm}/f_e)$$

The magnifying power marked on the objective lens is based on $16 \text{ cm}/f_o$, and the power of the ocular is calculated and marked on the basis of $25 \text{ cm}/f_e$. If it seems too approximate to neglect the "ones," consider an example of using a 10 power objective and a 10 power eyepiece. Multiplying these gives a total MP of 100. If we carry out the calculation not neglecting the "ones," we get $MP = (10 - 1)(10 + 1) = 99$. When you are looking into a microscope you will not notice the difference between a magnification of 99 and one of 100.

12–5–8   THE REFRACTING TELESCOPE

The refracting telescope uses lenses throughout, whereas reflecting telescopes use at least one image-forming mirror. There are telescopes that give an inverted image. These are usually used in astronomy because the inversion is not troublesome (it makes little difference if the star is seen upside down). There are also telescopes that have extra lenses or prisms (see Figure 12–11) and are convenient for terrestrial use. The astronomic type is basic, so it is the one that will be analyzed.

At one end of the telescope tube is an objective lens, which forms a real image of the object observed near the far end of the tube. This real image is examined with an eyepiece that is used with the eye close to it. It is used like a simple microscope to examine that image.

The magnifying power of a telescope expressed in the basic form is the angular size of the object when seen without the telescope (α of Figure 12–29(b)) to the angular size when seen with the telescope. It is usual when working with a telescope to have the virtual image seen in the eyepiece a long distance away. The "object" being examined by the eyepiece (the image in the tube) is then at the focal point of the eyepiece, f_e. If the image size is I, the angle β in Figure 12–29 is I/f_e.

The object being observed is at a very long distance (Figure 12–29 is not to scale), so the real image is formed at the focal distance f_o from the objective lens. The angle shown as α in Figure 12–29 is given by I/f_o. The angle α is shown outside the telescope lens, too, and it is the same as the angular size of

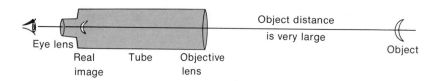

Figure 12–29 The refracting telescope.

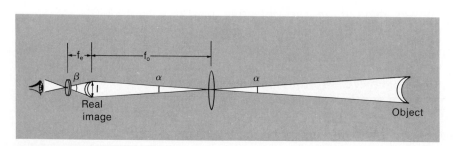

the object when seen with the naked eye. The magnifying power of the whole instrument is β/α, where

$$MP = \beta/\alpha = \frac{l/f_e}{l/f_o}$$

This can be manipulated to the form

$$MP = f_o/f_e$$

The magnifying power of a telescope is just the ratio of the focal length of the objective lens to the focal length of the eyepiece. The length of the telescope is just slightly more than the focal length of the objective lens, so the longer the telescope the higher the magnifying power possible. Telescopes frequently have interchangeable eyepieces, and the MP obtained will depend on the eyepiece used. For example, a certain telescope may have an objective lens of $f_o = 900$ mm and the available eyepieces may range from 25 mm to 6 mm in focal length. The magnifying power available will then range from 900 mm/25 mm or 36 times to 900 mm/6 mm or 150 times.

The longer the focal length of an objective lens, the larger the image in the tube will be; but also, the larger the image, the greater the area over which the light is spread. Put another way, the larger the image, the fainter it will be. To obtain a brighter image the objective lens would have to be increased in size. The light gathering ability (related to the diameter of the objective lens) of a telescope can be as important as the magnifying power.

12-5-9 FOCUSING WITH MIRRORS

Most of the world's large optical telescopes use a curved mirror rather than a lens as an objective. Curved reflecting surfaces also will focus radio waves. Radio telescopes and microwave communication systems use curved mirrors. Even sound can be focused by a curved mirror, and a curved reflector with a microphone at the focal point is often used in the recording of distant sounds such as bird calls.

The type of mirror to be analyzed will be that which is a portion of the surface of a sphere. In practice the mirrors used are sometimes paraboloids, but the spherical mirror shows the basic principles. Real images are formed by concave mirrors, so these will be analyzed.

In Figure 12–30, M is the cross section of a concave mirror that has a center of curvature at C. A ray of light is shown approaching the mirror, and an axis is drawn through the

A

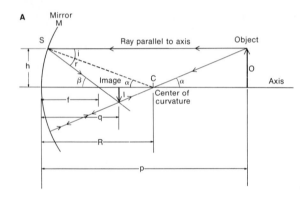

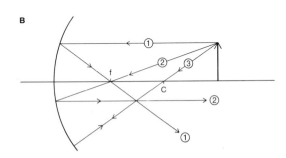

Figure 12–30 A principal ray diagram showing image formation by a concave mirror.

center parallel to this ray. The ray strikes the mirror at S, where the normal line is the line through the center. The angles of incidence, i, and of reflection, r, are equal by the law of reflection. The angle shown as α is equal to i and the angle β is equal to $i + r$ or 2α.

The ray cuts the axis at a distance from the mirror which is the focal length. The distance of S from the axis is h. If the angles are small, then in radians $\alpha = \dfrac{h}{R}$, and $\beta = \dfrac{h}{f} = 2\alpha$ or $\dfrac{h}{2f} = \alpha$. Then $\dfrac{h}{2f} = \dfrac{h}{R}$. The distance h cancels and leaves

$$f = R/2.$$

Parallel rays at any distance from the axis are focused at the distance f, the focal length. Also, the focal length is half the radius of curvature.

To analyze for image formation by a mirror, principal rays may be used. Three are shown in Figure 12–30(b). They are:
1. A ray parallel to the axis, reflected to go through the focal point.
2. A ray through the focal point, reflected to go parallel to the axis.
3. A ray through the center of curvature, reflected back along itself.

Using similar triangles in the same way as was done for lenses, it is easy to show that for a mirror

$$\frac{1}{p} + \frac{1}{q} = \frac{1}{f}$$

This is the same as for a lens but p and q, the object and image distances, are positive when they are in front of the mirror.

If the mirror is so wide that angles are not small, then the focusing will be spoiled. This phenomenon is called **spherical aberration.** The spherical aberration is corrected by using a non-spherical mirror. For parallel rays from distant objects, a parabolic mirror is used to correctly focus rays that enter parallel to the axis of the parabola. There is still some difficulty with off-axis rays. For diverging rays from a nearby object, the best focusing is obtained with a mirror that is an ellipsoid.

Mirrors have the advantage that all colors of light are reflected in the same way, while with lenses different colors are focused to different positions. Multiple lens systems consisting of different kinds of glass — achromats or apochromats — correct for this to a large extent.

DISCUSSION PROBLEMS

12-2 1. When Snell's law is applied to Ptolemy's data, as shown in Table 12–1, the values of sin i/sin r are not all the same (constant), though there does not seem to be the regular change that occurs in the ratio i/r. Even if the measured numbers obtained result in a constant, perhaps measuring to one more significant figure would show a variation. The topic to discuss is whether or not experimental data can ever show a law to be exactly true. In other words, can a law be proved by experiment, or can it only be shown to describe the data within a certain limit of precision?

12-2-2 2. In looking from air at an object in water, the distance to the object appears to be less than it really is. In looking from water to air, what apparent distance effect might you expect? Analyze it in a manner similar to the apparent depth analysis in the text. If you have the opportunity to wear goggles when you swim, observe the effect.

12-4-1 3. The index of refraction of the hot air immediately above a roadway or desert sand on a hot sunny day has a lower index of refraction than the cooler air above it. You may have noticed that at certain times the road or sand well ahead of you reflects sky light and appears like water. Explain the physics of the phenomenon.

12-4-2 4. If the front surface of the cornea of your eye is the main focusing mechanism, one would expect that with water next to the eye it would not be possible to focus. Test this, perhaps next time you go swimming. Discuss the necessity of goggles for seeing under water. Is it necessary that the glass surface of the goggles be flat? Fish see under water; how must their eyes differ from ours?

12-4-3 5. Make a careful drawing of a rainbow from observation to show the various colors as you see them, not as you expect them. Use colored pencils or crayons if possible. You may have to wait for rain, or you can produce a rainbow with a fine spray of water and a bright light or the sun behind you.

12-5-6 6. Angular sizes can be very useful, as in the measurement or estimation of distances to objects of known dimensions. In this way the distance of a person or of a plane, or of the height of an eagle, could be found. The accuracy in the distance would be the same as that in the measurement of the angle. In speaking of the separation of stars seen in the sky, or the size of a UFO in the sky, only the angular size can be used. Design an instrument or a device to help you measure angular size.

PROBLEMS

12-2   1. The finding of a law from experimental data is an important aspect of science. The data below represent measured values of the angle of incidence for which light reflected from a surface is completely polarized. At other angles of incidence there is only partial polarization. The angle for complete polarization seems to be related to the index of refraction.

Use the data below as well as thought and trial and error to discover the relation between θ and μ.

SUBSTANCE	ANGLE, θ, FOR COMPLETE POLARIZA- TION	INDEX OF REFRACTION, μ
water	53.1°	1.333
fluorite	55.1°	1.434
light crown glass	56.6°	1.517
heavy flint glass	58.7°	1.647
diamond	67.5°	2.417

12-2-1   2. For a light ray incident on the surface of a medium at 30° from the normal line, calculate the angle of refraction if that medium is: (a) water, $\mu = 1.333$; (b) diamond, $\mu = 2.417$.

12-2-1   3. This problem may not be practical, but it is a challenge. Find an angle of incidence i for which the angle of refraction r is just one half of i. Consider a ray going from air into glass for which $\mu = 1.60$.

12-2-2   4. The index of refraction of a piece of glass is to be measured, using the apparent depth method. A caliper is used to find the real thickness, which is 3.18 mm. A microscope focused through the glass onto a mark on the far surface reads 13.11 mm on a vertical scale as in Figure 12-31, and 14.70 mm when focused on the near surface. What are the apparent thickness and the index of refraction?

12-2-2   5. A fish looks vertically from water to air and sees an apparent distance effect somewhat like the apparent depth effect we see when looking the other way. Make a ray drawing or sketch to show this effect, and then analyze it starting with Snell's law to find how the apparent distance above the water surface is related to the real distance.

12-3   6. Find the speed of light in water of index 1.333. The speed in air is 3.00×10^8 m/s.

12-3   7. Find the speed of light in glass of $\mu = 1.50$ if the speed in air is 3.00×10^8 m/s.

12-3-1   8. A ray of light going from water ($\mu = 1.33$) to glass ($\mu = 1.60$) has an angle of 60° from the normal line in the water. What is the direction in the glass?

12-3-1   9. (a) A ray of light in air is incident at 45° on the surface of some water. Find the angle of refraction in the water. Use $\mu = 1.33$.

(b) A ray of light is incident at 45° on the surface of oil, $\mu = 1.55$, which is floating on

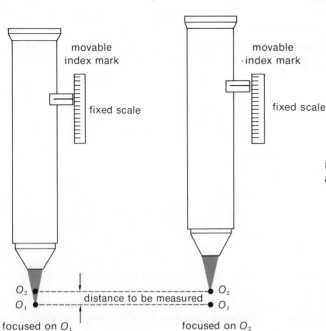

Figure 12-31 Measuring the apparent thickness of a piece of glass as described in Problem 4.

water. Find the angle of the ray from the normal line in the oil and in the water.

12-3-1   10. For silicate flint glass, the index of refraction for red light is 1.613 and that for blue light is 1.632. Find the angle between the red and the blue light in that glass when white light in air is incident on the glass surface at 60° from the normal line.

12-4-1   11. Calculate the critical angle for total reflection of a ray in glass, $\mu = 1.517$, in contact with water, $\mu = 1.333$.

12-4-1   12. What index of refraction will give a critical angle against air of 45°?

12-4-2   13. A ray of light is incident on a 60° prism at an angle of 45°. Make a diagram to show how the ray passes through the prism. Calculate the angle of deviation. The index of refraction of the glass is 1.673.

12-4-2   14. Calculate the minimum angle of deviation of a ray passing through a prism having an index of refraction 1.673. Make a sketch of such a ray. The prism angle is 60°.

12-4-2   15. (a) Show geometrically that the ray that suffers minimum deviation has an angle of refraction equal to half of the prism angle.

(b) What would be the angle of incidence for a ray striking a prism that has an angle of 60° if there is to be minimum deviation? The index is 1.673.

12-4-2   16. (a) Find the angle of minimum deviation of yellow light going through a 60° prism that is made of light crown glass ($\mu = 1.517$).

(b) Find the angle for a prism of heavy flint glass ($\mu = 1.647$) that would give the same minimum angle of deviation that was produced by the light crown glass prism of part (a).

12-4-2   17. In cold weather, ice crystals in the atmosphere result in a phenomenon somewhat like a rainbow. A ring forms around the sun with the brightness enhanced on either side of the sun; these bright areas are called sun dogs. The sun dogs result from light going at minimum deviation through the prism-like hexagonal ice crystals as in Figure 12-32. The crystal is like a 60° prism of ice.

(a) The inner, red edge of the dog occurs at 21.5° from the sun. What is the index of refraction of ice for red light?

(b) The blue occurs at 22.4°. What is the index for blue light?

12-4-3   18. A rainbow effect can be demonstrated using glass spheres rather than water drops. Find the angle to the bow formed by glass spheres of index 1.52.

12-4-4   19. What would be the index of refraction of the material of a sphere if light from a distant object is focused by one surface of the sphere onto the other surface?

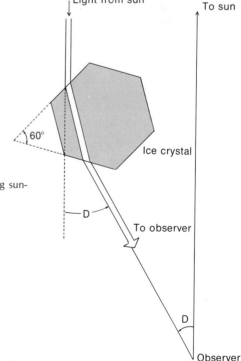

Figure 12-32 The path of light through an ice crystal in producing sun-dogs, described in Problem 17.

12–4–4   20. A solid clear plastic rod has one end ground to be a portion of the surface of a sphere 5.0 cm in radius. The index of the plastic is 1.50. A lamp is put 25 cm from the curved end. How far back in the rod will the rays from the lamp be brought to a focus?

12–5–2   21. Make principal ray diagrams to scale in order to find the distance from the lens to the image in each of the following cases:
(a) $f = 4$ cm, object distance $p = 14$ cm,
(b) $f = 4$ cm, object distance $p = 6$ cm,
(c) $f = 7$ cm, object distance $p = 14$ cm,
(d) $f = 4$ cm, object distance $p = 3$ cm.

12–5–2   22. Use a principal ray diagram to find the image position and size when an object 1 cm high is placed 4 cm from a lens of focal length 3 cm.

12–5–3   23. Solve for the image distance and the magnification in each situation of Problem 21, this time using formulae rather than scale drawings.

12–5–3   24. Solve Problem 22 using equations rather than a diagram.

12–5–4   25. A thin lens is used to cast a real image on a screen. The magnification is 3, and the separation of object and image is 1 meter.
(a) Where must the lens be?
(b) What is the focal length of the lens?

12–5–4   26. What must be the distance to a bird that is 10 cm long, in order to obtain an image 1 cm long if the camera has a lens of focal length 135 mm?

12–5–4   27. What is the focal length of a lens that casts an image at distance of 16 cm from it and magnified 25 times?

12–5–4   28. You wish to photograph a page that is 21 cm by 28 cm. Your camera has a 50 mm lens. The film is the standard 35 mm film, which gives an image size of 24 mm by 36 mm.
(a) Find the image and object distances needed to get the whole page on the film.
(b) Find the amount by which the lens must be moved forward from its standard position, a distance f from the film.

12–5–5   29. An object is 9.0 mm from a lens that has a focal length of 10.0 mm. Make a principal ray sketch to show the approximate position of the image, and then calculate the image position.

12–5–6   30. Two lines 3.6 mm apart are viewed from a distance of 6 meters. What is their angular separation (a) in radians and (b) in minutes of arc?

12–5–6   31. What is the angular size of a person 1.5 meters high, when viewed from (a) 3 meters, (b) 100 meters, (c) 1 kilometer?

12–5–6   32. A lens is held close to the eye to view a millimeter scale; the other eye looks at a wall 4 meters away. One millimeter seen through the lens is the same size as 20 cm viewed at a distance of 4 meters. What is the magnifying power of the lens? What is the focal length of the lens?

12–5–6   33. What must be the focal length of a lens for it to have a magnifying power of 20 when it is used as a simple microscope?

12–5–7   34. (a) What is the focal length of an objective lens marked $45\times$? (b) With that lens, what magnification would be obtained if the tube length were 24 cm?

12–5–7   35. A microscope is used with an 8 mm objective lens and an eyepiece marked $10\times$. The tube length is 16 cm. What magnifying power is obtained?

12–5–8   36. A telescope objective lens of focal length 1.22 meters is used to form an image of the moon in the telescope, and the image is measured to be 10.7 mm across. What is the angular size of the moon? Express the answer in degrees.

12–5–8   37. A telescope will usually have an objective lens of fixed focal length, but a variety of eyepieces may be used. Find the magnifying powers obtained when a telescope with an objective lens of 1.22 m focal length is used with an eyepiece of:
(a) 40 mm,
(b) 25 mm,
(c) 9 mm.

12–5–9   38. A concave mirror with a radius of curvature of 15 cm is used to make an image of an object that is 20 cm away.
(a) What is the focal length?
(b) Make a principal ray sketch to find the approximate image location.
(c) Calculate the image position.

12–5–9   39. A concave mirror with a radius of curvature of 4.0 cm forms an image of an object that is 2.1 cm away. (a) Where is the image? (b) What are the comparative sizes of the image and the object?

CHAPTER THIRTEEN

LIGHT: INTENSITY, ABSORPTION, AND SPEED

The topic of illumination is important in a society where people spend a lot of their working time indoors and where the length of the useful day is extended by "artificial" light. This area includes the topics of the units for the intensity of a light source and the illumination on surfaces.

Light intensity is decreased as it goes through an absorbing medium. The description of the decrease in light intensity with the thickness of a medium has some importance in chemical analysis by measurement of absorption. The light absorption process is very similar to that for the absorption of radiation from radioactive materials or of x-rays. The equations that describe the decay of radioactive material with time are the same as those that describe the change in light intensity with thickness. The general ideas in this section are very important in many subject areas to be encountered later, either in this book or in later experiences.

13–1 INTENSITY OF A LIGHT SOURCE

Though the old-fashioned candle (Figure 13–1) has been discarded as an illumination standard, it has not been forgotten. In times of power failure it is still a good friend. A more reproducible standard was devised and defined in such a way that the luminous intensity would be similar to that of a standard candle. The new intensity standard is based on the light emmission of a glowing surface of molten platinum. One international standard candle or **candela** of the new type is the luminous intensity of one sixtieth of a square centimeter of platinum at its fusion point (1769°C). The standard candle consists of a spe-

Figure 13-1 The original standard of illumination, now discarded.

cially constructed furnace with a conical hole at the top, as in Figure 13-2. At the bottom of the hole, where the area is just one sixtieth of a square centimeter, is platinum at the fusion point. At this temperature the glow produced is not unlike that of a single candle.

The intensity of a source of light does not refer to the total amount of light given off by the source. A candle emits light almost evenly in all directions, as shown in Figure 13-3, but the new standard candle emits light only into a small cone. The intensity of the new source in candelas was chosen to give approximately the same intensity of illumination on a surface at a given distance as would an old standard candle. We would say that the two sources emit the same "amount of light" or "light flux" into cones of similar size. What is needed to complete the definition of the intensity of a light source, and to allow us to deal with illumination, is a way of measuring the three-dimensional angle at the apex of a solid cone. *Stereo* is a prefix meaning "solid" or "three-dimensional," and the unit for such an

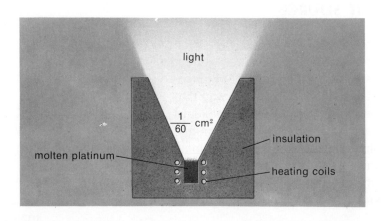

Figure 13-2 The modern "standard candle." The surface of the melted platinum has an intensity of 1 candela.

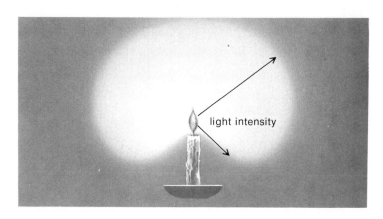

Figure 13–3 The variation in light given off in different directions by a source. This is described by specifying the amount of light flux going into a cone of a given size. The new standard candle (Fig. 13–2) emits almost the same amount of light into a given cone size as does the old candle.

angle is called a **steradian.** The unit used for the "amount of light flowing" or "light flux" is called a **lumen.** The symbol for lumen is lm. The intensity of a source in a given direction is measured by the number of lumens emitted into a solid angle of one steradian in that direction. A source of **one candela emits one lumen per steradian.**

Now, in order to use this quantity, the definition of the solid angle in steradians will have to be investigated.

13–1–1 SOLID ANGLES

To introduce the measurement of solid angles, consider small angles, so that in drawing arcs the curved arc length and the straight chord will be effectively equal.

In Figure 13–4(a) is shown a small solid angle designated by the lower case omega (ω). The cone is "rectangular"; at a radius r, chords or arcs have been drawn that are of the lengths Δy and Δx. The symbol Δ is the Greek capital delta, and here has the meaning "a small increment in." The size of ω depends on both of the angles shown as $\Delta\theta_y$ and $\Delta\theta_x$. If either of these angles is increased, ω will also be increased. In radian measure, $\Delta\theta_y = \Delta y/r$ and $\Delta\theta_x = \Delta x/r$. In Figure 13–4(b), $\Delta\theta_y$ has been doubled and the solid angle has also been doubled. In Figure 13–4(c), both $\Delta\theta_x$ and $\Delta\theta_y$ are doubled, and the solid angle is four times as big as in (a). It is evident that ω is proportional to the product of $\Delta\theta_y$ and $\Delta\theta_x$, or

$$\omega \propto \Delta\theta_y\Delta\theta_x$$

$$\omega \propto \frac{\Delta y}{r} \times \frac{\Delta x}{r} = \frac{\Delta A}{r^2}$$

The quantity $\Delta y\Delta x$ is the small area ΔA, which is the base of the cone measured "normally" to the radius lines. The solid angle

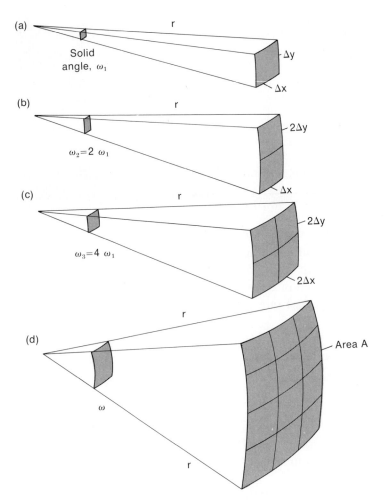

Figure 13–4 In (a) is a small solid angle in a rectangular cone with a base Δx by Δy. In (b), one side of the base has been increased to $2\Delta y$, and the solid angle in the apex of the cone has doubled. In (c), the area of the base is multipled by 4 and so is the solid angle at the apex. In (d), the large solid angle is given by the curved area of the base, A, divided by r^2.

is proportional to the small area at the bottom of those small cones divided by the square of the radius. For a constant solid angle, the area increases as the square of the radius; the ratio does not depend on the radius chosen, but only on the size of the solid angle:

if $$\omega \propto \Delta A / r^2$$

then $$\omega = k \Delta A / r^2$$

where k is a proportionality constant.

The unit for description of a solid angle is defined so that the constant k is unity; and in this unit, the steradian, we define

$$\omega = \frac{\Delta A}{r^2}$$

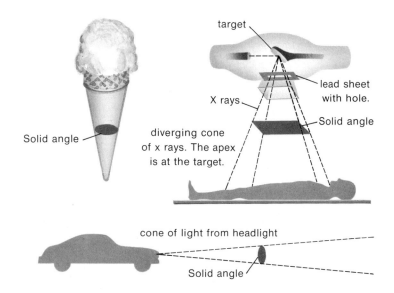

Figure 13–5 Examples of solid angles.

A large solid angle is made up of a lot of small solid angles, and the size of the large angle in Figure 13–4(d) is the total area on the curved surface, shown as A, divided by the square of the radius:

$$\omega = \frac{A}{r^2}$$

One steradian (also called one **unit solid angle**), the symbol for which is sr, is the solid angle for which the ratio $A/r^2 = 1$. For example, if $r = 6$ cm and $A = 36$ cm², then $A/r^2 = 1$.

Examples of solid angles are found in ice cream cones, diverging light beams, and diverging x-ray beams, as shown in Figure 13–5.

In working with small solid angles, the curved area (Figure 13–4(d)) to be evaluated can be approximated by the flat area formed by the chords, which is easily measured.

EXAMPLE 1

Find the solid angle at the apex of a cone that has a circular base 5 cm in diameter and sides 12 cm long. These are the approximate dimensions of an ice cream cone.

The area of the base is $\pi d^2/4$. This is the flat area, but it will be close to the curved area and the error will be recognized.

$$A = \pi \times (5 \text{ cm})^2/4$$
$$= 19.6 \text{ cm}^2$$
$$r = 12 \text{ cm}$$
$$\omega = A/r^2 = 19.6 \text{ cm}^2/12^2 \text{ cm}^2$$
$$= 0.14$$

The solid angle at the apex is 0.14 steradian or 0.14 sr.

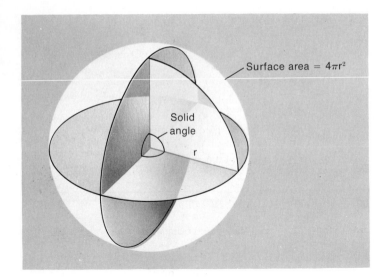

Figure 13–6 The solid angle in a whole sphere.

The largest possible solid angle is that in the center of a sphere, as shown in Figure 13–6. The curved area is the whole area of the surface of the sphere, which is $A = 4\pi r^2$. The solid angle ω is A/r^2, which is just 4π. That is, the solid angle in a sphere is 4π steradians.

The unit called the **square degree** is sometimes used. A square degree is a solid angle as in Figure 13–4(a), but with $\Delta\theta_x = \Delta\theta_y = 1°$. In radian measure, $1° = 0.01745$ radian, and a square degree is 0.01745^2 or 0.0003046 steradian. In a whole sphere, there are 4π steradians or $41,253$ square degrees.

13–1–2   LUMENS

The intensity of a light source measured in candelas is the same as the number of lumens going out into one steradian. The candela is defined in terms of the specific glowing platinum surface, and the lumen is defined in this manner: a source of brightness I candelas gives I lumens per unit solid angle or per steradian. In a solid angle of size ω, it gives $I\omega$ lumens.

EXAMPLE 2

Consider a small light source of 500 candelas (approximately 500 candlepower). At a distance of 1 meter, a hole with an area of 1 cm² is cut in a screen as in Figure 13–7. How many lumens go through the hole?

The solid angle ω is $1 \text{ cm}^2/(100 \text{ cm})^2$ or 10^{-4} steradians. The source is emitting 500 lumens per steradian, so in 10^{-4} steradians there are 500×10^{-4} lumens or 0.05 lumens.

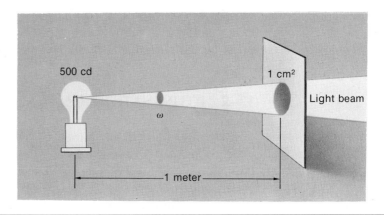

Figure 13–7 Given the source intensity, how many lumens of light flux go through the hole shown?

500 cd

1 cm²

Light beam

ω

1 meter

In general, $I = L/\omega$ or $L = I\omega$, where L is the number of lumens, I is the luminous intensity of the source in candelas, and ω is the solid angle.

In a whole sphere there are 4π unit solid angles or steradians. If a 1 candela source emits uniformly in all directions, giving 1 lumen per steradian, it emits a total of 4π or 12.57 lumens of light flux. Any real light source may have different luminous intensities in different directions.

Some representative source intensities are illustrated in Table 13–1.

The intensity of a source could also be described in terms of the rate at which energy is being radiated. The unit in this case would be the **watt per steradian.** Careful measurement has shown that for light emitted by a standard candle at 1769°C, one lumen is equivalent to 0.00146 watt or 1.46 milliwatts. One lumen per steradian is then 1.46 milliwatts per steradian. The temperature of the source is specified because the color varies with the temperature. A source at a lower temperature would be more reddish, while one at higher temperatures would be almost white; and for a spectral energy distribution differing greatly from the standard type, the unit of the lumen loses its meaning. On the other hand, the power radiated in watts/steradian would be a measurable quantity whatever the spectral distribution of energy, and even if the radiation were not in the visible spectrum range.

TABLE 13–1 Some representative source intensities in lumens per steradian or candelas

SOURCE	APPROXIMATE INTENSITY IN CANDELAS
40 watt light bulb	40
100 watt light bulb	130
40 watt fluorescent bulb	200
1000 watt street lamp	2500
firefly	Can you find out?

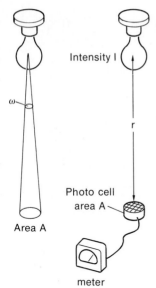

Figure 13–8 The intensity of illumination at a distance r from a source of intensity I is given by I/r^2.

If the source is not effectively a point, but perhaps in a long line like a fluorescent tube, the luminous intensity will often be expressed in lumens per meter of length of the source.

13–2 ILLUMINATION ON A SURFACE

Having defined the lumen of light flux, intensity of illumination on a surface is simply defined as the number of lumens falling on a unit area of surface. The units are **lumens per square meter,** and this term is also known by the name **lux.**

An old unit for illumination, the **foot candle,** was one lumen per square foot. One foot candle is equivalent to 10.8 lux (about 10). If an illumination is specified to be 100 foot candles, it would be about 1000 lux.

If L lumens fall on an area A, the illumination is simply L/A.

In terms of source intensity and distance, the illumination on a surface normal to the light beam is calculated quite simply for a small source. Let I be the intensity of a source in candelas and let the area be A at a distance r, as shown in Figure 13–8. The solid angle ω is given by A/r^2. The number of lumens in that solid angle is $I\omega = IA/r^2$, and this number of lumens is spread over the area A. The number of lumens per unit area is the illumination E, which is given by

$$E = IA/r^2A = I/r^2$$

This is the familiar **inverse square law** for a point source. For a source of intensity I in candelas, at a distance r in meters, the illumination in lumens per square meter or lux is simply I/r^2.

The expression given is the usual inverse square law, but the limitations for its application should be noted. First, the source must be very small, effectively a point compared with the distance r; second, the illuminated area must be "perpendicular" or normal to the direction of the light; and third, the light reflected onto the surface from walls and other surroundings must be negligible.

EXAMPLE 3

A light meter reading at 5.0 meters from a small source shows an illumination of 40 lux. What is the luminous intensity of the source in that direction?

Solve $E = I/r^2$ for I:

$$I = Er^2$$

where
$$E = 40 \text{ lux} = 40 \text{ lm/m}^2$$
$$r = 5.0 \text{ m}$$

$$I = 40\ \frac{lm}{m^2} \times 5.0^2\ m^2$$

$$= 1000\ lm$$

The unit of I has come out to be lumens; but luminous intensity, I, should be in lumens per steradian. The steradian is dimensionless, however. It is a unit of area divided by the square of a length. The steradian can then be inserted where it is appropriate, or even deleted where it is not.

The answer is that the luminous intensity of the source in that direction is 1000 lumens per steradian or 1000 candela.

If the source is not a point but a long line that has a luminous intensity of I candelas per unit length, the illumination on a surface at a perpendicular distance r is given by

$$E = 2I/r$$

The law is not inverse square for a long linear source (the length must be much longer than r), but merely an inverse law.

If the source of light is distributed evenly over an area that is large compared with the distance r, there will be no decrease in intensity with distance from the illuminating area. In terms of the luminous intensity I candelas per unit area, the illumination E is equal to I.

In practice, usually none of these limiting situations will apply perfectly, and it will often be necessary to measure the actual illumination rather than to calculate it. In any case, it is well to keep in mind that the common inverse square law for illumination has several limitations.

13-2-1 ANALYZING FOR A POWER LAW

In a real situation it is often satisfactory to use an inverse square law to calculate illumination, but there are also times when high precision is needed; and because there are factors that can cause deviations from an inverse square law, the actual law for that particular situation must be found. That is, there may be basically an inverse square law but also there may be a slight deviation from it. The reverse can happen in researching a new phenomenon. The actual law may be unknown and it must be found experimentally. In such a case, if the law is found to be described by a power close to -2, a basic inverse square law may be assumed and the reasons for the deviation sought. If the experimental result gives a power close to -1, a basic inverse law may be assumed and again the deviations from it sought.

Note that in an investigation such as this the scientist does not set out to *prove* that an inverse square law holds or does not hold. Rather, the object is to find what the actual power is. It may be -1.9, -2.1, -0.9, etc. Then the question is, "How do you go about solving for a power in a law?"

A power law would be described by an equation of the form

$$u = bv^n$$

One variable is u; perhaps it is intensity of illumination, in which case the symbol E may be substituted for u. The other variable is represented by v; perhaps it is a distance, so let it be r. The quantity b is just a constant, and n is the power sought. If an inverse square law applies, n will be -2. In this example the form of the equation will be written

$$E = br^n$$

To investigate the situation, a large number of readings will be obtained of the illumination E for as wide a range of r as is possible (or applicable). Some mathematical analysis will show how to solve for n.

Take the logarithm of both sides of the equation, and remember that $\log ab = \log a + \log b$. Also, $\log x^n = n \log x$. Then

$$\log E = \log b + n \log r$$

EXAMPLE 4

In a certain physical situation a meter was used to measure the illumination at various distances from a source. The data are reproduced below, with the distances in meters and the illumination in lux.

Distance r, meters	Illumination E, lux
1.0	20.5
2.0	5.1
3.0	2.5
4.0	1.35
5.0	1.00
6.0	0.70

The problem is to write the equation relating E and r. For illumination, a power law is expected; as suggested in the text, the logarithm of each quantity is found and then graphed. The logarithms follow (note the negative log value), and the graph is shown in Figure 13–9.

r	$\log r$	E	$\log E$
1.0	0.00	20.5	1.31
2.0	0.30	5.1	0.71
3.0	0.48	2.5	0.40
4.0	0.60	1.35	0.13
5.0	0.69	1.00	0.00
6.0	0.78	0.70	-0.15

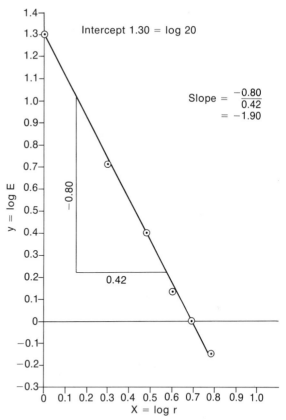

Figure 13–9 A plot of data concerning illumination and distance from the source. Both scales are logarithmic, and the result is that in this case the illumination varied inversely as the 1.9 power of the distance.

Intercept 1.30 = log 20

Slope $= \dfrac{-0.80}{0.42}$
$= -1.90$

$y = \log E$

-0.80

0.42

$X = \log r$

A slope triangle is shown on the line drawn among the points in Figure 13–9. The slope is −1.90, and this is the power n in the relation. The intercept is 1.30, which is the log of 20.0. The equation of the straight line is

$$y = 1.30 - 1.90\,x$$

or

$$\log E = \log 20.0 - 1.90 \log r$$

from which

$$E = 20.0\,r^{-1.90}$$

or

$$E = \frac{20.0}{r^{1.90}}$$

In this particular situation the power was measured experimentally to be −1.90. The deviation from −2.00 could be due to a factor such as reflected light or a source of large area.

The experimental readings will be the values of E and r. You can let the quantity log E be represented by y, and log r by x. Log b is the log of a constant, and that is just another constant; let it be a. Making these substitutions, the equation takes the form

$$y = a + nx$$

This is the familiar equation for a straight line. If a graph is made of y (or log E) against x (or log r), the experimental points would lie on a straight line or at least suggest a straight line. The slope of the line would be n, the power sought. The y intercept, a, would be the log of the constant b.

In Appendix 4 is a table of common logarithms. Instructions on the use of logarithms are included in Appendix 4.

13–2–2 OBLIQUE ILLUMINATION FROM A POINT SOURCE

If the light from a source falls obliquely onto a surface, it will be spread over a greater area than if the surface were normal to the beam. This is illustrated for a point source and a small area by Figure 13–10. The surface has been tipped to an angle θ away from the normal to the light beam. The solid angle is given by the normal area A' divided by r^2. The actual area shown is $A = \Delta x \Delta y$. The normal area can be seen, with reference to the diagram, to be given by $\Delta x \Delta y \cos \theta$ if Δy and Δx are small compared to r. The solid angle ω is A'/r^2 or $(\Delta x \Delta y \cos \theta)/r^2$. The number of lumens in this solid angle is $I\omega$ or

$$L = I\omega = I (\Delta x \Delta y \cos \theta)/r^2 = IA(\cos \theta)/r^2$$

The illumination on the actual area A is L/A or

$$E = L/A = I(\cos \theta)/r^2$$

The intensity of illumination varies as the cosine of the angle at which the surface is tipped away from the normal to the light beam. This is illustrated in Figure 13–11.

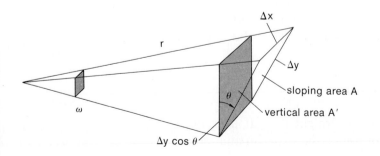

Figure 13–10 Finding a solid angle if the base is not normal to the axis of the small cone but has been tilted by an angle θ.

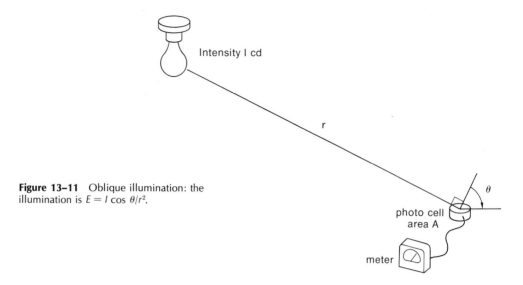

Figure 13–11   Oblique illumination: the illumination is $E = I \cos \theta / r^2$.

EXAMPLE 5

A table is to be illuminated by a single light of 100 cd over the center at a height of 1.5 meters. The table is 2 meters long. Find the illumination at the center of the table and also at the center of one end. The situation is shown in Figure 13–12.

(a) At the center the illumination is normal to the surface so:

$$E = I/r^2$$

where
$$I = 100 \text{ cd} = 100 \text{ lm/sr}$$
$$r = 1.5 \text{ m}$$

$$E = \frac{100 \text{ lm}}{1.5^2 \text{ m}^2 \text{ sr}}$$

$$= 44.4 \text{ lm/m}^2 = 44.4 \text{ lux}$$

(b) For the illumination at the end, first calculate the angle, which is shown as θ in Figure 13–11.

$$\tan \theta = 1 \text{ m/1.5 m} = 0.667$$
$$\theta = 33.7°$$
and
$$\cos \theta = 0.832$$

The square of the distance is

$$r^2 = (1.5^2 + 1^2) \text{ m}^2 = 3.25 \text{ m}^2$$

Use
$$E = \frac{I \cos \theta}{r^2}$$

$$= \frac{100 \text{ lm} \times 0.832}{3.25 \text{ m}^2}$$

$$= 25.6 \text{ lm/m}^2$$

(If r was solved for, it would be found as 1.80 m; then $\cos \theta = 1.5/1.8 = 0.833$. There would have been no need to calculate the angle.)

The answers are that at the center the illumination is 44.4 lux, while at the end it is only 25.6 lux.

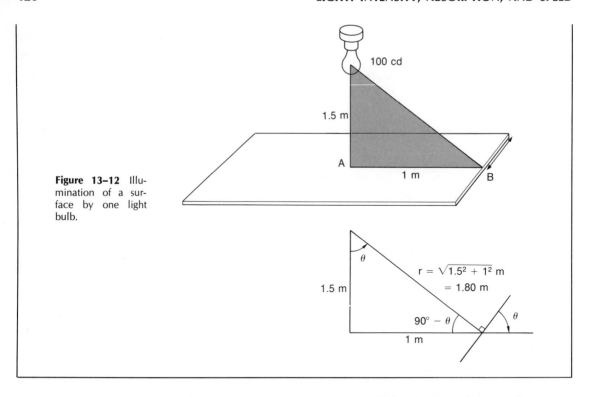

Figure 13-12 Illumination of a surface by one light bulb.

The phenomenon of oblique illumination is also important in considering the illumination and energy of sunlight incident on the ground at various latitudes. As one travels northward, especially in winter, the illumination on any horizontal surface is decreased considerably. Even in the summer, the energy of the sun is spread over a larger area in northern latitudes than it is near the equator, and the total energy available per unit area for plant growth is greatly reduced. At a latitude of 60° at noon in midsummer, the intensity of solar radiation on the ground is only one quarter of that at the equator. In midwinter, when the noon sun at 60° latitude is only 6½° above the southern horizon, illumination is reduced to 1.3 per cent of the value at the equator in summer. In summer, above the Arctic circle where the sun is up all day, the total energy per unit area per day actually exceeds that at the equator. Plant growth in an Arctic summer can be very rapid; but because of the cold soil layer just below the surface and the short growing season, the total vegetation per unit area is small compared with that in more southerly regions.

13-2-3 ILLUMINATION AND VISUAL ACUITY

Visual acuity can be described as sharpness of vision. It is a measure of the detail that can be seen. Visual acuity is measured by finding the separation of two objects that can just be seen as two. For example, two light sources may be seen as two when

TABLE 13–2 Illumination and comments on acuity in a variety of conditions

CONDITION	ILLUMINATION ON A SURFACE, lm/m^2	BRIGHTNESS OF A WHITE SURFACE, cd/m^2	SEEING CONDITIONS
very dark sky, no moon, heavy cloud	3×10^{-5}	10^{-5}	limit of vision
bright moon	10^{-1}	3×10^{-2}	acuity 1/10 normal
twilight	100	30	acuity drops at illumination below this
bright room	1000	300	normal acuity
bright sun	3×10^5	10^5	acuity decreased

they are nearby but only as one if they are sufficiently far away. This phenomenon is demonstrated by the headlights of a car, which appear as one source when the car is far away but can be seen separately as it gets closer. Similarly, two lines could be made so close together that the unaided eye would see them as only one. A magnifier may show them as two.

It is normal for people to be able to distinguish two objects as two if their angular separation is a minute of angle or more. Those with acute vision can see objects closer together than a minute. Those who can distinguish only objects separated by more than a minute of angle have less acute vision than normal. The measure of visual acuity is the inverse of the minimum angle of separation in minutes of objects that can be seen as separate. For example, normal visual acuity is 1/1 or 1. If a person can see only objects that are at least two minutes apart, that person's visual acuity is 1/2 or 0.5. If the limit is 0.67′, the acuity is 1/0.67 or 1.5. That person's eyes are 1.5 times as "sharp" as normal.

Visual acuity depends on the brightness of the objects being looked at, which in turn depends on the illumination. The brightness of a surface is designated by how much light is reflected to the viewer from the area in question compared to the illumination from a standard candle. Molten platinum (at the fusion point) emits 1 candela per 1/60 cm^2, which would be an area brightness of 600,000 candelas per square meter. A white, diffusely reflecting surface, illuminated to a level of 1 lm/m^2, would have a brightness when viewed from any direction of $1/\pi$ or 0.318 cd/m^2. If you are in a well lit room, the page of a book probably has a brightness of about 300 cd/m^2.

Normal visual acuity is obtained over a very wide range of brightness of the viewed object, from about 30 to 30,000 cd/m^2. At low levels of surface brightness the acuity is less than normal. The drop begins at about 30 cd/m^2, and at 0.03 cd/m^2 vision is only about a tenth that achieved at higher light levels. This phenomenon can be demonstrated by looking at the page of a book outdoors on a dark night. After your eyes have become adapted to the dark so that the page can be seen clearly, it will not be possible to read the printing. In very bright sunlight, acuity also drops. In Table 13–2 are some illumination levels under various conditions, giving a typical surface brightness for a white paper and comments on seeing conditions.

13–3 ABSORPTION

When a light beam passes through material, the intensity of the emerging light is less than the intensity incident on the medium. The percentage or fraction by which it is reduced is the **absorbance;** the percentage or fraction that passes through is the **transmittance.** The absorption process is of particular interest in the analysis of solutions to find the concentration of some absorbing material. To do this, the part of the spectrum absorbed by the substance being measured is determined, and then a measurement of the absorption (or transmittance) leads to a determination of the concentration of absorbing material. This requires an understanding of the absorption process.

The amount of absorption depends on the number of absorbing molecules in the light path. This number could vary either because of a variation in concentration or because of a variation in path length at a constant concentration. The end result is the same. The same difference in absorption is produced by the doubling of the concentration or by the doubling of the path length. The analysis is a little easier to explain when considering path length variations. In practice, the path length (the size of the container) is usually held constant and the concentration is varied; but doubling the concentration with a constant path length is the same as keeping the concentration constant and doubling the path.

Source

Detector

Figure 13–13 To consider the effect of absorption alone, have the source and detector fixed in position and insert various thicknesses, x, of absorber.

In this part of the work the source and measuring device will be considered to be at fixed positions while various absorber thicknesses are inserted between them, as in Figure 13–13. The variation of intensity with distance will be eliminated. It will also apply to situations such as sunlight. The distance to the sun is so great that changing a distance on earth has no effect.

One important characteristic of the absorption process is that a given amount of the absorber will transmit a certain fraction of the radiation entering it. A thickness of material that transmits half of the radiation falling on it will transmit exactly one half, no matter what the incident intensity may be. Such a

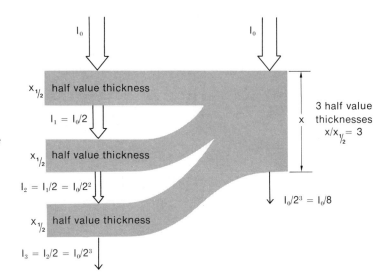

Figure 13–14 Absorption in successive half value thicknesses of material.

thickness is called a **half value thickness** or **half value layer.** The transmittance would be 0.50 or 50 per cent.

Consider a set of half value thicknesses, as in Figure 13–14. If an intensity I_0 falls on the first layer of thickness $x_{1/2}$ (indicating a half value thickness), the transmitted radiation is $I_0/2$. This is the intensity entering the second layer, which again transmits just half of that, or $I_0/2^2$. A third layer again cuts this by a half, to $I_0/2^3$. After passing through n half value layers, the intensity is described by $I = I_0/2^n$. The number of half value layers is found from the total thickness. If we have a total thickness x, and if we know the half value thickness $x_{1/2}$, then the number n of half value layers is $x/x_{1/2}$.

If the thickness x is an integral number of half value layers, so that $x/x_{1/2} = n$ is an integer, the intensity transmitted is easily calculated. However, if n turns out not to be an integer, then the intensity of transmitted light may be found using logarithms.

The equivalent situation, referring back to Figure 13–13, is that I_0 is the intensity measured with no absorber in place and I is the intensity when a thickness x is inserted. The measured intensity is described by

$$I = I_0/2^n$$

where $n = x/x_{1/2}$ or

$$I/I_0 = 2^{-n}$$

EXAMPLE 6

If the half value layer of a certain material is 3.0 cm and if the intensity with no absorber is measured to be 100 units, what would be the intensity if 15.0 cm of material were inserted?

One half value layer is 3.0 cm; 15.0 cm is therefore 15/3 or 5 half value layers.

$$I = I_o/2^n$$

where $I_o = 100$ and $n = 5$, so

$$I = 100/2^5 = 100/32 = 3.1$$

The intensity passing through 15.0 cm would be measured as 3.1 units.

EXAMPLE 7

If it is found that a thickness of 1.6 cm of a certain absorbing material reduces the light intensity to 1/16 of the value with no absorber, what thickness would reduce the intensity by half?

To do this you should recognize that 16 is a power of 2, namely 2^4. It is therefore given that

$$I/I_o = 1/2^4 = 2^{-4}$$

The quantity n, the number of half value layers, is 4 so one half value layer is 1.6 cm/4 = 0.4 cm = 4 mm. That is,

$$n = \frac{x}{x_{1/2}} = 4 \text{ and } x = 1.6 \text{ cm}$$

$$\frac{1.6 \text{ cm}}{x_{1/2}} = 4$$

from which

$$x_{1/2} = \frac{1.6 \text{ cm}}{4} = 0.4 \text{ cm} = 4 \text{ mm}$$

The half value layer is 4 mm.

The foregoing material is fine if the absorber is an integral number of half value layers. If it is not, logarithms may be used to calculate the intensity passing through the absorber. The logarithmic expression for the intensity transmitted is often the way that the absorption equations are expressed.

Let

$$I = I_o 2^{-n}$$

where $n = x/x_{1/2}$. Taking logs,

$$\log I = \log I_o - n \log 2$$

or

$$\log I = \log I_o - \frac{\log 2}{x_{1/2}} x$$

or

$$\log I/I_o = -\frac{\log 2}{x_{1/2}} x$$

EXAMPLE 8

If the light incident on the surface of a body of water is 1000 lux, and at a depth of 1 meter this is reduced to 800 lux, what would the intensity at 20 meters be expected to be?

First the half value layer must be calculated:

$$\log I/I_o = -\frac{\log 2}{x_{1/2}} x$$

where $\quad\quad I/I_o = 800 \text{ lux}/1000 \text{ lux} = 0.80$

so $\quad\quad\quad \log I/I_o = -0.097$

$\log 2 = 0.301$

$x = 1 \text{ m}$

Substituting these:

$$-0.097 = -\frac{0.301}{x_{1/2}} 1 \text{ m}$$

$$x_{1/2} = \frac{0.301}{0.097} \text{ m} = 3.1 \text{ m}$$

The half value layer is 3.1 m.

Now use

$$\log I/I_o = -\frac{\log 2}{x_{1/2}} x$$

Put

$$x = 20 \text{ m}$$

$$x_{1/2} = 3.1 \text{ m}$$

$$\log 2 = 0.301$$

Then

$$\log I/I_o = -\frac{0.301 \times 20 \text{ m}}{3.1 \text{ m}}$$

$$\log I/I_o = -1.94$$

so $\quad\quad\quad I/I_o = \text{antilog} -1.94$

$$= 0.011$$

Negative logs are more awkward to work with than positive logs. An alternative way to handle the problem is to write

$$\log I/I_o = -1.94$$

as $\quad\quad\quad -\log I/I_o = +1.94$

$$\log (I/I_o)^{-1} = 1.94$$

$$\log \frac{I_o}{I} = 1.94$$

Then $\quad\quad\quad I_o/I = 87$

and $\quad\quad\quad I/I_o = 1/87 = 0.011$

I_o was 1000 lux, so I at 20 m will be 11 lux.

13–3–1 ABSORPTION OF GAMMA RAYS

The absorption equations just derived apply to more than visible light. For example, they often apply to the radiation given out by radioactive materials. The radiations can be hazardous. If you have radioactive material about, your work area may be made safe by putting absorbing material around the radioactive substance. Lead is often used to absorb the gamma radiation given off by many radioactive materials. Each material will emit radiation of different energy and each material will then have its own half value thickness of absorber. For example, the radiation from radioactive cobalt-60 is cut in half by 1.0 cm of lead, while only 3 mm (0.3 cm) will reduce the radiation from iodine-131 by half.

EXAMPLE 9

If you must have some cobalt-60 stored nearby as in Figure 13–15, but with it there you find that the radiation at your workspace is 1000 times that considered to be a safe level by the authorities in your country, what thickness of lead should be put around the source?

Use
$$\log I/I_o = \frac{-\log 2}{x_{1/2}} x$$

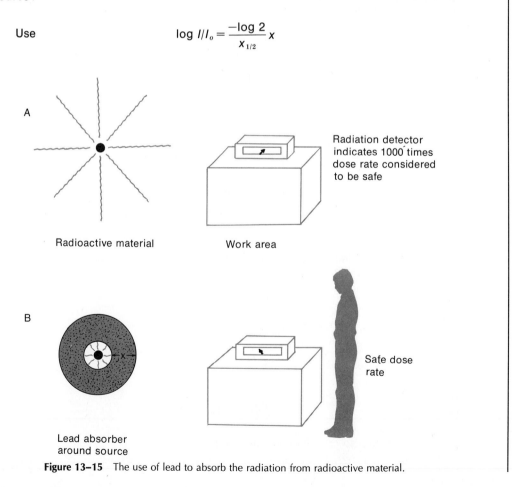

Figure 13–15 The use of lead to absorb the radiation from radioactive material.

$$x = \frac{-x_{1/2}}{\log 2} \log I/I_o$$

$$I/I_o = 1/1000$$

$$\log I/I_o = -3.00$$

$$x_{1/2} = 1.0 \text{ cm}$$

$$\log 2 = 0.301$$

$$x = -\frac{1.0 \text{ cm}}{0.301}(-3.00)$$

$$= 10.0 \text{ cm}$$

If you put 10 cm of lead around the radioactive cobalt-60, you may work unconcernedly at your desk. You may prefer to have someone experienced at the job put the lead there.

13-3-2 THE ABSORPTION COEFFICIENT

The ability of materials to absorb radiation is often described by an absorption coefficient. Materials that readily absorb (have a small half value layer) have a high absorption coefficient, while those that absorb only a little have a low absorption coefficient. Our absorption equations will be put into the standard form with the accepted type of absorption coefficient.

The fraction of radiation passing through a thickness x of material was expressed in terms of common logarithms (that is, logarithms to a base 10), and, writing $\log 2 = 0.3010$,

$$\log I/I_o = -\frac{0.3010}{x_{1/2}} x$$

By the definition of the logarithm this means that

$$I/I_o = 10^{-\frac{0.3010}{x_{1/2}} x}$$

Scientists often use a number other than 10 in such work; this number, given the symbol e, has the value $2.718. \ldots$ It is actually a transcendental number like π and it has special properties for such work. The number e can be represented by the series:

$$e = 1 + \frac{1}{1!} + \frac{1}{2!} + \frac{1}{3!} + \frac{1}{4!} \ldots$$

You can make your own calculation of e from this. Various powers of e are

$$e^1 \cong 2.718$$
$$e^2 \cong 7.389$$
$$e^3 \cong 20.09$$
$$e^4 \cong 54.60$$

Using fractional powers, one finds that

$$e^{2.303} \cong 10$$

This power can be found from the use of common logs:

$$e^a = 10$$

$$\log e^a = a \log e = \log 10 = 1$$

$$a = \frac{1}{\log e} = \frac{1}{0.4343} = 2.303$$

Substitute $e^{2.303}$ for 10 in the relation

$$I/I_o = 10^{-\frac{0.301}{x_{1/2}}x}$$

$$I/I_o = (e^{2.303})^{\left(-\frac{0.301}{x_{1/2}}x\right)}$$

$$I/I_o = e^{-\frac{0.693}{x_{1/2}}x}$$

This is the equation for the radiation intensity passing through a thickness x and in terms of a power of e. To find I/I_o the quantity $0.693 x/x_{1/2}$ is calculated, and then (from tables of powers of e) I/I_o is calculated.

Appendix 6 includes values of e^x for positive and negative powers of e. Powers of e are referred to as **exponentials**. An absorption process that follows the above law is called **exponential absorption**.

The quantity $0.693/x_{1/2}$ is called the **absorption coefficient** k. So

$$I/I_o = e^{-kx}$$

where $k = 0.693/x_{1/2}$, or $x_{1/2} = 0.693/k$.

EXAMPLE 10

It had been stated that for cobalt-60 gamma rays the half value layer of lead was 1.0 cm. What is the absorption coefficient?

Use
$$k = 0.693/x_{1/2} \quad \text{and} \quad x_{1/2} = 1.0 \text{ cm}$$
$$k = 0.693/1.0 \text{ cm}$$
$$= 0.693/\text{cm}$$

The absorption coefficient is 0.693 per centimeter. Note the inclusion of units with an absorption coefficient.

13-3-3 DEVIATIONS FROM EXPONENTIAL ABSORPTION

The logarithmic relation between transmittance and concentration of absorber applies only when the light measuring device is sensitive to only the portion of the spectrum being absorbed. For instance, if a blue-looking solution absorbs only the red portion of the spectrum but passes the blue portion of the spectrum without absorption, there will be deviation from exponential absorption if the whole spectrum is included in the measurement. This is because after the red portion of the spectrum is practically cut out, there will be very little more decrease in intensity with thickness because the blue light is not absorbed at all. This is perhaps an extreme case (as examples often are), but in working with light absorption the only part of the spectrum that should be analyzed is the part that is absorbed by the solution.

A similar situation arises if two parts of the spectrum have different absorption coefficients. Initially, the absorption coefficient measured would be the sum of the two; but as one part of the spectrum became completely absorbed, the slope of the transmittance curve would decrease to the value of the smaller coefficient.

13-4 THE SPEED OF LIGHT

The speed of light is so high that, before a successful measurement was made, there were many unsuccessful attempts, some by the greatest scientists. Whether the speed of light is infinite or just very high was pondered by the ancient Greeks. It was known that light traveled faster than sound because a distant event, perhaps a hammer blow, was seen before it was heard.

Galileo devised a system involving two stations many kilometers apart. At each station was an observer with a covered lantern. One person uncovered his lantern and, as soon as the distant person saw the light, his lantern was uncovered. The first person could then note the arrival time of the returning light. The experiment failed because the travel time was so much less than the reaction time of the observers. The best methods devised since have not been so different, however. An intermittent beam is sent to a distant station, using either a toothed wheel or a rotating mirror, and the light is returned by a mirror. Galileo's idea was basically good.

The first successful measurement of the speed of light was actually done by an astronomer who set out to do something else, something comparatively trivial. The process of looking for one thing but finding something else even more important frequently occurs in scientific work. The process is even given a name, **serendipity,** a name derived from the tales of the adventures of the three Princes of Serendip. It has been sometimes described as "you dig for water and you strike oil."

In the mid-seventeenth century the concept of the earth orbiting the sun was used by many astronomers because the picture such a system presented was much simpler to analyze than an earth-centered system. But there was *no direct evidence* for this and many, with good grounds, argued against it. The motions of the sun, moon, or stars, even the seasons and eclipses, can be readily explained on the basis that they, and not the earth, move. Two of the problems of the time were to demonstrate earth motion and to measure the speed of light.

Galileo had, some years previously, discovered four satellites of Jupiter, objects that revolve about that planet. Once each revolution they pass into the shadow of Jupiter, and these eclipses show on earth as a sudden disappearance of their light. In fact, these eclipses allowed a precise method of timing the periods of those satellites. Clocks were also being perfected, but the measurements of the periods of the satellites (the time between successive eclipses of the same satellite) were not consistent. The periods were at times longer than average, and at times less. Why?

Olaf Roemer (1644–1710), a Dane, was hired by a new observatory in Paris to make precise measurements of those periods. Roemer's data were as inconsistent as anyone else's. He did note, however, that the periods were always too short when Jupiter was in the east in the morning (a morning star) and were too long when Jupiter was in the west in the evening. Roemer's ultimate explanation can be described by reference to Figure 13–16(a). If indeed the earth moves in an orbit about the sun, then when Jupiter is seen in the morning, the earth would

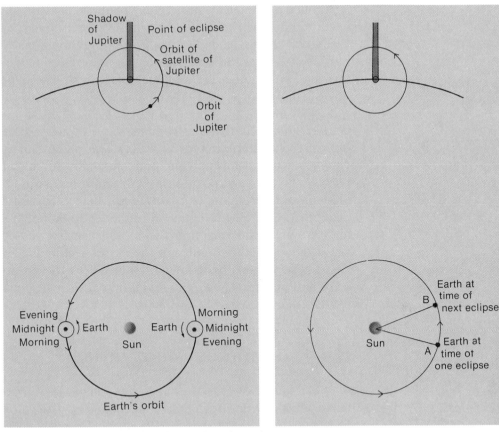

Figure 13–16 Viewing a satellite of Jupiter from the earth.

be approaching Jupiter. When Jupiter is seen in the evening, the earth is receding from Jupiter. As in Figure 13–16(b), as the earth approaches Jupiter an eclipse is seen when the earth is at *A*. The next eclipse is seen when the earth is at *B*. The last light from the second eclipse did not have to travel as far to reach the earth as did the light from the first eclipse. So the second eclipse is seen too early and the period as measured on earth is too short—in fact, it is too short by the time it took for the light to travel the distance *AB*. The size of the earth's orbit had been measured, so it was then no difficult task to find how far the earth had moved in that period. The amount by which the eclipse was shortened was measured, and the speed of light was merely distance/time. Applying Roemer's analysis, the actual periods of Jupiter's satellites could be found precisely. So not only did he achieve his goal, but his explanation was based on earth motion. Roemer was the first in history to find a phenomenon that required earth motion for its explanation and could not be explained without it. He was also the first to measure the speed of light. Roemer's achievements illustrate serendipity at its finest.

Since Roemer's time, other phenomena that depend on earth motion for their explanation have been found. Light from the stars shows a slight change in direction owing to earth motion. It is like the slanting rain seen out the side window of a moving car. Though the rain seen from a car may appear about 45° off the vertical, with starlight it is ±20 seconds of angle as the earth moves to and fro in its orbit. Starlight shows a slight change in wavelength as the earth moves toward a star and an opposite change as it moves away. The pattern of stars even changes when viewed from opposite ends of the orbit. The nearby stars change position by a fraction of a second of angle. (This is so small that it is not quite fair to say that the "pattern" changes.) This phenomenon is called parallax. Parallax was looked for by the ancient Greeks, and because they could not see it they rejected the idea of the earth orbiting the sun.

These various phenomena, all based on the observation of stars and planets, show convincingly that the earth moves in its orbit.

13-4-1   THE TOOTHED WHEEL METHOD

The speed of light on earth was not measured until 1849, when Fizeau, working in Paris, succeeded by using an apparatus of which the central part was a toothed wheel. As in Figure 13-17, a light beam is passed between two teeth (A and B) of a wheel to a distant mirror. The light is reflected back to go between two other teeth, C and D. The wheel is then set spinning, and the rate is increased until the returning light is blocked by tooth D, which moves to what was the space between C and D in the time that the light traveled $2d$. As the speed is increased further, the space between D and E would move into position to pass the returning beam.

In Fizeau's apparatus the rotating teeth produce a pulsating beam that replaced Galileo's covered lanterns.

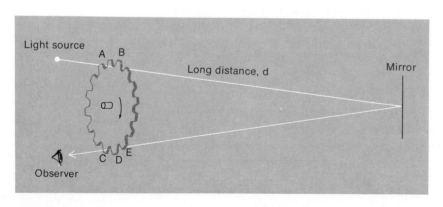

Light source

A   B

Long distance, d

Mirror

Observer

C D E

Figure 13-17   The toothed wheel used to measure the speed of light.

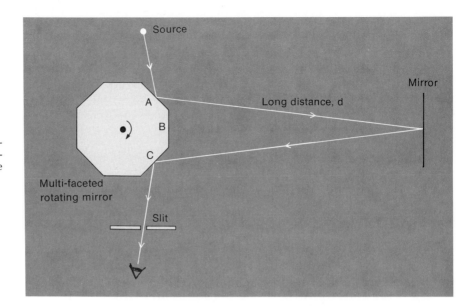

Figure 13–18 The rotating multi-sided mirror used to measure the speed of light.

13–4–2 THE ROTATING MIRROR

In 1850, Foucault again measured the speed of light, using a rotating mirror in place of a toothed wheel. Foucault's method was refined by the American scientist Albert Michelson, who made many measurements of the speed of light between 1879 and 1931.

Michelson first worked with a single rotating mirror like Foucault's, but his final apparatus consisted of a multi-faceted mirror as in Figure 13–18. With the mirrors stationary, the light was reflected from face A to the distant mirror M. The returning light reflected from the face C through a slit to the observer. When the mirror rotates, owing to the travel time of the light, face C is no longer in position to reflect the light through the slit. If the rotation speed is increased so that face B replaces face C in the travel time of the light, then the light will be seen through the slit. The travel time is found from the rotation speed and the number of faces.

Michelson's principal contribution in this field was in his painstaking attention to precision and detail in every way. His work stands as a monument to those who appreciate a masterpiece of precision.

DISCUSSION PROBLEMS

13–1–1   1. Give more examples of solid angles in the world about you.

13–2 2. The actual illumination on a desk below a lamp will depend on the amount of light reflected from the walls as well as on the source intensity. The size of the room will also influence the illumination. In a small room with white walls, what variation would you expect from an inverse square law? Outline your reasoning.

13–2–3   3. Investigate the explanation for the decrease in visual acuity at low levels of illumination.

13-2-3   4. What criterion would you use to establish a minimum level of illumination in a room?

13-4   5. Progress in science is made when a project gives unexpected results. Experiments that work as expected tell little that is new. Comment on these statements with respect to the measurement of the speed of light.

PROBLEMS

13-1-1   1. What is the solid angle in the beam of a flashlight if at 3 meters it makes a circle of light 1.0 meter in diameter on a wall? Assume that the flat area on the wall approximates sufficiently closely the area of the curved surface that should be used.

13-1-1   2. A horn used at a birthday party has a wide end of 3 cm radius, and it is 20 cm long. What is the solid angle at the apex?

13-1-1   3. A window 1 m by 1.5 m is viewed from a distance of 10 m. What is the solid angle formed by the window? If the viewing distance is increased to 30 m, by what factor does the solid angle change?

13-1-2   4. What is the total number of lumens emitted by a light of intensity 130 cd if it emits equally in all directions?

13-1-2   5. How many lumens of light flux fall on a page that is 21.6 cm by 28.0 cm lying on a table, when there is a light source of 130 cd 1.50 m above it?

13-1-2   6. A source of 1 cd is 2.00 m from an eye that has a pupillary diameter of 5 mm. How many lumens enter the eye?

13-2   7. What is the illumination on a page 1.50 m below a source of 130 cd?

13-2   8. What is the intensity of a street lamp that is 9.0 m above the ground if the illumination directly below it is 24.7 lux?

13-2   9. Find the expression that gives the illumination near a long linear light source.
Consider a source of length l, in which each unit of length is of intensity I candelas. Let the source be along the axis of a long cylinder, length l and radius r.
(a) Find the number of lumens emitted per unit length and by the length l.
(b) Find the area of the inside of the side walls of the cylinder.
(c) Calculate the number of lumens falling on each unit of area inside the cylinder. If the light escaping from the ends is neglected, that will give an expression for the intensity of il-

lumination at a distance r from a long source of I lumens per meter.

13-2-1   10. Find the exponent in the expression relating the two sets of data below. Also give the equation relating u and v.

u	v
3.0	1.12
4.6	2.5
5.6	3.9
7.3	6.4
8.0	7.5

13-2-1   11. Find the equation relating u and v in the following set of data. In the equation use the exponent that is found in the analysis and then say what the form of the basic relation probably is.

u	v
75	0.40
125	0.517
175	0.610
225	0.690

13-2-1   12. Find the equation relating u and v that applies to the following set of data. In the equation use the exponent that you actually find and then say what the form of the basic relation probably is.

u	v
5.05	20
1.30	30
0.274	50
0.103	70
0.035	100

13-2-2   13. A lamp of 130 cd is placed 1.00 meter above the center of a square table that is 1.00 meter on a side. Find the intensity of illumination:
(a) at the center of the table.
(b) at the middle of each side.

13-2-2   14. Find the illumination at the center of a table that is 3 meters long. Above each end at a height of 2 meters is a lamp of 200 cd.

13-2-2   15. Find the illumination at the center and also at the ends of a table 2.00 meters long. At a height of 1.67 meters above each end is a lamp of 130 cd.

13-2-2   16. A lamp of 200 cd is 1.5 m above a table. Where on the table will the illumination be just a half of what it is directly below the lamp?

13-2-2   17. (a) Find the illumination at

the center of a table when 1.5 m above it there is a long line of fluorescent lamps with an intensity of 165 cd/m.

(b) If the table is 1 meter wide, what is the illumination at the edge?

13-3   18.   If the half value thickness of a certain light-absorbing material is 1.00 mm, find the intensity I in terms of the intensity I_o with no absorber that would pass through:
(a) 2.00 mm.
(b) 3.00 mm.
(c) 3.32 mm.

The situation is as illustrated in Figure 13-13.

13-3   19.   Find the half value thickness for a given material if 8.0 cm reduce the intensity to 1/8 of the value that would occur with no absorbing material present.

13-3   20.   Find the thickness of absorbing material that causes an intensity to be given by $I_1 = 0.75\ I_0$ when it is known that a thickness of 1.00 cm results in an intensity given by $I_2 = 0.90\ I_0$.

13-3-1   21.   By how much would 0.5 cm of lead reduce the gamma radiation received from a piece of radioactive cobalt-60 when it is known that 1.0 cm reduces the radiation to a half?

13-3-1   22.   At a given position near where some radioactive material is stored, the radiation level is 20 times the tolerable amount. Find what thickness of lead must be used to reduce the intensity by the required amount. The half value layer for that situation is 6 mm.

13-3-2   23.   Find the absorption coefficient for a situation in which the half value layers are:
(a) 3 mm.
(b) 1.0 cm.
(c) 0.0693 cm.

13-3-2   24.   If a material has a linear ab-

sorption coefficient of 0.50/cm for a given radiation, find the fraction of the radiation that will pass through the following thicknesses:
(a) 1.00 cm.
(b) 1.00 mm.
(c) 5.0 cm.

13-3-2   25.   If a measured radiation intensity with no absorber is 100 units, find the intensity that would pass through (a) 2 cm and (b) 6 cm of material for which the absorption coefficient is 0.3/cm.

13-4   26.   If Galileo's method was to be used to measure the speed of light and the observers were on hills 7.5 km apart, what time interval would have to be measured? To find this, use the known value of the speed of light, 3×10^8 m/s.

13-4   27.   Find the time by which the observed period of one of the satellites of Jupiter would be changed from the true value when the earth moves toward Jupiter. Consider the satellite named Ganymede, which orbits Jupiter in 7 days, 3 hours and 43 minutes. Treat the orbit of the earth as a circle 1.50×10^{11} meters in radius; the time to orbit the sun is 1 year or 3.15×10^7 seconds. Light travels at 3×10^8 m/s.

13-4   28.   Calculate the speed of light using the following data. A wheel with 200 teeth is used with a mirror 7.5 km away. The light is sent between two teeth to the distant mirror and reflected back between two other teeth when the wheel is stationary. The returning light is blocked by a tooth when the wheel spins at 50 revolutions per second.

13-4   29.   Calculate the speed of light using the following data. The light is reflected from one face of an eight sided mirror to another mirror 35 km away. The returning light is reflected from another face of the mirror and through a slit. When the mirror rotates at 536 revolutions per second the returning light is also seen through the slit.

CHAPTER FOURTEEN

ELECTRIC CHARGES, FORCES AND FIELDS

Tremendous progress has been made since the days when the study of electricity was concerned mainly with static electricity and with current electricity turning motors and heating wires. The latter aspect of the subject is still important, especially in the everyday applications of electricity. However, a large number of devices have been developed using charges in motion but under the influencè of the forces of static charges.

The first topic will be the description of the forces between charges and the concept of the electric field; and then we will discuss how to set charged particles free to move and be used. In this there is much of the "magic" of modern scientific instruments.

The charges of importance are those on the electron, a negatively charged particle of very small mass, and on the proton, a positively charged particle with about 1836 times the mass of the electron. The positive charge on the proton is, as far as can be determined, identical in size to that on the electron, though of opposite sign, of course. The size of this charge is usually represented by $+e$ on the proton and $-e$ on the electron. The value of e is 1.60×10^{-19} coulomb. The coulomb is the unit for measurement of electric charge adopted for use in the SI units.

There is another unit for charge, one based on the centimeter-gram-second system of units, that is still occasionally encountered. It is referred to as an "electrostatic unit," abbreviated esu, also called a statcoulomb. One statcoulomb is 3.33×10^{-10} coulombs, and the electronic charge in these units is 4.80×10^{-10} esu.

Every atom has a nucleus in which most of its mass is concentrated, and this nucleus has a positive electric charge. The nucleus is considered to be a collection of protons, each with a charge $+e$, and neutrons, which have no charge at all. The nucleus is surrounded by electrons, which have a negative charge, and ordinarily the net charge on the atom is zero.

Objects become electrically charged by acquiring, in some way, an excess or a deficiency of electrons. An excess leads to a net negative charge and a deficiency leads to a net positive charge.

14-1 FORCES BETWEEN CHARGES

Charles Coulomb first found how the electric force depends on the amount of charge Q_1 and Q_2 on two objects and the distance between them. It was in 1784–85 that he discovered that the force depends on the product of the charges and varies inversely as the square of the distance r between them. This relation is known as Coulomb's law, and it is written as

$$F = \text{constant} \times Q_1 Q_2 / r^2$$

The proportionality constant depends on the units used. If the charges are measured in coulombs, the distance in meters, and the force in newtons (the SI units), the proportionality constant is found to be approximately 9×10^9. More precisely, and with units included, it is 8.998×10^9 N m^2/C^2. The unit represented by C is the coulomb. Rather unexpectedly, the constant in Coulomb's equation is related to the speed of light c, being described in these units by $10^{-7} c^2$. It is not just a chance numerical relationship but is a consequence of light being an electromagnetic phenomenon.

With the constant inserted, and in SI units, Coulomb's law is

$$F = 9 \times 10^9 \, Q_1 Q_2 / r^2$$

The electric forces can be either attractive or repulsive, depending on whether the charges are of like or of different sign. In fact, the Q's are put into Coulomb's law with their signs attached, and a negative value for F indicates attraction while a positive value indicates repulsion.

EXAMPLE 1

If two small objects are suspended by thin strings and then electrically charged, they may repel each other and hang apart as in Figure 14-1. The repulsive force is related to the angle of the string from the vertical (see Section 1-10). It would be consistent with such an experiment to obtain a force of 10^{-4} N with a separation of the objects of 5 cm or 0.05 m. In such a case, how much charge would be on the objects if the charges are equal and of the same sign?

Use Coulomb's law with $Q_1 = Q_2$, call it Q. The distance r is 0.05 m and $F = 10^{-4}$ N.

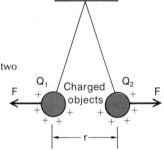

Figure 14–1 The repulsion between two charged objects.

$$F = 9 \times 10^9 \ \frac{\text{N m}^2}{\text{C}^2 \ r^2} \ Q^2$$

Solve for Q^2

$$Q^2 = \frac{Fr^2 \ \text{C}^2}{9 \times 10^9 \ \text{N m}^2}$$

Substitute the values

$$Q^2 = \frac{(10^{-4} \ \text{N}) \ (0.05^2 \ \text{m}^2)\text{C}^2}{9 \times 10^9 \ \text{N m}^2}$$

$$Q^2 = 2.8 \times 10^{-17} \ \text{C}^2$$

If you do not have a calculator that gives square roots directly, the expression should be put in terms of an even power of 10.

$$Q^2 = 28 \times 10^{-18} \ \text{C}^2$$
$$Q = 5.3 \times 10^{-9} \ \text{C}$$

The charge on each object will be 5.3×10^{-9} coulombs or 5.3 nanocoulombs.
The charges encountered in electrostatics are usually very small when expressed in coulombs.

Electric charge is one of the important properties of what we call matter. Another, already familiar, is inertia. The question of why matter has these properties has not been answered, but from the behavior of matter we have described two of the properties as inertia and electric charge. It is interesting also that the force laws for mass and for electric charge are of such similar forms, as can be seen when they are written as:

for electric charge $\qquad F_e = 9.00 \times 10^9 \ \dfrac{Q_1 Q_2}{r^2}$ in SI units

for gravitational mass $\qquad F_g = 6.67 \times 10^{-11} \ \dfrac{M_1 M_2}{r^2}$ in SI units

It is of interest to take two of the basic particles of nature, two protons, and to compare the electrical force to the gravitational force between them. For a proton

$$Q = e = 1.60 \times 10^{-19} \text{ C}$$
$$M_p = 1.67 \times 10^{-27} \text{ kg}$$

In the force laws $Q_1 = Q_2 = e$, so $Q_1 Q_2 = e^2$; and $M_1 = M_2$, so $M_1 M_2 = M_p^2$. Taking the ratio of the equations, the quantities r^2 cancel to leave

$$\frac{F_e}{F_g} = \frac{9.00 \times 10^9 \ e^2}{6.67 \times 10^{-11} \ M_p^2}$$

$$= \frac{9.00 \times 10^9 \times 1.60^2 \times 10^{-38}}{6.67 \times 10^{-11} \times 1.67^2 \times 10^{-54}}$$

$$= 1.2 \times 10^{36}$$

The electrical force is about 10^{36} times as strong as the gravitational force for those particles. Gravity is regarded as being an extremely weak force in the heirarchy of forces.

14-1-1   ELECTROMETERS

Electrometers, devices to measure electric charge, are often based on electrical forces.

When a person holds onto an electrostatic generator or stands near high voltage apparatus, the hair stands on end. Even brushing hair in dry air produces enough electricity that the charged hairs repel each other and stand out from the head. This repulsion is the principle of the electrometer.

Some modern electrometers differ very little in principle from the original gold leaf electroscopes of many years, even centuries, ago. The **gold leaf electroscope** consists of a metal can with a window; through an insulator on top, a metal rod is inserted. At the bottom of the rod is a metal plate with a piece of gold leaf attached to one side. The gold leaf is fastened only at the top so that it hangs loosely. Electric charge on the rod spreads over the metal plate and gold leaf. The repulsion between the charges causes the gold leaf to stand out at an angle, as in Figure 14-2(a). The greater the charge, the greater is the angle. This simple device is amazingly sensitive, being able to indicate a charge of about a billionth (10^{-9}) of a coulomb.

It is no surprise that this design has been improved on, though in some electrometers the principle is the same. One such device has replaced the gold leaf with a very thin U-shaped metalized quartz fiber, as in Figure 14-2(b). By making the fiber

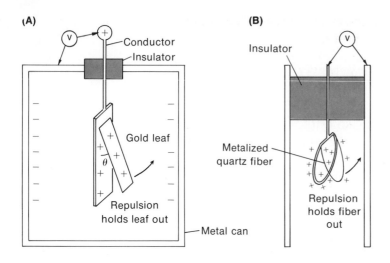

(A)

Conductor
Insulator
Gold leaf
θ
Repulsion holds leaf out
Metal can

(B)

Insulator
Metalized quartz fiber
Repulsion holds fiber out

Figure 14–2 The old gold leaf electro-scope (a), and its more modern version, the quartz-fiber electrometer (b).

very light, so thin that a microscope is required to see it, the sensitivity has been increased manyfold so that the detection of 10^{-12} or even 10^{-16} coulomb is possible. Such an electrometer is used in a pocket radiation measuring device, which is shown in Figure 14–3.

14–2 ELECTRIC FIELDS

You cannot push or pull an object, this book for instance, without "touching" it. The idea of exerting a force at a distance, without contact, is ridiculous except in the familiar cases such as electrically charged objects or even gravitational attraction. Our familiarity with these stops us from wondering "how can it happen?" This concept of action at a distance, which is how electric or gravitational forces seem to be able to operate, is puzzling.

One of the solutions to this is to say that somehow the electric charges (let us limit the discussion to electric forces) affect the space around them. We say that the space around a charge is pervaded with what we call an electric field. When another electric charge is put at a position where there is already an electric field, then there will be a force on that charge. The force is associated with the field at that position, not with the distant charge that produced the field. The field has a certain strength and it also has a direction; that is, electric field is a vector quantity.

The concept of a field not only gets rid of the troublesome idea of action at a distance, but it actually simplifies the analysis of many situations regarding electric forces. For example, the calculation of the force between just two charges of a given size is straightforward. The values of the known quantities are put

Direct-Reading Dosimeter

Figure 14–3 A pocket device for measuring radiation from radioactive materials or x-ray machines. At its heart is a quartz-fiber electrometer like that in Figure 14–2(b).

into the equation known as Coulomb's law and the unknown is solved for. However, practical situations usually involve more complicated arrays of charges, not simple point charges or spheres for which Coulomb's law can be applied very simply. It is in the complex situations that the concept of field can be very useful.

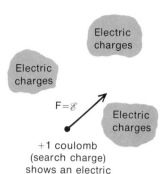

Figure 14–4 If an electric charge is placed at some point and there is a force on it, there is said to be an electric field there. The strength of the electric field is specified by the size of the force on a unit charge at that point.

When a charge is put at a place in space where there is an electric field, there will be a force on that charge. The size of the force will depend on the strength of the field at that point in space as well as on the size of the charge placed there. From this concept, the units used to measure a field are derived. The force on a unit positive search charge at any place in space is called the **field strength** at that point. This is illustrated in Figure 14–4. The direction of the force gives the direction of the field. If a charge of 1 coulomb is put at a certain point in space and there is force of, say, 5 newtons on it, the electric field at that point is said to have a strength of 5 newtons per coulomb. It does not matter where the charge or array of charges that causes this field may be. The net force on a unit charge (called a search charge) at that point measures the field there. Now, if a charge of 2 coulombs was placed at that point in the field, the force on it would be just twice as much as the force on the 1 coulomb charge.

From this the relation is deduced that if the field strength at some point is $\mathscr{E}$, the force on a charge q is given by the product of the field strength and the charge:

$$F = \mathscr{E} q$$

Also, if there is a force F on a charge q at a certain place, the electric field there is given by $\mathscr{E} = F/q$.

To calculate the field strength $\mathscr{E}$ at any point in space near a point charge of size Q, a charge q is imagined to be placed at that point. By Coulomb's law the force on q is described by

$$F = 9 \times 10^9 \ Qq/r^2$$
$$= 10^{-7} c^2 Qq/r^2$$

Knowing that $\mathscr{E} = F/q$, the q will cancel to give the field near a point charge Q as

$$\mathscr{E} = 9 \times 10^9 \ Q/r^2$$
$$= 10^{-7} c^2 Q/r^2$$

The **search charge** used in finding a field is always positive. If the search charge is near a positive charge Q, the force is repulsive and the direction of the field is radially away from Q.

(A)

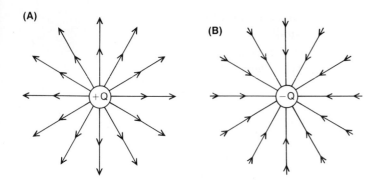

(B)

Figure 14–5 The pattern of the electric field in the space around a positive charge (a) and a negative charge (b).

If Q is negative, there will be attraction and the field is everywhere directed toward the negative charge. These fields can be pictured, as in Figure 14–5(a) and (b), in two dimensions where the lines drawn indicate the direction of the electric field in the space around the charges. The pictures lead to the idea of **field lines.**

EXAMPLE 2

Find the electric field at a distance of 0.1 m (10 cm) from a point charge of 0.02 microcoulombs (μC).

Use Coulomb's law with the search charge $Q_1 = +1$, and $Q_2 = 0.02 \times 10^{-6}$ C. Also, $r = 0.1$ m. Then the force F divided by Q_1 is the electric field:

$$\mathscr{E} = 9 \times 10^9 \frac{\text{N m}^2}{\text{C}^2} \times \frac{Q_2}{r^2}$$

$$Q_2 = 0.02 \times 10^{-6} \text{ C}$$

$$= 2 \times 10^{-8} \text{ C}$$

$$r = 10^{-1} \text{ m}$$

$$\mathscr{E} = 9 \times 10^9 \frac{\text{N m}^2}{\text{C}^2} \frac{2 \times 10^{-8} \text{ C}}{10^{-2} \text{ m}^2}$$

$$= 18 \times 10^3 \text{ N/C}$$

$$= 1.8 \times 10^4 \text{ N/C}$$

The electric field is directed away from the charge. If the charge of 0.02 μC was negative, the field would have been directed toward the charge.

EXAMPLE 3

Consider two charges of +0.02 μC and −0.02 μC as in Figure 14–6(a). They are 17.32 cm apart. The strength of the field is desired at the point P, which is 10 cm from each charge and at an angle of 30° as shown. If there was only the charge of +0.02 μC, the field would be that shown as $\mathscr{E}_1$ in the figure; the value, as found in the previous example, is 1.8×10^4 N/C. If there was only the charge of −0.02 μC, the field would be that shown as $\mathscr{E}_2$, again of magnitude 1.8×10^4 N/C. To find the net effect, remember that a field is a force per unit charge, so a field may be separated into components just as a force can be. As in Figure 14–6(b), $\mathscr{E}_1$ is found (using trigonometry) to be the equivalent of 1.56×10^4 N/C in the direction of the line joining the two charges and 0.9×10^4 N/C directed perpendicularly away from that line. Similarly, $\mathscr{E}_2$ has a component along the direction of the line between the charges of 1.56×10^4 N/C also, and a

(A)

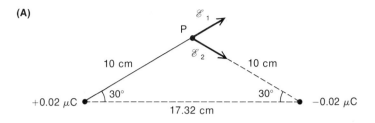

(B)

(C)

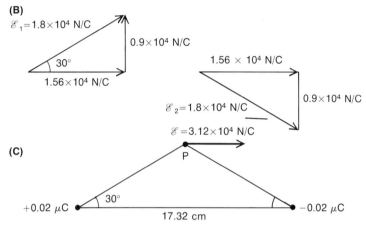

Figure 14–6 The calculation of an electric field resulting from two charges. (a) At point P the field from the two charges are shown as $\mathscr{E}_1$ and $\mathscr{E}_2$. These can be separated into components as in (b), and the net field is shown in (c).

component perpendicularly toward that line of 0.9×10^4 N/C. When these two fields occur together at the point P, the perpendicular components cancel and the other components add to give a net field at P of 3.12×10^4 N/C, as shown in Figure 14–6(c).

EXAMPLE 4

Find the force on an electron (charge 1.6×10^{-19} C) when it is at a position at which the field is 3.12×10^4 N/C. Such a field strength is not unlike that existing in a low voltage x-ray machine.

Use

$$F = Q\mathscr{E}$$

where

$$\mathscr{E} = 3.12 \times 10^4 \text{ N/C}$$
$$Q = 1.60 \times 10^{-19} \text{ C}$$
$$F = 3.12 \times 10^4 \text{ N/C} \times 1.60 \times 10^{-19} \text{ C}$$
$$= 5.0 \times 10^{-15} \text{ N}$$

This force of 5.0×10^{-15} N is extremely small, but so is the mass of an electron (it is only 9.1×10^{-31} kg), and the acceleration under such a force may be very high. Use Newton's second law to find it!

$$F = ma$$
$$a = F/m$$

where

$$F = 5.0 \times 10^{-15} \text{ N} = 5.0 \times 10^{-15} \text{ kg m/s}^2$$
$$m = 9.1 \times 10^{-31} \text{ kg}$$
$$a = \frac{5.0 \times 10^{-15} \text{ kg m/s}^2}{9.1 \times 10^{-31} \text{ kg}}$$

$$= 5.5 \times 10^{15} \text{ m/s}^2$$

This is indeed a large acceleration.

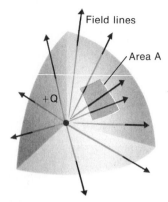

Figure 14-7 The representation of the three-dimensional field in the space near a positive charge. At any place, the strength of the field is represented by the number of lines penetrating a unit area such as A.

14-2-1   ELECTRIC FIELD LINES

Pictures such as those in Figure 14–5 are very useful because they not only show the direction of the field everywhere around a charge, but they also give an indication of the field strength in different places. The field is strongest near the charge where the lines are close together. When r is large, the field is weaker and this is where the lines are farther apart. The density of lines gives an idea of field strength, and is even a way to describe it. Of course, the field is not only in the plane of the paper but is in all directions. Figure 14–7 is a picture of an electric field in three dimensions. The lines are depicted as radiating outward through the surface of a sphere. A sphere of small radius would have more lines penetrating each unit area of its surface than would a sphere of large radius. The adopted convention is to imagine one line through each square meter of such a surface if the field strength is one unit (one newton per coulomb in the SI system). A field of 5 newtons per coulomb would be designated by five **lines per square meter** and so on. This is very artificial but it is, in fact, a very useful concept (invented, incidentally, by Michael Faraday, who is said to have invented this concept because he couldn't do math and the idea of field lines allowed him to solve a lot of problems in spite of that).

The total number of these fictitious lines emanating from a unit charge, one coulomb, can be calculated. At a distance r the field strength (with $Q = 1$) is given by $\mathscr{E} = 9 \times 10^9 / r^2$ (newtons per coulomb or lines per square meter). A sphere of radius r has a surface area given by

$$A = 4 \, \pi r^2$$

The number of lines cutting through each square meter of this surface is $9 \times 10^9 / r^2$. The total number of lines through the spherical surface around the unit charge is given by the number per square meter ($\mathscr{E}$) times the area A;

$$\mathscr{E} A = 4 \, \pi \times 9 \times 10^9$$

This is the number of fictitious lines of electric field emanating from a unit positive charge. From a charge Q the number of lines is

$$N = 4 \, \pi \times 9 \times 10^9 \, Q$$
$$= 1.13 \times 10^{11} \, Q$$

It would not matter if the surface surrounding the charge was not spherical, for if that number of lines emanates from the charge,

Q, that same number must go outward through any surface surrounding the charge. By similar reasoning the same number of lines would terminate on a negative charge, $-Q$.

Because of the factor of 10^{11}, the number of lines is often rather large and no attempt is made to draw that many in a diagram. However, the concept of field lines is very useful since the electric field strength is the number of lines passing through a unit area. This concept can simplify the analysis of many situations, and some of these will be illustrated in the next section.

EXAMPLE 5

In Example 1 the charge on each object was calculated to be 5.3×10^{-9} C. How many lines would one say emanate from such a charge?

The number is given by

$$N = 4\pi \times 9 \times 10^9 \ Q$$

where

$$Q = 5.3 \times 10^{-9}$$
$$N = 4\pi \times 9 \times 10^9 \times 5.3 \times 10^{-9}$$
$$= 599$$

Using the concept of field lines, one would say that 599 lines emanate from that charge.

EXAMPLE 6

Find the field strength, in terms of lines per square meter, 5 cm from a charge of 5.3×10^{-9} C. Then find the force on a like charge (5.3×10^{-9} C) at that position.

In the previous example it had been shown that 599 lines emanate from 5.3×10^{-9} C. The area of a sphere 0.05 m in radius is given by

$$A = 4\pi r^2$$
$$= 4\pi \times 0.05^2 \ \text{m}^2$$
$$= 0.0314 \ \text{m}^2$$

Through this area there are 599 lines, which would give the number per square meter, the electric field, as

$$\mathscr{E} = \frac{599 \ \text{lines}}{0.0314 \ \text{m}^2} = 19{,}000 \ \text{lines/m}^2$$

The field will be 19,000 lines/m^2 or 19,000 N/C. This can also be written as 1.9×10^4 N/C.

The force on 5.3×10^{-9} C will be

$$F = \mathscr{E}Q$$
$$= 1.9 \times 10^4 \ \text{N/C} \times 5.3 \times 10^{-9} \ \text{C}$$
$$= 1.00 \times 10^{-4} \ \text{N}$$

This answer checks with that of Example 1.

The picture may be clearer if the area is expressed in square centimeters. The area of a sphere of 5 cm radius is 314 cm^2. There are 599 lines through this area, or just under 2 per square centimeter, 1.9 to be precise.

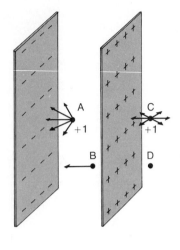

Figure 14–8 The field between and outside a pair of charged plates. If the plates are effectively infinitely large, the field outside the plates at positions such as C and D is zero. The field between the plates will be straight across, as at B.

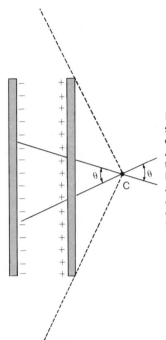

Figure 14–9 The reason why there is no field outside parallel plates. Although the charges on the nearby plate have a large effect because they are close, within a given angle there are more charges on the more distant plate.

14–2–2   FIELD BETWEEN PARALLEL PLATES

To show how to use the concept of field lines, and to introduce a situation that will be of later practical value, let us find the field between two large parallel plates with charges evenly distributed on the surfaces. One plate will have a positive charge, the other a negative charge. A battery is a good source of charge. Many electrical instruments or devices make use of such parallel or almost parallel plates. The concepts also apply to nerve sheaths and cell membranes, which accumulate positive ions on one side and negative ions on the other.

Such parallel plates are illustrated in Figure 14–8 with search charges of +1 unit at A, B, C, and D. The charge at A shows a few of the many forces toward or away from the individual charges on the plates. The sideways components of all these forces cancel, and the resultant force is straight across between the plates, as shown acting on the charge at B. The charges outside the plates at C and D have no net force at all on them if the size of the plates is very large compared to the plate separation. This is because the attraction to one plate will be canceled by the repulsion from the other. This may seem puzzling because one plate is further from the charge than the other. However, as shown in Figure 14–9, inside any given angle such as θ, the repulsion from the near plate is due to fewer charges than is the attraction to the more distant plate. It will be found that these cancel. This does not occur at the

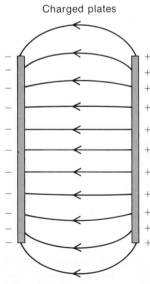

Charged plates

Figure 14–10 The pattern of the electric field lines between parallel plates.

edges of the plates, as shown by the charges outside the dotted lines. If the plates are large compared to their separation, the force outside the plates will be effectively zero.

From charge B it is seen that the direction of the field is straight across from the positive plate to the negative plate, as in Figure 14–10. At the edges of the plates the field lines do bend outward, an **edge effect,** but this can be ignored in this analysis.

All of the field lines emanating from the positive charges go directly across between the plates and terminate on the negative charges. The number of lines per unit area between the plates, as seen in Figure 14–11, is the same as the number leaving each unit area. The charge per unit area is called the **area density of charge** and is often represented by the Greek lower case sigma, σ. In terms of the total charge Q on a plate of area A, we say that $\sigma = Q/A$. The number of lines emanating from a charge σ is $4\pi \times 9 \times 10^9\ \sigma$, and this is also the field strength between the plates. That is, for parallel plates of area A, charged to have $+Q$ on one and $-Q$ on the other or a charge density σ, the electric field between the plates is given by

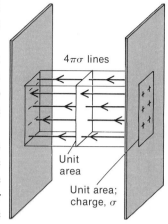

Figure 14–11 A diagram illustrating the calculation of the field strength between charged plates.

$$\mathscr{E} = 4\pi \times 9 \times 10^9\ \sigma$$
$$= 4\pi \times 9 \times 10^9\ Q/A$$

The units for $\mathscr{E}$ are expressed either as lines per m² or as N/C.

The factor of 10^9 indicates that fields would be very large. But the coulomb is actually a very large amount of charge ever to find in this type of situation. The charges encountered in real situations would ordinarily be much smaller, measured in units of microcoulombs. One microcoulomb (μC) is 10^{-6} coulomb.

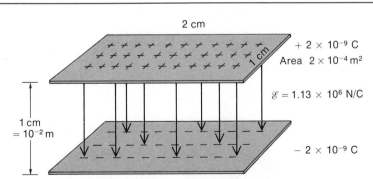

Figure 14–12 The charged parallel plates and the electric field between them.

EXAMPLE 7

Find the electric field between two metal plates as in Figure 14–12, each 1 cm × 2 cm, spaced 1 cm apart. There is a charge of 2×10^{-9} C on one and -2×10^{-9} C on the other.

The number of lines emitted from 1 C is 1.13×10^{11}. From 2×10^{-9} C there will be $2 \times 10^{-9} \times 1.13 \times 10^{11} = 2.26 \times 10^2 = 226$ lines.

The area in square meters is 10^{-2} m $\times$ 2×10^{-2} m $= 2 \times 10^{-4}$ m². The number of lines per square meter is the field strength. If from 2×10^{-4} m² there are 226 lines, then from 1 m² there would be $226/(2 \times 10^{-4}) = 1.13 \times 10^{6}$ lines.
The field is then 1.13×10^{6} N/C.

An example of a device in which this phenomenon and calculation is used is the **oscilloscope.** An oscilloscope tube is shown in Figure 14–13. A beam of electrons is "shot" down the tube between two sets of parallel plates to strike the phosphor on the end of the tube, where a glowing spot is produced. The electron beam can be deflected by charges on the parallel plates. The amount of deflection depends on the field between the plates. The force is toward the positive plate, of an amount $e\mathscr{E}$. This force causes the electron to accelerate in that direction while it is between the plates, so it emerges toward the face of the tube to hit it in a different position than it would strike if there were no charges on the plates.

The electron beam can be made to move around on the face of the tube by changing the charges on the two sets of plates; if the motion is sufficiently rapid, the result to the eye will be a line on the face of the tube. The amount of charge on the plates depends on the voltage applied to them. The voltage in turn may be a signal from something that is detecting heart or nerve impulses, and these are then displayed on the face of the tube.

14–3 VOLTAGE OR POTENTIAL

The term "voltage" is a common one, and basically it is related to work done when a charge is moved. This in turn depends on the force on the charge, which depends on the electric

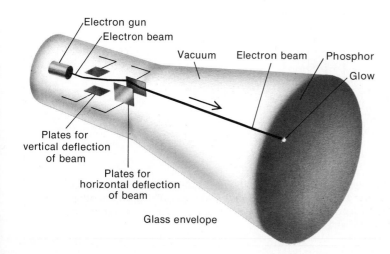

Figure 14–13 An oscilloscope tube. An electron beam passes between two sets of plates. The electric field between them changes the direction of the beam so that it can be made to hit anywhere on the phosphorescent face of the tube.

Electron gun
Electron beam
Vacuum   Electron beam   Phosphor
Glow
Plates for vertical deflection of beam
Plates for horizontal deflection of beam
Glass envelope

field. The term "potential" in electricity is almost synonymous with voltage. This is a convenient place to introduce the basic meaning of voltage.

14-3-1 THE VOLT

Basically, **voltage** refers to **potential energy per unit charge.** It is from this definition that voltage is frequently referred to as just **potential.** The term **volt** is a tribute to the memory of Alessandro Volta, who in 1800 invented the first of the devices that are now called "batteries."

In SI units, the unit of charge is the coulomb; also, in these systems energy is measured in joules. If, in moving a charge of one coulomb between two points, the work done is one joule, it is said that the **potential difference** between those two points is 1 volt. Across a voltage V, the work done on one coulomb is V joules; on q coulombs it would be qV joules.

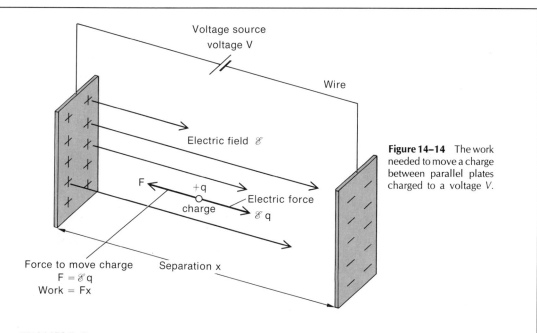

Figure 14–14 The work needed to move a charge between parallel plates charged to a voltage V.

EXAMPLE 8

Consider the hypothetical situation illustrated in Figure 14–14. A battery of voltage V (note the symbol for a battery) is connected to two metal plates so that one becomes positively charged and the other becomes negatively charged. The small object with a positive charge, $+q$, between the plates would be repelled from the positive and attracted to the negative plate. If you originally held the charge just outside the negative plate, you could apply a force F to move it across to just outside the positive plate.

If the charge was 1.6 C and the voltage was 6 volts, then how much work would be done in moving the charge between the plates; also, what would the potential energy of that charge be, with respect to the negative plate, when it was at the positive plate?

The work is qV or charge times voltage.

Work = 1.6 C × 6 V
      = 9.6 joules

The work done in moving the charge is 9.6 joules. This is the potential energy it has when it is at the positive plate. The P.E. of the 1.6 C is 9.6 joules. The P.E. per coulomb is 9.6 J/1.6 C = 6.0 J/C. This checks as being the voltage.

14-3-2   THE ELECTRON VOLT

In Chapter 4 the unit of energy of a particle, the electron volt, was introduced without describing its origin.

The potential energy of a charge q at a position where the voltage is V is just qV. This is the amount of work that would have to be done on a charge $+q$ in moving it from a negatively charged plate across to a positively charged plate. If that charge was released just outside the positive plate, it would accelerate toward the negative one and the potential energy qV would change to kinetic energy. Similarly, if a charge $-q$ were released at the negative plate, it would arrive at the positive plate with a kinetic energy again equal to qV. In practical situations, the charges dealt with are usually electronic charges of size e (negative for electrons and positive for protons).

If a charge e moves across a voltage V, the energy it acquires is eV. The charge e has been measured to be 1.60×10^{-19} coulomb. If V is 1 volt, the electronic charge will acquire 1.60×10^{-19} joule of energy. If it moves across 2 volts, it will acquire double this amount of energy. The quantity 1.60×10^{-19} joule occurs so frequently that it has been found convenient to use that quantity as a unit for measurement of energy per particle. The name given to it is the electron volt, abbreviated as eV, where

$$1 \text{ eV} = 1.60 \times 10^{-19} \text{ joule per particle}$$

The common multiples and submultiples of the electron volt are

$$1000 \text{ eV} = 1 \text{ keV} \qquad \text{(k is for kilo)}$$
$$1{,}000{,}000 \text{ eV} = 10^6 \text{ eV} = 1 \text{ MeV} \qquad \text{(M is for mega)}$$
$$10^9 \text{ eV} = 1 \text{ GeV} \qquad \text{(G is for giga)}$$
$$1/1000 \text{ eV} = 1 \text{ meV} \qquad \text{(m is for milli)}$$

The energy of a particle can be expressed in electron volts no matter how that energy was acquired. Charged plates and actual charged voltages are not necessary. Radium ejects alpha particles, some of which have an energy of about 4 million electron volts or 4 MeV. Electrons may be whirled in a betatron (a type of particle accelerator) until they acquire an energy of perhaps 100 MeV, even though such a voltage does not exist in the machine. A modern accelerator may give particles many GeV (one GeV is 10^9 eV). On the other end of the scale, electrons ejected by light shining on the sensitive surface of a phototube may have only about 1 eV. Slow neutrons, which are not charged, will move with about 0.02 eV of kinetic energy per particle.

14-4   ACCELERATING CHARGED PARTICLES

To study the relation between voltage and speed or energy of charges, first imagine two charged plates and a charge that is free to move about or to be pushed around between those plates, as in Figure 14-14. It will be assumed, at least at first, that there is a vacuum between the plates so that gas molecules do not interfere with the motion of the charges. If the movable charge is positive, pushing it toward the positive plate would require work. It would be like pushing a mass up a hill, giving it potential energy. If the charge were released at the positive plate, it would accelerate toward the negative plate. Like a mass sliding down a hill (Figure 14-15), it would gradually acquire

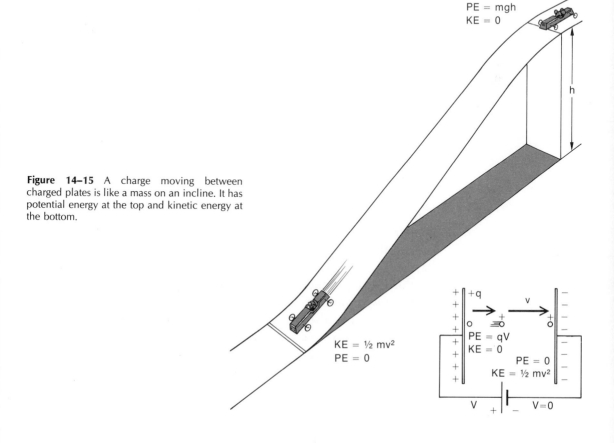

Figure 14-15 A charge moving between charged plates is like a mass on an incline. It has potential energy at the top and kinetic energy at the bottom.

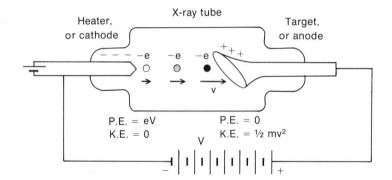

Figure 14–16 An x-ray tube, showing an electron with potential energy at the cathode and kinetic energy at the anode.

kinetic energy. If the moving charge was negative, the top of the "hill" would be the negative plate.

Many modern devices make use of a high voltage to accelerate particles to high speeds. One of these is the x-ray machine. In an x-ray tube, a high voltage V is put across the tube, as shown in Figure 14–16. At the negative end is the source of electrons, called the **cathode.** The positive terminal, called the **anode** or the **target,** is a piece of heavy metal embedded in a large mass of copper. The copper conducts heat away. Electrons released at the cathode accelerate to the anode and strike it with an energy eV. If V is 100,000 volts, the energy of each electron as it hits the target is 100,000 electron volts or 100 keV. When the electron strikes the heavy metal target, a part of its kinetic energy is changed to heat energy and a part changes to a form of electromagnetic radiation variously called x-rays, Roentgen rays, or **bremsstrahlung** (from the German *bremsen,* to brake as a car, and *strahlung,* radiation).

14–4–1 THE ELECTRON GUN

Many devices require a narrow beam of high speed electrons, and this beam is produced by an electron gun. The picture on a TV tube is produced by a beam of electrons striking a phosphor-coated screen, producing only a small bright dot; but this dot sweeps back and forth, up and down, varying all the while in intensity and producing a picture. Thirty complete pictures are produced per second, and the result to our eyes is a picture with motion. A cathode ray tube in an oscilloscope, as in Figure 14–13, also makes use of a moving beam of electrons striking a phosphor. The source of the electron beam is the electron gun at the back of the tube. An electron gun may be used also to inject a beam into a betatron or linear accelerator. These devices are sometimes used as sources of radiation for therapy or study.

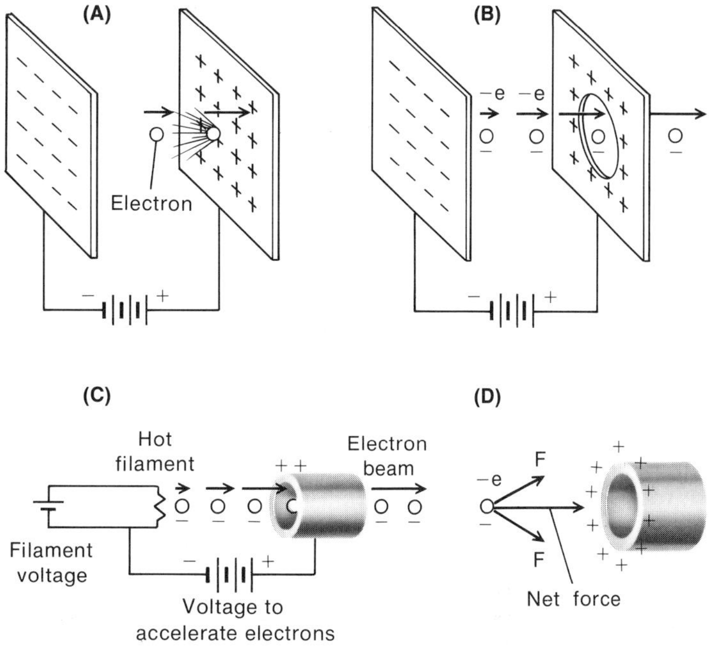

Figure 14-17 Electrons passing between parallel plates (a), through a hole in a plate (b), and from a hot filament to a charged ring or tube (c). The forces on an electron are directed toward the center of the tube (d).

The principle of the electron gun will be illustrated in a series of steps. First, if an electron is ejected from the negative plate of a charged capacitor, it will be attracted to the positive plate, arriving there at a speed depending on the voltage across the plates. The energy it has will be eV. The electric field is straight across between the plates as in Figure 14–11, so this is the direction of the acceleration of the electron. If a hole is drilled in the positive plate, some of the electrons will go through the hole and form a beam as in Figure 14–17. As they approach the hole, they are attracted to all sides of it, and the net force is straight through the hole.

As a refinement, consider the attraction of an electron emitted from a hot filament or cathode to a positively charged ring or tube, as in Figure 14–17(c).

The accelerating voltages used in the electron gun in the back of a cathode ray tube may range from about a thousand volts in a laboratory oscilloscope to 30,000 volts in the picture tube of a color television set. In fact, most color TV sets use three guns, one for each of three colors that combine to give the multicolored picture. The voltage used in television sets is such that some of the kinetic energy of the electrons is converted, as they hit the tube face, to x-rays. This is one of the reasons for having a heavy glass plate in front of the picture tube.

The source of the electrons for an electron gun is a hot wire or filament. The atoms of a hot solid may have enough energy to actually "bump" nearby electrons completely out of the metal. This is referred to as thermionic emission.

14–5   CALCULATION OF PARTICLE SPEEDS

The kinetic energy of a particle of mass m moving at a speed v is given by

$$\text{K.E.} = \frac{1}{2}\, mv^2$$

If the energy is acquired by a particle of charge e moving across a voltage V, the (kinetic) energy it acquires is eV, the same as would be required to push it the other way between the plates. Then,

$$\text{K.E.} = eV$$

Combining these expressions for kinetic energy,

$$\frac{1}{2}\, mv^2 = eV$$

from which

$$v = \sqrt{2eV/m}$$

In this expression, eV is the energy in electron volts. If electrons are being accelerated, then e is 1.60×10^{-19} C and the voltage used is V. Then the product eV is the energy in joules.

It is frequently useful, especially with particles other than electrons, to convert the energy to joules. If the energy is E electron volts, then in joules it is $1.60 \times 10^{-19}\, E$. Write

$$\frac{1}{2}\, mv^2 = 1.60 \times 10^{-19}\, E$$

$$v = \sqrt{2 \times 1.60 \times 10^{-19}\, E/m} \qquad (E \text{ is in electron volts})$$

EXAMPLE 9

Find the speed of an electron after it has accelerated across 1000 volts. Use $v = \sqrt{2eV/m}$, where

$$e = 1.60 \times 10^{-19} \text{ coulomb}$$
$$m = 9.11 \times 10^{-31} \text{ kg}$$
$$V = 1000 \text{ volts} = 10^3 \text{ V}$$

SI units are being used, and v will be in m/s.

$$v = \sqrt{\frac{2 \times 1.60 \times 10^{-19} \times 10^3}{9.11 \times 10^{-31}}}$$

$$= \sqrt{0.351 \times 10^{15}}$$

We bring this to an even power of 10 to carry out the square root operation:

$$v = \sqrt{3.51 \times 10^{14}}$$
$$= 1.87 \times 10^7$$

The speed will be 1.87×10^7 meters/second. This can be expressed in terms of the speed of light, c, which is 3.00×10^8 m/s.

$$v = \frac{1.87 \times 10^7}{3.00 \times 10^8} c$$

$$= 0.062 c$$

The electron is moving at $0.062 c$ or 6.2 per cent of the speed of light.

EXAMPLE 10

One of the particles often encountered is called an alpha particle. An alpha particle is actually a nucleus of helium, and its mass, almost 8000 times the mass of an electron, is 6.6×10^{-27} kg.

Find the speed of an alpha particle that has an energy of one million electron volts, or 1 MeV.

Use
$$v = \sqrt{2 \times 1.60 \times 10^{-19} \, E/m}$$

where
$$E = 10^6 \text{ eV}$$
$$m = 6.60 \times 10^{-27} \text{ kg}$$

SI units are used, so v will be in m/s.

$$v = \sqrt{\frac{2 \times 1.60 \times 10^{-19} \times 10^6}{6.60 \times 10^{-27}}}$$

$$= 7.0 \times 10^6$$

The speed will be 7.0×10^6 m/s. This is only 0.023 of the speed of light.

An interesting result is obtained if you use the formulae developed above to calculate the speed of an electron that has an energy of 1 million electron volts. Carrying it out as in Example 9, the result is that it would be moving at twice the speed of light. When the speeds are measured experimentally, this is found to be incorrect. The reason for our incorrect answer is that we have used the mechanics developed by Newton, and this is not valid near the speed of light. As objects approach the speed of light, the mass increases, and Newtonian mechanics does not take this into account. Rather, the mechanics developed by Einstein from the special theory of relativity in 1905 must be used. Relativity theory also explains why the mass increases, though this phenomenon was actually known before the development of the theory. In 1903 Marie Curie, in her doctoral thesis, presented results of M. Kaufmann, and her interpretation is as follows:

> It follows from the experiments of M. Kaufmann, that for the radium rays, of which the velocity is considerably greater than for the cathode rays, the ratio e/m decreases while the velocity increases . . . we are therefore led to the conclusion that the mass of the particle, m, increases with increase of velocity. . . .
>
> . . . we find that m approaches infinity when v approaches the velocity of light. . . .[*]

The mass was shown by Einstein to depend on the speed v according to

$$m = \frac{m_0}{\sqrt{1 - (v^2/c^2)}}$$

where m_0 is the mass when the particle is at rest and c is the speed of light.

In Table 14–1 are shown the results of the calculation of electron speeds at various energies, comparing the results from Newtonian mechanics and relativistic mechanics.

14–5–1 RELATIVISTIC SPEED

The calculation of the relativistic speed of a charged particle can be based on two concepts. The first is the relation between mass and energy. It is that the amount of energy E equivalent to a certain mass is given by

$$E = (\text{mass equivalent}) \times c^2, \quad \text{or} \quad \text{mass equivalent} = E/c^2$$

---

[*]M. Curie, *Radioactive Substances*, as reproduced by Philosophical Library Inc., New York, 1961, p. 42.

TABLE 14–1 **The speeds of electrons of various energies. The results from nonrelativistic mechanics and from relativistic mechanics are shown. The speed of light is 3.00×10^8 m/s.**

	SPEED IN m/s BASED ON:	
ENERGY IN ELECTRON VOLTS	*Nonrelativistic Mechanics*	*Relativistic Mechanics*
1	5.93×10^5	5.93×10^5
10	1.88×10^6	1.88×10^6
100	5.93×10^6	5.93×10^6
1,000	1.87×10^7	1.87×10^7
10,000	5.93×10^7	5.93×10^7
100,000	1.88×10^8	1.64×10^8
1,000,000	(5.93×10^8)	2.82×10^8

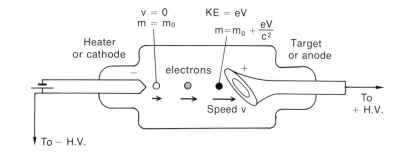

Figure 14–18 At relativistic speeds, a particle increases in mass as it gains energy.

Secondly, the mass of a particle increases with speed and is given by

$$m = \frac{m_0}{\sqrt{1 - (v^2/c^2)}}$$

The speed of light is c, and m_0 is the "rest mass" of the particle, the mass measured when it is not moving.

An electron at rest has a mass $m_0 = 9.11 \times 10^{-31}$ kg. The energy it acquires is eV, which is converted into the increased mass, as illustrated in Figure 14–18. The moving mass m is given by m_0 plus the mass equivalent to the energy eV that the particle was given. This added mass is given by E/c^2, where $E = eV$. Then:

$$\text{change in mass} = eV/c^2$$

so that

$$m = m_0 + (eV/c^2)$$

But also

$$m = \frac{m_0}{\sqrt{1 - (v^2/c^2)}}$$

Therefore,

$$m_0 + (eV/c^2) = \frac{m_0}{\sqrt{1 - (v^2/c^2)}}$$

The speed observed is v, and this equation may be solved for it. The speeds so calculated do coincide with those actually measured.

It is only a matter of algebra to solve for v in a general way or to put in the numbers for the various quantities (in SI units) and solve numerically for the speed v. The solution for the speed v is carried through as follows,
Solve for $\sqrt{1 - (v^2/c^2)}$ to get:

$$\sqrt{1 - (v^2/c^2)} = \frac{m_0}{m_0 + (eV/c^2)}$$

Divide the numerator and denominator of the right hand side by m_0 to get

$$\sqrt{1 - (v^2/c^2)} = \frac{1}{1 + \dfrac{eV}{m_0 c^2}}$$

Square both sides, and put the equation in the form

$$v^2/c^2 = 1 - \frac{1}{\left(1 + \dfrac{eV}{m_0 c^2}\right)^2}$$

Solve for v to get

$$v = c\sqrt{1 - \frac{1}{\left(1 + \dfrac{eV}{m_0 c^2}\right)^2}} \quad \text{or} \quad v/c = \sqrt{1 - \frac{1}{\left(1 + \dfrac{eV}{m_0 c^2}\right)^2}}$$

EXAMPLE 11

Find the speed of an electron that has accelerated across a million volts; that is, its energy is 1 MeV. Express it as a speed in m/s and as a fraction of the speed of light.

Use

$$v/c = \sqrt{1 - \frac{1}{\left(1 + \dfrac{eV}{m_0 c^2}\right)^2}}$$

where

$$e = 1.60 \times 10^{-19} \text{ C}$$
$$V = 10^6 \text{ volts}$$
$$eV = 1.60 \times 10^{-13} \text{ CV}$$
$$m_0 = 9.11 \times 10^{-31} \text{ kg}$$
$$c = 3.00 \times 10^8 \text{ m/s}$$
$$m_0 c^2 = 8.2 \times 10^{-14}$$
$$eV/m_0 c^2 = 1.95$$
$$1 + (eV/m_0 c^2) = 2.95$$

$$\frac{v}{c} = \sqrt{1 - \frac{1}{(2.95)^2}} = \sqrt{1 - \frac{1}{8.70}} = \sqrt{0.885}$$

$$= 0.941$$
$$v = 0.941 \times 3.00 \times 10^8 \text{ m/s}$$
$$= 2.82 \times 10^8 \text{ m/s}$$

That 1 MeV electron moves at 2.82×10^8 m/s. This is 94.1 per cent of the speed of light.

The expression for the speed of a particle with energy eV is shown above to be:

$$v = c \sqrt{1 - \frac{1}{\left(1 + \frac{eV}{m_0 c^2}\right)^2}}$$

As the accelerating voltage V gets very large, the term $1/[1 + (eV/m_0 c^2)]^2$ becomes very small. The quantity under the root sign is then one minus a very small quantity, and it is therefore always less than one, though it may be very close to it. No matter how large the voltage V is, the speed v will always be less than the speed of light c, which is therefore the limiting speed.

14-6   VOLTAGE AND ELECTRIC FIELD

Moving a charged particle between negatively and positively charged objects requires work, and the amount of work depends on the voltage. Also, work is force times distance: force is related to electric field, and therefore, somehow, voltage must be related to the electric field.

To relate the various quantities mentioned, consider the charged plates shown in Figure 14–19. The voltage across the plates is V, and the work required to move the charge q from one plate to the other is qV. The force F on the charge is $q\mathscr{E}$, where $\mathscr{E}$ is the electric field between the plates. The work done to move

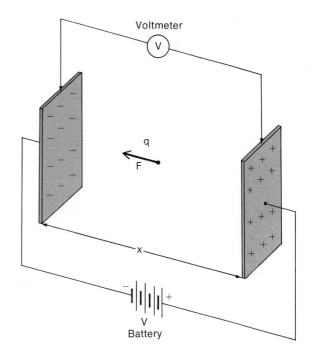

Figure 14–19   Diagram used in finding the relation between voltage and electric field.

the charge from one plate to the other is Fx (where x is the distance), and $F = q\mathscr{E}$. Therefore,

$$\text{work} = qV$$

and

$$\text{work} = q\mathscr{E}x$$

from which

$$q\mathscr{E}x = qV$$

Cancel q to get

$$\mathscr{E}x = V$$

or

$$\boxed{\mathscr{E} = V/x}$$

Unexpectedly, perhaps, this introduces another unit for expression of the strength of an electric field: a unit of voltage divided by a unit of distance. Two sets of units for electric field have already been dealt with. All three sets of units are identical and interchangeable, and in the SI system they are:

$$\text{volts/meter} = \text{newtons/coulomb} = \text{lines/meter}^2$$

EXAMPLE 12

In Example 7 and Figure 14–12, it was shown that two plates charged to 2×10^{-9} C would have a field between them of 1.13×10^6 N/C.
(a) Express this field in volts per meter.
(b) Find the voltage across the plates.
(c) What would be the force on a unit charge between the plates?
(d) What work in joules would be done in moving the unit charge from one plate to the other?

ANSWERS

(a) Volts per meter are the same as newtons per coulomb, so the field is 1.13×10^6 volts per meter.
(b) The electric field is given by $\mathscr{E} = V/x$, where x is the plate separation, 10^{-2} m in this example. The voltage V is $\mathscr{E}x$, so

$$V = 1.13 \times 10^6 \, \frac{\text{volts}}{\text{meter}} \times 10^{-2} \text{ m}$$
$$= 1.13 \times 10^4 \text{ volts}$$
$$= 11,300 \text{ volts}$$

The plates in that example would acquire that charge if 11,300 volts was applied. Though such parallel plates were illustrated as used in an oscilloscope, the voltages applied to the plates in an oscilloscope will be only of the order of 1 to 10 volts, not 11,300 as in this example.
(c) The force on a unit charge in a field of 1.13×10^6 N/C is just 1.13×10^6 N.
(d) If this unit charge is moved 1 cm (10^{-2} m) across between the plates, the work is force $\times$ distance or 1.13×10^6 N $\times 10^{-2}$ m $= 1.13 \times 10^4$ joules. The work per unit charge is the voltage, and this checks with part (b) above.

14–7   THE PARALLEL PLATE CAPACITOR

Consider a voltage V applied across two parallel plates of area A and separation x. A charge of $+Q$ will go on one plate, and $-Q$ on the other. The amount of charge on each plate depends not only on the voltage but also on the physical dimensions of the system. For a given system, a set of plates for example, a certain amount of charge will go on at a voltage of one volt, double that charge at two volts, and so forth. The amount of charge for each volt applied is referred to as the **capacity** of the system. If the charge is Q coulombs with a voltage V, the capacity is given by

$$C = Q/V$$

The units are **coulombs per volt;** one coulomb per volt is also referred to as a **farad** (after Michael Faraday). The term **capacitance** has the same meaning as capacity in electricity.

$$1 \text{ coulomb per volt} = 1 \text{ farad}$$
$$= 1 \text{ F}$$

In practice, a capacity of one farad is found to be so large that it is rarely realized. The size of a **capacitor** is commonly measured in **microfarads** (μF) or **picofarads** (pF), where

$$1 \ \mu\text{F} = 10^{-6} \text{ farad}$$
$$1 \ \text{pF} = 10^{-12} \text{ farad}$$

EXAMPLE 13

In Examples 7 and 8, a set of plates 1 cm $\times$ 2 cm spaced 1 cm apart was charged to 11,300 volts and held 0.02×10^{-6} coulomb. What was the capacity of the system? (C is used for capacity; the unit coulomb is written out.)
Use
$$C = Q/V$$
where
$$Q = 2 \times 10^{-8} \text{ coulomb}$$
$$V = 1.13 \times 10^4 \text{ volts}$$
$$C = \frac{2 \times 10^{-8} \text{ coulomb}}{1.13 \times 10^4 \text{ volts}}$$
$$= 1.77 \times 10^{-12} \text{ farad}$$
$$= 1.77 \text{ pF}$$

The capacity is 1.77 picofarads or 1.77×10^{-12} coulombs/volt.

14–7–1 THE DIELECTRIC CONSTANT

When a voltage is applied to a capacitor, a certain amount of charge will flow onto it. The amount of charge depends not only on the voltage but also on the physical dimensions: plate area and separation. Just as the capacity of a pail to hold

TABLE 14–2 Dielectric constant (specific inductive capacities), k. Data culled principally from *The Handbook of Chemistry and Physics,* **Chemical Rubber Company, Cleveland. The data are in approximate order of increasing** k.

MATERIAL	k
vacuum	1.0000 (by definition)
air	1.0006
liquid oxygen	1.5
paraffin	2.0–2.5
ice	2–3
polyethylene	2.3
rubber, hard	2.8
beeswax	2.75–3.0
nylon	3.5
quartz, fused	3.75–4.1
glass, Pyrex	3.8–5.1
diamond	5.5
cell membrane	(?) probably 5–10
silver bromide	12.2
ethanol	24.3
water, 100°C	55.3
water, 50°C	69.9
water, 20°C	80.4
water, 0°C	88.0

water depends on its dimensions, so does the capacity of a set of parallel plates to hold charge. One difference is that the electrical capacitor is also influenced by the material between the plates. That material must be an insulator, a **dielectric,** so that the charge stays on the plates and doesn't flow from one to the other. For example, if a certain set of plates has a certain capacity (charge per volt) with air between the plates and if glass is then placed between the plates, the capacity will be increased by about four times. The factor by which the capacity increases when a material is put between the plates is called the dielectric constant of the material. Some dielectric constants are listed in Table 14–2.

14–7–2 BIOLOGICAL CELLS AS CAPACITORS

Electrical phenomena in cells are of basic significance. In general, there is a net negative charge on the inside of the cell, while the fluid outside the cell contains positive ions. The charges are separated by the thin cell membrane, which makes the system, effectively, a capacitor (see Figure 14–20). The membrane is very thin compared to the cell size. Each cell is "wrapped" inside a parallel plate condenser; we may speak of the capacity of the system or, more usefully, the capacity per unit area of cell surface. Nerve cells are somewhat special. They are, effectively, long thin cylinders and electrical disturbances travel along them, constituting nerve impulses.

Here are some typical figures concerning cells: The cell membrane is about 0.1 μm thick, and the voltage across the membrane is about 0.090 volt or 90 millivolts (mV). One millivolt is a thousandth of a volt. These voltages are measured with extremely small-tipped probes, one in the fluid outside the cell and one penetrating the membrane, as shown in Figure 14–21. The probe that is put into the cell is made of a glass tube drawn out to have a tip less than 1 μm in diameter. It is filled with a conducting KCl solution that can be connected to the voltage measuring instrument by wires. The membrane is somewhat permeable, and positive sodium ions (Na^+) are attracted to the negative charges inside. Some of these ions do penetrate the barrier. This would soon neutralize the inside negative charge if there were not a mechanism in the cell to move these sodium ions out again. For each ion moved out the work required is the charge $+e$ times the voltage, 0.090 volt or 0.090 electron volt. This action of moving sodium ions out of the cell is referred to as sodium pumping. Though the energy per ion moved across that voltage is small, there are so many ions that it has been estimated that possibly 20 per cent of the energy we consume in a resting state is used in moving the sodium ions out of the cells across the cell wall voltage, that is, in operating what is referred to as the sodium pump.

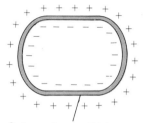

Cell membrane, thickness x

Figure 14–20 A biological cell as a capacitor. The plate separation is the thickness of the cell membrane, and the area is the surface area of the cell.

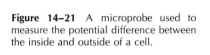

Figure 14–21 A microprobe used to measure the potential difference between the inside and outside of a cell.

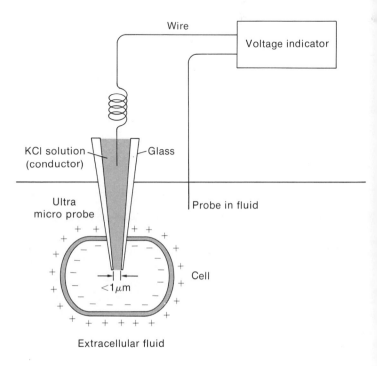

(A)

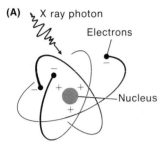

(B)

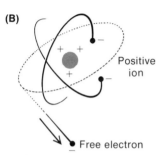

Figure 14–22 An x-ray photon ejecting an electron from an atom, and leaving the free negative electron and a positive ion.

14–7–3 CAPACITORS FOR RADIATION MEASUREMENT

X-rays were discovered by Wilhelm Roentgen in 1895; in March 1896 he reported:

> ". . . . I observed the following phenomenon: Electrified bodies in air, charged positively or negatively, are discharged if x-rays fall upon them; and this process goes on the more rapidly the more intense the rays are . . ."

The mechanism of this process is now well understood and it is exceedingly useful, for biological effects seem to be proportional to the amount of such an electrical effect.

One property of some types of radiation, such as ultraviolet, x-rays, and the various radiations from radioactive material, is that they may cause ionization of the air through which they pass. The radiation, in the process of being absorbed by the air (or other material), transfers energy to the electrons of the atoms. In many instances the electron is ejected from the atom as in Figure 14–22. If the radiation producing it was of high energy, such as an x-ray, the secondary electron will have an almost comparable energy. As it speeds through the air of the chamber it will knock other electrons free, leaving more positive ions also. If this process is allowed to occur in the air between the plates of a charged capacitor, as in Figure 14–23, the electrons will be attracted to the positive plate and the positive ions to the negative plate. When the electrons and ions reach the plates, a bit of charge on each plate will be neutralized, causing a reduction in the charge and voltage on the capacitor. The more intense the radiation, the faster the capacitor will lose its charge. Also, a low intensity of radiation for a long time will cause the same charge reduction as a high intensity for a short time. A capacitor with air between the plates can be a very sensitive and basic way to detect and measure radiations.

In Figure 14–24 is a plot of the voltage on a charged

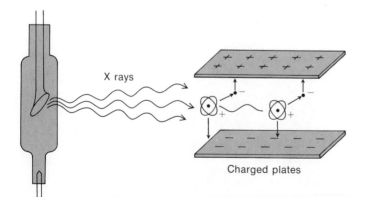

X rays

Figure 14–23 X-rays produce electric charges (+ and –) in the air between charged plates. These charges, free electrons and positive ions, are attracted to the plates where they neutralize some of the charge.

Charged plates

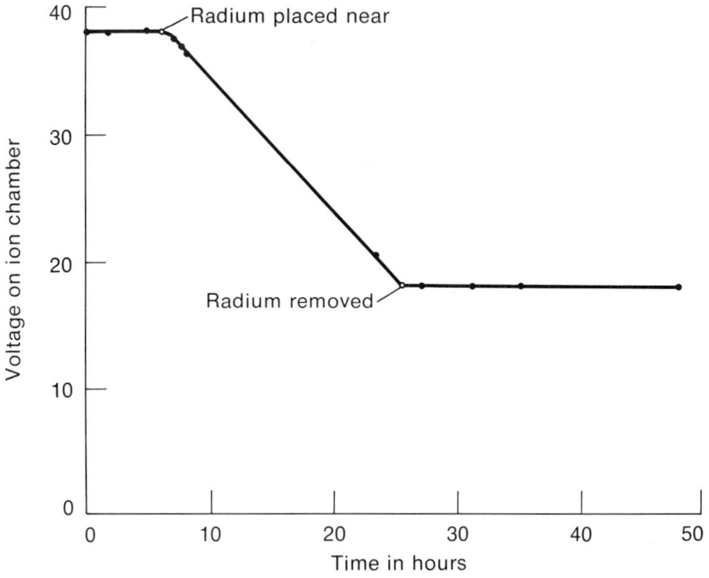

Figure 14–24 Actual voltage measurements on an air capacitor. The voltage remains steady until a radium sample is brought near; then it begins to discharge. Radium emits gamma rays, which are similar to high voltage x-rays.

(graph: Voltage on ion chamber vs. Time in hours; "Radium placed near" and "Radium removed" annotations)

capacitor as a function of time. The voltage was constant until a radium sample was brought near. Then the charge and voltage dropped until the radium was removed.

It is probably quite safe to say that the majority of the instruments for the measurement of the dose from x-ray or gamma ray sources are fundamentally air capacitors or, as they are sometimes called, ionization chambers. They were used even prior to 1900, and today they not only form the standards for radiation measurement but also are carried by almost every person who works near radioactive materials.

In Figure 14–3 a pocket type radiation dosimeter was illustrated. Inside it are an air chamber and a capacitor. A quartz fiber electrometer (Figure 14–2) is used to register the charge on the capacitor. When radiation such as x-rays or gamma rays hit the chamber, ionization is produced in the cavity and the charge is reduced. The image of the fiber moves across a scale to indicate the radiation dose received. A cross-section of such a dosimeter is shown in Figure 14–25.

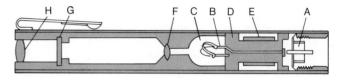

Figure 14–25 A pocket-type radiation dosimeter. The ionization occurs in the cavity at C. The charge or voltage across the cavity is indicated by the electrometer, *B. H, G,* and *F* form a microscope with which to view the electrometer.

A: Charging switch

B: Electrometer

C: Air cavity

D: Insulation

E: Capacitor (to vary total dose required to discharge the electrometer)

F: Objective lens of microscope (to focus electrometer fiber onto eyepiece scale)

G: Eyepiece scale

H: Eye lens of microscope

14–8 RADIATION DOSE

The unit developed for the measurement of exposure "dose" of radiation with x-rays or gamma rays is based on the ionization effect. It has been given the name **roentgen,** abbreviated r. One roentgen is the amount of radiation that produces in one cubic centimeter of dry air at 0°C and 760 mm pressure (standard temperature and pressure, abbreviated s.t.p.) one electrostatic unit of charge of each sign. Transformed to SI units, this is 3.33×10^{-4} coulombs/m³. This is not quite the exact wording of the formal definition, but it adequately describes the unit.

The roentgen describes a total radiation exposure irrespective of the time over which it occurs. In a single radiation session it may be necessary, for example, to given an exposure dose of about 200 r to a tumor. It does not matter if this is given in one minute or in ten, as long as the total radiation reaches the desired amount. It means only that if the radiation fell on air at the tumor position, 200 esu of charge would have been produced in each cm³ of the air. The output of the radiation-producing machine can be measured with an ion chamber, and would be expressed in the number or roentgens per unit time, that is, in roentgens per minute or r/s.

DISCUSSION PROBLEMS

14–1   1. Where would the excess electrons be located when a thick-walled hollow metal sphere is charged negatively?

14–2–2   2. Deduce the pattern of electric field lines between a point positive charge, Q, and a flat plane on which there is an evenly distributed charge of $-Q$. Sketch the field.

14–3   3. What would be the shape of the path of an electron that is projected into an electric field between parallel plates?

14–4–1   4. A modification of an electron gun could be designed to eject positively charged protons (hydrogen ions) from the back of a rocket. Compare the speeds obtainable with such a device, called an ion engine, to the speed of ejection of rocket propellant using chemical fuel. Discuss the possible advantage of a rocket ion engine.

14–5   5. The target of an x-ray tube must be cooled. The target is the metal block that is struck by the electron beam to produce x-rays. Use your knowledge of mechanical energy to explain the heating.

PROBLEMS

14   1. Calculate the total charge in coulombs on one mole of electrons. This quantity of charge is given the name of 1 faraday.

14–1   2. What two charges of equal size would have a force between them of 1 N when they are 1 m apart? Express the answer in coulombs and in esu.

14–1   3. Use Coulomb's law to calculate the force between two equal charges of 3.33×10^{-10} C when they are 0.01 m apart.

14–1   4. Find the force on a charge of $+5$ microcoulombs (μC) when it is midway between two other charges, which are 15 cm apart and which carry charges of -15 μC and $+15$ μC.

14–1   5. Find the force on a charge of 5 microcoulombs (μC) that is along a line between two charges, one of -15 μC, and the other of $+15$ μC, and 5.0 cm from the charge of -15 μC. The charges are 15 cm apart.

14–1   6. Find the charge on two small spheres that are suspended as in Figure 14–1 by threads 15 cm long, from points of support

1.0 cm apart. The spheres are 0.3 cm in radius and of foam plastic of density 15 kg/m³. The spheres carry equal charges and hang 4.5 cm apart (between their centers).

14-2 7. What charge must be put on a small sphere of mass 0.0306 g to cause it to be suspended by an upward electric field of 3000 N/C?

14-2 8. Find the mass that could be lifted if it were in an electric field of 1000 N/C and had a charge of 1.00 C on it.

14-2 9. Find the electric field that is horizontal past a small sphere that has a mass of 0.0303 grams, contains a charge of 1.0 μC, and when suspended by a thread 0.33 m long hangs at an angle of 15° from the vertical.

14-2 10. Find the electric field midway between two charges of 15 μC but of opposite sign and 0.15 m apart.

14-2 11. Find the speed acquired by an electron when it is accelerated in an electric field of 30,000 N/C over a distance of 0.005 m. The mass of an electron is 9.1×10^{-31} kg.

14-2 12. An electron with a speed of 7×10^6 m/s moves through a space 2.0 cm long where there is an electric field perpendicular to the initial direction of motion. The electric field is 1000 N/C. Find:
(a) the time spent in the field,
(b) the acceleration in the field,
(c) the speed acquired perpendicular to the initial speed,
(d) the direction of motion as it emerges from the field.
(The mass of an electron is 9.1×10^{-31} kg.)

14-2-1 13. (a) Find the number of field lines that come from a charge of 1.00 μC.
(b) Find the area of a sphere 1 m in radius around the charge in (a).
(c) Find the number of field lines going through each unit area of the sphere in (b).
(d) Use $\mathscr{E} = 9 \times 10^9 \ Q/r^2$ to find the field at 1.00 m from a charge of 1.00 μC, and compare the result to (c).

14-2-1 14. Using the concept of electric field lines, find the electric field at the surface of a sphere 1.00 cm in radius when it has a charge of one nanocoulomb distributed evenly over its surface.

14-2-1 15. Find the radius of a sphere that, when charged with one microcoulomb, will have an electric field of 3×10^6 N/C at its surface. If the sphere is in air, a field of that intensity will cause the charge to leak off into the air.

14-2-2 16. Find the total charge on a flat plate 0.30 m square when the field near the surface is outward from it with a strength of 1000 N/C.

14-2-2 17. Two parallel plates in air are to be charged, one positively and one negatively, to such an amount that the electric field between them is 80 N/C. Find the charge on each plate if the area of each plate is 0.0225 m².

14-3-1 18. (a) Find the voltage between two points if 1500 joules are used in moving 12.5 coulombs between those points.
(b) If 12.5 coulombs are moved per second between the points in (a), what is the power used?

14-3-1 19. A charge of 0.016 C is to be moved between two electrically charged plates that have a field of 300 N/C between them. The separation is 0.025 m.
(a) Find the force on the 0.016 C.
(b) Find the work done in moving it from the negative to the positive plate.
(c) Find the potential energy of the charge with respect to the negative plate when it is just outside the positive plate.
(d) Find the potential energy/charge.
(e) What is the voltage between the plates?

14-3-1 20. Find the voltage between two metal plates in air, which are charged to an amount that causes an electric field of 80 N/C between them:
(a) when the separation is 0.010 m,
(b) when the separation is 0.030 m.

14-3-1 21. Two metal plates in air are charged such that 15 J is used to move 3 C between them.
(a) What is the voltage between the plates?
(b) What is the electric field between the plates if the separation is 0.010 m?
(c) What is the electric field between the plates if the separation is 0.030 m?

14-5 22. (a) What is the potential energy of a charge of 1.6 C when it is just outside the positive plate of a pair of metal plates in air charged to 120 volts?
(b) Find the speed the charge in (a) would acquire when it reached the negative plate if it were released just outside the positive plate. The mass is 9.11 grams.
(c) Find the speed that a charge of −1.6 C would acquire if it were released outside the negative plate in a situation similar to that in (a) and (b).
(d) Does the separation of the plates affect the answers above?

14-5 23. Find the speed of the particles from a "gun" that shoots small metal pellets and is built on the principles of an electron gun.

An accelerating voltage of 1000 volts is used. The metal ball is charged to 1.60×10^{-10} C (this charge would be acquired by a ball of 1 mm radius connected to a 1400 volt battery). The mass of the ball is 1.2×10^{-5} kg (a sphere of aluminum 1 mm in radius).

Would the projectile from such a weapon be as dangerous to anyone as the voltage would be to the operator?

14–5   24.   At what accelerating voltage would electrons in an electron gun move at a tenth of the speed of light?

14–5   25.   What voltage would give electrons a speed of 7×10^6 m/s?

14–5   26.   (a) What speed would be acquired by protons accelerated by 10,000 volts in a device like an electron gun? It would be a proton gun.

(b) What momentum would be given to 1000 moles of protons (about 1 kg of hydrogen) accelerated by 10,000 volts in a proton gun? The mass of a proton is 1.67×10^{-27} kg. The charge is $+e$.

(c) If that proton gun served as a rocket engine, what speed would it give to a 1000 kg rocket after ejecting 5.8 kg of protons? (See Discussion Problem 4.)

14–5–1   27.   Use relativistic mechanics to find the speed of an electron accelerated through 200,000 volts.

14–5–2   28.   What voltage would be required to accelerate an electron so that its mass became twice the rest mass?

14–5–1   29.   Use relativistic mechanics to find the speed of an electron accelerated through 2 million volts.

14–6   30.   What voltage would have to be put onto two metal plates 0.010 m apart so that the electric field between them would be 1000 N/C?

14–7   31.   What must be the capacity of an object such that will acquire 4×10^{-4} C when 6 volts are applied?

14–7   32.   Find the charge that would flow onto a capacitor with a rating of 50 μF when 60 volts are applied to it.

14–7   33.   Find the capacity of two metal plates in air, each of which is 2 cm $\times$ 3 cm, that are separated by 1.00 mm of air. To do this work in steps.

(a) When V volts are applied, what would be the electric field between the plates? Express the field in volts/meter and in lines/m².

(b) What is the total number of electric field lines between the plates?

(c) What charge on each plate would give rise to the number of lines found in (b)?

(d) Using the above results, find the charge per volt, which is the capacity in farads.

14–7   34.   In Problem 33 the capacity of a set of parallel plates was calculated but only one quantity, the voltage, was not given as a number.

Calculate the capacity of parallel plates in air, each of area A and separated by a distance x. Follow the same steps outlined in Problem 33.

14–7–1   35.   A capacitor is charged to 90 volts and is then found to have 3×10^{-6} C on each plate (positive on one and negative on the other, of course).

(a) With the battery still connected, glass of $k = 4.0$ is slipped between the plates. Describe what happens to the charge on the plate.

(b) If the battery is disconnected before the glass is inserted, what happens? What changes?

14–8   36.   (a) How much free charge, of either sign, would be produced in 2 cubic centimeters of standard air by 100 milliroentgens of radiation?

(b) What size of capacitor, charged to 40 volts, would have to be connected to the air chamber in (a) in order that 10 r would completely discharge it?

CHAPTER FIFTEEN

CURRENT ELECTRICITY

This chapter will deal with phenomena arising from the motion of charges. The main subject will have to do with the flow of electricity in wires and the common concept of electric current. Moving charges also give rise to what are called magnetic fields. Moving charges interact with the devices known as magnets, or more precisely, with magnetic fields, so magnetic phenomena will be included.

15–1 CURRENT

A movement of electric charge is called an electric current, and a flow of one coulomb per second is given the name of one **ampere** (A) or simply one **amp.** If a charge Q flows in a time t, the current I in amps is given by $I = Q/t$.

Some materials allow electric current to flow through them more easily than others. These are called good conductors. Other materials do not allow current to flow through them, and these are called insulators. There is, of course, a reasonably continuous gradation in ability to conduct charge, so categorization between insulators and conductors cannot always be done with certainty. It is perhaps better to regard all materials as being conductors; some, like the metals, are extremely good conductors, while others, those ordinarily called insulators, are very poor conductors. In ordinary electrical work, materials like plastics or waxes are regarded as being non-conductors or insulators. It is principally in dealing with extremely small currents such as are encountered in radiation measuring instruments that the current through so-called insulators becomes important.

In dealing with ordinary electrical devices such as light bulbs, heaters, or general lab equipment, the current supplied will be measured in amperes, or perhaps in units as small as milliamperes (1 milliamp = 1 mA = 1/1000 amp). The current to

469

the heater or filament of an x-ray tube will be several amps; the current flowing through the tube to produce x-rays will be measured only in milliamps.

The current obtained from photoelectric cells is usually in the range of millionths of an amp or microamps (1 micro-amp = 1 μA = 10^{-6} amp). This is actually a very small current.

The current associated with measurements in biological systems or with ion chambers for measurement of x-rays or gamma radiation may be 10^{-10} or 10^{-16} amp. Measurement of such exceedingly small currents has special problems associated with it!

EXAMPLE 1

Current in amps is flow of charge in coulombs per second. If a current I flows for a time t, the charge Q that passed is the product of amount per second times the number of seconds, or $Q = It$.

(a) If 1 amp flows for 1 minute, how many coulombs have gone by?

$$Q = It = 1 \text{ coulomb/s} \times 60 \text{ s}$$
$$= 60 \text{ coulombs}$$

(b) If 1 microamp flows for an hour, how many coulombs go by?

$$I = 10^{-6} \text{ coulomb/s}$$
$$t = 3600 \text{ s} = 3.6 \times 10^3 \text{ s}$$
$$Q = It$$
$$Q = 10^{-6} \times 3.6 \times 10^3 \text{ coulombs}$$
$$= 3.6 \times 10^{-3} \text{ coulombs}$$

15-1-1 THE NATURE OF A CURRENT

The atoms of metals, which are the best conductors of electricity, have electron arrangements like those shown in Figure 15-1. Most of the electrons are in full shells but one, two, or three are in larger orbits. These outer electrons are attracted to the positive charge on the nucleus, but most of this attractive force is canceled by the repulsion of all the inner electrons. As a result, the outer electrons are referred to as being *loosely bound.*

When the atoms are packed together in an array to form a solid as in Figure 15-2, those outer electrons lose their association with a particular atom. They are as much attracted to the "neighbors" as to the one they were originally with. As a result, they are really free to move inside the metal. If an electric field is applied across the metal, these **conduction electrons,** as they are called, will all move under the force resulting from the field. As in Figure 15-3, the electrons will be pushed to one side of the metal. The accumulated electrons result in a nega-

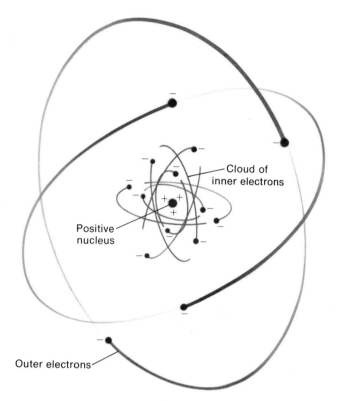

Figure 15-1 The conductors have loosely bound outer electrons.

tively charged region at one end, and the atoms that are left behind are deficient in electrons and have a net positive charge. The motion of the electrons formed a momentary current.

The end result would have been the same if it had been positive charges that moved, though of course in the opposite direction. When the subject of electricity was first being developed and no one knew which kind of charge really moved, it was assumed that the positive charges moved. A current was then regarded as being a motion of positive charges. The end result is the same, anyway, in most cases. We now know better, but still the conventional way to designate a current in a metal is as a motion of positive charges. When the end result is the same, it is usual to accept this (but be smug in your knowledge of reality) rather than to battle with convention.

If a wire is connected between two charged plates as in Figure 15-4, the electrons will flow through and neutralize the positive charges. There will initially be a large current, but as the charge is drained the current will become smaller. (This is the type of situation in which there is an exponential decrease.) The current flowing at any given voltage depends on the physical characteristics of the wire. Some materials conduct electricity more easily than others: they are said to have low resistivity. Also, a large diameter wire would conduct more current for a given voltage than would a thin one. These factors,

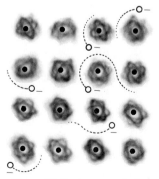

Figure 15-2 In a metal, the conduction electrons are free to move among the atoms.

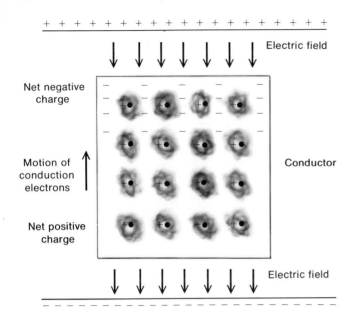

Net negative charge

Motion of conduction electrons

Net positive charge

Conductor

Electric field

Electric field

Figure 15–3 When an electric field is put across a conductor, the conduction electrons are pulled to one side, leaving a net positive charge behind.

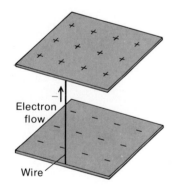

Electron flow

Wire

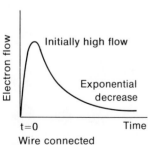

Electron flow

Initially high flow

Exponential decrease

t=0   Time
Wire connected

Figure 15–4 The flow of electrons through a wire to discharge a pair of charged plates.

size and type of wire, together give to that wire the characteristic called resistance.

A battery is similar to a charged capacitor. At one pole is an accumulation of positive charge, and at the other is negative charge. If a wire is put across the battery, a current goes through the wire; but new charge is generated by the battery so that the battery voltage is reasonably steady and the current is constant.

Conduction through ionic solutions is slightly different. In a solution containing positive and negative ions, both move. The positive ions drift in one direction, the negative ions in the other.

Semiconductors are different again. Metallic solids consist of an array of atoms with conduction electrons free to move about; insulating solids consist of arrays of atoms with all electrons tightly bound. The semiconductors are crystalline solids of certain materials with traces of impurity. Germanium and silicon are two of the common materials used for semiconductors. If the impurity is an atom with one more electron than the basic crystalline material, that extra negative electron will not fit into the crystal pattern. An *n* type (negative) semiconductor is obtained. When an electric field is placed across the material, the extra electrons will hop from atom to atom through the crystal, giving a flow of current. If the impurity has one less electron per atom than the basic crystal material, there will be an electron deficiency at one point in the crystal. If an electric field is put across the crystal, an electron from a neighboring atom will jump to that vacant space. An atom with a net positive charge (an electron deficiency) is left behind. Such a semiconductor is a *p* (or positive) type. That *hole* where there is an electron deficiency will quickly be occupied by an electron from the next atom, so the positive hole moves under the influence of the field. In a p type semiconductor the motion of the hole left by the electrons is identical to the motion of a positive charge.

15-2 OHM'S LAW

If a voltage is applied across a piece of material, be it a wire, block, membrane, or any other shape, a current will flow. The size of the current will depend on the voltage. Also, the higher the resistance of the object, the lower the current; that is, the current is inversely proportional to resistance. Letting I be the intensity of the current, V the voltage, and R the resistance, the statements above can be put as a proportionality:

$$I \propto V/R$$

and with a proportionality constant k,

$$I = kV/R$$

The units adopted for general use are such that the constant k is 1. Then, with I in amps, V in volts, and R in a unit called an ohm, the relation is

$$I = V/R \quad \text{or} \quad V = IR \quad \text{or} \quad R = V/I$$

Any of these three forms of the equation is called **Ohm's law.** The last, $R = V/I$, is often used as a definition of **resistance** or impedance: it is the ratio of voltage to current. Of all the relations in electrical circuit work, Ohm's law is by far the most often used.

A source of steady voltage such as a battery is often designated by a long and short line as in Figure 15–5. A resistance is shown by a zig-zag line, while wires of negligible resistance are just straight lines. Meters are shown by circles, like dial faces with their function indicated by I or V. A current meter or **ammeter** is connected so that all the current in the wire goes through it to be registered. A **voltmeter** is connected *between* the two points between which the voltage is to be found. The connections for the voltmeter may be shown as actually connected wires, or by arrows to indicate that the voltage reading can be obtained just by touching the leads to the two points between which the voltage (or potential difference) is to be found. The meter connections could be as in Figure 15–5(b).

The symbol for ohms is the upper case Greek omega, written as Ω.

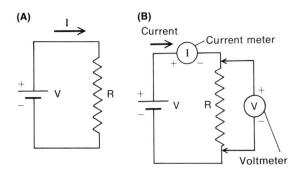

Figure 15–5 An electric circuit and the use of a current meter and a voltmeter.

EXAMPLE 2

With a circuit as in Figure 15–5, if the voltmeter reads 120 volts and the ammeter reads 0.83 amp, what is the resistance?

$$V = 120 \text{ volts}$$
$$I = 0.83 \text{ amp}$$
$$R = V/I$$
$$= 120/0.83 \text{ (with these units, } R \text{ is in ohms)}$$
$$= 145 \text{ ohms}$$

The resistance is 145 ohms.

EXAMPLE 3

If in a simple circuit, as in Figure 15–5, the resistance is known to be 1500 ohms and the voltage is 1.5 volts, what will be the current?

$$V = 1.5 \text{ volts}$$
$$R = 1.5 \times 10^3 \text{ ohms}$$
$$I = V/R$$
$$= 1.5/1.5 \times 10^3 \text{ amps}$$
$$= 1.0 \times 10^{-3} \text{ amp}$$
$$= 1.0 \text{ milliamp}$$

The current will be 1.0 milliamp.

EXAMPLE 4

Consider a current of 3 amps flowing in succession through a 30 ohm resistance, a 20 ohm resistance, and then a 10 ohm resistance, as in Figure 15–6. Find the voltage across each resistor, shown as V_1, V_2, and V_3 in the figure, and then the total battery voltage V.

If a current I flows through a resistor R, the voltage between the ends of the resistor is given by Ohm's law, $V = IR$. In all of the resistors, $I = 3$ amps.

$$\text{When } R = 30\,\Omega, \quad V_1 = 3 \times 30 = 90 \text{ volts}$$
$$\text{When } R = 20\,\Omega, \quad V_2 = 3 \times 20 = 60 \text{ volts}$$
$$\text{When } R = 10\,\Omega, \quad V_3 = 3 \times 10 = 30 \text{ volts}$$

The total voltage is the sum, 180 volts, so this is what must be supplied by the battery.

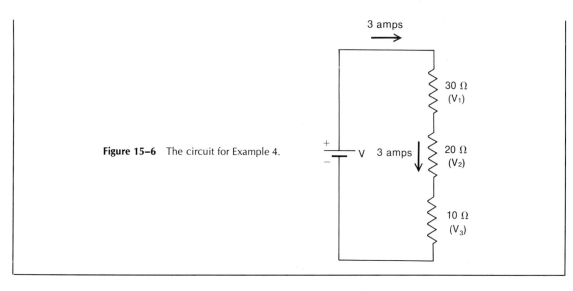

Figure 15-6   The circuit for Example 4.

In the above example, the resistors are connected in such a way that the current that goes through the first one continues through each in turn. The current in each is the same. Resistors connected in this way are referred to as being **in series.** The effective resistance is the numerical sum of the individual resistors or

$$R = R_1 + R_2 + R_3 + \ldots$$

EXAMPLE 5

Find the current when a voltage of 180 volts is applied across a resistor of 60 ohms. This is based on Example 4, but the three resistors are replaced by one resistor that is the sum of the three (30 + 20 + 10 = 60).

$$V = 180 \text{ volts}$$
$$R = 60 \text{ ohms}$$
$$I = V/R = 180 \text{ volts}/60 \text{ ohms}$$
$$= 3 \text{ amps}$$

This is the current that was used in Example 4, and this example shows the validity of replacing several resistors that are in series by one that is their sum.

15-2-1   RESISTORS IN PARALLEL

A **parallel** connection is illustrated in Figure 15-7. A voltmeter that is connected across either resistor is effectively across the voltage source. The current I from the source divides, with I_1 going through one resistor and I_2 through the other so that $I = I_1 + I_2$. By Ohm's law, $I = V/R$ and the resistors R_1 and R_2 could be replaced by one as in Figure 15-7(b), one that

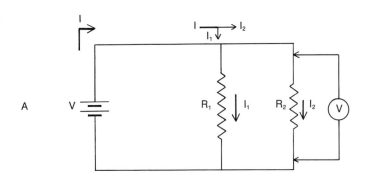

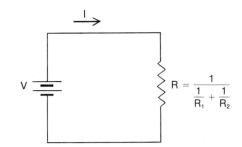

Figure 15–7 (a) A circuit with two resistors in parallel. (b) An equivalent circuit, with R chosen to have such a value that, for the same voltage as in (a), the current is also the same as in (a).

for the same voltage would pass the same current. To find the one resistor that is equivalent to the two in parallel, use

$$I = I_1 + I_2$$

Put $I_1 = V/R_1$ and $I_2 = V/R_2$ so that

$$I = V\left(\frac{1}{R_1} + \frac{1}{R_2}\right)$$

In a circuit with a voltage V and one resistor of value R, the current is given by

$$I = \frac{V}{R} = V\left(\frac{1}{R}\right)$$

The current would be the same as in the circuit with R_1 and R_2 in parallel only if R were chosen of such a size that

$$\frac{1}{R} = \frac{1}{R_1} + \frac{1}{R_2}$$

That is, the two resistors in parallel could be replaced by one that is of a value given by the above relation.

EXAMPLE 6

A voltage source of 100 volts is connected to two resistors in parallel as in Figure 15–7. R_1 is 440 ohms and R_2 is 880 ohms. Find
(a) the current in each resistor,
(b) the total current from the source,
(c) the one resistor that would be the equivalent of those two in parallel.
(a) The voltage across each is 110 volts. By $I = V/R$,

$$I = V/R_1 = (110/440) \text{ amp} = 0.25 \text{ amp}$$
$$I_2 = V/R_2 = (110/880) \text{ amp} = 0.125 \text{ amp}$$

(b) The total current from the source is $(0.25 + 0.125)$ amp $= 0.375$ amp.
(c) The equivalent resistor is found from

$$\frac{1}{R} = \frac{1}{R_1} + \frac{1}{R_2}$$

Solve for R to get

$$R = \frac{R_1 R_2}{R_1 + R_2}$$

$$= \frac{440 \times 880}{440 + 880} \text{ ohms}$$

$$= 293 \text{ ohms}$$

The equivalent resistor has a value of 293 ohms.
Check this by solving a simple circuit for I when $V = 110$ volts and $R = 293$ ohms.

$$I = V/R = (110/293) \text{ amp}$$
$$= 0.375 \text{ amp}$$

The check is satisfactory.

15–3 POWER IN AN ELECTRIC CIRCUIT

When a charge Q is moved across a voltage V, the energy that must be supplied is QV. If this occurs in a time t, the rate at which energy is used is QV/t, and the rate of use of energy is the definition of power. With Q in coulombs and V in volts, the energy is in joules. If the time t is in seconds, the power is in joules/second or watts. Also, Q/t is the current, I. Therefore, the relation giving the power P in watts is

$$P = IV$$

In words, power $=$ current $\times$ voltage, or watts $=$ amps $\times$ volts.
Sometimes it is convenient to express power in terms of resistance and current, or resistance and voltage. To do this, we use Ohm's law and substitute for the unwanted quantity. For

example, in $P = IV$ we substitute for V from Ohm's law, $V = IR$. This results in

$$P = I^2R$$

According to this, watts are also equal to amp^2 × ohms. In a similar way (do this yourself), it can be shown that

$$P = V^2/R$$

This shows that, also, watts are equal to volts^2/ohms.

EXAMPLE 7

What is the current through a light bulb with a rating of 100 watts at 120 volts? Solve $P = IV$ to get $I = P/V$. Substitute the values and find

$$I = 100/120 = 0.83$$

The current will be 0.83 amp.

EXAMPLE 8

In Example 2, we found that a device drawing 0.83 amp at 120 volts would have a resistance of 145 ohms. Just as a check, use $P = V^2/R$ with $V = 120$ volts and $R = 145$ ohms:

$$P = (120^2)/145 = 14400/145$$
$$= 100$$

The power is still 100 watts.

EXAMPLE 9

A typical x-ray tube may operate at a voltage of 100 kilovolts and with a current through the tube of 5 mA. The power is dissipated mostly as heat in the target that the electrons hit, and partly as x-rays. What is the total power?

$$I = 5 \times 10^{-3} \text{ amp}$$
$$V = 100 \text{ kV} = 10^5 \text{ volts}$$
$$P = IV$$
$$= 5 \times 10^{-3} \times 10^5$$
$$= 5 \times 10^2 = 500$$

The power is 500 watts.

This would lead to heating of the target, and perhaps even melting at the point of impact of the electrons, unless means were provided to carry the heat away.

15–4 RESISTANCE AND RESISTIVITY

The resistance of a given piece of material, wire, membrane, or whatever it may be, depends not only on its atomic or molecular structure but also on its physical dimensions.

In Figure 15–8 we show a piece of material put between two charged plates. The current I depends, of course, on the voltage. The bigger the area shown as A, the bigger the total current; but on the other hand the longer the length l, the lower the electric field and hence the current. In a proportionality equation,

$$I \propto \frac{A}{l} V$$

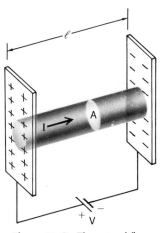

Comparing this with Ohm's law, $I = V/R$, it can be said that the resistance R is given by

$$R \propto l/A$$

Figure 15–8 The rate of flow of charge depends directly on the area and inversely on the length of the conductor.

For a given length l and cross-sectional area A, different materials will have different resistances. This difference can be taken into account in a proportionality constant. This constant, often represented by ρ (rho), is called the resistivity. Putting this in, the resistance of an object is given by

$$R = \rho l/A$$

The resistivity of a material is found by measuring the various quantities of length, area, and resistance and solving for ρ. Numerically, ρ is the resistance of a piece of material for which A and l are both one unit (for example, a unit cube of the material).

The units for resistivity ρ can be found by solving for ρ. This gives

$$\rho = \frac{RA}{l}$$

R is in ohms, A is in m², and l is in meters. One unit, m, will cancel, leaving the units of ρ as ohm meters or Ω m. Frequently the unit ohm centimeters may be encountered. One Ω cm is a hundredth of an Ω m.

Resistivities of some materials are shown in Table 15–1. These include some of the best conductors, the metals; some insulating materials; and some biological materials.

TABLE 15–1   Resistivity of some materials. Metals,
insulators, liquids, and some biological materials are
included. The sources of the data are varied and in
some cases are an average from more than one source.
The resistivities of the biological materials are from
A. M. Gordon and J. W. Woodbury in T. C. Ruch and
H. D. Patton (Eds.), *Physiology and Biophysics,*
W. B. Saunders Co., 1966.

MATERIAL	RESISTIVITY, ρ $\Omega\,m$
silver	1.59×10^{-8}
copper	1.75×10^{-8}
gold	2.44×10^{-8}
aluminum	2.82×10^{-8}
brass	7×10^{-8}
iron	10×10^{-8}
nichrome	100×10^{-8}
Bakelite micarta	5×10^{8}
glass, plate	2×10^{11}
nylon	4×10^{12}
epoxy, cast resin	$10^{14}–10^{15}$
ceresin	5×10^{16}
polystyrene	$10^{15}–10^{17}$
water	10^{5}
alcohol	3×10^{3}
glycerin	6.4×10^{6}
olive oil	5×10^{10}
squid axon membrane	1.33×10^{7}
axoplasm	2
cytoplasm	1

EXAMPLE 10

Find the resistivity of a wire that is 10 meters long and circular in cross-section with a diameter of 1 mm (10^{-3} m), and which has a resistance of 0.4 ohm. From the calculated resistivity, identify the material from Table 15–1.

From $R = \rho l/A$, solve for ρ to get

$$\rho = RA/l$$

Substitute

$R = 0.4\ \Omega$
$A = \pi d^2/4$
  $= \pi \times 10^{-6}\ m^2/4$
  $= 7.85 \times 10^{-7}\ m^2$
$l = 10\ m$

so

$$\rho = \frac{0.4\ \Omega \times 7.85 \times 10^{-7}\ m^2}{10\ m}$$
$$= 3.14 \times 10^{-8}\ \Omega\,m$$

The resistivity is $3.14 \times 10^{-8}\ \Omega$ m, and from Table 15–1 the material is probably aluminum.

15–5 THE POTENTIOMETER

The name "potentiometer" actually means a meter to read potential or voltage. It is an ingenious circuit used to *compare* voltages. All so-called measurements of anything, such as length or mass, are comparisons with an accepted unit, so the word meter for measurement, rather than perhaps comparator for comparison, is not incorrect.

This particular device has been chosen for discussion because a large number of pieces of equipment in a clinical lab or a chemistry lab or a biology lab use a potentiometer type of circuit. Also, that circuit illustrates many general aspects of electric circuits.

15–5–1 VOLTAGE AT ANY POINT IN A CIRCUIT

When a mass is lifted up from the earth, it is given potential energy. In electrical work, voltage is potential energy per unit charge and thus is analogous to elevation. The arbitrary zero level in an electric circuit, as shown in Figure 15–9, is often chosen at the negative pole of the battery; any other part of the circuit can then be assigned a voltage level. This is what would be read by a voltmeter whose negative end is connected to the $V = 0$ position, as also shown in Figure 15–9. The values of the resistances, the voltage, and the current are those used in Example 4. The voltage is highest at the positive side of the battery and decreases through the circuit. The battery is like the water pump shown in Figure 15–10, pumping water to an elevation of 180 m. The first resistor is like a pipe bringing the water down 90 m, and the next brings it down to 30 m. The last

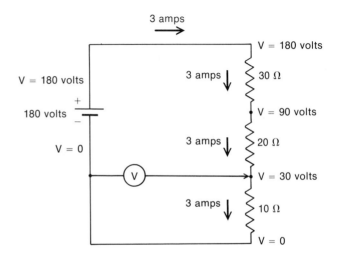

Figure 15–9 The designation of the potential at any point in a circuit.

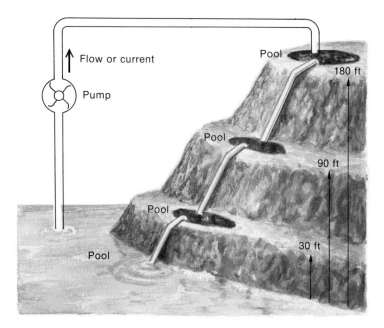

Flow or current

Pump

Pool
180 ft

Pool
90 ft

Pool

Pool
30 ft

Pool

Figure 15–10 An analogy between water level and electrical potential. This physical analogy is comparable to the electric circuit in Figure 15–9.

one brings it to the pool at the bottom, the zero potential level. It might be superfluous to say it, but water flows downhill; it will not flow in a horizontal channel. That is, water will not flow through a pipe if it is at the same level or potential at both ends. Similarly, electric charge will not flow between two points that are at the same potential. Electrical potential is like height, which determines gravitational potential energy. Potential or voltage is a bit like water pressure.

To illustrate this further, consider the two circuits in Figure 15–11. Each of these is the same as the one in Figure 15–9, and they are connected at the zero potential level. If a wire and current meter are used to join the two at different potential levels as shown by Figure 15–11(b), charge would flow and register in the meter. However, if a wire and current meter join them as in Figure 15–11(c), no charge would flow through the meter. This would be like joining the two ponds in Figure 15–12. They are at the same level, so no water flows between them.

15–5–2 A VARIABLE VOLTAGE SOURCE

To go one step further, consider a voltage source connected by low-resistance wires to a long, high-resistance wire as in Figure 15–13. The battery is shown to have a voltage V_0 and the resistance wire is of length l. The voltmeter is shown with a movable contact. If the total resistance of the wire is R_0, then the resistance of the portion between the voltmeter contacts is $(x/l)R_0$. Then, by Ohm's law, the reading of the voltmeter at a

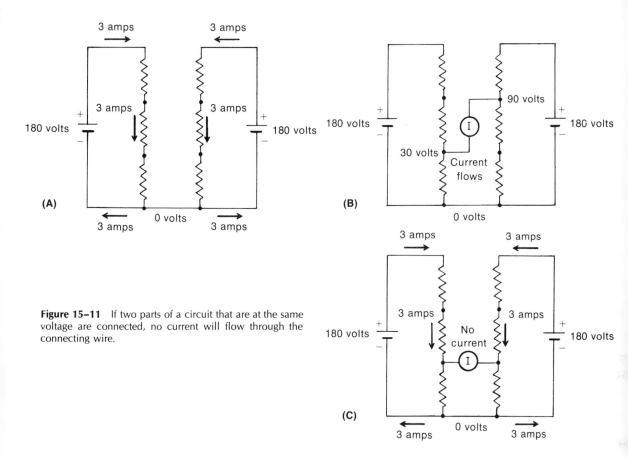

Figure 15–11 If two parts of a circuit that are at the same voltage are connected, no current will flow through the connecting wire.

position x will be $(x/l)V_0$. There will be a direct relation between the position x and the voltage V.

An arrangement such as that in Figure 15–13 can be used to obtain a variable voltage. As in Figure 15–14, a resistor with a sliding contact, referred to as a **rheostat,** is connected across a voltage source V_0. The voltage across the output terminals can be varied from zero to a maximum of V_0.

Figure 15–12 A physical analogy to the connecting wire in the center of Figure 15–11 (c).

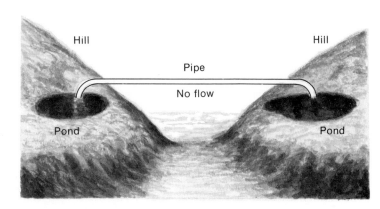

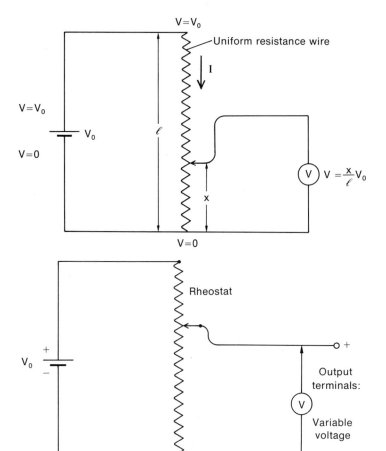

$V = V_0$

Uniform resistance wire

I

$V = V_0$

V_0

ℓ

$V = 0$

V $V = \dfrac{x}{\ell} V_0$

x

$V = 0$

Figure 15–13   The voltage along a uniform resistance wire increases uniformly with distance.

Rheostat

V_0 $+$ $-$

$+$

Output terminals:

V

Variable voltage

$-$

Figure 15–14   A resistance wire with a sliding connection, called a rheostat, can be used to provide a variable voltage when needed.

15–5–3 COMPARING VOLTAGES

In the circuit shown in Figure 15–15(a), a battery marked V_1 is connected through a sensitive current meter to the variable position on the resistance wire. As long as V_1 is less than V_0, there will be a point on the resistance wire where the voltage is just equal to V_1. When the sliding contact is in that position, no current will flow from V_1; the current meter will read zero. If that **null point** (point of zero current) is at x_1, then the voltages V_1 and V_0 compare as the lengths x_1 and l or

$$\frac{V_1}{V_0} = \frac{x_1}{l} \qquad \text{or} \qquad V_1 = \frac{x_1}{l} V_0$$

Comparing voltages is done by comparing the lengths.

If a different voltage V_2 is used, as in Figure 15–15(b), then

$$V_2 = \frac{x_2}{l} V_0$$

(A)

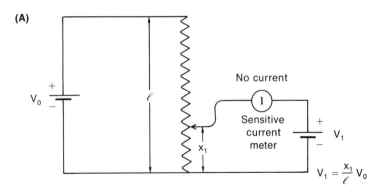

Figure 15–15 A potentiometer circuit used to compare voltages by comparing distances.

(B)

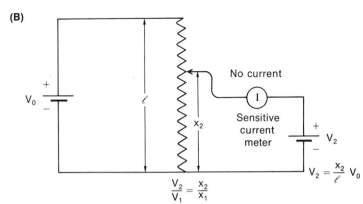

The voltages V_1 and V_2 compare as the lengths x_1 and x_2. This can be shown by dividing the above equations to get

$$V_1/V_2 = x_1/x_2$$

A device used to compare voltages in this way is called a potentiometer.

The slidewire need not be linear as shown. It may be in a large arc, almost a full circle, as shown in Figure 15–16. The sliding contact is connected to a knob, and the balance position is read on a circular dial.

There are advantages in comparing voltages in this way. One is in precision. A common pointer-type voltmeter will be good to probably 2 per cent; a voltmeter with an accuracy of 0.5 per cent is an extremely expensive laboratory instrument. However, with a resistance wire even 1 meter in length, if the balance point is found to within even half a millimeter, an accuracy of about a tenth of one per cent or less is obtained.

A further advantage is that at the time of measurement no current is being drawn from the voltage source being measured. Drawing current from a source often changes its steady state voltage. If you want the voltage at zero current, a potentiometer will give it. The moving pointer type of voltmeter, however,

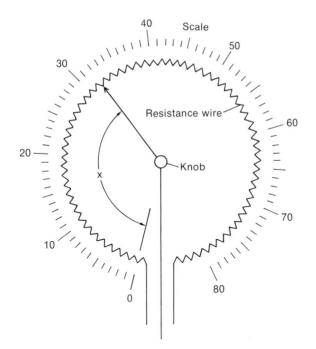

Figure 15–16 The resistance wire of a commercial potentiometer is often arranged in a circle so that the position of the contact is controlled by a knob.

ordinarily draws current to operate the mechanism. This will frequently change the voltage being measured; it does not give the voltage at zero current.

15–6 ELECTRIC CURRENT AND MAGNETIC FIELD

When an electric charge is still, there is an electric field around it. When the charge moves, a magnetic field also appears. Associated with any current is a magnetic field; they are inseparable. It is for this reason that magnetic effects are included here in the chapter on current electricity. Because of the magnetism associated with a current, there is an interaction between a current-carrying wire and a magnetic field much like the one that occurs between an ordinary bar magnet and an external magnetic field. Electrical measuring instruments of the moving coil (and pointer) type make use of this effect.

Magnetic fields are associated not only with wires through which a current flows, but also with charged particles moving through space. This effect is made use of in devices that produce extremely high energy particles and x-rays; these devices have some advantages in treatment of deep tumors. These effects also occur in television tubes and even in the mechanism producing northern lights. Phenomena and devices based on the interaction between moving particles and magnetic fields are dealt with in Chapter 16. In this chapter the interaction between magnetic fields and current-carrying wires is considered.

15-6-1 DETECTING A MAGNETIC FIELD

A basic way to detect a magnetic field is to put a current-carrying wire in it. If there is a force on the wire, then a magnetic field is present. However, a more familiar way is to use a small bar magnet, such as is used for a compass needle. The compass needle lines up along the magnetic field. The explanation of this is usually given in terms of the concept of magnetic poles. A bar magnet contains an N and an S (or North and South) pole. The direction of a magnetic field is in the direction of the force on an N pole.

As in Figure 15-17, if the bar magnet is inclined to the field, a torque will be exerted to move it toward the field direction. Magnetic poles don't exist; they are just a convenient way to analyze what happens.

15-6-2 PATTERNS OF MAGNETIC FIELD

The pattern of a magnetic field around a bar magnet is shown in Figure 15-18. The magnetic field lines come out of the N pole and go into the S pole. In Figure 15-19 are some patterns of magnetic fields. In (a) is the ordinary bar magnet or dipole, and in (b) is a conventional type of horseshoe magnet. Parts (c) and (d) illustrate variations of the horseshoe magnet; in (c), if the poles are close together compared to the width, the magnetic field lines are straight across just like the electric field lines between parallel plates (Section 14-2-2) and a uniform field results. In (d) the pole faces are curved and a circular piece of iron is put in. The field lines go from the N pole perpendicularly across to the iron and then from the iron to the S pole. This is the configuration commonly used in electric meters.

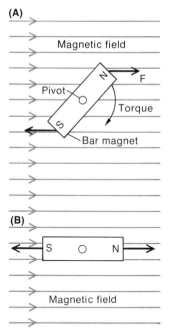

Figure 15-17 A small bar magnet in a magnetic field experiences a torque that tends to cause it to line up with the field direction.

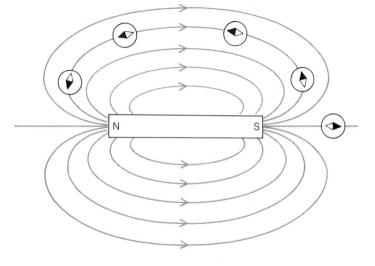

Figure 15-18 A small compass indicating the field pattern around a bar magnet.

(A)

(B)

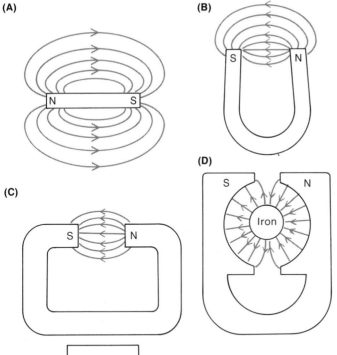

Figure 15-19 The magnetic field pattern around a bar magnet (a) and a horseshoe magnet (b). Large flat pole pieces (c) give a space of uniform field. The special arrangement in (d) gives a radial pattern of lines. This latter configuration is used in electric meters.

(D)

(C)

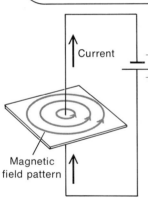

Figure 15-20 The pattern of the magnetic field around a current-carrying wire.

15-6-3 MAGNETIC FIELD AROUND A CURRENT

The pattern of the magnetic field around a current is unlike any of those illustrated for bar or horseshoe magnets. The magnetic field lines are curved as in Figure 15–20. The direction of the field is *around* the current. Considering the flow to be that of positive charges, the direction of the field is such that if you grasp the wire with your *right hand* with the thumb in the direction of the current, the fingers curl around in the direction

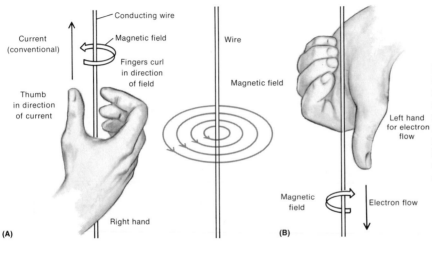

Figure 15-21 (a) The *right hand rule* to help remember the relation between magnetic field direction and current direction. (b) The *left hand rule* is used to indicate the magnetic field direction if electron flow, rather than conventional current, is considered.

of the field. This is shown in Figure 15–21(a). If you consider the flow to be made up of negative charges, the field direction is found in a similar way but using the left hand, as in Figure 15–21(b).

Arrows are placed on the lines to indicate a field direction or a current direction, but there is a problem in drawing them because they exist in three dimensions. To assist, another convention has developed, using ⊙ to indicate the point of an arrow coming out of the paper and ⊕ or + to indicate the tail of an arrow going into the paper. These are shown in Figure 15–22.

15–6–4 THE FIELD AROUND A LOOP AND A COIL

The pattern of magnetic field around a single wire consists of concentric circles. For a loop like that in Figure 15–23(a), the field pattern can be drawn as it would be on a plane cutting the loop. The current goes down on one side and up on the other, as in Figure 15–23(b), and basically the magnetic field is circular around each wire. But in the center of the loop, the fields from the two sides of the loop add, and the field is strengthened. The pattern of field lines comes out from one face of the loop, goes around the outside, and goes in through the other face. It is very similar to the pattern around a bar magnet, as shown in Figures 15–18 and 15–19. The loop is like an exceedingly short bar magnet.

A series of loops forming a coil or a solenoid has the pattern of magnetic field shown in Figure 15–24. Between adjacent coils most of the field cancels because from one wire the field is outward and from the next it is inward.

A spinning charged particle is like a loop of current. Electrons and protons do have magnetic fields associated with

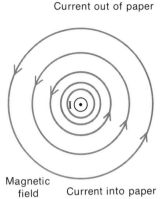

Current out of paper

Magnetic field

Current into paper

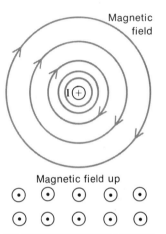

Magnetic field

Magnetic field up

Current

Magnetic field down

Figure 15–22 (a) and (b) Indicating current direction by points and tails of arrows. (c) The magnetic field lines are shown by a similar convention.

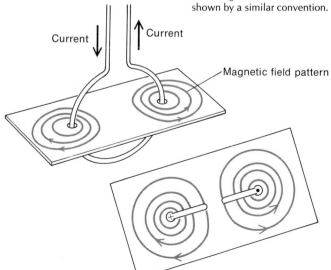

Current   Current

Magnetic field pattern

Figure 15–23 The magnetic pattern around a loop of wire that is carrying a current. The pattern is similar to that from a very short bar magnet.

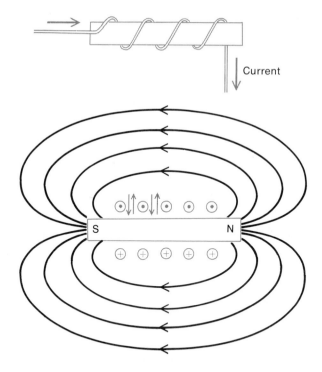

Current

Figure 15–24 The magnetic field pattern around a series of loops, called a coil or solenoid. The pattern is very much like that around a bar magnet.

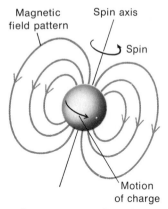

Magnetic field pattern   Spin axis

Spin

Motion of charge

Figure 15–25 A charged particle that is spinning is like a loop of current, and it has a magnetic field.

them and with their spins. Even a neutron has a magnetic field, which is food for thought. The spinning proton has a field like that shown in Figure 15–25.

Ordinarily, the particles (protons, neutrons, and electrons) in an atom are arranged such that the magnetic fields cancel, leaving a net magnetic field of zero. This is not always the case. In iron the particles may be aligned so that the magnetic fields of some of the outer electrons add to produce the net effect of the common magnet.

There is a magnetic field around the earth, which (as in Figure 15–26) is very similar to that around a loop of current or a short bar magnet—it is like a dipole. A compass needle on the surface of the earth turns to have the N end pointing north. This means that the magnetic pole at the geographic north of the earth is an S pole. The axis of the magnetic dipole field at the earth is inclined to the axis of rotation and does not pass through the geometrical center of the earth. The north geomagnetic pole is at about 73° north latitude and 100° west longitude, which is in Arctic Canada. The magnetic pole wanders slowly, and the polarity of the earth has even reversed occasionally throughout geologic time.

The earth's magnetism is probably a result of electric currents in the metallic core of the earth.

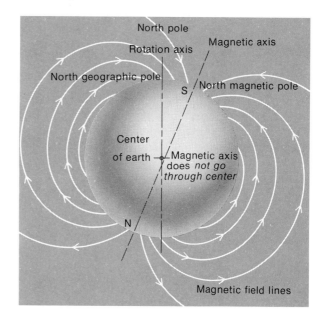

Figure 15–26 The magnetic field around the earth.

15–6–5   A CURRENT IN A MAGNETIC FIELD

The field around a current-carrying wire is shown in Figure 15–27(a). The (positive) current is shown coming out of the page, and the magnetic field lines B are counterclockwise around it. If this current-carrying wire is placed in a uniform magnetic field as in Figure 15–27(b), the resulting field pattern is the combination of those from (a) and (b). The field has a downward component to the left of the wire and an upward component to the right, as in Figure 15–27(c). Below the wire, the field from the wire adds to the uniform field; above the wire the two fields cancel. The resultant field pattern is given in Figure 15–27(d).

What happens in this case is that there is a force on the wire as in Figure 15–27(e), which is such that the field lines tend to straighten. This is not the actual cause, but it describes the phenomenon. The three things, current direction, magnetic field direction, and direction of the force or thrust, are mutually perpendicular.

This force on a wire carrying a current in a magnetic field is what turns electric motors and what causes a pointer in an electric meter to move.

To find the direction of the force on the wire, a simple trick is often used. The thumb and first two fingers of the *left* hand are placed perpendicular to each other, as in Figure 15–28. The first finger is put in the direction of the magnetic field, and the center finger in the direction of the current; the thumb then points in the direction of the thrust on the wire. This can be an aid to

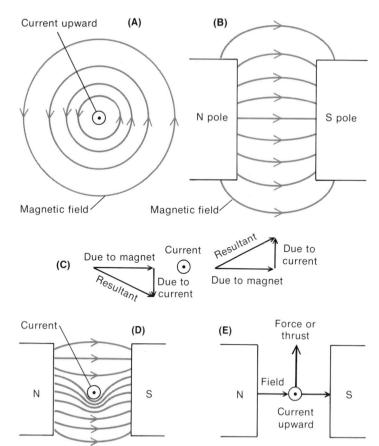

Figure 15–27 The magnetic field from a current in a wire (a) and the field of a magnet (b) combine as shown in (c) to give the field pattern in (d). A force is exerted on the current (e).

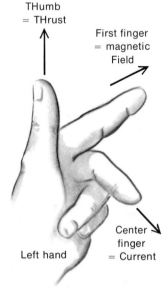

Figure 15–28 The left hand motor rule to help remember the direction of the force on a current in a magnetic field. (Use the right hand if electron flow, rather than conventional current, is considered.)

memory as well as good exercise at times, for it often involves almost standing on your head! It is a way to express what is called the **left hand motor rule.**

15–6–6 FORCE ON A CURRENT IN A MAGNETIC FIELD; UNITS FOR MAGNETIC FIELD

The size of the force on a current in a magnetic field depends on the magnitude of the current I, the strength of the magnetic field B, and also on the length l of the wire. The force is directly related to each of these:

$$F \propto BIl$$

or

$$F = kBIl$$

where k is a proportionality constant. The units in the SI system are newtons for force, amperes for current, and meters

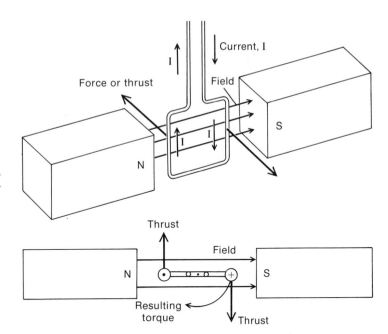

Figure 15-29 The forces and torque on a current-carrying loop of wire in a magnetic field: (a) pictorial view; (b) schematic.

for length. The units for B are devised to make the proportionality constant just 1. Then

$$F = Bll$$

or

$$B = F/Il$$

The units for magnetic field strength B are newtons per ampere meter. This unit is also given the name of one **tesla,** after a scientist who did a lot of work in this area. It is also called one **weber per square meter,** where the lines of magnetic field are measured in **webers.**

An old unit for a magnetic field was the **gauss.** A gauss is a small unit, there being 10 thousand (10^4) gauss in one tesla or weber/m^2.

15-6-7 FORCE ON A LOOP OF WIRE IN A FIELD

When a coil or loop of wire is put into a magnetic field as in Figure 15-29, the forces on the two sides of the coil are in opposite directions. This results in a torque that tries to turn the coil. This is the configuration that is the key to electric meters and motors!

15-7 ELECTRIC METERS

Moving coil type electric meters are probably the most common of all. They make use of the torque on a coil in a

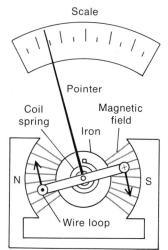

Scale

Pointer

Coil spring

Magnetic field

Iron

N

S

Wire loop

Figure 15–30 The essential parts of a meter movement.

magnetic field. The construction is often like that in Figure 15–30. Curved pole pieces and an iron insert as in Figure 15–19(d) give a magnetic field that is uniform over a considerable arc and is also directed to the center.

The force on the wires depends on the magnetic field strength, on the length of the wire in the field, and on the current.

The field strength is a constant, though with time it may change; an old meter is not so reliable as a new one, and it should be checked against a standard. Magnetic materials used in modern meters give high field strengths and therefore make sensitive meters.

The length of the wire in the field is fixed, of course; but in making the meter, the length is made large by using many turns (in other words, a coil, not just a single loop of wire).

Thus, field strength and wire length are fixed in the meter construction. The only variable is the current through the coil, and the torque that causes it to turn depends directly on the current. The meter is made so that as the coil turns, a spring is twisted to give a restoring force. The coil turns until the torque

---

EXAMPLE 11

Consider an electric meter of the moving coil type as illustrated in Figures 15–29 and 15–30. The problem is to find the force constant of the spring, given the following data:

The field strength is 0.3 tesla (N/A m).
The length of the coil in the field is 0.015 m.
There are 40 turns on the coil.
The radius of the coil is 0.010 m (this is the distance from the axis to the wires on which the force acts).
The coil is to turn through 57.3° (1 rad) with a current of 0.001 amp.
First find the total length of the wire in the field on one side of the coil. This length l is 40 × 0.015 m or 0.6 m.
The force on one side of the coil is given by Bll or

$$F = 0.3 \frac{N}{A\,m} \times 0.001\ A \times 0.6\ m$$

$$= 0.00018\ N = 1.8 \times 10^{-4}\ N$$

This is the force shown in Figure 15–29.
The torque due to such a force on each side of the coil with a radius of 0.010 m is:

$$\text{Torque } L = 2 \times 1.8 \times 10^{-4}\ N \times 10^{-2}\ m$$
$$= 3.6 \times 10^{-6}\ N$$

The angle turned through will be given by $L = k\theta$, where k is the torque constant. Then $k = L/\theta$ and $\theta = 1$ rad. Therefore

$$k = 3.6 \times 10^{-6}\ N/rad$$

The restoring spring in a moving coil meter will be a very fine, delicate thing. To minimize friction, a precaution that is necessary when the forces are so small, a meter will often be made with jeweled bearings as in a good watch.

is balanced by the spring. The deflection of the pointer is directly proportional to the current!

This is what I have set out to show: that electric meters of the moving coil type give a deflection that depends on the current through the coil. They are all basically current meters, even though some of them may be labeled voltmeters. How can this be?

15-7-1   VOLTMETERS

A sensitive current meter can be transformed to give a deflection that depends on the voltage applied to it, by adding a resistance in series with the coil. If the resistance is R and a voltage V is applied, the current through it (as in Figure 15-31) is given by Ohm's law,

$$I = V/R$$

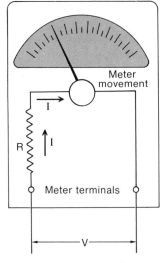

Figure 15-31 A voltmeter consists of a sensitive meter movement in series with a resistance. The current in the meter depends on the voltage applied to the terminals.

The current, and hence the meter deflection, is proportional to the voltage. The current meter has been transformed to a voltmeter! An ideal voltmeter would read a voltage without drawing any current. The instrument that comes closest to this state is the electrostatic voltmeter, of which the gold leaf electroscope and quartz fiber electroscope are examples. There are also others that have extremely high resistance and hence draw very little current. They usually go under the name of electrometers. The ordinary voltmeter has many uses anyway, but the fact that it draws current must often be remembered. The higher the resistance of the meter, the less current it will draw for a given voltage. When a potentiometer is balanced, it draws no current.

There is always some resistance in the wires of the coil, and the total voltmeter resistance includes this.

EXAMPLE 12

Consider that you have a meter movement (coil, pointer, and so forth) that gives a full scale deflection for 100 microamps through the coil. You want, however, a voltmeter that will read up to only 1 volt. What resistance must be added? (In this case, assume zero resistance for the coil.)

You want to make the resistance such that when $V = 1$ volt, you have $I = 100$ μA. Then with 1 volt applied to the meter terminals, full scale deflection will result. A mark labeled "1 volt" can then be put at the end of the scale, and the rest of the scale can be marked in fractions of a volt.

Use Ohm's law, $V = IR$ or

$$R = V/I$$
$$V = 1 \text{ volt}$$
$$I = 100 \ \mu\text{A} = 1 \times 10^{-4} \text{ amp}$$
$$R = 1/(1 \times 10^{-4}) \text{ ohms}$$
$$= 10^4 \text{ ohms}$$
$$= 10,000 \text{ ohms}$$

The circuit will be assembled as in Figure 15–31 with the resistance $R = 10,000$ ohms (actually, 10,000 ohms less the resistance of the coil).

EXAMPLE 13

With the meter of Example 12, work out a general expression for the resistance for any full scale voltage desired.

Use $R = V/I$ with $I = 10^{-4}$ amp to get

$$R = V/10^{-4}$$
$$= 10,000 \ V$$

For each volt of the full scale reading, the resistance must be 10,000 ohms. For example, for a full scale reading of 100 volts, the resistance must be 100 times 10,000 ohms or a million ohms. Such a meter is said to have a sensitivity of 10,000 ohms per volt.

15–7–2 METER SENSITIVITY

If a current I is required to give full scale deflection of a meter movement, then in order for it to read full scale for a voltage V, the resistance must be

$$R = V/I \quad \text{or} \quad R = \left(\frac{1}{I}\right) \times V$$

The resistance per full scale volt is just the inverse of the full scale current. The less current it requires, the more sensitive is the movement and the higher the resistance per volt. Meter sensitivity is expressed in ohms per volt. The higher this figure is, the less current the meter requires.

Meters will ordinarily be marked with their sensitivity as in Figure 15–32.

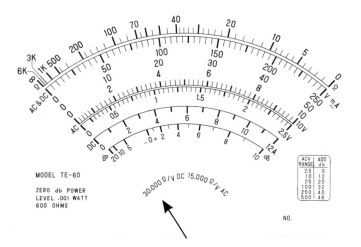

Figure 15–32 The sensitivity of a meter in ohms per volt is marked on the meter face (arrow).

MODEL TE-60

ZERO db POWER
LEVEL .001 WATT
600 OHMS

ACV RANGE	ADD db
2.5	0
10	12
25	20
100	32
250	40
500	46

NO.

15-7-3 MULTI-RANGE VOLTMETERS

A voltmeter may be made with a variety of full scale ranges by using a multi-pole switch that connects different resistances in series with the meter movement. This is illustrated in Figure 15-33. Each switch setting will give a different full scale voltage reading.

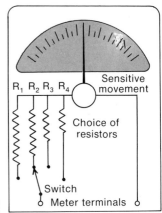

Figure 15-33 A voltmeter adapted to read various ranges of voltage.

15-7-4 CURRENT METERS

A basic, sensitive meter movement can be adapted to read any range of current that is desired, just as it can be adapted to read any range of voltage. This is done by putting a resistance across (in parallel with) the meter coil to shunt a fraction of the current away from the coil. For example, if a meter movement deflects full scale for a current of 1 milliamp and a meter that reads 1 amp full scale is wanted, then a low resistance must be put across the coil as in Figure 15-34. Its value must be chosen such that if 1 amp flows into the meter terminal, 0.999 amp will be bypassed through the shunt resistance R_s and only 0.001 amp or 1 mA will go through the meter coil. It must be 999 times easier for the current to go through the shunt, so its resistance must be 1/999 of the resistance of the coil.

The resistance of the fine wire of a meter coil may be, for example, 400 ohms. Continuing the example above, the shunt resistance would be 1/999 of this or very close to 0.4 ohm. The net resistance of a current meter is very low, though not always negligible.

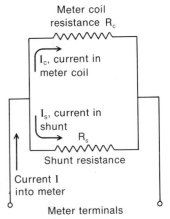

Figure 15-34 The use of a shunt resistance to convert a sensitive current meter to one of any desired range.

15-7-5 MULTIRANGE CURRENT METERS

By provision of various shunt resistances, a sensitive meter movement may be adapted to measure a wide range of current. As in Figure 15-35, a switch may be provided to put in different shunt resistances and give different ranges.

15-7-6 ELECTRIC MOTORS

If you have a coil in a magnetic field, such as those shown in Figures 15-29 and 15-30, rotation of the coil is caused by putting a current through the coil. The coil rotates until the torque produced by the current is balanced by the spring. If the spring were not there, the rotation would continue, and that is the beginning of the idea of an electric motor. The problem to be

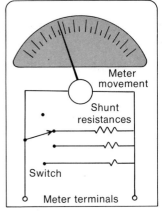

Figure 15-35 A meter adapted to read various ranges of current by choosing different shunt resistances.

Coil

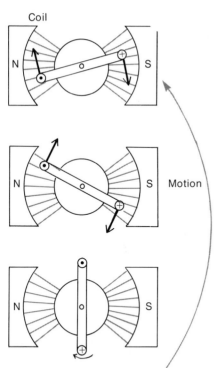

Motion

No force - Equilibrium position
At this instant change
the direction of the
current in the coil

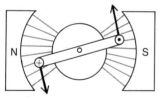

Forces are in opposite direction
if current is not switched

Figure 15-36   The electric motor.

overcome is the fact that when the coil turns 180°, the force is
in the direction opposite to the rotation, as shown in Figure
15-36.

This problem is not insurmountable; just arrange for the
current to change direction each time the coil passes the
equilibrium position, as shown to the right in Figure 15-36.
Such a reversal can be accomplished in more than one way.
Some are:

(a) Hire a small, fast person to interchange the battery
wires at the appropriate time. This never did work.

(b) Put a split ring on the axis, with the current fed to the
ring by means of two "brushes" connected to the battery terminals. In one orientation of the coil the current goes in one half
of the ring, through the coil, and out the other half. But as the

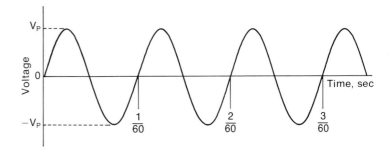

Figure 15–37 The way in which an alternating voltage changes with time.

coil and axis turn, the brushes move onto the other halves of the ring and the current is switched in the coil.

(c) Use a source of voltage that switches direction periodically. This is what is called alternating voltage (or current). The electricity commonly supplied to homes, labs, and factories is alternating, switching direction completely 50 or 60 times a second. This type of voltage is shown in Figure 15–37. The motor speed must be synchronized exactly with the frequency of the voltage. Electric clocks use synchronous motors so they keep in step with the electricity fed to them.

Electric motors needed an introduction because motors, tiny or large, play such a big role in our technology.

15–8 GENERATING ELECTRICITY

When an electric charge moves in a magnetic field, there is a force on it. A wire has electric charges associated with the atoms, and many of the electrons are free to move. If a wire is moved across a magnetic field, the electrons in the wire also move across the field. Therefore, there is a force on them as shown in Figure 15–38, the direction being given by the motor rule described in Figure 15–28. The electrons move to one end of the wire; it becomes negatively charged, leaving the other end positive. The wire is like a battery—there is a voltage between the ends. An external circuit (a light bulb, perhaps) can be connected to the ends of the wire and a current will flow.

This phenomenon of induction of a voltage in a conductor moving in a magnetic field is the method by which almost all the electricity used in our society is generated. The practical devices use a coil turning a magnetic field, just as for a meter or a motor. The force to turn the coil may be provided by water power, steam from a fossil-fueled boiler, or steam generated in a nuclear reactor.

Most generators are made to provide alternating current or voltage because each time the coil goes by the equilibrium position (as for the motor in Figure 15–36), the direction of the

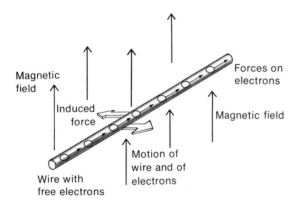

Figure 15–38 The motion of electrons in a conductor that is moved across a magnetic field.

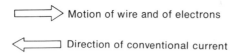

Motion of wire and of electrons

Direction of conventional current

induced voltage changes. So alternating current (ac) is easy to produce with a generator.

15-8-1 LENZ'S LAW

Looking back to Figure 15–38, the electrons moving along the wire are seen to produce a current in the wire. If an external circuit draws the electrons away, the flow along the wire is continuous. This flow of electrons within the external magnetic field results in a force, which — by the motor rule — is seen to oppose the motion of the wire. To keep the wire moving, energy must be expended to overcome this induced force. This is the energy that is transferred to the current to be used in the external circuit.

The process can be broken into steps for analysis.

(a) The wire is moved across the magnetic field.
(b) The free electrons in the wire are forced in one direction.
(c) These moving electrons result in a current.
(d) The force on the wire resulting from this current opposes the motion of the wire.

Statement (d) is an example of Lenz's law, which in a more general form is that **the voltage induced in a physical system by a change in conditions is always in such a direction that the current set up by the voltage will oppose that change.**

In our example, the position of the wire was changed and a force resulted to oppose that change.

Rather than moving the wire in the field, the field could be moved across the wire. The "motion" of a magnetic field can also result from a change in the magnitude of the magnetic field through the coil. If the magnetic field is made to increase, the current in the coil will produce its own magnetic field that will oppose the increase. If the field decreases, the induced field in the coil will tend to stop the decrease.

Is it any wonder that Lenz's law is often called the law of natural cussedness?

15-8-2 TRANSFORMERS

A transformer is a device that changes the value of an alternating voltage. It is based on the phenomenon that a changing magnetic field in a conducting coil induces a voltage in that coil. The construction may be like that in Figure 15–39. As a current flows in the **primary coil,** a magnetic field is set up in the iron. This goes through the **secondary coil,** inducing a voltage in it. The size of the induced voltage depends on the ratio of the number of turns in the two coils. It is a direct ratio, so

$$V_2/V_1 = N_2/N_1$$

The voltage in the secondary may be greater or less than that in the primary, depending on the **turns ratio** N_2/N_1. The voltage may be stepped up, stepped down, or kept the same. In cases in which the voltage is kept the same, the transformer may be used to isolate the secondary circuit from the primary for some reason.

Whereas transformer theory has not been discussed in detail, these devices are mentioned because of their wide use in laboratories. Most laboratory instruments do not require the standard ac line voltage, often 120 volts; so a transformer is used to provide what is needed. A transformer does not work with direct (non-alternating) current. In fact, if a transformer-

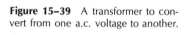

Figure 15-39 A transformer to convert from one a.c. voltage to another.

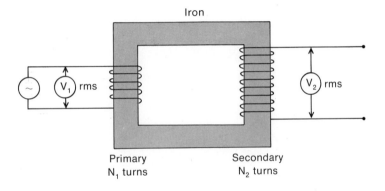

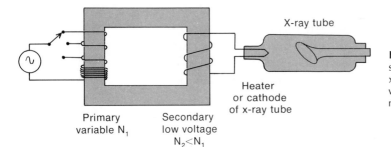

Figure 15–40 An example of the use of a step-down transformer. The heater of an x-ray tube requires a low voltage, which can be varied by changing the number of turns used in the primary.

X-ray tube

Heater
or cathode
of x-ray tube

Primary
variable N_1

Secondary
low voltage
$N_2 < N_1$

operated instrument is connected to a dc source, the first indication of malfunction may be smoke production. The dc current is not limited by the "back voltage" effect obtained with ac in a coil. A large current will then flow; the heating effect may be appreciable and the net effect may be disastrous.

X-ray tubes make use of both step-down and step-up transformers. At one end of the x-ray tube is a heating coil, which requires usually from about 4 to 6 volts. A step-down transformer is used to obtain this. The purpose of the heat is to evaporate off electrons that will accelerate across the tube and produce x-rays. The hotter the filament, the greater the number of electrons and the more intense the x-radiation. The x-ray intensity is controlled by controlling the heater voltage. This is achieved by a switch that selects the number of turns in the primary, as in Figure 15–40; hence, the turns ratio and the secondary voltage are variable. The control could be on the secondary, but that goes directly to the x-ray tube, which may be at a dangerously high voltage. The transformer also serves to isolate the primary wires (and the operator) from the high voltage. The high voltage across the x-ray tube, often in the region of a hundred thousand volts, is provided by a transformer with about a thousand times as many turns on the secondary as on the primary. This is illustrated (not exactly) in Figure 15–41. The high voltage may be controlled in the same manner as the heater control voltage. Sometimes the voltage applied to the primary will be controlled,

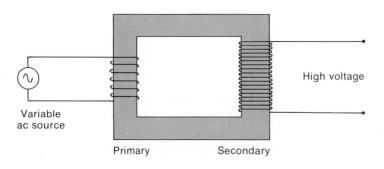

High voltage

Figure 15–41 An example of the use of a step-up transformer: obtaining the high voltage required for an x-ray tube.

Variable
ac source

Primary   Secondary

perhaps by another transformer that changes the voltage only a little by selecting the number of secondary turns.

There is frequently a problem in that a certain device may require a direct current such as that from a battery, not an alternating current. Hand calculators run on batteries, and in order to charge them (if they are rechargeable) current must be fed into them in one direction. The current from the wall plug is alternating and would push current in for 1/120 second and take it out for the next 1/120 second. That would do no good. Also, x-ray tubes need the accelerating current and voltage always in one direction so that the electrons will go only one way.

Alternating voltage would be useless in these cases, and the situation must be rectified. The devices that change ac to dc are called **rectifiers.**

15–9 RECTIFIERS AND DIODES

Some materials are unusual in that they have different resistances for current in different directions. If even a small voltage is applied in one direction, current will flow easily; but very little current will flow in the other direction when the voltage is reversed. These materials go under the name of rectifiers or diodes. The distinction in the names is on the basis of either the use or the amount of current. If the device is used to supply dc power to an instrument, the term rectifier will be used; while if only a small current is involved and the device is part of an electronic circuit, it will be called a diode. In either case, rectifier or diode, it is a device that allows current to flow in only one direction. The symbol for either is a line and a triangular arrowhead that shows the direction of current flow.

Diodes may be made from copper covered with a layer of copper oxide in contact with a fluid. Current will flow easily from the solution to the copper, but not in the reverse direction. Other examples of rectifying materials are selenium and germanium crystals.

In Figure 15–42 are shown two rectifier circuits, one with a battery connected in one direction and the other with the battery reversed. In one case current flows, and a resistor must be included to limit it. In the other case no current flows.

A similar circuit with an alternating voltage source is shown in Figure 15–43(a). The pattern of voltage applied is in Figure 15-43(b); current flows only when the applied voltage is in the proper direction, as in part (c). It flows only half the time, but when it does it is always in the same direction. It is a **pulsating** but **direct** current. In part (a) of the figure the instantaneous voltage across the resistance R depends on the current flowing,

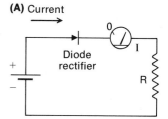

(A) Current

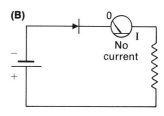

(B)

Figure 15–42 A diode or rectifier connected so that current will flow, and with the battery reversed so that no current flows.

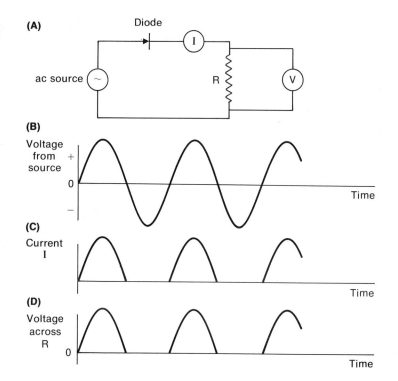

Figure 15–43 A rectifier circuit, showing also the applied voltage, the current, and the voltage across the load resistor. The load voltage is direct but pulsating.

according to Ohm's law, $V = IR$. The voltage across R therefore varies in the same manner as the current, as in Figure 15–43(d). The applied alternating voltage in part (b) is changed to the direct pulsating voltage in part (d).

The load resistance shown as R in Figure 15–43 may not necessarily be a plain resistance, but could be some device being used. For instance, it could be an x-ray tube, which does offer a high resistance to current flow.

Rectification like that shown in Figure 15–43 allows current to flow only during half of the cycle (or wave of voltage), and it is called **half wave rectification.** The basic idea of a circuit for an x-ray tube using half wave rectification is shown in Figure 15–44. Note the symbol for a transformer used in this figure.

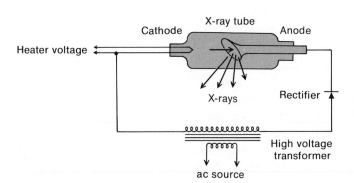

Figure 15–44 The basic concept of the high voltage circuit, including the rectifier, for an x-ray tube.

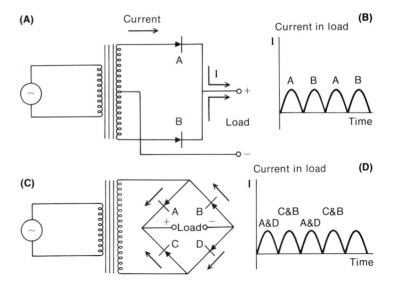

Figure 15–45 Circuits used for full wave rectification.

Combinations of two or even four rectifiers, as in Figure 15–45(a) and (c), will allow current to flow during both halves of the cycle; this is called **full wave rectification.** No matter what the direction of polarity of the source, the current always flows in the same direction through the load. The voltage and current are shown in parts (b) and (d) of the figure. It is still pulsating, but is an improvement on half wave rectification.

15–9–1 SMOOTHING CIRCUITS

The pulsating character of the current from a rectifier is sometimes undesirable. It would be preferable to have a steady current in the load and a steady voltage across it. A method to achieve this, though not perfectly, is to insert a capacitor across the load as in Figure 15–46. When electrons flow through the

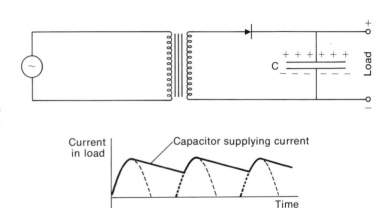

Figure 15–46 The use of a capacitor to smooth the voltage and current in a rectified circuit.

rectifier, they flow into the capacitor as well as through the load. The capacitor becomes charged. When the electrons cease to flow through the rectifier, the capacitor provides a current through the load; it partially discharges in the interval until the next pulse occurs. The next pulse brings the charge on the capacitor to the peak value again. The result is a fairly steady current in the load, or fairly steady voltage. The variation in voltage on the load may be as in Figure 15–46. This is referred to as a "ripple" in the voltage. The amount of the ripple depends, to a large extent, on the size of the capacitor. Increasing the size would reduce the ripple; adding a coil and even another capacitor would reduce the ripple even more. These "smoothing circuits" are of various designs and will not be elaborated upon. The smoothing must be designed to give an output voltage with a ripple small enough to be acceptable with the instrument for which it is used.

15-10 AMPLIFICATION

Electrical signals that one measures in the study of nerve impulses, effects with light, sound detectors, and similar phenomena are often small; progress in topics in biophysics has been directly related to progress in the ability to amplify and therefore study these electrical signals. The importance of electrical amplification in a biological system was expressed very well by Lord Adrian in a lecture at Oxford in 1946. His statement, in a small book that has had the rare quality of maintaining its value over the years and hence being reprinted (in 1967) is:*

> We know now that the impulse, the wave of activity which travels down a nerve-fiber when it is stimulated artificially, is always accompanied by an electric effect. It is small enough, but not too small to measure, and it is from records of these electrical changes that we have most of our information about the events which take place in the nerves and the brain. The transmitted change is accompanied by an electrical effect because the region which is active develops a negative potential with respect to the inactive regions beyond it. But the potential difference which can be measured by electrodes at two points on the nerve does not amount to more than a few hundredths of a volt and it does not last for more than a few thousandths of a second. The flow of current in and around the active region is therefore very small indeed, and the advance of knowledge about it has depended mainly on the progressive improvement in instruments for measuring very small and very brief currents.

*The Physical Background of Perception by Lord Adrian (Oxford University Press; 1946 and 1967).

Instruments recording the effect directly, without amplification, culminated in Einthoven's string galvanometer, developed at the beginning of this century, but after the war of 1914 valve amplification increased the sensitivity about a thousandfold and made it possible to use instruments which would show the exact time course of the change in spite of its extreme brevity.

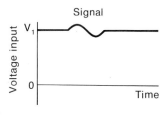

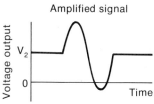

Figure 15–47 A small voltage change that constitutes a signal, and the amplified signal. The steady values V_1 and V_2 are of no consequence.

The idea of amplification is that the small voltage or current *changes* that constitute the information or signal are made into larger voltage or current *changes*. It is the changes that are amplified, not the voltages or currents themselves. As in Figure 15–47, although the changes in voltage are increased, the steady values V_1 and V_2 are of no consequence.

The original signal will often be in the form of a voltage, yet voltage and current signals are related by Ohm's law. A change in voltage can cause a change in current through a resistor; alternatively, a change in current through a resistor will cause a corresponding voltage change.

Among the most important devices that amplify electrical signals are **transistors** and what could be called **tubes** or **valves.** I do not intend to describe the inner working of a transistor, but only its behavior as seen from the outside. In Figure 15–48 we show the signal applied as a small voltage change to one part of the amplifying device. This results in a change in current through the other part. It is like pushing on the side of a flexible pipe to control the flow of water (Figure 15–49). If the water is not under a high pressure, even a small push could make a large change in the flow rate. The changing current in the amplifying device can be put through a resistor, shown by R in Figure 15–48. If the current is initially I_1, then the voltage across R will be given by $V_1 = I_1R$. With a change in current to $I_1 - \Delta I$ (a decrease, for instance), the voltage V_2 across R becomes:

$$V_2 = (I_1 - \Delta I)R$$
$$= I_1R - \Delta IR$$
$$= V_1 - \Delta V_2$$

where $\Delta V_2 = \Delta IR$. A change in voltage across R has resulted from the change in current through the device.

Figure 15–48 The principle of amplification. The device may be a transistor or a tube; a small change in voltage at the input results in a large current change in the resistance R.

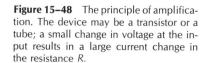

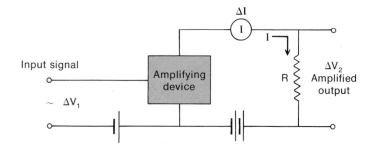

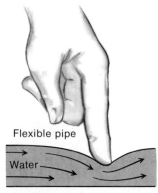

Flexible pipe

Water

Input signal changes current
(finger)   (water flow)

Figure 15–49 An analogy to an amplifying device is a flexible tube carrying water. Pressure applied to the side changes the flow rate.

Nearly all amplifiers work on this principle. A small input voltage change results in a relatively large current change. By use of a load resistor, this current change is made into a large voltage change. Amplifying devices such as transistors or tubes have this characteristic of amplifying voltage changes. If the input signal is a tiny current, it may be put through a resistor to make a voltage to apply to the amplifier. Then a change in the current results in a change in the voltage signal, which in turn causes a larger change in current through the amplifying device.

15–11 ALTERNATING CURRENT AND VOLTAGE, ac

A problem that comes to mind immediately is how to express the size of an alternating voltage or current. Because of the periodic change in sign, the average value of each is zero. The needle of an ordinary meter would be kicked one way for 1/120 of a second, and then the other way; back and forth, changing direction 120 times per second. The needle would not move at all because of its relatively large inertia. The peak values on either side of zero could be used as an expression of size or value. This is sometimes done, though it is not the peak value that is important in calculations such as that for the power dissipated in a resistance.

Power can be expressed as

$$P = VI \quad \text{or} \quad P = V^2/R \quad \text{or} \quad P = I^2R$$

With a resistance R, at the moment that the voltage has a value V the power being dissipated is V^2/R. Even if V is negative, V^2 is positive so the power is positive. The power changes with time, but the average power is found by taking the average (or mean) value of V^2 and dividing by R. The value of V that is used to

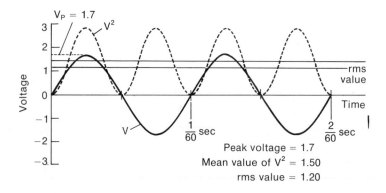

Peak voltage = 1.7
Mean value of V^2 = 1.50
rms value = 1.20

Figure 15–50 An a.c. voltage, V, with a peak value of 1.7 volts. The way in which V and V^2 vary with time is shown, as well as the average of V^2 and the square root of the average of V^2. The last quantity is called the root mean square or r.m.s. value.

calculate the power using the formula $P = V^2/R$ is the square root of the mean of the square. This is abbreviated as the **root mean square** or **rms** value. It is this rms value that is commonly quoted when referring to an ac voltage. In Figure 15–50 is shown the manner in which ac voltage varies with time. In that example it goes from a maximum of 1.7 volts to a minimum of -1.7 volts. The quantities V^2, the mean of V^2 or $\overline{V^2}$, and the rms value or $\sqrt{\overline{V^2}}$ are also shown.

The same process of reasoning is used with ac current. Power can also be described by $P = I^2R$. Again it is the rms value of I that is important in quoting the value of the current that can be used in the calculation of power.

With the common sinusoidal pattern of variation of current and voltage, the rms value is $1/\sqrt{2}$ of the peak value. The voltage commonly distributed in labs and homes in North America is referred to as 120 volts ac. This means that the rms value is 120 volts. The peak value is $\sqrt{2}$ times this or 170 volts. The voltage actually varies from $+170$ to -170, averaging to zero, but the value useful for power calculations and some others is 120 volts.

15–11–1 CAPACITORS WITH ac AND dc

A capacitor consists of two plates separated by a non-conductor. If, as in Figure 15–51(a), a direct voltage is applied to a capacitor, a current flows immediately on the closing of the switch S. But as the capacitor becomes charged, the voltage builds up as in Figure 15–51(b) and the current drops, approaching zero as in Figure 15–51(c).

If an alternating voltage is applied to the capacitor as in Figure 15–52(a), the capacitor will charge up one way when the source voltage builds up in one direction. This is shown in Figure 15–52(b). But when the source voltage begins to decline, the capacitor begins to discharge; then, with the voltage reversal, it charges the other way. This repeats during every cycle; with an alternating voltage applied to a capacitor, an alternating current flows in the circuit. This alternating current is not in phase with the voltage. The larger the capacity, the larger the flow of charge or current, and vice versa. In some ways a capacitor in an ac circuit is like a resistor in a dc circuit.

15–11–2 COILS WITH ac AND dc

The subject of a changing magnetic field across a conductor (or a loop of wire) inducing a voltage in the conductor has not

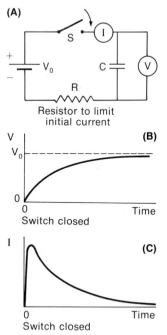

Figure 15–51 A direct voltage applied to a capacitor. When the switch S is closed (a), the voltage across the capacitor rises (b). The current, initially large as in (c), drops exponentially.

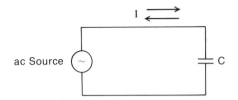

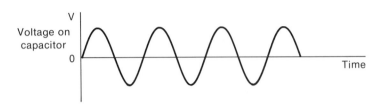

Voltage on capacitor

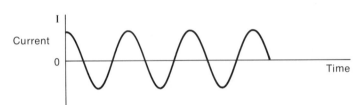

Current

Figure 15–52 A capacitor in an a.c. circuit.

been dealt with. However, it is a topic that has at some time been shown to most of us. It is the principal of the electric generator. The situation of a current beginning to flow into a coil is not dissimilar. The coil becomes a magnet, building up a magnetic field. But this magnetic field is changing across the very coil that is producing it. A voltage is induced in that coil to impede the current entering (Lenz's law).

As a result of the phenomenon just mentioned, a coil offers far more impedance to the flow of an alternating current than to the flow of a direct current. Such a coil is sometimes called an **inductor** when used in an electrical circuit.

DISCUSSION PROBLEMS

15–5   1. Any measurement in physics is ultimately a comparison with an agreed standard. Describe how this is so with length and with mass.

15–6–5   2. A beam of positively charged particles is projected into a magnetic field. The beam is like a current-carrying wire, without the wire to confine the moving charges. Make a diagram to show the force that is now on each moving charge and deduce the subsequent path of the beam in the magnetic field.

15–6–5   3. Show that the force on an individual particle, with charge e and speed v moving across a magnetic field of strength B, is

given by $F = Bve$. To do this, consider that there are n charges per unit length of a wire, and those in a length numerically equal to the speed v pass a given point per second. Then the current $I = nve$. Find the force on a length l and then the force on each moving charge.

15–6–6   4. Two current-carrying wires are parallel but carry current in opposite directions. The current in each wire interacts with the magnetic field of the other wire. In what directions are the resulting forces?

15–6–6   5. Show that the radius of curvature of the path of a particle with charge e and mass m, moving at a speed v in a magnetic field B, is given by $r = mv/Be$. Use the result given in Problem 3, that $F = Bve$.

PROBLEMS

15-1   1. (a) Find the total charge that flows in 1 minute through a light bulb in which the current is 0.833 amp.

(b) Find the current when 6.66×10^{-9} C flows past a point in one hour.

15-1   2. How many electrons pass across an x-ray tube per second if 5 milliamps flow?

15-2   3. The following situations involve Ohm's law.

(a) What current flows when a voltage of 120 volts is connected across 360 ohms?

(b) 1000 volts is applied to 10^9 ohms. What current flows?

(c) A current of 3 amps flows through 18 ohms. What is the voltage?

(d) 0.50 milliamp flows through 470,000 ohms. What is the voltage?

(e) A voltage of 120 volts is applied across a resistor and 0.040 amp flows. What is the resistance?

(f) Find the resistance of a lamp when it is found that with 12 volts applied to it, 1.75 amps flow.

15-2   4. (a) Draw a circuit showing 120 volts connected across three resistors in series. The values of the resistors are 30 ohms, 40 ohms, and 1130 ohms.

(b) Calculate the current flowing from the voltage source in (a).

(c) Calculate the voltage across each of the resistors in (a).

15-2   5. Find the resistance that must be put in series with a resistor of 400 ohms such that, when 120 volts are applied across them both, 0.010 amp flows.

15-2   6. What resistance must be put in series with a resistor of 10,000 ohms such that when 12.0 volts are applied to them both, the voltage across the resistor added is 1.00 volt?

15-2-1   7. In a certain situation a voltage V is applied to a resistor R, and a current I flows through R. When 435 ohms are put in series with R and the same voltage is applied, the current through R is reduced to 0.61 of the value before the 435 ohms were added. What is the value of R? (This trick can be adapted to find the resistance of an electric meter.)

15-2-1   8. Resistors of 1 Ω, 2 Ω, 3 Ω and 4 Ω are connected in parallel and connected to a voltage source of 12.0 volts.

(a) Draw the circuit.

(b) Find the current through each resistor.

(c) Add the results in (b) to find the current from the voltage source.

(d) Find the equivalent resistor that would allow that current when $V = 12.0$ volts.

(e) Find the equivalent resistance using the formula for resistors in parallel.

(f) Using the results from (e), find the total current that flows. Does it check with (c)?

15-2-1   9. Find the total current that flows when 120 volts are applied across three resistors in parallel, their values being 666 Ω, 444 Ω and 437,000 Ω.

15-2-1   10. Find the effective resistance of an infinite number of resistors in parallel, and then find the total current that would flow through them when 1.5 volts are applied. The values of the resistors in ohms are 1, 2, 4, 8, 16, etc., each being double the resistance of the previous one.

15-2-1   11. Find the current that would flow when 12.0 volts are applied to an infinite series of resistors that have values in ohms of 0!, 1!, 2!, 3!, 4!, 5!, etc., continuing the series. 0! is defined as 1, and also 1! = 1.

15-2-1   12. In a certain circuit 1.00 amp flows toward two resistors that are in parallel. One of them is of 41 ohms and 0.10 amp is flowing through it. What must be the value of the other resistor?

15-3   13. What is the maximum voltage that can be put across a resistor of one million ohms if anything more than half a watt of power in it will result in what can conservatively be described as overheating? Please do this problem theoretically, not experimentally.

15-3   14. How many 150 watt light bulbs can be put in parallel on a 120 volt circuit before a 15 amp circuit breaker will be tripped?

15-3   15. Resistance changes with temperature. A 60 watt light bulb, for instance, may have a resistance of 15 ohms when it is cold, but when 120 volts are applied to it the power dissipated in it is 60 watts. Find the ratio of the resistance when it is hot to the resistance when it is cold.

15-3   16. A heating coil in a pipe is to raise the temperature of water flowing at the rate of 0.41 kg/minute by 5°C. The heater is to run from a 12 volt battery.

(a) What power is needed?

(b) What must be the resistance?

15-3   17. What length of nichrome wire 1 mm in diameter would have a resistance of 10 ohms?

15-4   18. How much current would flow through a sheet of glass 3.0 mm thick and 15.0 cm square when both sides are silvered to make contact and 900 volts are applied to it?

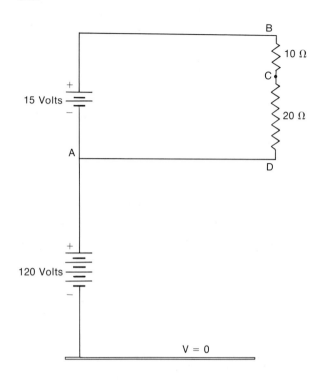

Figure 15–53 The circuit diagram for Problem 22.

15–4   19. A wire 100 m long and 3 mm in diameter has a coating of nylon 0.10 mm thick. The wire, except for the ends, is in a conducting fluid. With 10.0 volts between the wire and the fluid, what current would flow through the coating?

15–4   20. Find the comparative masses of wires of aluminum and copper, each 400 meters long and each of such a size that the resistance is 1.00 ohm. Which is the better conducter of the two, copper or aluminum, when size is considered and when weight is considered?

15–5   21. The positive terminal of a 90 volt battery is connected to resistors of 400 ohms, 300 ohms, 200 ohms, and 100 ohms that are in series, and then back to the negative battery terminal. Calculate the current and then, with the negative pole assigned a zero potential, find the potential at the ends of each of the four resistors.

15–5–1   22. What are the voltages at the points marked A, B, C, and D of Figure 15–53? The zero voltage reference point is shown.

15–5–3   23. Find the current through the 10,000 ohm resistor of Figure 15–54 when the

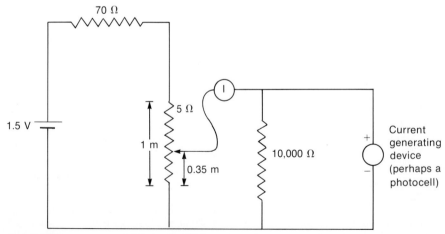

Figure 15–54 The circuit diagram for Problem 23.

sensitive current meter shows zero current. The 5 ohm resistor is a one-meter resistance wire, and the null reading is obtained at 0.35 meter from the end.

15-6-5   24.   Consider a wire in which there is a current and, consequently, a magnetic field around it as in Figure 15–20. Another wire is placed near that wire, parallel to it, and a current in the second wire is in the same direction as that in the first. There will be an interaction between the current in each wire and the magnetic field around the other wire. In what direction will the force on each wire be?

15-6-6   25.   Find the horizontal magnetic field when there is a vertical force of 3.3×10^{-3} N on a horizontal wire 0.67 m long carrying a current of 15 amps.

15-7   26.   Find the torque on a rectangular coil as in Figure 15–29. The coil is 2.5 cm square, has 100 turns, is in a field of 0.50 tesla, and carries a current of 100 microamps.

15-4   27.   A current meter is to have a square coil 3.0 cm on a side with 4 turns. It is to have a torque constant of 3.6×10^{-3} N/rad and deflect 57.3° with a current of 5.0 amps. What must be the magnetic field?

15-7-1   28.   Find the resistance of the coil of a voltmeter if it reads 9.15 volts when a (nominal) 9 volt battery is connected across it, and only 3.25 volts when a resistance of 10,000 ohms is connected in series with the meter and the same 9 volt battery.

15-7-1   29.   What resistance must be put in series with a voltmeter that has a full scale reading marked 10 volts but is to be changed to give it a full scale reading with 50 volts applied? The resistance of the meter at 10 volts is 200,000 ohms.

15-7-1   30.   Find the current required to give full scale deflection of a voltmeter that is marked 20,000 ohms per volt.

CHAPTER SIXTEEN

ATOMS AND NUCLEI

16–1 ATOMS

A one-line definition of an atom is not as simple as it was in the days before it was possible to take the atom to pieces, to add to it, or to make new ones of unusual particles. Instead of starting with a definition, some of the characteristics of atoms and the structural features that are important in different situations will be described.

The existence of atoms was inferred from the study of material under different situations. The fact that the amounts of various materials entering into chemical reactions is in ratios of fairly small whole numbers was explained by Dalton about 1807 on the basis of matter consisting of atoms. Chemists identified different elements, and each element was believed to consist of a different type of atom. The list of elements grew until by 1931 the existence of 90 of them had been demonstrated.

The number of **naturally occurring elements** on earth is still only 90. Behind the use of the number 90, rather than the number 92 that may be familiar, lies a scientifically interesting tale. Dmitri Mendeleev, a Russian chemist (1834–1907), devised a scheme of ordering and numbering the elements. On that scheme all elements were assigned a number, called the atomic number, and the highest, 92, was eventually assigned to uranium. Also on the basis of the scheme, the existence of new elements, ones that had not been found up to that time, was predicted and it became a challenge to find them. People, places, and institutions gained fame if they were involved in finding new elements. Some of the heavier elements were known to exist only fleetingly in radioactive decay chains as uranium or thorium gradually transformed to lead. Radium is an example of one of the longer-lived members of such a chain. It is unstable and decays with what is called a "half-life" of 1600 years. Radium occurs in nature only as a product of decaying uranium. Of the elements having atomic numbers lower than that of lead, the last found was number 59, praseodymium, in 1931. The list was then complete except for elements numbered 43 and 61. For these the search continued. Exotic minerals from faraway places were analyzed. Occasionally a discovery was reported but proved false. The count of the list of the elements occurring naturally on earth stood, and still stands, at 90.

Both of those missing elements have now been made in the laboratory and their characteristics show that they will never be found in nature on earth. Both are radioactive and have fairly short half-lives. The longest-lived form of element 43, now called **technetium,** decays to half of its original population in 1.5 million years, and element 61, **promethium,** has a half-life of only 5.5 years. When the elements of which the earth is made were formed, these and other radioactive species were probably included. The longer-lived radioactive elements, uranium and potassium, for example, still exist; but the short-lived ones have disappeared. Spectral lines of technetium have, however, been identified in the light from some stars. Where could this element come from?

Technetium is no longer a rare curiosity, for in the nuclear medicine laboratory it ranks as one of the most useful radioactive isotopes. A large number of hospital isotope laboratories or nuclear medicine departments have their own technetium generators.

16–2 MODELS OF ATOMS

Atoms are known from the behavior of matter; in different situations, different atomic properties are dominant. When the pressure-volume-temperature relations in a gas (kinetic theory) are being analyzed, it is often sufficient to consider atoms as hard spheres (Chapter 6). Gases such as helium and neon are monatomic, and the pressure is described by the model of the collision of hard spheres with the container wall. Some crystals take shapes described by the packing of spheres in different ways. In these instances, in which phenomena of nature are satisfactorily explained by the concept of an atom as a sphere, it is quite legitimate to deal with it as such. The internal structure of the atom affects those phenomena by only a small amount. Other things, such as the absorption of light by solids or conduction of electricity, do require that the electrons attached to the atoms be considered.

In the investigation of emission of light from the atoms of a gas, the outer electrons of the atom are involved. Many of the features of the spectra are described by considering the atom as having a central portion, the nucleus, which has a net positive

(A)

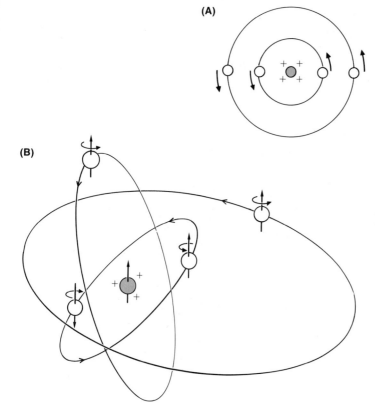

(B)

Figure 16–1 An atomic model used to explain the main features of atomic spectra (a) and to explain some of the details of the spectra (b).

charge, and electrons in orbit about it [Figure 16–1(a)]. Such a model of an atom describes most features of the spectra, but the details of the spectra do require more elaborate models, such as that in Figure 16–1(b). The inner electrons are important in the production of some types of x-rays.

Radioactivity is a property associated with the tiny central nucleus, and in studying this phenomenon the structure of that nucleus must be considered. The electrons are of minor importance in radioactivity. In any given situation, the atomic model that will best explain or describe the features being studied is used.

One of the chief areas of interest is spectral analysis, and it has been from attempts to understand spectra that the electronic structure of atoms has been revealed. That will be the next topic.

16–3 ATOMIC SPECTRA

When the light from a glowing gas is analyzed with a spectroscope, a series of discrete spectral lines is seen as in Figure

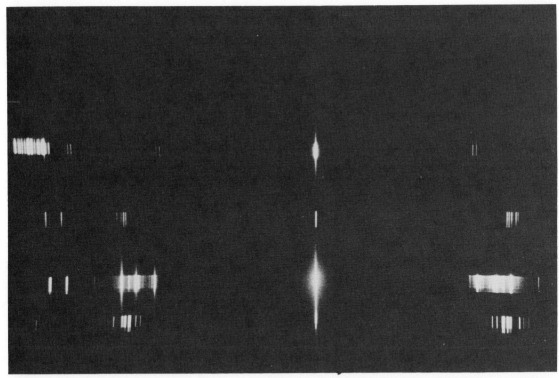

Figure 16–2 The line spectra of various elements. These spectra were obtained with a grating and are, from top to bottom, those of neon, helium, mercury, and argon.

16–2, rather than a continuous spectrum. Each element emits its own set of lines, and these spectra can be used for analysis. But what is the origin of these lines? Why do the gases emit only certain colors or wavelengths and not others? The spectral lines were seen and worked with throughout the 1800's and into the early 1900's, though their originating mechanism was obscure. Tables of wavelengths from each element were compiled. In order to gain an understanding of them, some form of order was looked for in what was initially apparent chaos.

A string on a musical instrument can be made to vibrate with only certain wavelengths or frequencies (these are related by wavelength = wave speed/frequency), and an organ pipe of a given length will produce only certain frequencies. In each case there is a fundamental frequency, and there are also harmonics, which are integral multiples of the fundamental. The wavelengths of the harmonics of a sound are simple integral fractions of the wavelength of the fundamental, and it was natural to look for such a simple relation among the wavelengths of spectral lines. Some prominent wavelengths of the lines in the hydrogen spectrum are shown in Table 16–1, and a spectrogram of hydrogen is shown in Figure 16–3.

TABLE 16–1 Wavelengths of some of the lines observed in the hydrogen spectrum*

LINE	WAVELENGTH, nm
H_α	656.84
H_β	486.136
H_γ	434.048
H_δ	410.176
H_ϵ	397.0
H_ζ	388.9
H_η	383.5
H_θ	379.8

*The lower case Greek letters used here are, in order, alpha (α), beta (β), gamma (γ), delta (δ), epsilon (ϵ), zeta (ζ), eta (η), and theta (θ).

There is, apparently, an order, noticeable principally in the picture, but there is no simple ratio between the wavelengths. It was apparent that the phenomenon of emission of these wavelengths was not simple resonance. But what order was there? So long as the order in the spectral patterns could not be described, there could be no clue as to what lay behind them — to the nature of the atom that gave rise to them. A pattern had to be found!

Johann Balmer, in Germany in 1885, made a major contribution to deciphering the spectra when he found that the wavelengths of the series of lines of the hydrogen spectrum

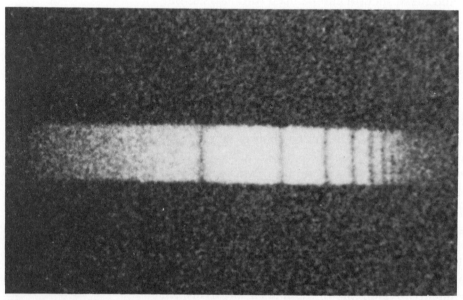

Figure 16–3 A spectrum of hydrogen. An order in the spacing of the lines is apparent. This spectrum is, for variety, that of the star Sirius. Lab spectra are the same except that they have bright lines on a dark background. This spectrum seems to emphasize the order.

could be described by an equation of the form

$$\lambda = \text{constant} \times n^2/(n^2 - 4)$$

The constant is 364.705 nanometers; n could take the values 3, 4, 5 . . . . Each value of n results in the calculation, very precisely, of a line of the hydrogen spectrum. This series of spectral lines became known as the Balmer series. Balmer had found a description of the order in one of the series of spectral lines. Other series were soon found, ones that had 1^2 and 3^2, or 9, in place of the 4, or 2^2, in Balmer's equation.

One other piece of information was necessary to complete the groundwork for the explanation of line spectra. This resulted from the work of two men, Max Planck (1900) and Albert Einstein (1905). Planck introduced the concept of radiant energy occurring in discrete amounts, and Einstein showed that light, in its interaction with electrons (the photoelectric effect), behaves as though it occurs in discrete bundles, **quanta of radiation** or **photons.** In accordance with Planck's ideas, the energy of a quantum depends on the part of the spectrum to which it belongs. This is characterized by the wavelength λ, or the frequency f (these are related by $f = c/\lambda$). The energy of a photon is proportional to the frequency, or $E \propto f$. This can be put into an equality using a constant of proportionality. This constant is called Planck's constant, symbol h. Then the energy carried by a photon is given by

$$E = hf$$
$$\text{or } E = hc/\lambda$$

The value of h is 6.63×10^{-34} joule seconds. The shorter wavelengths, violet for example, have more energy per photon than do the longer wavelengths of red light. In Table 16-2 is shown the energies associated with various colors or wavelengths, in joules and in electron volts.

TABLE 16-2 **Energies associated with photons of various wavelengths or colors**

WAVELENGTH, NANOMETERS	COLOR	ENERGY, JOULES	ENERGY, ELECTRON VOLTS
650	red	3.05×10^{-19}	1.91
540	green	3.68×10^{-19}	2.29
450	blue	4.41×10^{-19}	2.75
410	violet	4.84×10^{-19}	3.02

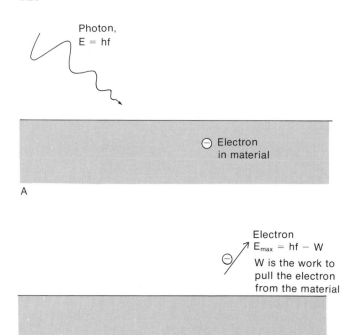

Photon,
E = hf

⊖ Electron
in material

A

Figure 16–4   The photoelectric effect.

Electron
$E_{max} = hf - W$
W is the work to
pull the electron
from the material

⊖

B

16–3–1   THE PHOTOELECTRIC EFFECT

The return to a particle theory of light after more than a hundred years of amply demonstrating its wave properties could not be done lightly. The next demonstration of the particle property of light was described in 1905 by Einstein. For this work Einstein was awarded a Nobel prize in 1921.

When light falls on some materials, electrons are ejected from the material. The energy of the light is transferred to some of the electrons. This is the phenomenon on which various light measuring devices (photometers) are based. In such devices the electric current (number of electrons flowing) depends on the intensity of the light. A bit of thought would lead to the conclusion that an increase in light intensity would increase the energy of the electrons ejected but not necessarily the number. For a wave of given frequency, an increase in intensity of the wave increases the amplitude of the electric field. The energy that could be passed to an electron should be greater for a greater intensity. This was not found to be the case. The number of electrons ejected was increased with increased intensity, but the electron energy did not change. Einstein found that the energy of the electrons was related to the frequency of the light, or the inverse of the wavelength. Red light, for example, resulted in less energetic electrons than blue light. The relation he found was that the observed maximum electron energy, E, was given by

$$E = hf - W$$

The quantity W is called the **work function** of the material. It is the work done in pulling the electron from the material. This work ranges from the value W upward, depending on where the electron was in the material when it absorbed the light. The energy E is a limiting value, and E does not change with intensity, only with frequency. The process referred to as the photoelectric effect is illustrated in Figure 16–4.

The photoelectric effect is actually an inelastic collision. The photon hits the electron and is absorbed by it. All of the energy of the photon is transferred to the electron. In Section 7–7–1 it was shown that, in an inelastic collision between two particles, momentum is conserved but the total kinetic energy is reduced. If the kinetic energy is to be conserved, a third particle must be there to take some momentum. The photoelectric effect does not occur with a free electron but only with an electron bound to an atom or an extended piece of matter. This third component can allow conservation of momentum.

16–3–2 THE COMPTON EFFECT

An elastic collision may occur between a photon and a free electron, as contrasted to the inelastic process that is the photoelectric effect. This elastic process is referred to as the Compton effect. In the Compton effect a photon with energy hf_1 or hc/λ_1 approaches an electron as in Figure 16–5(a). After an interaction the electron moves away with part of the energy; a photon also still exists, but with a reduced energy, hf_2 or hc/λ_2. In this reaction both momentum and energy are conserved just as in the two-particle system.

The Compton effect is important at photon energies in the region of x-rays and gamma rays, but not for visible light.

16–4 NEILS BOHR'S MODEL OF THE ATOM

An atomic model that would describe the positions of spectral lines of various elements was developed by Niels Bohr in Denmark. By 1915 the model was developed to such a degree that most of the features of the spectra could be accounted for. It was a triumph of theoretical and empirical work and catapulted Bohr to a position among the leading physicists of our

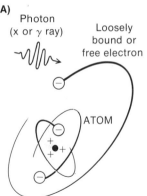

(A)

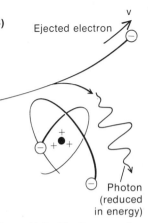

(B)

Figure 16–5 The Compton effect. Only part of the energy of the photon is transferred to the free electron.

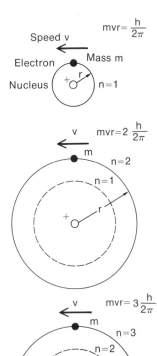

Figure 16–6 Allowed orbits of the electron in a hydrogen atom.

time, a position that he maintained by producing more and more simplifying concepts in various areas of atomic and nuclear physics.

The atomic model developed by Bohr uses the now familiar orbiting concept. In the center of the atom is the relatively small but heavy positively charged nucleus (discovered by Rutherford in 1906). The electrons orbit the nucleus just as the planets orbit the sun, but they are held in their orbits by the force of electrical attraction. One of the important concepts that Bohr introduced was that only certain orbits are possible for the electrons. In these **allowed orbits** the action in one revolution has to be an integral number times Planck's constant. By **action** is meant the product of momentum and distance. That is, in the allowed orbits:

$$mv \times 2\pi r = nh$$

For the smallest orbit, n has the value 1; it then becomes 2, 3, and so forth for successively larger orbits. It is customary now to put that equation in the form

$$mvr = nh/2\pi = n\hbar$$

The quantity mvr, momentum times radius, is angular momentum. In the allowed orbits, the *angular momentum* can be only an integer times $h/2\pi$. The quantity $h/2\pi$ occurs so frequently that it is represented by a single symbol, $\hbar$ (read h-bar or h-stroke). That is, in the first orbit the angular momentum is $\hbar$. In the second orbit it would have to be $2\hbar$, and so on. In the nth orbit the angular momentum would be $n\hbar$, where n is called the **principal quantum number.**

These orbits are shown in Figure 16–6. The idea introduced is that of a physical quantity, in this case the angular momentum, which cannot take on a continuous range of values but for which only certain values can occur. It is as though a wheel could be spun only at certain definite speeds but not at those in between. Changes in speed of such a wheeled vehicle (because of certain allowed wheel revolution speeds) could occur only in sudden jumps. Natural phenomena on the very small scale behave in this way. The allowed steps are so small that on the scale of ordinary objects these small jumps are not noticeable, so the idea that variations can be continuous has arisen. Bohr's work marked the beginning of the change from ordinary concepts of nature.

To see how the Bohr model of the atom describes spectra, consider the hydrogen atom. In Figure 16–7 are illustrated some possible circular orbits for hydrogen. Ordinarily each hydrogen

(A)

(B)

Figure 16–7 An electron that gains energy must move to a larger allowed orbit, and when it falls back to a lower orbit it radiates energy.

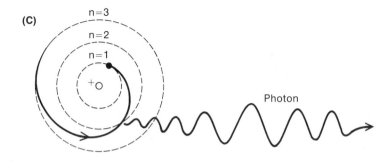

atom has the electron in the lowest allowed orbit. In this orbit the electron has the least energy. If energy is given to the electron, it may be raised to another orbit as in part (b). When an atom has extra energy like this, it is said to be in an **excited state.** The electron usually does not remain for long in the higher orbits, but will spontaneously fall back to a lower orbit. The energy that the electron must get rid of in such a transition is radiated as a photon of light [Figure 16–7(c)]. The energy of the photon is the same as the energy difference between the orbits. Thus, the line spectra were an indication that only certain orbits were allowed, since the frequency, and thus the energy, of the light from each transition is always the same.

The model developed by Bohr will be analyzed as far as finding the sizes of the allowed orbits. This will give experience in the analysis of such a model. The first step in this type of work is to consider the simplest case, which will be the hydrogen atom with the electrons in circular orbits.

This model of the atom is illustrated in Figure 16–8. The electron orbits at a speed v and a radius r. The electron mass is small compared with the mass of the nucleus. The force required to hold the electron in the circle is described by $F = mv^2/r$, this force being provided by the electrical attraction

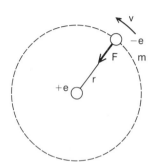

Figure 16–8 An electron orbiting a proton to form a hydrogen atom.

between the electron and the nucleus. Both charges are of a magnitude e. This force is found from

$$F = 10^{-7}c^2 Q_1 Q_2/r^2 \qquad \text{(see Section 14-1)}$$

with Q_1 and Q_2 each being of size e. Then

$$F = 10^{-7}c^2 e^2/r^2$$

By Bohr's quantum condition, in the nth orbit

$$mvr = nh/2\pi$$

This can be solved for v, which can be substituted in the equation $F = mv^2/r$. But also $F = 10^{-7}c^2 e^2/r^2$. These equations can then be solved for the radii of possible orbits, and the result is that:

$$r = \frac{n^2 h^2}{4\pi^2 10^{-7}c^2 me^2} = \frac{n^2 \hbar^2}{10^{-7}c^2 me^2}$$

The values of the various quantities are:

$$n = 1, 2, 3 \ldots .$$
$$h = 6.6262 \times 10^{-34} \text{ joule seconds}$$
$$c = 2.9979 \times 10^8 \text{ m/s}$$
$$m = 9.1095 \times 10^{-31} \text{ kg}$$
$$e = 1.6022 \times 10^{-19} \text{ coulomb}$$

Substitution of these values gives the possible radii for different values of n. Using nanometers, the radii of the possible orbits for the hydrogen atom are described by

$$r = 0.0529 \, n^2 \text{ (nanometers)}$$

where n is the principal quantum number and can take integral values 1, 2, 3 . . . . The radii calculated from this equation were used to make Figure 16–9.

The above calculations are for hydrogen, which has only a charge of $+e$ on the nucleus. The calculation could be carried out for other elements that would have a charge Ze on the nucleus, where Z is the atomic number. The force would be correspondingly greater, and it is possible to show that the allowed orbital radii are smaller by the atomic number, Z. In helium, for instance, $Z = 2$, and the first orbit is just half the size that it would be for hydrogen.

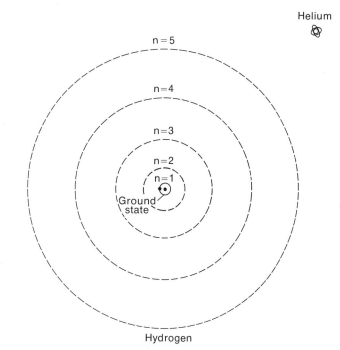

Helium

n = 5

n = 4

n = 3

n = 2

n = 1

Ground
state

Hydrogen

Figure 16–9 Radii of the possible orbits of the electron in a hydrogen atom, and the electrons in the lowest orbit of helium.

Atomic radii calculated with this formula do correspond with atomic sizes found in other ways.

16–4–1 OTHER ATOMIC MODELS

The Bohr model of the atom was so successful for the explanation of line spectra that it was one of the milestones in modern physics. Some modifications were required to account for atoms with more than one electron and to explain other features. One of these features is that most spectral lines are not single, but consist of two or more components close together. These features were explained by considering elliptic orbits, spinning electrons, orbits inclined at various angles, and so on. The analysis of spectra has, in fact, given most of the information we now have concerning atomic structure. However, some basic questions remained. Why were only certain orbits allowed? Why did the orbiting electron not radiate its energy? This would be expected because the orbiting electron would produce an oscillating electric and magnetic field, like a radio wave. And radio waves had been worked with for years.

The explanation came with the work of Prince Louis de Broglie while he was a graduate student in 1924. De Broglie postulated that just as photons exhibited both wave properties and particle properties, perhaps electrons, for which we ordinarily see the particle behavior, also would exhibit wave properties. He developed the theory mathematically; with that to go on, the wave properties of electrons were demonstrated experimentally in 1927, by Davisson and Germer in the U.S. and by G. P. Thompson in England. The wave properties are not limited to electrons but are exhibited by all particles. Again it is the case that in some situations particle properties predominate, and in other situations the wave properties are most important. De Broglie showed that we should look for the situations or experiments in which the wave properties could be seen.

Erwin Schrödinger applied de Broglie's wave concepts to atomic theory. To state the result simply, in the allowed orbits the circumference is an integral number of wavelengths. In the first orbit there is one wave, in the second orbit two waves, and so on. The wave, as it travels around the orbit, would be something like a standing wave in a rope, and it can be expected that there would be something unique about such orbits.

So what is an electron like? It has properties of mass (inertia), electric charge, and spin, and also properties of a wave. These are incompatible. But the particle model is just that, a *model* that describes some of the properties of electrons. The wave model describes other properties. An electron is neither of these things. Just as it was seen that light is neither a wave nor a particle, an electron also is neither, but it does show properties of each. Atomic models are as vague as concepts of the particles of which they are made. The models, however, have been very *useful,* and it is because they are useful that they are retained.

One of the uses of the wave model of the electron has been that it led to the idea of an **electron microscope.** A light microscope is limited by the wavelengths of light. Because of diffraction, detail much small than the wavelengths cannot be seen. But electrons of high energy have very short wavelengths, and electron microscopes with magnifications over a hundred times greater than those of light microscopes are used to observe the structure of bacteria and to see viruses. Field emission electron microscopes show the arrangement of atoms on a solid.

16–4–2 ATOMS WITH MORE THAN ONE ELECTRON

The hydrogen atom has only one electron, normally in the lowest orbit (for which the principle quantum number is 1.) Helium has two protons in the nucleus, and therefore two outer electrons to balance the two positive charges. Both of these electrons go into the orbit for $n = 1$, but because of the greater nuclear attraction the radius is much smaller than for hydrogen. The two electrons differ in that the spin of one of them is in approximately the same direction as the angular momentum in the orbit, while the spin of the other is in the opposite direction. These two electrons are not just tiny balls but actually occupy the whole sphere around the atom, like a fuzzy shell about the nucleus. This first shell is referred to as the K shell, and it is filled by those two electrons.

Lithium has three electrons, and the third must be in an orbit for which $n = 2$. Inside this orbit there is a shell of two negative electrons and a nucleus with three positive protons, the total attractive force of which is that of one positive charge. In many ways the orbit of the third electron is similar to that of hydrogen, and lithium is chemically related to hydrogen. The shell for $n = 2$, called the L shell, does not become full until it has a total of 8 electrons in it. Including those in the K shell, an atom with a full L shell has a total of 10 electrons. There will be a corresponding number of protons, 10, and the element of atomic number 10 is neon.

The noble gases, including helium and neon, are characterized by having filled shells, all the electrons in the outer shell being relatively tightly bound (in neon by an excess charge of $+8$ inside that shell). The alkali metals, such as lithium and sodium, are characterized by having only one electron in the outer shell, with a single net positive charge inside that shell, the inner electrons screening the effect of all but one of the protons.

16–5   THE NUCLEUS

Sir Ernest Rutherford must have been very surprised when in 1911 he found the approximate size of an atomic nucleus based on experiments performed by himself, Hans Geiger, and Sir Ernest Marsden. The existence of atoms had been accepted for over half a century, and in most phenomena studied they behaved almost like hard spheres about 10^{-8} cm across. Rutherford, by studying the deflection of alpha particles projected at thin metal foils, found that almost the whole of the mass (actually about 99.95 per cent of it) and the positive charge were concentrated in a space less than a ten thousandth of this size. Rutherford reported his findings in the unnecessarily cold, unimpassioned phraseology of scientific writing in this way:

"Considering the evidence as a whole, it seems simplest to suppose that the atom contains a central charge distributed through a very small volume, and that the large single deflexions are due to the central charge as a whole, and not to its constituents."

He had discovered the nucleus!

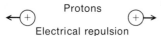
Protons
Electrical repulsion

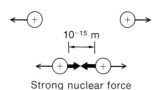
10^{-15} m
Strong nuclear force

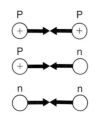
P   P

P   n

n   n

Figure 16-10 Nuclear forces.

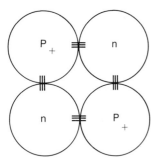
P+   ≣   n

n   ≣   P+

Figure 16-11 An alpha particle.

16-5-1   NEUTRONS AND PROTONS IN NUCLEI

Subsequent work has shown that the nucleus can be considered as a collection of protons and neutrons. These two kinds of particles are almost identical, differing mainly in the presence of the positive charge on the proton. The masses of these particles are almost the same, the proton having 1836 and the neutron 1837 times the mass of the electron. In Table 16-3 are shown the masses of neutrons, protons, and electrons. The masses are given in kilograms and in what are called atomic mass units. The u is based on a standard of the isotope carbon-12 being 12 units. Protons and neutrons are referred to as **nucleons,** and they attract each other with a very strong force when they get within about 10^{-15} meter of each other. This force is neither electrical nor gravitational. It is another kind of force, called a **nuclear force.** How this nuclear force depends on distance has not been found. It is known that it is not described by an inverse square law but that it operates only over a very short range. Inside this range of action it is *hundreds* of times stronger than the electrical forces resulting from the charges on protons. At short range the electrical forces are almost negligible; protons attract protons, protons attract neutrons, and neutrons attract neutrons. Some of these ideas are illustrated in Figure 16-10.

Another characteristic of the nuclear force is that a nucleon can strongly attract only a limited number of other nucleons. This results in a subgrouping of two protons and two neutrons in tightly bound units, called **alpha particles** (Figure 16-11). Though they are really a group of four, they are so tightly bound that they behave as one. Alpha particles do attract each other and other nucleons to form larger nuclei. The alpha particle alone is also a nucleus of a helium atom.

Different elements are distinguished by different chemical properties. Chemical properties depend on the arrangement of outer electrons, which is determined by the number of electrons. The atom is electrically neutral, there being one negative orbiting electron for each positive proton in the nucleus, so the element is basically determined by the number of protons in the nucleus. All atoms having one proton behave as hydrogen. All atoms with eight protons behave as oxygen, and atoms with 92

TABLE 16-3   The masses of protons, neutrons and electrons.

PARTICLE	MASS IN kg	MASS IN u
proton	1.6725×10^{-27}	1.0073
neutron	1.6748×10^{-27}	1.0087
electron	9.1091×10^{-31}	0.00055

protons are uranium. The neutrons in the nucleus do not affect chemical properties, but they do affect the mass. Ordinary oxygen actually consists of three different types. Each has 8 protons, but some oxygen atoms have 8 neutrons, some have 9, and some even have 10. It is said that there are three naturally occurring **isotopes** of oxygen.

The weight of an atom depends on the total number of nucleons (protons plus neutrons), and this is called the **mass number**, A. The number of protons is called the **atomic number**, Z. The difference between A and Z is the number of neutrons. The three isotopes of oxygen referred to have mass numbers of 16, 17, and 18. These nuclear species or **nuclides** are called O^{16}, O^{17}, and O^{18}. There are several notations in common use. The atomic number may or may not be put as a subscript before the symbol for the element. The mass number is put as a superscript either before or after the symbol. It is actually redundant to include both the atomic number and the symbol, since one determines the other. The species of nucleus with 8 protons and 8 neutrons (total 16 particles) may be written as

$$\,^{16}_{8}O \quad \text{or} \quad \,_{8}O^{16} \quad \text{or} \quad O^{16} \quad \text{or} \quad \,^{16}O$$

The first of these is the officially recognized form, though any of the others may be encountered. The third form is used in this book, since the expression "O sixteen" is used in talking.

16–5–2   THE CHART OF THE NUCLIDES

Combinations of protons and neutrons will now be examined in an orderly way.

One proton alone, with one orbiting electron, forms the hydrogen atom in its most common form. It is called hydrogen one, H^1, and forms 99.985 per cent of ordinary, natural hydrogen. The remaining 0.015 per cent of natural hydrogen also has a neutron in the nucleus. Since the proton and neutron are of almost identical mass and by comparison the electron mass is almost neglible, this type of hydrogen atom, H^2, weighs twice as much as the more common H^1. It is called **heavy hydrogen** and also is given the name **deuterium**, symbolized by D. These are shown in Figure 16–12.

A hydrogen atom can also be formed with two neutrons and a proton. This is hydrogen three, H^3, also called **tritium.**

The various combinations of neutrons and protons can be illustrated with a series of rows of boxes. In each box along one row is one proton. In the next row there are two protons, then three, and so on as illustrated in Figure 16–13. The atoms in the

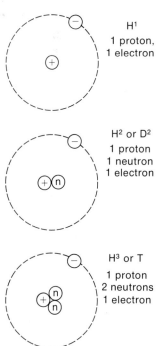

H^1
1 proton,
1 electron

H^2 or D^2
1 proton
1 neutron
1 electron

H^3 or T
1 proton
2 neutrons
1 electron

Figure 16–12 Isotopes of hydrogen.

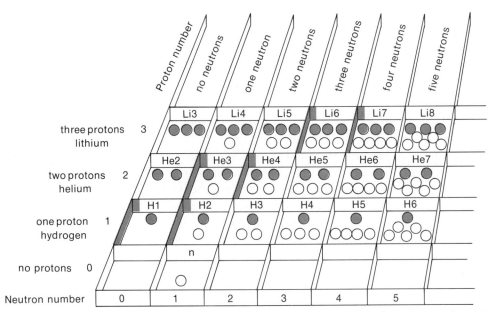

Figure 16–13 Combinations of neutrons and protons. Not all combinations shown here exist even fleetingly in nature. The nuclides in the dark-sided boxes are stable and are found in nature.

row with one proton will all be hydrogen atoms, the row with two protons will contain helium atoms, and the row with three protons will contain lithium atoms. In the first column of boxes nothing is added, but in the second column a neutron is put in each box, in the third column two neutrons are added, and so on. Such an array of boxes allows all combinations of neutrons and protons: all nuclear species or nuclides are represented. Each element is represented by a row and each possible **isotope of that element** is in one of the boxes along that row. In each of the squares of a chart showing these nuclides, information concerning that isotope can be inserted. Such a chart, devised initially by Emile Segré, is sometimes called a **Segre chart** or a **chart of the nuclides.** In Figure 16–14 a portion of such a chart is shown with some of the data on isotopes of hydrogen, helium, and the other elements up to neon.

The chart of the nuclides is to a physicist what Mendeleev's periodic table is to a chemist.

16–6 RADIOACTIVE NUCLEI

Nuclei made of any number of protons and neutrons could be imagined. Only certain combinations are stable, however, and for the lighter elements there must be *approximately* equal numbers of neutrons and protons. If a nucleus is formed with an

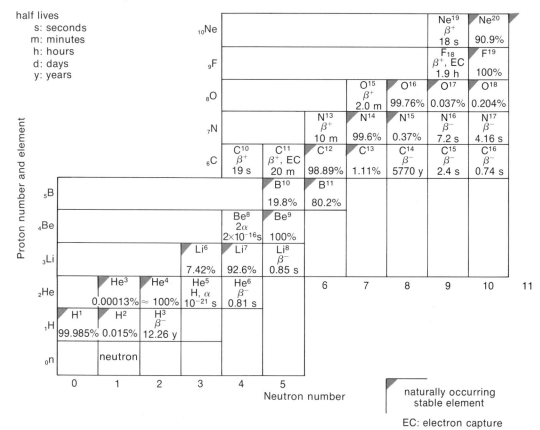

Figure 16–14 A portion of the chart of the nuclides. Data concerning each nuclide or isotope can be put into the appropriate box. On this chart the percentage abundance is shown for stable nuclides, and decay method and half-life for radioactive nuclides.

excess of neutrons, one of those neutrons will undergo a transformation to obtain a more "acceptable" or stable nucleus. The neutron will change into a proton, which has a positive charge, and an ordinary electron with a negative charge. The electron will be ejected from the nucleus, and such an electron is called a **beta particle** or a β^- particle. This is one of the forms of radioactivity [see Figure 16–15(a)].

This does not mean that the neutron consisted of a proton and an electron, for if a nucleus with an excess of protons is formed, one of the protons will transform to a neutron and eject a positively charged electron, called a **positron** or a **beta plus** (β^+) particle [Figure 16–15(b)]. The electrons that occur in ordinary matter are all negative. These positive electrons are not ordinary at all. They are an example of what is called antimatter. When a positive electron and a negative electron meet, they annihilate each other. The mass disappears and two high-energy gamma rays are produced as in Figure 16–16.

(A)

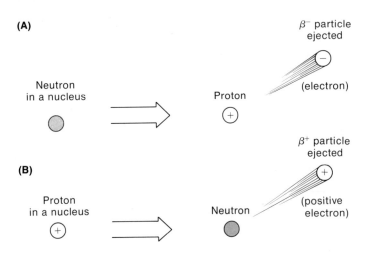

(B)

Figure 16–15 (a) An excess neutron in a nucleus may transform to a proton (+) and an electron (−), ejecting the electron (which is called a β^- particle). (b) An excess proton in a nucleus transforms to a neutron and a positive electron (β^+), which is ejected.

Beta radioactivity may be of two forms, β^- or β^+. The β^+ activity has high-energy **annihilation radiation** associated with it. In both of these beta emission processes, a second particle is also emitted, a **neutrino** with β^+ emission and an antineutrino with β^- emission. The neutrinos have no charge and no rest mass, but spin and carry energy. They react very little with matter, and they will not be mentioned further for they are of no importance in the areas of concern to us. However, as particles they are fascinating, and in some areas of study they are extremely important.

Tritium is an example of a nucleus with an excess neutron. One of the neutrons will change to a proton and a β^- particle will be ejected. The energy of the β^- is up to 18.1 keV. This leaves a nucleus of helium, the special type of mass number 3, He³. The tritium nucleus does not necessarily decay immedi-

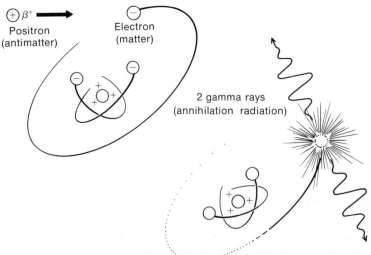

Figure 16–16 The fate of a positron. It meets an electron and annihilates it. Two gamma rays result.

ately. Some atoms last a long time, others not so long. The decay occurs at such a rate that, if initially there is a certain number of atoms, after 12.26 years half of them will have changed to He³. It is said that the half-life of H³ is 12.26 years. The decay of tritium to He³ is shown in Figure 16–17.

Accompanying either β^+ or β^- emission, there is frequently an emission of excess energy in the form of gamma (γ) rays. These are high-energy radiation of a form similar to light. The gamma rays behave like waves or particles (photons). Their energy is frequently in the MeV range.

Radioactive elements are changed to other elements after β^- or β^+ emission, as illustrated in the above examples. If a nucleus decays by β^- emission, it "loses" a neutron but "gains" a proton to change it to the isotope diagonally above and to the left on the chart of the nuclides. If β^+ emission occurs, the resulting nuclide is to the lower right. In Figure 16–18 is illustrated the portion of the chart near carbon. Arrows indicate the change in the nuclear species for the appropriate decay.

16-6-1 ALPHA PARTICLE EMISSION

Some radioactive materials emit alpha particles, which are clusters of 2 protons and 2 neutrons: a helium nucleus. The result is a nuclide with 2 less protons and 2 less neutrons than the original. Alpha particle emission occurs principally with the heavy elements; uranium, thorium, and radium, for example. The alpha particles are about 7294 times as massive as beta particles, which are electrons. For a given energy the alpha particles move very slowly compared to beta particles (about $1/\sqrt{7294}$ or 1/85 as fast). They also have double the charge. As an alpha particle moves through matter, perhaps tissue, it spends

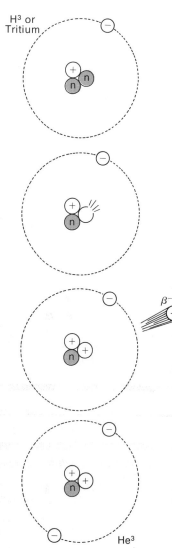

Figure 16–17 The decay of tritium to He³.

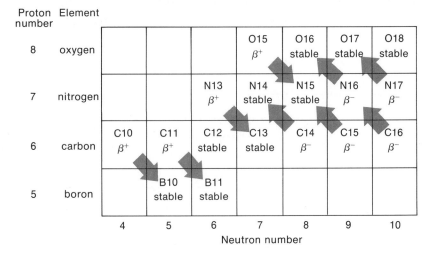

Figure 16–18 The transformations due to β^- and β^+ processes, shown on a chart of the nuclides near carbon.

Proton number	Element	4	5	6	7	8	9	10
8	oxygen				O15 β^+	O16 stable	O17 stable	O18 stable
7	nitrogen			N13 β^+	N14 stable	N15 stable	N16 β^-	N17 β^-
6	carbon	C10 β^+	C11 β^+	C12 stable	C13 stable	C14 β^-	C15 β^-	C16 β^-
5	boron		B10 stable	B11 stable				

Neutron number

EXAMPLE 1

What is the result after carbon-14 emits a β^- particle?

Carbon-14 is a well-known radioactive material used extensively in biological investigation. Carbon has atomic number 6; it has 6 electrons orbiting a nucleus that has 6 protons. The total number of nucleons in C^{14} is 14, so there must be 8 neutrons. The common form of carbon is C^{12}, which has 6 protons and 6 neutrons in the nucleus. Carbon-14 has an excess of neutrons, and it is one of those cases in which a neutron will transform to a proton and a negative electron, a β^-, which is emitted.

Original state of C^{14}:	6 p, 8 n, 6 electrons
During radioactive decay:	1 neutron changes to proton + electron
Final state:	7 p, 7 n, 7 electrons

An atom with 7 electrons and 7 protons is nitrogen. The atom that results is N^{14} (there are still 14 nucleons). The process can be written this way:

$$C^{14} \longrightarrow N^{14} + \beta^-$$

Carbon-14 changes to nitrogen-14 after β^- emission. The energy of the β^- ejected from C^{14} can be anything up to a maximum of 156 keV. There are no gamma rays. An antineutrino is also emitted, but it has been omitted from the equation so as not to clutter the concepts.

Most C^{14} atoms will remain as such for a long time before decaying to N^{14}, but the process is completely unpredictable for an individual atom. In a sample of C^{14} the decay rate is such that half the atoms change to N^{14} in 5770 years, or alternately, each year 0.012 per cent of them decay to N^{14}.

EXAMPLE 2

What results after copper-64 emits a β^+ particle?

Copper-64 is a positron emitter that has been sometimes used in clinical studies.

Copper has atomic number 29, so it has 29 protons. Cu^{64} has $(64 - 29)$ neutrons, or 35 neutrons. One proton changes to a neutron and a β^+ particle.

Original state of Cu^{64}:	29 p, 35 n, 29 electrons
In radioactive decay:	A proton changes to a neutron and a positron. The positron annihilates an electron.
Resulting atom:	28 p, 36 n (total 64), 28 electrons

An atom with 28 protons or atomic number 28 is nickel. The process has been this:

$$Cu^{64} \longrightarrow Ni^{64} + \beta^+ + \gamma$$

Copper-64 changes to nickel-64 after β^+ emission. In this case a gamma ray is also emitted. The positron may have an energy up to 0.573 MeV. Cu^{64} is unusual in that sometimes it will emit a β^- particle rather than a β^+.

85 times as long in the vicinity of each molecule, pulling on its electrons with double the force of a beta particle. The impulse given to the electrons of each molecule is about 170 times as much, and the electrons are very likely to be pulled away. Since chemical bonds are due to electrons, the alpha particles passing through tissue do a lot of damage at the molecular level. The alpha particles transfer their energy very quickly to the electrons of the material through which they pass, and as a result they do

not travel very far before coming to rest. The total range of an alpha particle in tissue would typically be the order of a tenth of a millimeter. They do not, therefore, penetrate clothing or skin. However, some of the alpha emitters are easily taken into the body; for instance, one of those occurring in the chain of decay from uranium is a gas. Some of the others, if ingested, are deposited in the bones where they remain almost permanently. Alpha emitters are therefore extremely dangerous, and handling them must be done with respect, and with knowledge of special techniques.

16–7 ACTIVITY OF A SAMPLE

The radioactivity of a sample of material is described by the number of atoms disintegrating per second in that sample. A sample in which an average of one atom disintegrates (emits an alpha or a beta particle) per second is said to have an activity of 1 **becquerel,** 1 Bq. The becquerel (disintegrations per second) is named in honor of Henri Becquerel, who in 1896 found that uranium steadily emits radiation; it was the discovery of radioactivity.

An older unit for measurement of radioactivity, the curie (Ci), is still encountered. This unit is named in honor of Marie Curie, who numbered the discovery of radium among her many achievements. Radium was the principal radioactive material from the time of its discovery for almost 50 years until radioactive isotopes became available from nuclear reactors about 1946. The radioactivity of uranium was discovered by Henri Becquerel in 1896, but Marie Curie noted that uranium ore was more radioactive than the uranium taken from it. She proceeded to isolate this radioactive material and found radium as well as several of the radioactive daughter products of radium. She actively pursued the study of these materials, their properties and uses, until almost the time of her death in 1934. Born Marie Sklodowska in Poland in 1867, she worked with her husband, Pierre Curie, at the Sorbonne in Paris. In Figure 16–19 she is shown on a Polish stamp issued in her honor. One of her daughters, Irène, married Frédéric Joliot, and together they did pioneer work on artificial radioactive materials. Another daughter, Ève, was a journalist and wrote a fascinating biography of her mother.* The claim that there was a third daughter, the famous milli, is not true.

One curie was initially defined as the number of atoms of radium that disintegrate per second in a sample consisting of one

---

*Ève Curie, *Madame Curie,* Pocket Books, Inc., New York, 1969.

Figure 16–19 Marie Curie on a commemorative Polish stamp. She was a Pole, born Marie Sklodow-ski, but she did her famous work in Paris.

gram of radium. This number was later measured to be 3.7×10^{10} disintegrations per second. The subdivisions or multiples of this are often encountered. Some of these, with an indication of the areas of use, are:

1 **Megacurie** = 1 MCi = 3.7×10^{16} dis/s = 3.7×10^{16} Bq. This is the unit often used to describe the radioactivity associated with nuclear reactors.

1 **kilocurie** = 1 kCi = 3.7×10^{13} dis/s = 3.7×10^{13} Bq. This is the activity used in cancer therapy units in hospitals.

1 **curie** = 1 Ci = 3.7×10^{10} dis/s = 3.7×10^{10} Bq.

1 **millicurie** = 1 mCi = 3.7×10^{7} dis/s = 3.7×10^{7} Bq. Sources for research, for implantation into tumors, or to be given as therapy doses are often in the millicurie range.

1 **microcurie** = 1 μCi = 3.7×10^{4} dis/s = 3.7×10^{4} Bq. Activities at this level are used in diagnostic tests or in many areas of research.

1 **nanocurie** = 1 nCi = 37 dis/s = 37 Bq.

1 **picocurie** = 1 pCi = 0.037 Bq = 2.22 dis/m. This is a very

low level of activity, sometimes found in natural materials and sometimes encountered in research.

The activity of a sample of material depends on the type of material and also on the amount. If in a certain sample there are *x* disintegrations per second, in a sample twice as big there would be 2*x* disintegrations per second. Cut the sample in half, and in one of the halves there would be only *x*/2 dis/s.

EXAMPLE 3

Consider a small sample of radioactive material as in Figure 16–20. The sample has an activity of 1 μCi, and 10 cm from it is a circular detector 2 cm in diameter. The sample is assumed to emit particles equally in all directions. How many particles will enter the detector per minute?

There will be 3.7×10^4 particles emitted per second in a solid angle of 4π steradians. The number per unit solid angle is $(3.7 \times 10^4)/(4\pi) = 2944$ per second per steradian.

The area of the detector is $A = \pi \times 2^2 \text{ cm}^2/4 = 3.14 \text{ cm}^2$. The solid angle to the detector is given by

$$\omega = A/r^2$$
$$= 3.14 \text{ cm}^2/10^2 \text{ cm}^2$$
$$= 0.0314 \text{ steradian}$$

The number of particles in this solid angle is:
0.0314 steradian $\times$ 2944 per second per steradian = 93 per second

There will be 93 per second or 5580 per minute entering the detector.

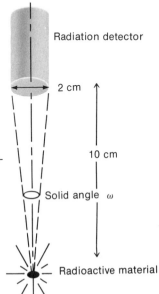

Figure 16–20 A sample of radioactive material and a detector. Only the radiation in the solid angle ω will enter the detector.

16–8   HALF-LIVES

If you have a pure sample of radioactive material, all of the same kind of atom, it will not remain that way. Each time an atom disintegrates (sends out a beta or an alpha particle), one of the atoms changes into a different element. The atoms are changing at whatever the disintegration rate is. The amount of the original material gradually decreases and the amount of the daughter material builds up. The time required for the amount of original material to decrease to half of its initial value is called the half-life, $t_{1/2}$, of that material. When the half-life is reached, the disintegration rate has also dropped to half of the initial rate (Section 16–6). It will take another half-life for the amount to drop by half again. The amount then remaining will again drop by half in another half-life, and so on. If the amount (number of atoms, perhaps) is N_0 at time 0, then the number remaining at various times is:

$$
\begin{aligned}
\text{at } t &= 0, & N &= N_0 \\
\text{at } t &= t_{1/2}, & N_1 &= N_0/2 \\
\text{at } t &= 2\,t_{1/2}, & N_2 &= N_1/2 = N_0/2^2 = N_0/4 \\
\text{at } t &= 3\,t_{1/2}, & N_3 &= N_2/2 = N_0/2^3 = N_0/8
\end{aligned}
$$

The amount at various numbers of half-lives is shown graphically in Figure 16–21.

This process is mathematically the same as the description of light absorption in terms of half value layers. It is the same as the cooling of a hot object for which the cooling rate depends on the temperature excess. It is even the same as the decay of sound in a room after the source of sound is cut off.

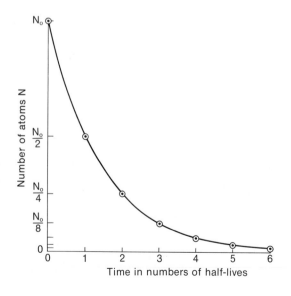

Figure 16–21   The amount of radioactive material remaining after various numbers of half-lives.

TABLE 16–4   The half-lives of a few radioactive materials

ELEMENT	ISOTOPE	HALF-LIFE
hydrogen	H-3 (tritium)	12.3 y
carbon	C-14	5730 y
cobalt	Co-60	5.3 y
iodine	I-131	8.07 d
radium	Ra-226	1602 y
uranium	U-235	7.1×10^8 y
uranium	U-238	4.51×10^9 y

The amount of radioactive material in terms of the number of half-lives, n, is

$$N = N_0/2^n$$

where $n = t/t_{1/2}$ (t is time)
so $N = N_0 2^{-t/t_{1/2}}$
Writing $2 = e^{0.693}$ (as was done for light absorption)
$N = N_0 e^{-0.693 t/t_{1/2}}$ ($e = 2.71828\ldots$)

The quantity $0.693/t_{1/2}$ is called the **decay constant** for the radioactive material, often called λ, and is the fraction disintegrating per unit time. Then

$$N = N_0 e^{-\lambda t} \quad \text{where } \lambda = 0.693/t_{1/2}$$

Both of these latter equations are often used in the description of the amount of radioactive material as a function of time. The half-lives of some common isotopes are listed in Table 16–4.

EXAMPLE 4

Some commercial signs for use in such places as exit indicators in aircraft are made to glow by the radiations from tritium (H^3) acting on a phosphor. How much would the amount of tritium, and hence the light output, drop in 6.2 years? The half-life of tritium is 12.3 years.

This problem can be done in several ways, as was described for light absorption. Let us use the exponential in the form

$$N/N_0 = e^{-0.693 \, t/t_{1/2}}$$
$$t = 6.2 \text{ years}$$
$$t_{1/2} = 12.3 \text{ years}$$
$$0.693 \, t/t_{1/2} = 0.349$$
$$e^{-0.349} = 0.71$$

The activity will drop to 0.71 or 71 per cent of the initial value.

16–9   RADIOACTIVE CHAINS

Sometimes the daughter nuclide produced after a radio-active decay is itself radioactive. There may even be a succession of decays, a radioactive chain. One example of such a multiple process concerns the notorious strontium-90.

Strontium-90 ($_{38}Sr^{90}$) decays with the emission of a β^- particle. The maximum β^- particle energy is 0.546 MeV, and the half-life is 28.9 years. The daughter is yttrium-90 ($_{39}Y^{90}$). Y^{90} is also radioactive, also emitting a β^- particle. The maximum energy of the Y^{90} beta particle is 2.27 MeV, high indeed. The half-life of Y^{90} is 64 hours; that is, half of the atoms of Y^{90} will decay within 64 hours after they have been formed. When Sr^{90} has been taken up by bone, which may happen if it is ingested, it will decay to Y^{90}, which in turn decays, sending electrons with 2.27 million electron volts of energy through the bone and surrounding cells. The daughter of Y^{90} is stable zirconium-90.

There are no gamma rays associated with the strontium-yttrium-zirconium chain. The process can be written as follows:

$$Sr^{90} \longrightarrow Y^{90} + \beta^- \text{ (0.546 MeV, 28.9 years)}$$
$$Y^{90} \longrightarrow Zr^{90} + \beta^- \text{ (2.27 MeV, 64 hours)}$$

Strontium-90 itself may result from the formation of Kr^{90} or Rb^{90}. Figure 16–22 shows the part of the chart of the nuclides near Sr^{90}. The decay chain from Kr^{90} to Zr^{90} is shown. All of these nuclides are among the hundreds of nuclides that result

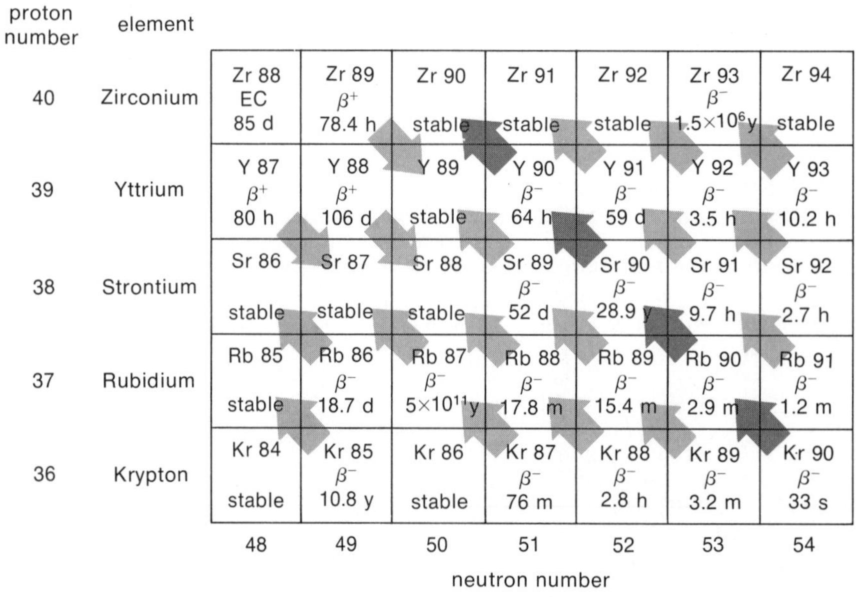

proton number	element							
40	Zirconium	Zr 88 EC 85 d	Zr 89 β^+ 78.4 h	Zr 90 stable	Zr 91 stable	Zr 92 stable	Zr 93 β^- 1.5×10⁶y	Zr 94 stable
39	Yttrium	Y 87 β^+ 80 h	Y 88 β^+ 106 d	Y 89 stable	Y 90 β^- 64 h	Y 91 β^- 59 d	Y 92 β^- 3.5 h	Y 93 β^- 10.2 h
38	Strontium	Sr 86 stable	Sr 87 stable	Sr 88 stable	Sr 89 β^- 52 d	Sr 90 β^- 28.9 y	Sr 91 β^- 9.7 h	Sr 92 β^- 2.7 h
37	Rubidium	Rb 85 stable	Rb 86 β^- 18.7 d	Rb 87 β^- 5×10¹¹y	Rb 88 β^- 17.8 m	Rb 89 β^- 15.4 m	Rb 90 β^- 2.9 m	Rb 91 β^- 1.2 m
36	Krypton	Kr 84 stable	Kr 85 β^- 10.8 y	Kr 86 stable	Kr 87 β^- 76 m	Kr 88 β^- 2.8 h	Kr 89 β^- 3.2 m	Kr 90 β^- 33 s
		48	49	50	51	52	53	54

neutron number

Figure 16–22   Decay processes around strontium-90.

from uranium fission. Strontium-90 is long lived (28.9 years), and it occurs in the fall-out from nuclear weapons. Its life is such that it can make its way up the food chain to humans.

Strontium-90 has earned a bad reputation because if it is taken into the body, it will lodge in the bone. This is because it is chemically somewhat similar to calcium. The body will take calcium to the bones in preference to strontium, and the amount of strontium taken up by bone is highest if there is a calcium deficiency. Once in the bone, the strontium sits until it decays by emission of a weak beta particle. Shortly afterward, the daughter yttrium-90 decays by emitting a high energy beta. The material in and near joints and the red-cell-forming bone marrow are very easily damaged by such radiation.

16–9–1 URANIUM SERIES

All of the elements of atomic number greater than that of lead (82) (except for bismuth-209) are radioactive. The naturally occurring element with the highest atomic number is uranium (92), and it also is radioactive. Uranium emits an alpha particle, the resulting nucleus is radioactive, and there is a whole chain of disintegrating elements till finally lead is reached. All the elements between lead and uranium occur in nature only as part of the chain of decay from uranium to lead. Different isotopes of uranium result in different chains. In Table 16–5 is shown one of the decay series, the one that involves the most abundant isotope of uranium, U-238. The particle emitted in each decay is shown, and also the half-life. Most of them also involve the emission of gamma rays, though these are not indicated. It is the gamma rays from radium that make it useful for cancer therapy; so in some areas they are important, while in some cases they are dangerous.

In Table 16–5, elements beyond uranium have been included. When the elements of which our earth is made were formed, these "transuranium" elements were probably formed also. Within days after the element formation process stopped, the fermium-100 would have almost completely disappeared, for its half-life is only 3.23 hours. A few hundred years later the californium-250 would be gone, and then in a million years or so it would have been hard to find any curium-96. The plutonium-94 would have been significant for many millions of years, the half-life being a third of a million years. But the elements have existed long enough that no Pu-94 is found today.

Uranium-238, with a half-life of 4.51×10^9 years, does remain. It is interesting that another of the radioactive chains involves uranium-233, which has a half-life of only 1.6×10^5 y, and no elements of this chain are to be found. The end result of

TABLE 16–5   One of the uranium series of radioactive elements

ELEMENT	ATOMIC NUMBER	ISOTOPE	PARTICLE EMITTED	HALF-LIFE	DAUGHTER
fermium	100	Fm-254	α	3.23 h	Cf-250
californium	98	Cf-250	α	13.1 y	Cm-246
curium	96	Cm-246	α	4710 y	Pu-242
plutonium	94	Pu-242	α	3.87×10^5 y	U-238
uranium	92	U-238	α	4.51×10^9 y	Th-234
thorium	90	Th-234	β^-	24.1 d	Pa-234
protactinium	91	Pa-234	β^-	1.1 m	U-234
uranium	92	U-234	α	2.47×10^5 y	Th-230
thorium	90	Th-230	α	7.8×10^4 y	Ra-226
radium	88	Ra-226	α	1600 y	Rn-222
radon	86	Rn-222	α	3.8 d	Po-218
polonium	84	Po-218	α	3.05 m	Pb-214
lead	82	Pb-214	β^-	26.8 m	Bi-214
bismuth	83	Bi-214	β^-	19.8 m	Po-214
polonium	84	Po-214	α	16.4 μs	Pb-210
lead	82	Pb-210	β^-	22 y	Bi-210
bismuth	83	Bi-210	β^-	5.01 d	Po-210
polonium	84	Po-210	α	138 d	Pb-206
lead	82	Pb-206	stable		

this chain is bismuth-209, natural bismuth. It is apparent that the elements have not existed for tens of billions (10^9) of years, or the U-238 would also have gone.

Uranium-234, with a half-life of 2.47×10^5 years, does appear in nature, but it amounts to only 0.0057 per cent of natural uranium. This number is just the ratio of the half-lives of U-234 and U-238 ($2.47 \times 10^5 / 4.51 \times 10^9 = 5.5 \times 10^{-5} = 0.0055$ per cent). The U-234 exists only as part of the decay chain.

The natural radium in nature, radium-226, is associated with uranium ore and is there only as part of the decay chain. The amount is in the ratio of the half-life of radium to that of U-238. That is $1602/4.51 \times 10^9$ or 3.6×10^{-7}. It is only 0.36 grams per tonne of uranium. The task facing Madam Curie in her original work, which was incidentally work toward her doctorate degree, was to separate the small amount of radium from large masses of ore. The task was even more difficult because radium had not yet been discovered. She saw that there was something in the ore that was not uranium but was also radioactive. When this material was separated out, its chemical properties were studied as well as its spectrum. It was found to be a new element!

The method used to date the time of formation of minerals on earth is based on the knowledge of this decay chain. If a mineral containing uranium is formed and if it can be established that it did not initially contain lead, then as time passed the decaying uranium would form lead-206 in the mineral. The longer the mineral existed, the greater would be the amount of lead in it. The age of the mineral can be found from the ratio of lead to uranium.

Another of the uranium series has U-235 at its head, those of higher atomic number having disappeared. U-235 has a half-life of only 0.71×10^9 years, contrasted to 4.51×10^9 years for U-238. Because U-235 has such a short life, it has decayed faster than U-238 and there is not as much of it left now. U-235 comprises only 0.72 per cent of natural uranium. When the elements were formed, the amount of U-235 was probably much closer to the amount of U-238.

In Figure 16–23 is a plot of the amounts of U-235 and U-238 present at various times in the past. The vertical scale is marked logarithmically. The present-day total uranium is given

Figure 16–23 The amounts of uranium-238 and uranium-235 at various times in the past. At 6×10^9 years ago the amounts were equal.

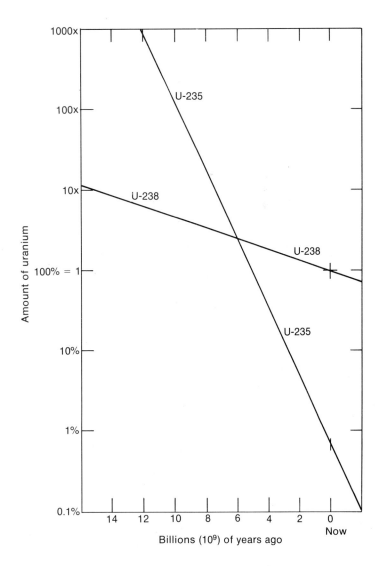

a value of 1. Six billion (6×10^9) years ago the amounts of U-235 and U-238 would have been equal. Ten billion years ago there would have been about 25 times as much U-235 as U-238. This ratio at the time of element production is highly unlikely.

There are two interesting points. Combining this type of reasoning with other theoretical ideas concerning the relative amounts of materials produced, the age of the elements is considered to be about 4.5 to 5×10^9 years. That gives the probable time in the past at which the elements of earth were formed. Astronomers, however, can give evidence based on such things as motion of galaxies and ages of stars that these objects have existed for 10 to 15×10^9 years. It seems that the elements of which the earth is comprised were formed at a time much later than that. How? Where?

16–10 NATURALLY OCCURRING RADIOACTIVE ISOTOPES

In addition to the series of heavy radioactive elements associated with uranium, there are others that occur in nature. These fall into three categories: those of long life that remain from the initial element formation; those of short life that are being formed now, principally by cosmic ray bombardment of the elements in the upper atmosphere; and those put there by man, chiefly as a consequence of nuclear bomb explosions.

16–10–1 LONG-LIVED RADIOACTIVE MATERIALS

Some of the long-lived naturally occurring radioactive materials are listed in Table 16–6 with a few of their properties. These all have lives long enough to have lasted since the time of element formation.

TABLE 16–6 Some long-lived naturally occurring radioactive materials. There are at least 10 more, but these are illustrative.

Element	Nuclide	% of Natural Element	Particle in Decay	Half-Life, Years
potassium	K-40	0.0118	β^- or β^+	1.28×10^9
rubidium	Rb-87	27.85	β^-	5.0×10^{10}
indium	In-115	95.72	β^-	5×10^{14}
cerium	Ce-142	11.07	α	5×10^{16}
neodymium	Nd-144	23.85	α	2.1×10^{15}

The most interesting of these is potassium-40, with a half-life of 1.28×10^9 years. Natural potassium consists of 93.08 per cent K-39 and 6.91 per cent K-41, both of which are stable, and 0.0118 per cent of the radioactive K-40. Potassium is one of the elements of the body, so the human body has developed over millions of years in the presence of radioactivity.

Potassium is abundant in the earth, and the energy released in the radioactive process (the beta particle and gamma ray average a total of 0.7 MeV for each disintegration) is probably one of the main sources of heat in the earth, along with the various uranium series. Without the K-40, the earth would probably long ago have become a cold, solid body. In the distant past there was even more of the K-40, and it could have been one of the main energy sources that caused the earth to be liquid at one stage after its formation even if it had formed initially as a cold body.

16–10–2 SHORT-LIVED RADIOACTIVE MATERIALS IN NATURE

The principal short-lived radioactive materials that occur in nature are H-3 (tritium) and carbon-14. Tritium has a half-life of only 12.33 years, and C-14 has a half-life of 5730 years. Neither of these could have been produced with the elements billions of years ago, but both are being continually produced by cosmic ray bombardment of the atmosphere. The cosmic rays are principally high-energy protons, most of which come from outer space and bombard the earth continuously. Elements in the upper atmosphere are battered about, and two of the products are H-3 and C-14.

The H-3 finds its way into the water on earth and, although it decays, there is continual replenishment. If water is cut off from circulation, such as in a deep ocean pool or even in a bottle (perhaps of fermented grape juice), the amount of tritium in it will slowly decay compared with water in circulation with the atmosphere and rain. The time that the water has been out of circulation could be measured by determining the amount of tritium left compared with the amount in circulating water.

This procedure was useful until the explosion of hydrogen bombs in the atmosphere. These bombs produced tritium that was in excess of the total amount naturally occurring in the world and, to put it politely, fouled up an otherwise valuable research technique.

The carbon-14 finds its way via the CO_2 of the air into growing things such as plants or trees. Once the organism dies, the amount of C-14 per unit mass of carbon begins to decrease and the age of the material can be determined by the amount remaining (the half-life is 5730 years). Such a calculation would have to be made on the assumption that the amount of C-14 in the atmosphere has been constant over the period involved, perhaps up to 10,000 years. Comparison of the dates computed from the C-14 amount to known dates has told of small, but significant, variations in cosmic ray intensity over the past several thousand years.

16–11 PAIR PRODUCTION

The existence of an electron with a positive charge was described in Section 16–5, as was the process by which it annihilates an ordinary electron. When they interact, they disappear as particles, and their entire particle mass is transformed to the energy in the gamma rays that result. It is like a positive charge and a negative charge neutralizing. The ordinary matter of the ordinary negative electron neutralizes the anti-matter of the positron, and the result is no matter at all.

The energy associated with the mass of an electron or of a positron (they are both 9.11×10^{-31} kg) is found from the relation

$$E = mc^2$$
$$m = 9.11 \times 10^{-31} \text{ kg}$$
$$c = 3.00 \times 10^8 \text{ m/s}$$
$$E = 9.11 \times 9.00 \times 10^{-15} \text{ joule}$$
$$= 81.9 \times 10^{-15} \text{ joule}$$

To express this in electron volts, use 1 eV $= 1.602 \times 10^{-19}$ joule; then the energy from the conversion of one electron mass is 0.51 MeV (million electron volts). The total energy in the two gamma rays is 1.02 MeV.

An almost reverse process also occurs. Rather than a pair of particles changing to gamma radiation, a high-energy gamma ray may change into a pair of particles. The gamma ray disappears as such! When you speak of a material absorbing gamma rays, you mean that if you put a certain number in you get fewer out. The difference, you say, is due to absorption. One of the absorption processes is **pair production**. The process occurs only near a nucleus.

As in Figure 16–24(a), the gamma ray approaches a nucleus. In (b) the gamma ray has disappeared and a β^+ particle and β^- particle appear. Part of the energy, equal to 1.02 MeV, goes into producing the mass of the particles (0.51 MeV each), and the remaining energy of the gamma ray will remain as kinetic

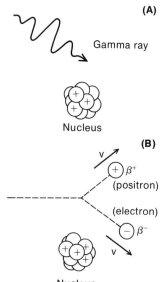

(A)

Gamma ray

Nucleus

(B)

v

β^+ (positron)

(electron)

β^-

v

Nucleus

Figure 16–24 The process of pair production.

Matter | Antimatter

Figure 16–25 The system of nucleons.

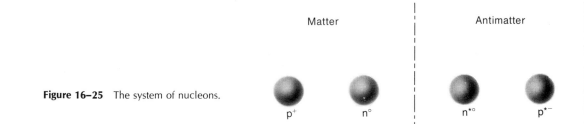

p^+ n° $n^{*\circ}$ p^{*-}

energy of the two particles. The energy needed to produce a particle is found from $E = m_0 c^2$, where m_0 is its rest mass. The β^+ particle will eventually be annihilated.

This process cannot occur unless the gamma ray energy is at least enough to create the two particles, that is, at least 1.02 MeV. This pair production process occurs only for high-energy gamma rays.

There are three methods by which absorption of high-energy radiation such as gamma rays or x-rays takes place:

1. The photoelectric effect (dominant at low energy).

2. The Compton effect (nearly all energies in the x-ray region.

3. Pair production (at high energy only).

All of these processes result in high-speed particles in the absorbing medium.

16–11–1 PROTONS AND ANTI-PROTONS

It is not only electrons that have an anti-matter counterpart; protons also do. The proton is 1836 times as massive as an electron so it requires 1836 times the energy to produce such a pair, that is, 1836×1.02 MeV or 1873 MeV. This can also be expressed as 1.87 GeV. The proton-antiproton pair could not be produced until accelerators of 2 billion electron volts (2 GeV) were made. The antiproton not only is anti-matter, but it has a negative charge in contrast to the positive charge of the ordinary proton. Antiprotons will annihilate protons, with the emission of two gamma rays, each of about 1 GeV.

Antineutrons also occur. They are of about the same mass as the antiproton. The proton, neutron, antineutron, and antiproton form a group as in Figure 16–25. They are a part of a larger group of large-mass particles called baryons.

It is not hard to imagine that in the beginning energy produced particles of matter and anti-matter. The positive protons as we know them combined with negative electrons to form atoms of matter, and the negative protons combined with the positive electrons to form anti-matter atoms. If somehow these separated, perhaps stars and galaxies formed—some of matter and some of anti-matter. The concept is intriguing, the only difficulty being that to date no evidence of anti-matter galaxies or stars has been found. Doing that is not easy because atoms of anti-matter would emit the same spectral lines as atoms of matter. It is by their light that we know the stars.

16–11–2 ELEMENTARY PARTICLES

Using high-energy machines, physicists have made a large number of particles other than the electron pairs and nucleons (p^+, n, n*, p*$^-$). (The asterisk indicates anti-matter.) The particles fall into groups when arranged according to their various properties. The various groups have been given names like mesons, muons, and baryons. The masses are of only certain discrete values, and attempts are being made to explain just why those masses occur. It is a problem similar to the unravelling of the order in spectral lines and then the explanation of the origin of spectra.

PROBLEMS

16–3   1.   Use Balmer's formula to calculate the wavelengths of the first four spectral lines in the Balmer series in the hydrogen spectrum, and tabulate them with the measured values shown in Table 16–1. What is the average difference between the experimental and theoretical values? Express it as a percentage and use the absolute values of the differences to calculate the average.

16–3   2.   Find the energy in electron volts associated with each of the first four lines in the Balmer spectrum of hydrogen.

16–3   3.   What wavelengths are associated with photon energies of 1, 10, 100, 1000, and 10,000 electron volts? Express the answers in nm.

16–3–1   4.   What would be the maximum wavelengths of light that could eject an electron from the materials listed below with their work functions in eV? The work function depends on the crystalline state and surface condition of the material, so the values are just representative. As well as the limiting wavelength, describe the colors that could eject electrons and those that could not. If the energy is not in the visible part of the spectrum, is it beyond the red or beyond the violet?

silver	3.7	eV
barium	2.5	eV
cesium	1.9	eV
gold	4.3	eV
iron	3.9	eV
potassium	2.0	eV
sodium	2.06	eV

16–3–1   5.   Find the maximum energy of the electrons ejected from silver and from potassium by green light of 540 nm. The work function of silver is 3.7 eV and that of potassium is 2.0 eV. Express the electron energy in eV.

16–4   6.   Planck's constant is so small that we do not ordinarily see that some quantities, angular momentum for example, occur not as continuous variables but only as discrete quantities. Find the first three possible speeds for a vehicle in a world where h has the value 6.62 J s rather than 6.62×10^{-34} J s. The quantity $h/2\pi$ would be 1.05 J s. The wheels of the vehicle are hoops of mass 15 kg and radius 0.20 m. The angular momentum of each wheel may have values only of $nh/2\pi$, where $n = 1, 2, 3, \ldots$.

16–4   7.   Work through all the steps from the quantum condition for the orbits of the hydrogen atom to the calculation of the radius of the first allowed orbit, $n = 1$. This orbit is also called the ground state of the atom.

16–4   8.   Starting with the quantum condition, carry through the steps to find the radius of the orbit for which $n = 1$ for a helium atom. Both electrons of helium ordinarily occupy the orbit for which $n = 1$.

16–5–2　9. List all the various carbon isotopes shown in the chart in Figure 16–14, writing the number of neutrons, protons, and electrons in each.

16–5–2　10. List the stable isotopes for the elements from hydrogen to carbon, showing the number of protons and neutrons in the nucleus of each.

16–6　11. If a neutron were to be knocked out of each of many O^{16} nuclei, what would the resulting nuclides be? After an hour or so, what would there be?

16–6　12. Describe what happens if you start with a pure sample of carbon-16. What do you have for the first few seconds? Then what do you have after a long period of time?

16–6–1　13. Uranium-238 emits an alpha particle. The resulting nucleus emits a β^-; then there is another β^- and two more alpha particles. This is part of the whole process of the decay of uranium. Make up a portion of the chart of the nuclides to show these events and the nuclides that exist after each particle emission. The elements needed are:

元素 92, uranium

　　　　element 92, uranium
　　　　element 91, protactinium
　　　　element 90, thorium
　　　　element 89, actinium
　　　　element 88, radium

16–6–1　14. Find what results when a neutron is added to a nucleus of U^{238}, the resultant nucleus emits a β^-, and again the daughter emits a β^-. The nuclide that then occurs has a long half-life. Show the process on a small portion of a chart of the nuclides and identify the resulting isotope. Some needed information is:

　　　　element 92, uranium
　　　　element 93, neptunium
　　　　element 94, plutonium

16–7　15. Find the efficiency with which a certain radiation detector counts radioactive disintegration. The efficiency is the ratio of the number of counts produced by the instrument in a given time to the number of particles incident on the counter in that time. The face of the counter is a 2.5 cm diameter circle placed 7.5 cm from a small source of 37,000 Bq. The counter registers 190 counts per second.

16–7　16. Detectors of radioactivity also respond to the ubiquitous cosmic rays, so that detection of small amounts of radioactive material is limited by the cosmic ray counts. The lower limit for a given counter is often considered to be that amount of radioactivity that produces a counting rate equal to the cosmic ray count. Find the minimum detectable limit for a counter with an efficiency of 80% placed at a position where the solid angle to the sample is 2.5 sr. The cosmic ray count is 35 per minute. Express the answer in picocuries as well in Bq.

16–8　17. How much iodine-131, which has a half-life of 8 days, would you have a month after you received a shipment of 100 mCi?

16–8　18. In approximately how many half-lives will the radioactivity of a sample drop to 0.1% of the initial value?

16–8　19. A bottled drink is claimed to be 8 years old. The concentration of tritium in the bottle is measured to be 80% of the concentration in a water sample taken from a river in the location where the drink was bottled. What is the actual time since the drink was made?

16–9　20. The decay chain from strontium-92 to zirconium-92 is shown in Figure 15–22. The half-life of Sr^{92} is 2.7 h. The Sr^{92} decays to Y^{92} which, as it is formed, begins to decay with a half-life of 3.5 h to stable Zr^{92}. Sketch a graph showing the amounts of each of the three nuclides individually and the total amount of all three over the first day after the sample was pure Sr^{92}. You are not expected to calculate the curves, only to apply thought and make a reasonable curve.

16–9–1　21. What percentage of natural potassium would have been the isotope K^{40} at a time 4.5×10^9 years ago? K^{40} forms 0.0118% of potassium today. The half-life of K^{40} is 1.28×10^9 y.

CHAPTER SEVENTEEN

SPECIAL RELATIVITY

The purpose of this chapter is to outline some of the physical phenomena that require the special theory of relativity for their explanation. Most of the effects of relativity are important only at very high speeds, but one of the consequences, the equivalence of mass and energy, is important in power production from nuclear reactors.

Toward the end of the nineteenth century, some phenomena were seen that were very troublesome to physicists. In electromagnetic theory an absolute reference frame, an ether pervading space, seemed necessary but it could not be detected.

Albert Einstein in 1905 published his special theory of relativity. He seemed to base it on no physical experiments but only on the idea of how the laws of physics would appear to people moving at different speeds in space. There are no markers in empty space to show who is doing the moving. There could then be no privileged system, and the world must seem the same to each of them. In particular, the speed of light would be the same to the observers in each system, no matter how they appeared to move. Also, the laws of physics, such as conservation of momentum, would have the same form in all systems. Einstein's conclusions seemed to solve some of the problems that had been arising in physics and also showed many other things to look for ($E = mc^2$, for one).

Einstein's new way of looking at the universe showed him to be one of the most remarkable thinkers of all time. The special theory of relativity has done much to shape science and the world since. Einstein published three scientific papers in 1905. One was on Brownian motion, one on the photoelectric effect, and one on special relativity. He was, at the time, an obscure 26 year old patent clerk in Switzerland. The paper on the photoelectric effect earned him a Nobel prize, but the work on relativity was so controversial that the Nobel prize committees never did recognize it as the most outstanding of all.

Einstein, it has been said, read little but thought a lot.

17-1   MOTION

In a modern jumbo jet cruising at perhaps 12,000 meters altitude and at 900 kilometers per hour, the passengers experience no sensation of motion. The motion can be detected only by looking out the window and using things outside, clouds or the ground, as a reference. It is possible to detect motion at constant velocity only by the use of an external reference frame of some sort, or some other object.

You may have had the experience of waiting in a line of stalled traffic; when the vehicle beside yours moves, you get the immediate sensation that you are doing the moving, but in the direction opposite to which the other vehicle moves. That is a case of detecting the relative motion of two objects but not being sure which is doing the moving.

If you were in a spacecraft cruising at constant speed and, on looking out a porthole, saw another craft move toward you and then away, you might have considerable difficulty determining the absolute speeds of the crafts. In fact, could it be done?

The question to ask is, inside a vehicle (a reference frame) that is either stationary or moving at a constant velocity, is there a way, without reference to outside objects, to measure the motion?

On earth, the air, the ground, and other objects form reference frames; so imagine the craft out in space. Without reference to outside objects such as stars, how could you measure the motion of the craft?

One method suggested for the detection of motion is to measure the speed of light in the direction of the motion and perpendicular to it.

The methods of measuring the speed of light use a beam that is sent to a distant mirror and from which it is reflected back. The time is measured using a device such as a rotating mirror. If an eight-sided mirror is used, the reflected light is seen (see Figure 13-18) when the speed of rotation is such that the multisided mirror turns an eighth of a revolution in the time of travel of the light out to the distant mirror and back. This apparatus was described in Section 13-4 as it would be for a stationary system. The speed of light was measured in that way on the earth, which is not a stationary system; there is ample evidence that the earth moves about the sun. The measurement of the speed of light by the rotating mirror method is not quite precise enough to be expected to be affected by earth motion. There is a device that is sensitive enough, so the analysis will be carried out to show the measurement of the difference between the speed of light in the direction of motion and that perpendicular to it.

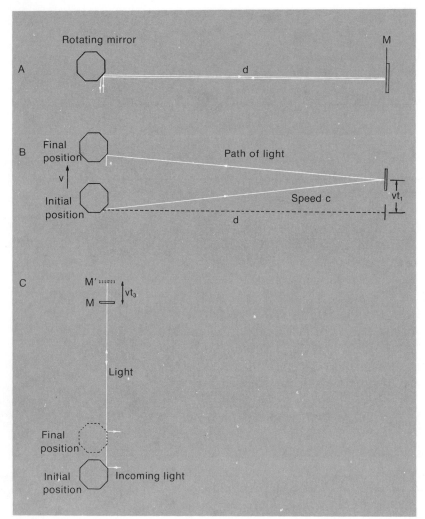

Figure 17–1 Measuring the speed of light. (a) A stationary system. (b) Measurement at right angles to the motion of the moving system. (c) Measurement along the direction of motion of the system.

17–2 THE SPEED OF LIGHT IN A MOVING SYSTEM

In Figure 17–1(a) is depicted an apparatus used to measure the speed of light as it would appear to the observer working with it. The rotating mirror method is shown with a mirror a long distance, d, away. The light is shown reflecting back from the same face that it went out from. This is not as in Section 13–4, but it is as valid.

In Figure 17–1(b) is the same system as in (a), but set in motion perpendicular to the path to the distant mirror. The actual path of the light is along the diagonals shown and is longer than the path in (a). The speed of the apparatus is V and the speed of light is c.

Let the time required for the light to go to the mirror be t_1. In this time the distant mirror moves a distance Vt_1.

The length of the actual outward path is given from the triangle shown in the figure and is $\sqrt{d^2 + V^2 t_1^2}$. It is also given by ct_1 so

$$ct_1 = \sqrt{d^2 + V^2 t_1^2}$$

or

$$c^2 t_1^2 = d^2 + V^2 t_1^2$$

from which, by a bit of algebra,

$$t_1 = \frac{d}{c\sqrt{1 - V^2/c^2}}$$

The time required to come back is the same, so the time out and back is

$$t_1 + t_2 = \frac{2d}{c\sqrt{1 - V^2/c^2}}$$

Thus, if the speed of light is being measured by this method in a north-south direction on earth, so that the apparatus is perpendicular to the direction of the motion of the earth in its orbit, the calculated speed based on $2d/(t_1 + t_2)$ would not be the true speed. Rather,

$$\frac{2d}{t_1 + t_2} = c\sqrt{1 - V^2/c^2}$$

This will be less than the true speed c.

The speed of light may also be measured along the direction of motion, as shown in Figure 17–1(c). The light leaves the rotating mirror, and after a time t_3 gets to the distant mirror, which has moved to the position shown as M'. The total distance the light travels is $d + Vt_3$ and is also ct_3. From this,

$$ct_3 = d + Vt_3$$

from which

$$t_3 = d/(c - V)$$

The light gets back to the rotating mirror in a time t_4. It has traveled a distance less than d by an amount Vt_4. From this,

$$d - Vt_4 = ct_4$$

and

$$t_4 = d/(c + V)$$

The total travel time, $t_3 + t_4$, is

$$t_3 + t_4 = \frac{d}{c-V} + \frac{d}{c+V}$$

This can be manipulated to the form

$$t_3 + t_4 = \frac{2dc}{c^2 - V^2}$$

The calculated speed given by $2d/(t_3 + t_4)$ when the measurement is along the direction of motion would be given by

$$\frac{2d}{t_3 + t_4} = c(1 - V^2/c^2)$$

This is not the same as would be obtained by a stationary piece of equipment or if the measurement were perpendicular to the motion.

The apparatus could be set up to do the two measurements simultaneously, one beam going along the direction of motion of the earth and the other perpendicular to it. Then the distance could be adjusted so that the two beams take the same travel time: that is, $t_1 + t_2 = t_3 + t_4$. The distances would then not be the same. Let the perpendicular distance be $d_\perp$ and the distance parallel to the motion be $d_\parallel$. Then

$$t_1 + t_2 = \frac{2d_\perp}{c\sqrt{1 - V^2/c^2}}$$

$$t_3 + t_4 = \frac{2d_\parallel}{c(1 - V^2/c^2)}$$

If $t_1 + t_2 = t_3 + t_4$, then the right-hand sides above are equal. The 2 and the c cancel, and $(1 - V^2/c^2)$ can be written as $(\sqrt{1 - V^2/c^2})^2$. Then

$$d_\parallel = \sqrt{1 - V^2/c^2}\, d_\perp$$

The distance in the direction of motion would be less than the distance perpendicular to the motion if the travel times of the light were the same (Figure 17–2).

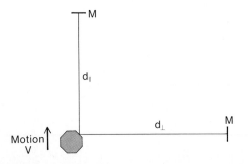

Figure 17–2 The paths perpendicular to the motion and along it. The distances are not the same, but the travel times of the light are the same.

17–3 THE MICHELSON-MORLEY EXPERIMENT

A piece of apparatus that is sufficiently precise to detect the difference in the measured speed of light in the direction of earth orbital motion and perpendicular to it is the Michelson interferometer (Section 11–4–3). If this interferometer (Figure 11–6) is set up to produce white light fringes, it means not that the lengths of the two arms are equal but that the travel times of the light in the two directions are the same. If those travel times are the same, then the light waves return with the vibrations in step.

When the Michelson interferometer is set up on earth with one arm east-west and the other north-south, then when white light fringes occur the distance to the two plane mirrors would not be expected to be the same. The east-west arm should be shorter than the north-south arm, for the motion of the earth in its orbit is in an east-west direction. This is shown in Figure 17–3(a).

In 1887, Albert Michelson and his co-worker, Edward

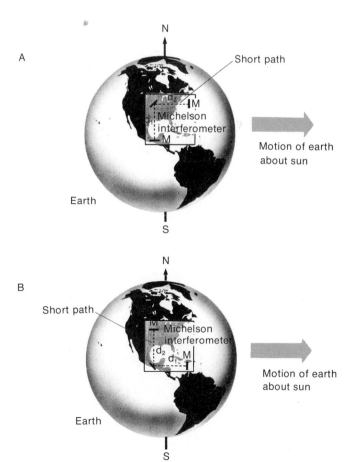

Figure 17–3 The Michelson-Morley experiment. The apparatus is adjusted as in (a) to have equal travel times on the two paths. It is then rotated to the position shown in (b). The apparatus is not drawn to scale.

Morley, set up an interferometer as in Figure 17–3(a) and adjusted it for white light fringes. Then they carefully rotated it through 90° so the arm that was expected to be the longest ($d_\perp$ of Figure 17–3) was in the direction of earth motion as in Figure 17–3(b). The travel times of the light on the two paths would be expected to change, and this would be shown by a change in the interference of the returning light waves. It didn't work! The fringes stayed the same. The travel time of the light in the path along the direction of earth motion was the same as in the direction perpendicular to it. Perhaps the earth does not move after all. The Michelson-Morley experiment has been repeated, refined, and repeated even more, but no earth motion has been detected.

The earth moves in its orbit at 30 km/s, while light travels at 3×10^5 km/s. The earth-sun system is also a part of a large collection of about 10^{11} stars that form a spiral galaxy. This galaxy, with the earth near the outer edge, rotates once in about 200 million years, and because of this motion the earth also has a speed of 75 km/s. Even this does not affect the Michelson-Morley experiment. Whatever unknown speed the galaxy may have through space has no effect, either.

17–3–1 THE LORENTZ CONTRACTION

One explanation of Michelson and Morley's result was put forward by H. A. Lorenz about 1904. His idea was that all objects contract in the direction of motion by the factor $1/\sqrt{1 - V^2/c^2}$. This would mean that the east-west arm of the interferometer was initially the shorter. Then when the instrument was turned that arm became longer while the other one shortened. This idea explained the effect, but it was not a satisfying explanation. There would be no way to check it because all measuring sticks would also shorten in the direction of motion.

17–3–2 EARTH MOTION

The Michelson-Morley experiment showed no motion of the earth through space, yet Roemer's work with the satellites of Jupiter, Bradley's work on the aberration of starlight, and the measurements showing that nearby stars periodically changed in direction (parallax) all demonstrated the motion of the earth in its orbit. The demonstrations of earth motion were all with reference to other objects in space. The Michelson-Morley experiment was to detect the motion of the earth with respect to space itself. This could not be done, for there are no markers, no reference points, in empty space. That means that nothing can move with respect to space; objects move only relative to each other.

17-4 THE SPEED OF LIGHT AS A CONSTANT

The Michelson-Morley experiment shows that light spreads out to form a spherical wavefront about the source. The light from the beam splitter traveled the same distance out to the mirrors in the same time. The experiment was done on an object

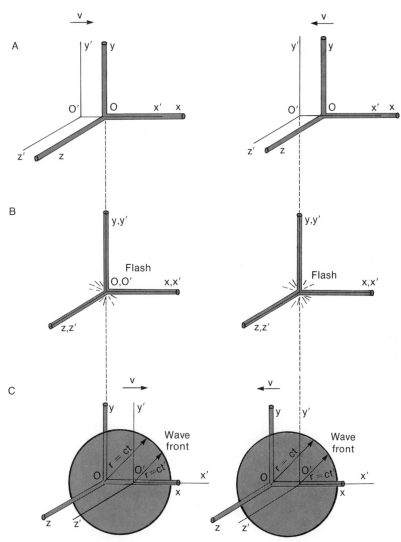

Figure 17-4 Moving reference frames (a). A light flashes when the origins coincide (b), and the wavefront shown in (c) is a sphere about both O and O'.

moving with respect to other space objects (the earth is a moving spaceship). If the experiment were done on another spaceship, the results would be the same — the wave would spread as a sphere about the source. The speed of light would be the same in all directions measured in either spaceship.

Consider two reference frames moving through space as in Figure 17–4. Instead of showing spaceships, only sets of mutually perpendicular rods (x, y, and z axes) are shown. One frame is the xyz frame and the other, called the primed system, is marked by x', y', and z' axes. In the two sets of drawings, one shows the primed axes moving and the other shows the unprimed axes moving. In space, no one knows which is moving.

As the two origins coincide, let a light be flashed as in (b); the wavefront then spreads out to form a sphere about the origin as in (c). But where is the center of the sphere? Is it at O or is it at O'? According to the idea of the constancy of the speed of light in all directions in any system, the wavefront would be a sphere about each origin as seen by the observer in each system.

As in Figure 17–5, the equation of the sphere is, in the unprimed system,

$$x^2 + y^2 + z^2 = r^2 = c^2 t^2$$

or
$$x^2 + y^2 + z^2 - c^2 t^2 = 0$$

In the primed system,

$$x'^2 + y'^2 + z'^2 = c^2 t'^2$$

or
$$x'^2 + y'^2 + z'^2 - c^2 t'^2 = 0$$

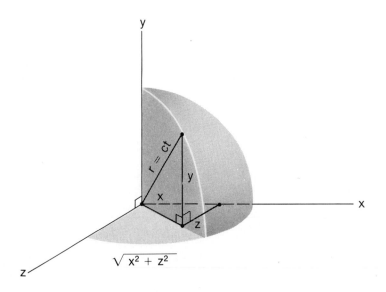

Figure 17–5 The equation of a spherical wavefront is found from $x^2 + y^2 + z^2 = c^2 t^2$.

Even the time t has been primed, because the problem is to make the same sphere a sphere about *each* observer at the time as seen by that observer.

To show why time must be different in the two systems, consider that each has two mirrors set up at the same measured distance from the origin in each system. One is along the x axis and one is along the y axis. A light flashes when the origins coincide, and a short time later the system is as in Figure 17–6.

The observer in each system sees the light reach his two mirrors at the same time, but the observer in the unprimed system sees the light reach the mirror on his (unprimed) x axis before he sees it reach the mirror on the primed axis, because (as in Figure 17–6) the mirror in the primed system has moved beyond his. Quite obviously, the observers are going to have arguments about time. The time cannot be the same in the two systems.

The problem is simultaneously to make

$$x^2 + y^2 + z^2 - c^2t^2 = 0$$

and

$$x'^2 + y'^2 + z'^2 - c^2t'^2 = 0$$

To do this, the manner in which the primed quantities are related to the unprimed quantities must be found. Since both expressions are equal to zero, equate them:

$$x^2 + y^2 + z^2 - c^2t^2 = x'^2 + y'^2 + z'^2 - c^2t'^2$$

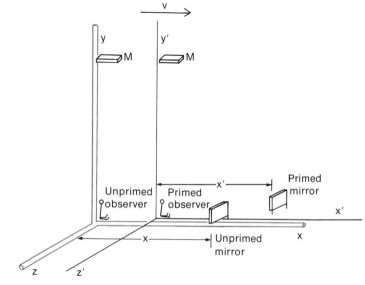

Figure 17–6 The observers in the two systems would dispute the simultaneous arrival of the light at the mirrors.

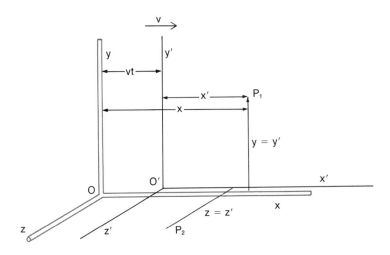

Figure 17-7 The relation between co-ordinates in two systems.

If the motion is in the x direction, then (as in Figure 17-7) points P_1 and P_2 are always at the same y and z distances. Let $y' = y$ and $z' = z$. Then

$$x^2 - c^2 t^2 = x'^2 - c^2 t'^2$$

For an observer in the unprimed system it would be expected that $x' = x - vt$, and on ordinary reasoning it would be expected that $t' = t$. These two equations are known as Galilean transformations. If this were so, then $x^2 - c^2 t^2 = x'^2 - c^2 t'^2$ would yield $x^2 = (x^2 - vt)^2$. But this is not so unless $v = 0$, so the Galilean transformations cannot be valid. They do apply closely in ordinary, low-speed situations, but they are at variance with the concept of the speed of light being the same in different systems moving with respect to each other.

Apparently, it is not true that $x' = x - vt$ and $t = t'$. No matter how much that makes sense in our ordinary experience, there is something wrong with it.

If x' is not equal to $x - vt$, then let it be given by $x' = \alpha (x - vt)$, where α is to be found. Also, t is not equal to t'. Let t' depend not only on t but also on position; that is, let $t' = \beta t + \gamma x$, where β and γ are to be found. After a long algebraic procedure, it is seen that

$$\alpha = \beta = 1 \left/ \sqrt{1 - \frac{v^2}{c^2}} \right.$$

and
$$\gamma = (v/c^2) \sqrt{1 - \frac{v^2}{c^2}}$$

These factors will make true the relation

$$x^2 - c^2 t^2 = x'^2 - c^2 t'^2$$

Substitute

$$x' = x \Big/ \sqrt{1 - \frac{v^2}{c^2}}$$

and $$t' = t \Big/ \sqrt{1 - \frac{v^2}{c^2}} + (vx/c^2) \sqrt{1 - \frac{v^2}{c^2}}$$

It is then only an exercise in algebra to show that the right-hand side reduces to $x^2 - c^2 t^2$, the same as the left side.

17–5   RELATIVISTIC TRANSFORMATIONS

It is on the basis of the type of reasoning above that it is said that times and distances are not the same to observers in systems moving with respect to each other. The only postulate on which this is based is that the speed of light is the same in all directions in any reference system. The deductions from this, though we have not worked them out, involve differences in lengths, differences in time intervals, and as a result of these, differences in the speed of an object when viewed by two different observers. The results are outlined below.

17–5–1   LENGTH

A length l_0 in a stationary system will be seen as a shorter length if it moves. The shortening will be in the direction of motion. The observed length l will be given by

$$l = l_0 \sqrt{1 - v^2/c^2}$$

This is the Lorentz contraction already mentioned.

17–5–2   TIME

Time intervals will be lengthened in a moving system and will be given by

$$t_2 - t_1 = (t_2 - t_1)_0 / \sqrt{1 - v^2/c^2}$$

This time dilation, as it is called, has been experimentally seen. Some of the elementary particles have a very short half-life, but when they are made to move at a high speed in an accelerator they appear to last for a much longer time than when they move slowly. An atomic clock has been flown around the world in the direction of the earth's rotation; on return to the starting point, it had lost time compared with a clock in the laboratory. When the clock was flown westward, against the earth's rotation, it gained time compared with the laboratory clock.

EXAMPLE 1

If a clock is put into a satellite and made to orbit at 8 km/s for a year, how much time would it lose? The speed of light is 300,000 km/s.
The time interval registered by the clock is $t_2' - t_1'$, and on earth it is $t_2 - t_1$:

$$(t_2' - t_1') = (t_2 - t_1) \sqrt{1 - v^2/c^2}$$
$$v = 8 \text{ km/s}$$
$$c = 300,000 \text{ km/s}$$

The ratio v/c is a small quantity. Write the surd as an exponent and use the binomial expansion to only one term because of the smallness of v^2/c^2.

$$\sqrt{1 - v^2/c^2} = (1 - v^2/c^2)^{1/2}$$

$$= 1 - \frac{1}{2}(v^2/c^2) + \ldots \text{ smaller terms}$$

Then
$$t_2' - t_1' = (t_2 - t_1)\left(1 - \frac{1}{2}\frac{v^2}{c^2}\right)$$

$$= (t_2 - t_1) - \frac{1}{2}\frac{v^2}{c^2}(t_2 - t_1)$$

The last term above is the amount by which the clock is slow. In that expression $t_2 - t_1$ is 1 year or 3.15×10^7 seconds.
The time loss of the clock is

$$\frac{1}{2}\frac{v^2}{c^2} \times 3.15 \times 10^7 \text{ s} = \frac{8^2 \text{ km}^2/\text{s}^2}{2 \times 9 \times 10^{10} \text{ km}^2/\text{s}^2} \times 3.15 \times 10^7 \text{ s}$$

$$= 0.0112 \text{ s}$$

The clock will lose just 0.011 second.
It is not yet practical to put a sufficiently precise clock through the rigors of a rocket launching to test this. It will probably be possible soon.

EXAMPLE 2

This is the twin paradox.
One twin stays on earth while his brother goes off in a spaceship at 0.99 of the speed of light for a period of what seems like 10 years on earth. What will be the time interval to the traveling twin? This will be the amount he ages.

$$(t_2 - t_1) = (10 \text{ years}) \sqrt{1 - \frac{0.99^2 c^2}{c^2}}$$

$$= 10 \times 0.141 \text{ years}$$
$$= 1.41 \text{ years}$$

The traveling twin will have aged by only 1.41 years while his brother on earth aged 10 years.

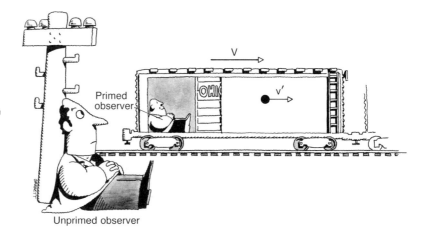

Figure 17–8 The velocity of an object viewed by two observers.

17–5–3   SPEED

The differences in observed distances and times lead also to velocities not being what would ordinarily be expected.

In Figure 17–8, an observer in a moving vehicle throws a ball, which he sees to move at a speed v'. One would expect an outside observer to see the ball move at a speed $v = v' + V$, where V is the speed of the vehicle. By relativity, however, the outside observer will see only a speed given by

$$v = \frac{v' + V}{1 + \dfrac{v'V}{c^2}}$$

If the speeds v' and V are small compared to the speed of light, c, the denominator will be almost exactly 1 and then v will be given by $v' + V$ as we expect.

EXAMPLE 3

Consider a vehicle moving at 0.7 of the speed of light with respect to an outside observer. An object in the vehicle is projected at 0.9 of the speed of light as seen by an observer in the vehicle. How fast does the outside observer see the object move?

$$v = \frac{v' + V}{1 + \dfrac{v'V}{c^2}}$$

where

$$v' = 0.9\,c$$
$$V = 0.7\,c$$
$$c = \text{speed of light}$$

$$v = \frac{0.9\,c + 0.7\,c}{1 + \dfrac{(0.9\,c)\,(0.7\,c)}{c^2}}$$

$$= \frac{(0.9 + 0.7)c}{1 + 0.63}$$

$$= 0.98\,c$$

The outside observer sees it move only at 0.98 of the speed of light.

EXAMPLE 4

An observer in a moving vehicle shines a light beam forward. The speed of the light is measured by an outside observer. The speed of the vehicle, V, can be any value, and $v' = c$.

The observed speed is

$$v = \frac{v' + V}{1 + \dfrac{v'V}{c^2}}$$

$$= \frac{c + V}{1 + \dfrac{cV}{c^2}}$$

$$= \frac{c + V}{1 + \dfrac{V}{c}} = \frac{c(c + V)}{c + V} = c$$

The speed will be seen by the outside observer to be just the speed of light, c.

The relativistic velocity transformation is borne out by the experimental finding that nothing has been found to exceed the speed of light.

17–5–4 MASS

If an observer in a moving system performs a collision experiment, and the masses can be measured before and after, it will be found that conservation of momentum holds. The velocities measured by an outside observer, however, as in Figure 17–9, will not be the same and the conservation of momentum will not be found to apply. This is contrary to the postulate that physical laws must be the same in all systems. It is found that conservation of momentum will apply if mass depends on speed according to the relation

$$m = \frac{m_0}{\sqrt{1 - \dfrac{v^2}{c^2}}}$$

where m_0 is the mass when $v = 0$. It is called the *rest mass*.

If the laws are to be the same, then the mass must be a function of speed. This had been found even before the theory of relativity was put forward. Apparently Einstein was not

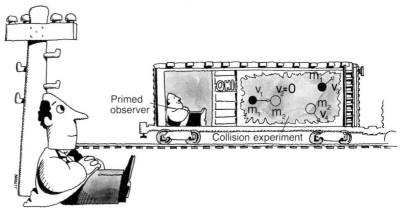

Figure 17-9 A collision experiment viewed from two systems.

Primed observer

Collision experiment

Unprimed observer

aware of it, but it is an example of how his theories answered already existing problems.

When a force is applied to a mass m, as in Figure 17-10, it accelerates as described by Newton's second law. As the object increases in speed, the mass also increases and the acceleration drops. As the speed gets close to the speed of light, the mass gets very large and the acceleration becomes very small. As v approaches the speed of light, the mass approaches infinity and the acceleration approaches zero. It would not be expected that a particle that has a rest mass m_0 not equal to zero could be accelerated to the speed of light.

m_0

$v = 0$

F

a

$$m = \frac{m_0}{\sqrt{1 - v^2/c^2}}$$

v

F

a

Figure 17-10 Accelerating an object that has increasing mass.

17-6 MASS AND ENERGY

When a force is applied to a mass and work is done on it, the mass not only moves but increases in mass. When the problem is worked through, it is found that the work done on the object, which is the kinetic energy given to it, is given by

$$E = \left(\frac{m_0}{\sqrt{1 - v^2/c^2}} - m_0 \right) c^2$$

$$= \text{increase in mass times } c^2$$

The energy apparently exists in the object in the form of increased mass. In words:

Energy equals change in mass times the square of the speed of light.

This is ordinarily written as $E = mc^2$, but in that equation m is the *change* in mass. This relation between mass and energy was found as a consequence of relativity theory. It has been used in various parts of the book.

17-7   GENERAL RELATIVITY THEORY

In 1916, Einstein brought out a general theory of relativity, which is in reality a theory of gravity. It also has had a great effect on modern science. Today the general theory is not the only theory of gravity, and gravity is a very active area of research in many laboratories of the world.

All that will be said about it is that Sir Arthur Eddington once gave a lecture on general relativity. After the lecture, questions were sought and someone asked, "Is it true that there are only three people in the world who understand general relativity?" After some thought, Sir Arthur asked, "Who is the third?"

Most of the material of this introductory physics book was actually in that category at some time. As time passes the concepts become clearer and easier to explain. As the author, I hope that you have understood many of the concepts, even of special relativity theory.

PROBLEMS

17-2   1.   Equipment is set up to measure the time required for a sharp sound to go to a wall 15.00 m away and back to a microphone beside the source. The speed of sound is 341.3 m/s at that time and place.

(a) Find the time needed for the sound to go to the wall and back if the equipment is stationary.

(b) Find the time that would be measured if the equipment were on a vehicle moving at 35 m/s parallel to the wall at a distance of 15.00 m.

(c) What is the difference in the measured time, expressed as a percentage?

(d) What is the actual path traveled by the sound in (b)?

17-2   2.   A sound source, a microphone, an electronic timer, and a reflector are mounted on a moving vehicle. The source and the microphone are near the front and the reflector is 15.00 m away at the back. The speed of sound is 341.3 m/s at that time and place.

(a) Find the time it takes for the sound to go to the reflector and back when the vehicle is moving at 35 m/s forward.

(b) Find the time it takes for the sound to go to the reflector and back when the vehicle is moving at 35 m/s backward.

17-2   3.   Find an expression for the speed of light when it is measured perpendicular to the direction of motion of the measuring equipment. The known quantities are the speed, V, of the equipment, the distance, d, to the reflector, and the travel time of the light, $t_1 + t_2$.

17-2   4.   What percentage error occurs when the speed of light is measured in an east-west direction on the earth and the speed of the earth in its orbit is neglected? The earth moves at 30 km/s and the speed of light is 300,000 km/s. This error is expected on the basis of ordinary, non-relativistic ideas but actually does not exist.

17-3-1   5.   On the basis of the Lorentz contraction, find the length of a meter bar, or the amount by which it shortens, when it moves at the following speeds in the direction of the bar:

(a) $V = c/2$,

(b) $V = c/10$,

(c) $V = 30$ km/s.

17-5-3   6.   A certain spaceship has the capability of sending a small rocket forward at half the speed of light. You see this spaceship come past you traveling at half the speed of light, and it ejects the rocket forward from it. How fast do you see that rocket move?

17-6   7.   When work is done on an object to cause it to move at a speed v, the energy acquired, according to Newtonian mechanics, is given by $E = mv^2/2$. In relativistic mechanics the energy acquired by an object made to move at a speed v is given by

$$E = \left\{ \frac{m_0}{\sqrt{1 - v^2/c^2}} - m_0 \right\} c^2$$

Use a binomial expansion to show that this relativistic expression reduces to the Newtonian expression if v is much smaller than c. This shows that Newtonian mechanics is quite valid at speeds much less than the speed of light.

POWERS OF 10

In scientific work it is common to express large or small numbers as a number between 1 and 10 multiplied by a power of 10. The powers may be positive or negative. The methods of writing numbers in this way, and multiplying and dividing numbers expressed as powers of 10, are shown below:

Positive Powers	Negative Powers
$1 = 10^0$	$0.1 = 10^{-1}$
$10 = 10^1$	$0.01 = 10^{-2}$
$100 = 10^2$	$0.001 = 10^{-3}$
$1000 = 10^3$	$0.0001 = 10^{-4}$

The exponent of 10 is the number of "places" after the 1.

The negative exponent is the same as the number of places the decimal point is moved to get 1.

Examples of scientific notation:

$$2000 = 2 \times 1000 = 2 \times 10^3$$
$$2,300,000 = 2.3 \times 1,000,000 = 2.3 \times 10^6$$

In the latter example, the exponent 6 is the number of "places" the decimal is moved to the left to obtain the number 2.3.

$$0.002 = 2 \times 0.001 = 2 \times 10^{-3}$$
$$0.0000023 = 2.3 \times 0.000001 = 2.3 \times 10^{-6}$$

In the latter example the number in the negative exponent is the number of "places" the decimal is moved to the right to obtain 2.3.

In multiplying, $10^a \times 10^b = 10^{a+b}$

In dividing, $10^a / 10^b = 10^{a-b}$

UNITS

In the Système International d'Unités (SI units) there are seven base units. There are standard multiples and submultiples of these base units. All other units are expressed in terms of the base units. These other derived units have accepted names in many cases.

Base units

Quantity	Unit	Symbol
length	meter	m
mass	kilogram	kg
time	second	s
temperature	kelvin	K
amount of substance	mole	mol
luminous intensity	candela	cd
electric current	ampere	A

Units that are multiples or submultiples of the base units are described by a standard set of prefixes, given in the following table.

SI prefixes

Prefix	Symbol	Factor by Which the Base Unit is Multiplied
exa	E	10^{18}
peta	P	10^{15}
tera	T	10^{12}
giga	G	10^{9}
mega	M	10^{6}
kilo	k	10^{3}
hecto	h	10^{2}
deca	da	10
–	–	1
deci	d	10^{-1}
centi	c	10^{-2}
milli	m	10^{-3}
micro	μ	10^{-6}
nano	n	10^{-9}
pico	p	10^{-12}
femto	f	10^{-15}
atto	a	10^{-18}

These prefixes are used with the base unit except for mass, in which case the prefix is used as though the gram is the base unit. For example, a thousandth of a kilogram is a gram, not a millikilogram.

Some derived units with special names

Quantity	Unit	Symbol	Definition
area	hectare	ha	$1\ ha = 10^4\ m^2$
fluid volume	liter	ℓ	$1\ \ell = 10^{-3}\ m^3$
force	newton	N	$1\ N = 1\ kg\ m/s^2$
pressure	pascal	Pa	$1\ Pa = 1\ N/m^2$
mass	tonne	t	$10^3\ kg$
energy	joule	J	$1\ J = 1\ kg\ m^2/s^2$
power	watt	W	$1\ W = 1\ J/s$
electric charge	coulomb	C	$1\ C = 1\ A\ s$
electrical potential	volt	V	$1\ V = 1\ J/C$
electric resistance	ohm	Ω	$1\ \Omega = 1\ V/A$

Conversion factors to change non-SI units to SI units

QUANTITY:	TO CHANGE:	TO:	MULTIPLY BY:
Length	inches	cm	2.540
	feet	m	0.3048
	yards	m	0.9144
	miles	km	1.6093
Area	square miles	km^2	2.590
	acres	hectare, ha	0.4047
	square feet	m^2	0.09290
Volume	cubic feet	m^3	0.02832
	cubic yard	m^3	0.7646
Mass	pound (avoirdupois)	kg	0.4536
	tons (2000 lb)	tonnes	0.9072
	grains	milligrams	64.80
	ounces (troy or apoth.)	grams	31.10
	ounces (avdp.)	grams	28.35
Density	g/cm^3	kg/m^3	1000
	lb/ft^3	kg/m^3	16.02
Speed	mi/h	m/s	0.4470
	mi/h	km/h	1.6093
Force	pound-force	newtons, N	4.448
	kilogram-force	N	9.806
	dynes	N	10^{-5}
Pressure	pound-force per square inch	N/m^2 or Pa	6895
	torr	Pa	133.3
	mm of Hg	Pa	133.3
	bar	kPa	100.00
	millibar	kPa	0.1000
	atmosphere	kPa	101.3
	$dynes/cm^2$	Pa	0.1000
Energy	Btu	joules, J	1055
	calories	J	4.187
	Calories (kcal)	J	4187
	kilowatt hour	J	3.6×10^6
	ergs	J	10^{-7}
Power	Btu/h	watts, W	0.2931
	horsepower	W	745.7
	kcal/day	W	0.04846
Luminous intensity	candlepower	candela	1.00
Illumination	foot candles	$lumens/m^2$ or lux	10.76

PROPERTIES OF MATERIALS

A. Solids

Substance	Density kg/m³	Specific Heat Capacity J/kg °C	Melting Point °C	Latent Heat of Liquefaction kJ/kg	Thermal Conductivity W/m °C	Electrical Resistivity Ω m
aluminum	2700	891	660	399	200	2.8×10^{-8}
antimony	6600	200	630	102	18	4.1×10^{-7}
copper	8930	380	1083	180	384	1.7×10^{-8}
glass (crown)	2500 to 2700	670	~1100	–	10	$>10^6$
gold	19,300	129	1063	64	290	2.4×10^{-8}
ice	920	2100	0	333	2.1	80
iron (cast)	7000 to 7700	500	~1100	100 to 140	48	$\sim 8 \times 10^{-7}$
iron (wrought)	7800 to 7900	480	1500	100 to 140	48	1.4×10^{-7}
lead	11,340	125	327	25	35	2.1×10^{-7}
marble	2500 to 2800	880	–	–	3.0	
nickel	8800	440	1455	59	–	7.8×10^{-8}
osmium	22,570	130	2970	141	–	9.5×10^{-8}
platinum	21,400	135	1769	114	69	1.1×10^{-7}
plutonium	19,800	(108)	639	(40)	8	1.4×10^{-6}
quartz	2660	710	~1750	560 to 1100	9.2	$>10^6$
silver	10,500	208	961	111	420	1.6×10^{-8}
uranium, α form	19,000	96	1132	–	30	3×10^{-7}

B. Liquids

Substance	Density at 0°C kg/m³	Specific Heat Capacity J/kg °C	Freezing Point °C	Boiling Point °C	Latent Heat of Vaporization kJ/kg	Viscosity Pa s	Index of Refraction
alcohol (ethyl)	792	2500	−115	78.3	861	0.0012	1.362
carbon bisulfide	1293	1000	−110	46.2	355	0.000367	1.632
helium	147		–	−270.8			
olive oil	920	1960	–	–	–	0.099	1.46
water	1000	4180	0	100	2250	0.001	1.333
water (sea)	1025	3930	(−9)	(104)	–	–	1.343
water (heavy)	1105	4300	3.8	101.4	2330	–	–

C. Gases

Values are for 0°C and 101.3 kPa.

SUBSTANCE	MOLECULAR WEIGHT kg/kmol	DENSITY kg/m³	MEAN SPEED OF MOLECULES m/s	BOILING POINT °C	C_p J/kg	$\gamma = C_p/C_v$
air	28.8	1.293	447	−190	991	1.402
ammonia, NH_3	17.03	0.771	578	−33.5	2170	1.336
argon, Ar	39.95	1.784	380	−189.2	530	1.667
carbon dioxide, CO_2	40.02	1.977	361	−78.5	840	1.306
carbon monoxide, CO	28.01	1.250	450	−191.5	1045	1.401
ethylene, C_2H_4	28.05	1.260	449	−102.7	1522	1.264
hydrogen, H_2	2.016	0.0899	1693	−252.5	14,300	1.41
helium, He	4.003	0.1785	1207	−268.6	5230	1.63
krypton, K	83.80	3.736	263	−152.3	247	1.33
methane, CH_4	16.04	0.717	300	−164	2220	1.313
nitrogen, N_2	28.01	1.2518	454	−195.2	983	1.41
oxygen, O_2	32.00	1.429	425	−183.0	1010	1.40
radon, Rn	222.0	9.73	161	−61.8	94	~1.67
xenon, Xe	131.3	5.887	210	−107	158	~1.67

D. Radioactive isotopes

ISOTOPE		HALF-LIFE	PARTICLE AND ENERGY MeV	GAMMA RAY ENERGIES MeV
tritium	$_1H^3$	12.26 y	β^- 0.0181	—
carbon	$_6C^{11}$	20.5 m	β^+ 0.96	—
	$_6C^{14}$	5.770 y	β^- 0.156	
sodium	$_{11}Na^{24}$	15.0 h	β^- 1.39	2.753, 137
phosphorus	$_{15}P^{32}$	14.3 d	β^- 1.71	—
sulfur	$_{16}S^{35}$	86.7 d	β^- 0.167	
potassium	$_{19}K^{40}$	1.3×10^9 y	β^- 1.32, EC*	1.32
cobalt	$_{27}Co^{60}$	5.27 y	β^- 0.31, 1.48	γ, 1.332 + γ, 1.1724
strontium	$_{38}Sr^{89}$	50.4 d	β^- 1.46	—
	$_{38}Sr^{90}$	28 y	β^- 0.54	(daughter is $_{39}Y^{90}$)
yttrium	$_{39}Y^{90}$	64.2 h	β^- 2.27	0.02% have γ's
technetium	$_{43}Tc^{99\,m}$	6.0 h		γ, 0.140, 0.142, 0.020

(technetium-99m is an excited state of Tc⁹⁹, and it decays to Tc⁹⁹)

	$_{43}Tc^{99}$	2.1×10^5 y	β^- 0.29	—
iodine	$_{53}I^{131}$	8.05 d	β^- 0.61 \ 0.25 \ 0.81	γ, range 0.728 to 0.08
promethium	$_{61}Pm^{147}$	2.5 y	β^- 0.225	(longest lived of all Pm isotopes)
gold	$_{79}Au^{198}$	64.8 h	β^- 0.96 \ 96%	γ, 0.412 (99%)
radon	$_{86}Rn^{222}$	3.82 d	α 4.586 \ 99%	(daughters emit many gamma rays and α particles)
radium	$_{88}Ra^{226}$	1622 y	α 4.78 \ 95%	(daughter is Rn^{222})
uranium	$_{92}U^{235}$	7.13×10^8 y	α 4.12 \ to \ 4.56	γ, range 0.07 to 0.38 (forms 0.72% of natural uranium)
	$_{92}U^{238}$	4.51×10^9 y	α 4.2	γ, 0.048 (forms 99.27% of natural uranium)

*EC = electron capture, a process in which an orbital electron is captured and annihilated by the nucleus with emission of a neutrino; atomic number is decreased by 1 and atomic mass does not change.

LOGARITHMS, INSTRUCTIONS AND TABLES

Logarithms are frequently introduced as a method for assisting in multiplication. If $y = ab$, then to find y, look up the logs of a and b; add them, and the sum is the logarithm of y. The antilog of this is the value y.

If $\quad\quad\quad\quad\quad y = ab$
then $\quad\quad\quad \log y = \log a + \log b$

This arises from the nature of a logarithm. The log of a number is the power to which the *base* is raised to get that number. The base for common logs is 10.

If $\quad\quad a = 10^c$, c is the log of a
and $\quad\quad b = 10^d$, d is the log of b
then $\quad\quad y = ab$
$\quad\quad\quad\quad y = (10^c)(10^d)$
$\quad\quad\quad\quad\quad = 10^{c+d}$

Let $g = c + d$, so that $y = 10^g$. Then g is the log of the product y, and y is the number whose log is g.

If y is a number a raised to a power n, the meaning is that a is multiplied by itself n times. It does not matter whether or not n is an integer. If n is $1/2$, the interpretation is a square root, for example.

If $\quad\quad\quad y = a^n$
then $\quad\quad y = a \times a \times a \times a \ldots (n \text{ times})$
$\quad\quad\quad \log y = \log a + \log a + \log a \ldots (n \text{ times})$
or $\quad\quad \log y = n \log a$

This relation holds for any exponent n.

The tables of logarithms are constructed for numbers between 1 and 10. Since the logs of all numbers from 1 to 10 lie between 0 and 1, these tables give what is called the **mantissa,** the portion to the right of the decimal point in the logarithm. The number to the left of the decimal point in the logarithm, the **characteristic,** depends on the position of the decimal point in the number in question; it shows the whole (integral) power of 10 that must be multiplied by the antilog of the mantissa. It is zero for numbers from 1 to 10; 1 for numbers from 10 to 100; -1 for numbers from 0.1 to 1; and so forth. In this type of work, in which numbers less than one are frequently used, the logarithms of such numbers cannot be expressed in the usual form. To show the method that must be used, the number should be expressed in terms of powers of 10. For example, find the log of 0.02. To do this, express it as

$$0.02 = 2 \times 10^{-2}$$
$$\log 0.02 = \log (2 \times 10^{-2})$$
$$= \log 2 + \log 10^{-2}$$
$$= \log 2 - 2 \log 10$$

but since $\qquad 10^1 = 10, \log 10 = 1$

therefore $\qquad \log 0.02 = -2 + \log 2$

and from the tables $\log 2 = 0.301$

Then $\qquad \log 0.02 = -2 + 0.301$
$$= -1.699$$

Natural logarithms use the base e, where $e = 2.71828. \ldots$ The principal use of natural logs is in the analysis of absorption or decay phenomena and some growth processes. The natural log of a number is the power to which e is raised to get the number in question. To use the tables to find the natural log of a number not included in the main body of the table, first express the number in the notation using powers of 10. Then make use of the idea that when quantities are multiplied, their logs are added. For example, let us find the natural log of 6500.

$$6500 = 6.5 \times 10^3$$
$$\ln 6.5 = 1.87$$
$$\ln 10^3 = 6.91$$
$$\ln 6.5 + \ln 10^3 = 8.78$$

Therefore $\qquad \ln 6500 = 8.78$

What is the natural log of 0.0135?

$$0.0135 = 1.35 \times 10^{-2}$$
$$\ln 1.35 = 0.300$$
$$\ln 10^{-2} = -4.605$$
$$\ln 1.35 + \ln 10^{-2} = 0.300 - 4.605 = -4.305$$

Therefore $\qquad \ln 0.0135 = -4.305$

Common logs of numbers from 0.01 to 10.9

x	0.00	0.01	.02	.03	.04	.05	.06	.07	0.8	.09
0.0		−2.00	−1.70	−1.52	−1.40	−1.30	−1.22	−1.15	−1.10	−1.05
0.1	−1.00	−.959	−.921	−.886	−.854	−.824	−.796	−.770	−.745	−.721
0.2	−.699	−.678	−.658	−.638	−.620	−.602	−.585	−.569	−.553	−.538
0.3	−.523	−.509	−.495	−.481	−.469	−.456	−.444	−.432	−.420	−.409
0.4	−.398	−.387	−.377	−.367	−.357	−.347	−.337	−.328	−.319	−.310
0.5	−.301	−.292	−.284	−.276	−.268	−.260	−.252	−.244	−.237	−.299
0.6	−.222	−.215	−.208	−.201	−.194	−.187	−.180	−.174	−.167	−.161
0.7	−.155	−.149	−.143	−.137	−.131	−.125	−.119	−.114	−.108	−.102
0.8	−.097	−.092	−.086	−.081	−.076	−.071	−.066	−.060	−.056	−.051
0.9	−.046	−.041	−.036	−.032	−.027	−.022	−.018	−.013	−.009	−.004
1.0	0	.004	.009	.013	.017	.021	.025	.029	.033	.037
1.1	.041	.045	.049	.053	.057	.060	.064	.068	.072	.076
1.2	.079	.083	.086	.090	.093	.097	.100	.104	.107	.111
1.3	.114	.117	.121	.124	.127	.130	.134	.137	.140	.143
1.4	.146	.149	.152	.155	.158	.161	.164	.167	.170	.173
1.5	.176	.180	.182	.185	.188	.190	.193	.196	.199	.201
1.6	.204	.207	.210	.212	.215	.217	.220	.223	.225	.228
1.7	.230	.233	.236	.238	.241	.243	.246	.248	.250	.253
1.8	.255	.258	.260	.262	.265	.267	.270	.272	.274	.276
1.9	.279	.281	.283	.286	.288	.290	.292	.294	.297	.299

x	0.0	0.1	0.2	0.3	0.4	0.5	0.6	0.7	0.8	0.9
2	.301	.322	.342	.362	.380	.398	.415	.431	.447	.462
3	.477	.491	.505	.519	.531	.544	.556	.568	.580	.591
4	.602	.613	.623	.633	.643	.653	.663	.672	.681	.690
5	.699	.708	.716	.724	.732	.740	.748	.756	.763	.771
6	.778	.785	.792	.799	.806	.813	.820	.826	.833	.839
7	.845	.851	.857	.863	.869	.875	.881	.886	.892	.898
8	.903	.908	.914	.919	.924	.929	.934	.940	.944	.949
9	.954	.959	.964	.968	.973	.978	.982	.987	.991	.996
10	1.0	1.004	1.009	1.013	1.017	1.021	1.025	1.029	1.033	1.037

Natural logs, $\log_e x$ or $\ln x$

x	0.00	0.01	0.02	0.03	0.04	0.05	0.06	0.07	0.08	0.09
0.0		−4.61	−3.91	−3.51	−3.22	−3.00	−2.81	−2.66	−2.53	−2.41
0.1	−2.30	−2.21	−2.12	−2.04	−1.97	−1.90	−1.83	−1.77	−1.71	−1.66
0.2	−1.61	−1.56	−1.51	−1.47	−1.43	−1.39	−1.35	−1.31	−1.27	−1.24
0.3	−1.20	−1.17	−1.14	−1.11	−1.08	−1.05	−1.02	−0.99	−0.97	−0.94
0.4	−0.92	−0.89	−0.87	−0.84	−0.82	−0.80	−0.78	−0.76	−0.73	−0.71
0.5	−0.69	−0.67	−0.65	−0.63	−0.62	−0.60	−0.58	−0.56	−0.54	−0.53
0.6	−0.51	−0.49	−0.48	−0.46	−0.45	−0.43	−0.42	−0.40	−0.39	−0.37
0.7	−0.36	−0.34	−0.33	−0.31	−0.30	−0.29	−0.27	−0.26	−0.25	−0.24
0.8	−0.22	−0.21	−0.20	−0.19	−0.17	−0.16	−0.15	−0.14	−0.13	−0.12
0.9	−0.106	−0.094	−0.083	−0.073	−0.062	−0.051	−0.041	−0.030	−0.020	−0.010
1.0	0	0.010	0.020	0.030	0.039	0.049	0.058	0.068	0.077	0.087
1.1	0.095	0.104	0.113	0.122	0.131	0.140	0.148	0.157	0.166	0.174
1.2	0.182	0.191	0.199	0.207	0.215	0.223	0.231	0.239	0.247	0.255
1.3	0.262	0.270	0.278	0.285	0.293	0.300	0.307	0.315	0.322	0.329
1.4	0.336	0.344	0.351	0.358	0.365	0.372	0.378	0.385	0.392	0.299
1.5	0.405	0.412	0.419	0.425	0.432	0.438	0.445	0.451	0.457	0.464
1.6	0.470	0.476	0.482	0.489	0.495	0.501	0.507	0.513	0.519	0.525
1.7	0.531	0.536	0.542	0.548	0.554	0.560	0.565	0.571	0.577	0.582
1.8	0.588	0.593	0.599	0.604	0.610	0.615	0.612	0.626	0.631	0.637
1.9	0.642	0.647	0.652	0.658	0.663	0.668	0.673	0.678	0.683	0.688

x	0.0	0.1	0.2	0.3	0.4	0.5	0.6	0.7	0.8	0.9
2	0.693	0.742	0.79	0.83	0.88	0.91	0.96	0.99	1.03	1.06
3	1.10	1.13	1.16	1.19	1.22	1.25	1.28	1.31	1.34	1.36
4	1.39	1.41	1.44	1.46	1.48	1.50	1.53	1.55	1.57	1.59
5	1.61	1.63	1.65	1.67	1.69	1.70	1.72	1.74	1.76	1.77
6	1.79	1.81	1.82	1.84	1.86	1.87	1.89	1.90	1.92	1.93
7	1.95	1.96	1.97	1.99	2.00	2.01	2.03	2.04	2.05	0.07
8	2.08	2.09	2.10	2.12	2.13	2.14	2.15	2.16	2.17	2.19
9	2.20	2.21	2.22	2.23	2.24	2.25	2.26	2.27	2.28	2.29
10	2.30									

$$\log_e 10 = 2.303 \qquad \log_e 10^{-1} = -2.303$$
$$\log_e 10^2 = 4.605 \qquad \log_e 10^{-2} = -4.605$$
$$\log_e 10^3 = 6.908 \qquad \log_e 10^{-3} = -6.908$$
$$\log_e 10^4 = 9.210 \qquad \log_e 10^{-4} = -9.210$$

TRIGONOMETRIC FUNCTIONS

θ is in degrees.

θ	$\sin \theta$	$\cos \theta$	$\tan \theta$	θ	$\sin \theta$	$\cos \theta$	$\tan \theta$
0	·0000	1·000	·0000	45	·7071	·7071	1·0000
1	·0175	·9998	·0175	46	·7193	·6947	1·0355
2	·0349	·9994	·0349	47	·7314	·6820	1·0724
3	·0523	·9986	·0524	48	·7431	·6691	1·1106
4	·0698	·9976	·0699	49	·7547	·6561	1·1504
5	·0872	·9962	·0875	50	·7660	·6428	1·1918
6	·1045	·9945	·1051	51	·7771	·6293	1·2349
7	·1219	·9925	·1228	52	·7880	·6157	1·2799
8	·1392	·9903	·1405	53	·7986	·6018	1·3270
9	·1564	·9877	·1584	54	·8090	·5878	1·3764
10	·1736	·9848	·1763	55	·8192	·5736	1·4281
11	·1908	·9816	·1944	56	·8290	·5592	1·4826
12	·2079	·9781	·2126	57	·8387	·5446	1·5399
13	·2250	·9744	·2309	58	·8480	·5299	1·6003
14	·2419	·9703	·2493	59	·8572	·5150	1·6643
15	·2588	·9659	·2679	60	·8660	·5000	1·7321
16	·2756	·9613	·2867	61	·8746	·4848	1·8040
17	·2924	·9563	·3057	62	·8829	·4695	1·8807
18	·3090	·9511	·3249	63	·8910	·4540	1·9626
19	·3256	·9455	·3443	64	·8988	·4384	2·0503
20	·3420	·9397	·3640	65	·9063	·4226	2·1445
21	·3584	·9336	·3839	66	·9135	·4067	2·2460
22	·3746	·9272	·4040	67	·9205	·3907	2·3559
23	·3907	·9205	·4245	68	·9272	·3746	2·4751
24	·4067	·9135	·4452	69	·9336	·3584	2·6051
25	·4226	·9063	·4663	70	·9397	·3420	2·7475
26	·4384	·8988	·4877	71	·9455	·3256	2·9042
27	·4540	·8910	·5095	72	·9511	·3090	3·0777
28	·4695	·8829	·5317	73	·9563	·2924	3·2709
29	·4848	·8746	·5543	74	·9613	·2756	3·4874
30	·5000	·8660	·5774	75	·9659	·2588	3·7321
31	·5150	·8572	·6009	76	·9703	·2419	4·0108
32	·5299	·8480	·6249	77	·9744	·2250	4·3315
33	·5446	·8387	·6494	78	·9781	·2079	4·7046
34	·5592	·8290	·6745	79	·9816	·1908	5·1446
35	·5736	·8192	·7002	80	·9848	·1736	5·6713
36	·5878	·8090	·7265	81	·9877	·1564	6·3138
37	·6018	·7986	·7536	82	·9903	·1392	7·1154
38	·6157	·7880	·7813	83	·9925	·1219	8·1443
39	·6293	·7771	·8098	84	·9945	·1045	9·5144
40	·6428	·7660	·8391	85	·9962	·0872	11·43
41	·6561	·7547	·8693	86	·9976	·0698	14·30
42	·6691	·7431	·9004	87	·9986	·0523	19·08
43	·6820	·7314	·9325	88	·9994	·0349	28·64
44	·6947	·7193	·9657	89	·9998	·0175	57·29
				90	1·000	·0000	∞

EXPONENTIAL FUNCTIONS

Table of e^x

x	0.0	0.1	0.2	0.3	0.4	0.5	0.6	0.7	0.8	0.9
0	1.00	1.105	1.221	1.350	1.492	1.649	1.822	2.014	2.226	2.460
1	2.718	3.004	3.320	3.699	4.055	4.482	4.953	5.474	6.050	6.686
2	7.389	8.166	9.025	9.974	11.02	12.18	13.46	14.88	16.44	18.17
3	20.09	22.20	24.53	27.11	29.96	33.12	36.60	40.45	44.70	49.40
4	54.60	60.34	66.69	73.70	81.45	90.02	99.48	109.9	121.5	134.3
5	148.4	164.0	181.3	200.3	221.4	244.7	270.4	298.9	330.3	365.0
6	403.4	445.9	492.7	544.6	601.8	665.1	735.1	812.4	897.8	992.3
7	1097	1212	1339	1480	1636	1808	1998	2208	2441	2697
8	2981	3294	3641	4024	4447	4915	5432	6003	6634	7332
9	8103	8955	9897	10938	12088	13360	14765	16318	18034	19930
10	22026	24343	26903	29733	32860	36316	40135	44356	49021	54176

Table of e^{-x}

The negative number indicates a multiplying power of 10.

x	.0	.1	.2	.3	.4	.5	.6	.7	.8	.9
0	1.000	0.905	0.819	0.741	0.670	0.607	0.549	0.497	0.449	0.407
1	0.368	0.333	0.301	0.273	0.247	0.223	0.202	0.283	0.165	0.150
2	0.135	0.122	0.111	0.100	0.091	0.082	0.074	0.067	0.061	0.055
3	0.050	0.045	0.042	0.037	0.033	0.030	0.027	0.025	0.022	0.020
4	1.83–2	1.66–2	1.50–2	1.36–2	1.23–2	1.11–2	1.01–2	9.1–2	8.2–3	7.4–3
5	6.7–3	6.1–3	5.5–3	5.0–3	4.5–3	4.1–3	3.7–3	3.35–3	3.03–3	2.74–3
6	2.48–3	2.24–3	2.03–3	1.84–3	1.66–3	1.50–3	1.36–3	1.23–3	1.11–3	1.01–3
7	9.1–4	8.3–4	7.5–4	6.8–4	6.1–4	5.5–4	5.0–4	4.5–4	4.1–4	3.7–4
8	3.35–4	3.04–4	2.75–4	2.49–4	2.25–4	2.03–4	1.84–4	1.67–4	1.51–4	1.36–4
9	1.23–4	1.12–4	1.01–4	9.1–5	8.3–5	7.5–5	6.8–5	6.1–5	5.5–5	5.0–5
10	4.54–5									

e.g., $e^{-5} = 6.7 \times 10^{-3} = 0.0067$

$e^{-10} = 4.54 \times 10^{-5}$

MATHEMATICAL CONSTANTS AND FORMULAE

$$\pi = 3.14159\ldots$$
$$e = 2.71828\ldots$$
$$1\ \text{rad} = 57.296°$$
$$1° = 0.017453\ \text{rad}$$

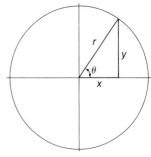

Figure A-1

Triangle; base a, altitude h: area $= ah/2$

Circle; radius r or diameter d:
circumference $= 2\pi r = \pi d$
area $\qquad = \pi r^2 = \pi d^2/4$

Sphere; radius r or diameter d:
surface area $= 4\pi r^2 = \pi d^2$
volume $\qquad = 4\pi r^3/3 = \pi d^3/6$

Bionomial expansion:

$$(1 + x)^n = 1 + nx + \frac{n(n-1)x^2}{2!} + \frac{n(n-1)(n-2)x^3}{3!} + \ldots\ldots$$

Solution of quadratic equation:

If $ax^2 + bx + c = 0$

then $x = \dfrac{-b \pm \sqrt{b^2 - 4ac}}{2a}$

Trigonometric relations:
As in Figure A–1
$\sin\theta = y/r$
$\cos\theta = x/r$
$\tan\theta = y/x$
$\sin^2\theta + \cos^2\theta = 1$
$\sin(a + b) = \sin a\cos b + \cos a\sin b$
$\cos(a + b) = \cos a\cos b - \sin a\sin b$

579

MISCELLANEOUS PHYSICAL AND ASTRONOMICAL CONSTANTS

Atomic mass unit: $\quad$ $u = 1.660 \times 10^{-27}$ kg

Mass of proton: $\quad$ $M_p = 1.673 \times 10^{-27}$ kg

Mass of neutron: $\quad$ $M_n = 1.675 \times 10^{-27}$ kg

Mass of electron: $\quad$ $M_e = 9.109 \times 10^{-31}$ kg
$$= 0.000549 \text{ u}$$

Electronic charge: $\quad$ $e = 1.602 \times 10^{-19}$ C

Avogadro's number $\quad$ $N = 6.023 \times 10^{23}$ particles/mole

Boltzmann's constant $\quad$ $k = 1.381 \times 10^{-23}$ J/K particle

Universal gas constant $\quad$ $R = 8.314$ J/mol K
$$= 8314 \text{ J/kmol K}$$

Universal gravitation constant $G = 6.67 \times 10^{-11}$ N m²/kg²

Speed of light $\quad$ $c = 2.998 \times 10^{8}$ m/s

Astronomical Unit $\quad$ A.U. $= 1.495 \times 10^{11}$ m

Earth: Mean radius: 6.36×10^6 m
Equatorial radius: 6.378×10^6 m
Polar radius: 6.357×10^6 m
Mean distance from sun: 1.495×10^{11} m
Standard surface gravity: $g_0 = 9.80665$ m/s²
Mass: 5.983×10^{24} kg

Planetary system: Mean radii of orbits in A.U.

Mercury	0.387	Saturn	9.539
Venus	0.723	Uranus	19.18
Earth	1.000	Neptune	30.06
Mars	1.524	Pluto	39.44
Jupiter	5.203		

ANSWERS TO ODD-NUMBERED
NUMERICAL PROBLEMS

CHAPTER 1

1. (a) 23.5 N,   (b) 3.87×10^{-3} kg, 3.87 g,
    (c) 3.8×10^{-2} N,   (d) 3800 dynes.

3. 735 N.

5. (a) 636 km,   (b) 1.83 km,   (C) 91.4 cm,   (d) 90,000 m,
    (e) 1530 mm,   (f) 0.084 mm,   (g) 84 μm.

7. At a distance, x_2, of 1.5 m.

9. 1750 N.

11. (a) 2.33 m,   (b) 1.167 m.

13. 0.98 m from the end.

15. The center of mass moves from 0.375 m from one support,
    past the center, to 0.375 m from the other support.

17. (a) $P = 89$ N,   (b) $T = 261$ N.

19. (a) $F_x = 47.6$ N, $F_y = 27.5$ N
    (b)   3.0 N,   22.0 N
    (c)   −50.6 N,   42.4 N
    (d)   0   −92.0 N

21. (a) 41 N,   (b) 50 N.

23. at 30°, 260 N m; at 45°, 212 N m; at 60°, 150 N m; at 80°,
    52 N m.

25. 1°, 0.10 g; 2°, 0.20 g; 3°, 0.30 g. The value of g has no
    effect.

27. (a) 0.55797, 0.558
    (b) 0.881917, 0.88
    (c) 0.441235, 0.441

CHAPTER 2

1. 1930 km/h, 536 m/s.

3. In 25 seconds, $s = 566$ m at 45° from initial direction,
$\bar{v} = 22.6$ m/s in the direction of s.
In 50 seconds, $s = 800$ m diametrically across the track,
$\bar{v} = 16$ m/s in the direction of s.
In 1 rev, $s = 0$, $\bar{v} = 0$.

5. 0–3 s, 1.2 m/s²,
3–4 s, 1.2 m/s²,
4–6 s, 1.6 m/s²,
6–9 s, 1.8 m/s²,
9–12 s, 0.1 m/s²,
20–24 s, −3.4 m/s².

7. 39 m/s.

9. $t = 10$ s $+ 5$ s $+ 2.5$ s $+ 1.25$ s $+ \cdots\cdots$

$$= 10 \text{ s} \left(1 + \frac{1}{2} + \frac{1}{4} + \frac{1}{8} + \cdots\cdots \text{ summed to an infinite number of terms}\right)$$

$$= 10 \text{ s} \left(\frac{1}{1 - \frac{1}{2}}\right) = 20 \text{ s}.$$

11. 3.2° east of north, 179 km/h.

13. (a) 26.4 m/s, (b) 125 s, (c) 0.42 m/s².

15. 1.33 m/s².

17. 122.5 m, 0.

19. (a) 24.5 m/s, (b) 15 m/s, (c) 30.6 m.

21. On earth, 24 m/s. On the moon, 9.71 m/s.

23. 25 s.

25. 304 km.

27. 1.71 m/s².

29. 1.63 km/s, 115 min. 38 s.

CHAPTER 3

1. 18,500 kg/m³, not pure gold.

3. water 1000 cm³/kg, 10.0 cm cube; aluminum 370 cm³/kg, 7.18 cm cube; lead 88.0 cm³/kg, 4.48 cm cube; osmium 44.4 cm³/kg, 3.54 cm cube; cement about 351 cm³/kg, 7.05 cm cube; quartz 377 cm³/kg, 7.23 cm cube.

5. 1.0 g.

7. 6.0×10^{16} kg/m^3.

9. 1.56×10^3 N.

11. (a) 0.28 m/s^2, box down, man up;
    (b) 6.07 m/s^2, box up, man down.

13. 1.9×10^7 m/s.

15. 9.8 m/s^2, 38.1° east of north.

17. 0.63 kg m/s.

19. 25 N in a direction opposite to the velocity.

21. 1 bullet, 0.010 m/s; 100 bullets, 1.0 m/s.

23. (a) 500 N, (b) 451 N, (c) 90.2 m/s^2, (d) 90.2 m/s,
    (e) 45.1 m,   (f) 415 m, (g) 460 m.

25. 0.7 m/s downward.

27. 8.64 km west.

29. 8.6 rev/min (0.90 rad/s).

31. 23.6 m/s^2.

33. 10 m/s, not speeding.

35. (a) 184 N, (b) 1.0 m/s^2.

CHAPTER 4

1. At closest approach, speed is 3.4% higher.

3. 465 years.

5. Mean R^3/T^2 is 1.011×10^{13} m^3/s^2. The moon is the highest.

7. 3.00 km/s.

9. (a) 0.19 km/s, (b) 1.04 km/s.

11. 0.400 years $= 146$ days.

13.

angle $\theta°$	θ, rad	$\sin \theta$	$\tan \theta$
1°	0.0175	0.0175	0.0175
2°	0.0349	0.0349	0.0349
3°	0.0524	0.0523	0.0524
4°	0.0698	0.0698	0.0699
5°	0.0873	0.0872	0.0875
6°	0.1047	0.1045	0.1051
7°	0.1222	0.1219	0.1228
8°	0.1396	0.1392	0.1405
9°	0.1571	0.1564	0.1584
10°	0.1745	0.1736	0.1763

15. 0.105 rad/s, 1.45×10^{-4} rad/s.

17. 628 rad/s.

19. 420 rad.

21. 3.9 m/s.

23. 8.8 rad/s².

25. The disc will have double the speed.

27. 32 kg m²/s.

29. 7.4×10^{-7} N.

31. 1.62 N/kg.

CHAPTER 5

1. 1.74×10^3 J.

3. 7.73×10^6 J/day, 89.5 J/s.

5. (a) 1860 kcal/day, (b) 26.6 kcal/kg, (c) 26.6°C.

7. 5.8×10^8 J.

9. 29°C.

11. (a) 3.2×10^{12} J, (b) 3.75×10^7 watts.

13. (A) 9.8 J, (b) 127 J, (c) 2090 J.

15. (a) 4000 J, (b) 613 J, (c) 0.85.

17. (a) 3130 J, (b) 0.

19. 5.9×10^6 m/s.

21. v (hoop) / v (sphere) = 0.775.

23. (a) −460°, (b) 37°C = 310 K, (c) −40°F = 233 K.

25. (a) 32°C, (b) 72°C.

27. 43.6°C.

29. 61°C.

31. 6 minutes.

33. 0.033 watts/m °C.

35. Conduction through a concrete wall is 3.3 times that through wood.

37. 15.7 W/m².

39. 10 m².

41. 3.8 m.

CHAPTER 6

1. 8.5×10^6 Pa.

3. 709 kPa.

5. 138 kPa.

7. 2400 N.

9. 14 cm.

11. $-156°C$.

13. 13.3 N.

15. (a) 6.07×10^{-21} J, (b) 478 m/s.

17. 2.8×10^{15} particles/m³.

19. 2.2×10^{-16} g/cm³.

21. Radon is 7.7 times as dense as air.

23. The fastest would have 1/4 the mass.

25. (a) 14.6 K, (b) 8.7 K.

27. (a) $5\,R/2 = 20.78$ J/mole K,
    (b) 7 R/2 = 29.09 J/mole K,
    (c) 1.40.

CHAPTER 7

1. 1.0×10^8 N/m².

3. 6.7×10^{10} N/m².

5. (a) 1.33×10^7 N/m², (b) 6.7×10^{-4}, (c) 6.7×10^{-3} cm.

7. 0.51 N/cm.

9. 0.0116 rad, 0.66°.

11. (a) 0.175, (b) 4.7×10^9, (c) 4790 N m,
    (d) 479 N m/degree, 2.7×10^4 N m/rad.

13. 0.013.

15. 1.25 J.

17. (a) 30 J, (b) 30 J, (C) 39 m/s.

19. 7.2 kg.

21. 5 m/s and 30 m/s.

23. 0.89.

25. 4.5×10^{-3} J.

27. (a) 568 Pa, (b) 284 Pa, (c) 5.7 Pa, (d) 4.1 Pa.

29. In the large chamber the tension is 1.6 times that in the small.

31. 7.5×10^{7} N/m.

33. 0.046 N/m.

CHAPTER 8

1. 23.5 mi/U.S. gal.

3. 77.6 kg.

5. (a) 0.05074 kg, (b) 0.4975 N, (c) 101.3 kPa.

7. 10.3 m.

9. 1104 kg/m³.

11. 184 kPa.

13. (a) 1000 N down, 1199 N up, (b) 200 N.

15. 4715 kg, 4.715 tonnes, 10,400 lb.

17. 900 kg/m³.

19. (a) 96 N, (b) 96 N.

21. 0.042 N s/m.

23. 12 s.

25. (a) 5000 J, (b) 12,000 J.

27. $A_2 = (1/3.4)A_1$.

29. 0.59.

CHAPTER 9

1. (a) 0.50 m, (b) 2.0 rad/s, (c) 1.0 m/s, (d) -2.0 m/s²
3. (a) 3.9 m/s, (b) 6.17 m/s²,
   (c) at points of maximum displacement,
   (d) toward the center, (e) outward.

5. 3140 m/s, 9.9×10^{7} m/s².

7. (a) 0.41 rad, (b) 23.6°, 156.4°, 203.6°, 336.4°.

9. 0.0049 N/m.

11. (a) 0.2484 m, (b) 0.9936 m.

13. 3.15 m.

15. 306 m/s.

17. 6.8×10^{-9} N m/rad.

19. 1:31.6.

21. 343 m/s.

23. $v = \sqrt{\text{tension/mass per unit length}}$

25. 8.0×10^{10} N/m².

27. 2600 km.

29. 3.4×10^{9} N/m².

CHAPTER 10

1. Long wavelength limit, 34 m to 17 m; short wavelength limit, 2.3 cm to 1.7 cm.

3. 0.0289 kg/mole.

5. 223 m/s.

11. 3750 Hz.

13. 256 Hz, 768 Hz, and 1280 Hz. Third and fifth harmonics.

15. 92 db.

17. 3.2×10^{-4} W/m².

19. 61.3 db.

21. zero db.

23. 0.000 000 000 000 000 79 watts.

25. 10:1.

29. I_0, 1.5 I_0, 1.75 I_0, 1.875 I_0 etc. to 1.998 I_0; 1.998 I_0, 0.998 I_0, 0.498 I_0, 0.248 I_0 to 0.002 I_0.

31. 0.21 s.

33. 0.019.

CHAPTER 11

1. 7.5×10^{14} Hz, 4.3×10^{14} Hz, 3×10^{8} Hz.

3. 16,980.

5. 20.5 cm, 23.55 cm.

7. (a) 450 nm, 900 nm, 1350 nm, 1800 nm, 2250 nm, 2700 nm.
    (b) 540 nm, 1080 nm, 1620 nm, 2160 nm, 2700 nm, 3240 nm.
    (c) 2700 nm.

9. (a) (i) 280 nm, 0.28 μm, 0.00028 mm,
            (ii) 200 nm, 0.20 μm, 0.00020 mm,
            (iii) 160 nm, 0.16 μm, 0.00016 nm.
    (b) No, for $\lambda = 400$ nm, the thickness would have to be 320 nm.

11. 61%, 89%.

13. 100 nm, 250 nm, 125 nm and shorter in U.V.

15. 0.000 5461 mm, 546.1 nm.

21. 6.3°, 1.1 cm.

23. 654 nm.

25. 846.4 nm, 11,810/cm.

27. 5.53 μm, 180/mm.

29. A linear structure produces lines, spacing 115 nm.

31. 546 nm.

CHAPTER 12

1. $\tan \theta = \mu$.

3. 73.74°.

5. $d' = \mu d$.

7. 2.00×10^8 m/s.

9. (a) 32.1°, (b) 27.1° in oil, 32.1° in water.

11. 61.5°.

13. 58.7°.

15. (b) 56.8°.

17. red, 1.306; blue, 1.317.

19. 2.00.

23. (a) 5.6 cm, (b) 12 cm, (c) 14 cm, (d) −12 cm.

25. (a) 25 cm from the object, (b) 18.75 cm.

27. 6.15 mm.

29. −90 mm.

31. (a) 28°, (b) 0.86°, 52′, (c) 0.086°, 5.2′.

33. 1.3 cm.

35. 200.

37. (a) 31,   (b) 49,   (c) 136.

39. (a) 42 cm,   (b) $I/O = 20$.

CHAPTER 13

 1. 0.087 sr.

 3. 0.015 sr, 1/9.

 5. 3.5.

 7. 58 lux.

 9. (a) $4\pi I$ and $4\pi Il$,   (b) $A = 2\pi rl$,   (c) $E = 2\,I/r$.

11. $u = 2.63\ v^{0.51}$, $u = c\sqrt{v}$.

13. (a) 130 lux,   (b) 93 lux.

15. 59 lux at each location.

17. (a) 220 lux,   (b) 198 lux.

19. 2.67 cm.

21. 0.707.

23. (a) 0.23/mm,   (b) 0.69/cm,   (c) 10/cm.

25. (a) 0.55,   (b) 0.17.

27. 62 s.

29. 3.0×10^5 km/s.

CHAPTER 14

 1. 96,400 C.

 3. 1.0×10^{-5} N.

 5. 338 N.

 7. 0.10 μC.

 9. 80 N/C.

11. 7.3×10^6 m/s.

13. (a) 1.13×10^5 lines,   (c) 12.6 m²,   (c) 9000 lines/m², (d) 9000 N/C.

15. 3 mm.

17. 15.9 picocoulombs.

19. (a) 4.8 N,   (b) 0.12 J,   (c) 0.12 J,   (d) 7.5 J/C,
     (e) 7.5 volts.

21. (a) 5 volts,   (b) 500 N/C,   (c) 167 N/C.

23. 0.16 m/s.

25. 138 volts.

27. 2.1×10^8 m/s.

29. 2.94×10^8 m/s.

31. 67 μF.

33. (a) 1000 V lines/m^2,   (b) 0.6 V lines,
     (c) 5.3×10^{-12} V coulombs,
     (d) 5.3×10^{-12} coulombs/volt, 5.3 pF.

35. (a) The charge increases by 4 times.
     (b) The voltage decreases to 1/4.

CHAPTER 15

1. (a) 50 C,   (b) 1.85×10^{-12} amps.

3. (a) 0.33 amps,   (b) 1.0 μA,   (c) 54 volts,   (d) 235 volts,
    (e) 3000 ohms,   (f) 6.9 ohms.

5. 11,600 ohms.

7. 680 ohms.

9. 0.45 amps.

11. 32.6 amps.

13. 707 volts.

15. 16 times.

17. 7.9 m.

19. 0.024 μA.

21. 0.090 A; 0.9 v, 27 v, 54 v, 90 v.

23. 3.5 μA.

25. 3.3×10^{-4} tesla.

27. 0.2 tesla.

29. 800,000 ohms.

CHAPTER 16

1. 656.47, 486.273, 434.173, 410.293, less than 0.04%.

3. 1240 nm, 124 nm, 12.4 nm, 1.24 nm, 0.124 nm.

5. None from silver, 0.30 from potassium.

9. C^{10}:6p, 4n, 6e.
   C^{11}:6p, 5n, 6e
   C^{12}:6p, 6n, 6e
   C^{13}:6p, 7n, 6e
   C^{14}:6p, 8n, 6e
   C^{15}:6p, 9n, 6e
   C^{16}:6p, 10n, 6e

11. O^{15}, N^{15}.

13. The nuclides involved are U^{238}, Th^{234}, Pa^{234}, U^{234}, Th^{230}, Ra^{226}.

15. 74%.

17. 7.4 mCi.

19. 4 years.

21. 0.13%.

CHAPTER 17

1. (a) 0.08790 s,   (b) 0.08837 s,   (c) 0.53%,   (d) 30.16 m.

3. $c = \dfrac{2d}{(t_1 + t_2)} \sqrt{1 + \dfrac{V^2(t_1 + t_2)^2}{4d^2}}$

5. (a) $\ell = 0.866$ m,   (b) $\ell = 0.9950$ m,
   (c) shortened by 5×10^{-15} m.

INDEX

Boldface numbers indicate definitions or major discussions.

PHYSICS
with Modern Applications

Leonard H. Greenberg
University of Regina
Regina, Saskatchewan

1978
W. B. SAUNDERS COMPANY
Philadelphia, London, Toronto

W. B. Saunders Company: West Washington Square
Philadelphia, PA 19105

1 St. Anne's Road
Eastbourne, East Sussex BN21 3UN, England

1 Goldthorne Avenue
Toronto, Ontario M8Z 5T9, Canada

Cover illustration is a reproduction of "Opus 1972", enamel on steel by Virgil Cantini.   Courtesy of the artist.

Physics with Modern Applications   ISBN 9-7216-4247-0

Last digit is the print number: 9   8   7   6   5   4   3   2   1

PREFACE

AIMS OF THE BOOK

There are many reasons for acquiring knowledge of basic physics, no matter what field of specialization you may be in. Concrete uses of physical principles and analysis in various fields of science, including biology, geology, and medicine, are given prominence in this book. These uses are in the understanding of phenomena and in instrumentation. But physics is useful not only to scientists; non-scientists often attempt to emulate the methods of the physicist because, as a science, physics has been so successful. To use our methods you must know what they are; know what laws in physics are and how they arise; know how theories are used; and see how we arrive at basic principles in nature.

All of us have a richer life when we understand phenomena of nature and scientific accomplishments. What causes the colors in soap bubbles? Why is sound carried so well downwind? What are the principles of space travel? What is radioactivity?

THE MATERIAL AND ORDER OF PRESENTATION

Probably the most successful of all sciences, from the standpoint of the accuracy of the description of natural processes, has been that dealing with motion. This includes the effect of forces. The book begins with the study of forces on objects that are not moving, and then proceeds to the description of motion. These two topics are combined in the study of force and motion in Chapter 3. The studies go from a description of the origin of the concepts to applications in today's world.

In Chapter 4 rotational motion is described. This begins with Kepler's description of orbits, for Kepler's laws are generally regarded as being the first of the scientific laws. These laws are shown to apply to the description of the paths of interplanetary spacecraft. Rotational dynamics on earth is not neglected.

The quantity we call energy is discussed in Chapters 5 and 6. There is no need to stress the importance of energy in today's world. The concepts developed are also important in the topics in the remainder of the book.

The basic material developed to this point is used in the study of the distortion of solids by forces (Chapter 7) and the properties of fluids at rest and in motion (Chapter 8). These topics have their own value but also are necessary for the discussion of vibration and waves (Chapter 9). The particular vibration called sound is worthy of separate treatment in Chapter 10, and phenomena in optics which result from the wave nature of light are discussed in Chapter 11. Optical instruments, such as microscopes, are of such wide use in many fields of science that the subject of ray optics is introduced (Chapter 12). Some characteristics of light—its intensity, absorption, and speed—are discussed in Chapter 13. The absorption phenomenon for light is described by the same equations that describe such things as the absorption of gamma rays, the decay of sound intensity in a room, and the cooling of a hot object. Unifying ideas such as this show how a little basic knowledge encompasses a broad range of application.

Phenomena concerning electricity (stationary charges, charged particles moving in electric fields, current electricity and its link to magnetism) are covered in Chapters 14 and 15.

The treatment of atomic models in Chapter 16 makes use of the material on electric forces. This includes discussions of the nature of radioactivity, of matter, and of antimatter.

The last chapter is an attempt to show the view of the world that requires the ideas of Einstein's special relativity for its proper description. The reasons for the contractions of the observed lengths of moving objects, the dilation of time, and the increase in mass with speed are described though not rigorously developed. The origin of the relation $E = mc^2$ is described in a similar way. It is not a rigorous treatment of relativity, but mainly an indication that the ideas have a firm foundation.

BACKGROUND NECESSARY

No background in physics is assumed, but physics at the level of the twelfth year of school in most educational systems is certainly an advantage.

Physicists go from basic ideas to application to individual situations by the use of mathematical reasoning. The rules of mathematics apply in the natural world. The more sophisticated your mathematics, the more powerful your abilities in physics.

To use this book, the ability to manipulate equations by the methods of algebra is a necessity. Basic trigonometry is also used and a knowledge of logarithms is assumed. Calculus is not used; but to go further in any branch of physics, calculus methods would have to be learned.

UNITS

Throughout the book, the units used are those of the Système International or SI. It used to be my practice to teach in diverse systems of units, and the physics often became almost secondary to the mastering of the different systems. The decision to work only in SI units simplified the presentation and clarified many concepts. At least in science we should work in only one system throughout the world. It has to start with you, the student. You will not be alone in those units, for more and more scientific journals are requiring SI units in the reports they publish.

ACKNOWLEDGMENTS

The final manuscript was prepared after obtaining a large number of helpful suggestions from Dr. Stan Williams of the Iowa State University of Science and Technology, from Dr. Robert E. Simpson of the University of New Hampshire, and from Dr. Bob Henson of the University of Missouri. To them I say thank you. They cannot be blamed, however, for faults that remain.

My wife, Ann, was understanding about all the late nights I spent at my desk. She also contributed much through discussion of the material and did some of the typing. Nevertheless, it is the staff of the publisher who transform the typescript to a book. Among those who performed this were Jay Freedman, Tom O'Connor, John Hackmaster, Lorraine Battista, and at the top of the list Joan Garbutt and John Vondeling. They were all wonderful and patient people to work with. There were others too, but space is limited, and I hope they realize that I have appreciated them.

In conclusion, I hope that you who read and study the book will gain by it.

L. H. GREENBERG

CONTENTS